Cut here and keep for reference

ALGEBRA

Arithmetic Operations

$$a(b + c) = ab + ac$$

$$\frac{a}{b} + \frac{c}{d} = \frac{ad + bc}{bd}$$

$$\frac{a + c}{b} = \frac{a}{b} + \frac{c}{b}$$

$$\frac{\frac{a}{b}}{\frac{c}{d}} = \frac{a}{b} \times \frac{d}{c} = \frac{ad}{bc}$$

Exponents and Radicals

$$x^m x^n = x^{m+n}$$

$$\frac{x^m}{x^n} = x^{m-n}$$

$$(x^m)^n = x^{mn}$$

$$x^{-n} = \frac{1}{x^n}$$

$$(xy)^n = x^n y^n$$

$$\left(\frac{x}{y}\right)^n = \frac{x^n}{y^n}$$

$$x^{1/n} = \sqrt[n]{x}$$

$$x^{m/n} = \sqrt[n]{x^m} = \left(\sqrt[n]{x}\right)^m$$

$$\sqrt[n]{xy} = \sqrt[n]{x}\sqrt[n]{y}$$

$$\sqrt[n]{\frac{x}{y}} = \frac{\sqrt[n]{x}}{\sqrt[n]{y}}$$

Factoring Special Polynomials

$$x^2 - y^2 = (x + y)(x - y)$$

$$x^3 + y^3 = (x + y)(x^2 - xy + y^2)$$

$$x^3 - y^3 = (x - y)(x^2 + xy + y^2)$$

Binomial Theorem

$$(x + y)^2 = x^2 + 2xy + y^2 \qquad (x - y)^2 = x^2 - 2xy + y^2$$

$$(x + y)^3 = x^3 + 3x^2 y + 3xy^2 + y^3$$

$$(x - y)^3 = x^3 - 3x^2 y + 3xy^2 - y^3$$

$$(x + y)^n = x^n + nx^{n-1}y + \frac{n(n-1)}{2}x^{n-2}y^2$$

$$+ \cdots + \binom{n}{k}x^{n-k}y^k + \cdots + nxy^{n-1} + y^n$$

where $\binom{n}{k} = \frac{n(n-1)\cdots(n-k+1)}{1 \cdot 2 \cdot 3 \cdot \cdots \cdot k}$

Quadratic Formula

If $ax^2 + bx + c = 0$, then $x = \dfrac{-b \pm \sqrt{b^2 - 4ac}}{2a}$.

Inequalities and Absolute Value

If $a < b$ and $b < c$, then $a < c$.

If $a < b$, then $a + c < b + c$.

If $a < b$ and $c > 0$, then $ca < cb$.

If $a < b$ and $c < 0$, then $ca > cb$.

If $a > 0$, then

$$|x| = a \quad \text{means} \quad x = a \quad \text{or} \quad x = -a$$

$$|x| < a \quad \text{means} \quad -a < x < a$$

$$|x| > a \quad \text{means} \quad x > a \quad \text{or} \quad x < -a$$

GEOMETRY

Geometric Formulas

Formulas for area A, circumference C, and volume V:

Triangle
$$A = \tfrac{1}{2}bh$$
$$= \tfrac{1}{2}ab \sin \theta$$

Circle
$$A = \pi r^2$$
$$C = 2\pi r$$

Sector of Circle
$$A = \tfrac{1}{2}r^2\theta$$
$$s = r\theta \quad (\theta \text{ in radians})$$

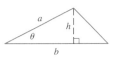

Sphere
$$V = \tfrac{4}{3}\pi r^3$$
$$A = 4\pi r^2$$

Cylinder
$$V = \pi r^2 h$$

Cone
$$V = \tfrac{1}{3}\pi r^2 h$$
$$A = \pi r \sqrt{r^2 + h^2}$$

Distance and Midpoint Formulas

Distance between $P_1(x_1, y_1)$ and $P_2(x_2, y_2)$:

$$d = \sqrt{(x_2 - x_1)^2 + (y_2 - y_1)^2}$$

Midpoint of $\overline{P_1 P_2}$: $\left(\dfrac{x_1 + x_2}{2}, \dfrac{y_1 + y_2}{2}\right)$

Lines

Slope of line through $P_1(x_1, y_1)$ and $P_2(x_2, y_2)$:

$$m = \frac{y_2 - y_1}{x_2 - x_1}$$

Point-slope equation of line through $P_1(x_1, y_1)$ with slope m:

$$y - y_1 = m(x - x_1)$$

Slope-intercept equation of line with slope m and y-intercept b:

$$y = mx + b$$

Circles

Equation of the circle with center (h, k) and radius r:

$$(x - h)^2 + (y - k)^2 = r^2$$

SPECIAL FUNCTIONS

Power Functions $f(x) = x^a$

(i) $f(x) = x^n$, n a positive integer

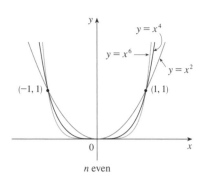

n even

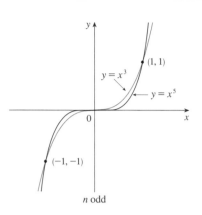

n odd

(ii) $f(x) = x^{1/n} = \sqrt[n]{x}$, n a positive integer

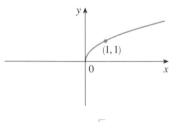

$f(x) = \sqrt{x}$

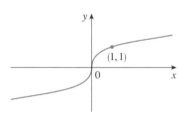

$f(x) = \sqrt[3]{x}$

(iii) $f(x) = x^{-1} = \dfrac{1}{x}$

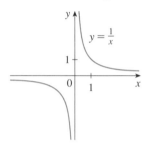

Exponential and Logarithmic Functions

$\log_a x = y \iff a^y = x$

$\ln x = \log_e x, \quad \text{where} \quad \ln e = 1$

$\ln x = y \iff e^y = x$

Cancellation Equations

$\log_a(a^x) = x \qquad a^{\log_a x} = x$

$\ln(e^x) = x \qquad e^{\ln x} = x$

Laws of Logarithms

1. $\log_a(xy) = \log_a x + \log_a y$

2. $\log_a\left(\dfrac{x}{y}\right) = \log_a x - \log_a y$

3. $\log_a(x^r) = r\log_a x$

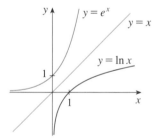

$\displaystyle\lim_{x \to -\infty} e^x = 0 \qquad \lim_{x \to \infty} e^x = \infty$

$\displaystyle\lim_{x \to 0^+} \ln x = -\infty \qquad \lim_{x \to \infty} \ln x = \infty$

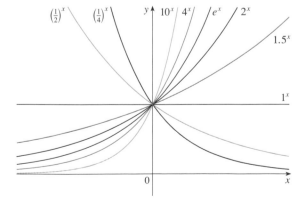

Exponential functions

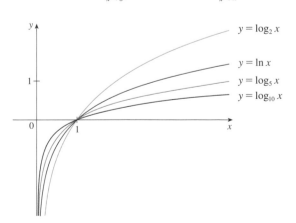

Logarithmic functions

DIFFERENTIATION FORMULAS

Constant function $\dfrac{d}{dx}(c) = 0$

Power Rule $\dfrac{d}{dx}(x^n) = nx^{n-1}$

Power Rule with Chain Rule $\dfrac{d}{dx}[g(x)]^n = n[g(x)]^{n-1}g'(x)$

Reciprocal function $\dfrac{d}{dx}\left(\dfrac{1}{x}\right) = -\dfrac{1}{x^2}$

Reciprocal function with Chain Rule $\dfrac{d}{dx}\left[\dfrac{1}{g(x)}\right] = -\dfrac{g'(x)}{[g(x)]^2}$

Natural exponential function $\dfrac{d}{dx}(e^x) = e^x$

Natural exponential function with Chain Rule $\dfrac{d}{dx}[e^{g(x)}] = e^{g(x)}g'(x)$

Exponential function $\dfrac{d}{dx}(a^x) = a^x \ln a$

Natural logarithmic function $\dfrac{d}{dx}(\ln x) = 1/x$

Natural logarithmic function with Chain Rule $\dfrac{d}{dx}[\ln g(x)] = \dfrac{g'(x)}{g(x)}$

Constant Multiple Rule $\dfrac{d}{dx}[cf(x)] = cf'(x)$

Sum Rule $\dfrac{d}{dx}[f(x) + g(x)] = f'(x) + g'(x)$

Difference Rule $\dfrac{d}{dx}[f(x) - g(x)] = f'(x) - g'(x)$

Product Rule $\dfrac{d}{dx}[f(x)\,g(x)] = f(x)\,g'(x) + g(x)\,f'(x)$

Quotient Rule $\dfrac{d}{dx}\left[\dfrac{f(x)}{g(x)}\right] = \dfrac{g(x)\,f'(x) - f(x)\,g'(x)}{[g(x)]^2}$

Chain Rule $\dfrac{d}{dx}[f(g(x))] = f'(g(x))\,g'(x)$

Chain Rule If $y = f(u)$, $u = g(x)$, then $\dfrac{dy}{dx} = \dfrac{dy}{du}\dfrac{du}{dx}$.

INTEGRATION FORMULAS

$\displaystyle\int cf(x)\,dx = cF(x) + C$

$\displaystyle\int [f(x) \pm g(x)]\,dx = F(x) \pm G(x) + C$

$\displaystyle\int c\,dx = cx + C$

$\displaystyle\int x^n\,dx = \dfrac{x^{n+1}}{n+1} + C \quad (n \neq -1)$

$\displaystyle\int (1/x)\,dx = \ln|x| + C$

$\displaystyle\int e^x\,dx = e^x + C$

$\displaystyle\int e^{kx}\,dx = \dfrac{1}{k}e^{kx} + C$

$\displaystyle\int a^x\,dx = \dfrac{1}{\ln a}a^x + C$

The Substitution Rule If $u = g(x)$ is a differentiable function and f is continuous on the range of g, then

$$\int f(g(x))\,g'(x)\,dx = \int f(u)\,du$$

Integration by Parts If $u = f(x)$ and $v = g(x)$, then $du = f'(x)\,dx$ and $dv = g'(x)\,dx$, and

$$\int u\,dv = uv - \int v\,du$$

Brief Applied Calculus

The cover image depicts global business communication. The image, called *Globe Encounters*, comes from New York Talk Exchange, a creation of the Senseable City Lab at MIT and exhibited at MoMA, the Museum of Modern Art, in 2008. It visualizes the volumes of Internet data flowing between New York City and cities around the world. The size of the glow on a particular city location corresponds to the amount of IP traffic between that place and New York City.

Brief Applied Calculus

JAMES STEWART
McMaster University and University of Toronto

DANIEL CLEGG
Palomar College

BROOKS/COLE
CENGAGE Learning™

Australia · Brazil · Japan · Korea · Mexico · Singapore · Spain · United Kingdom · United States

BROOKS/COLE
CENGAGE Learning

Brief Applied Calculus, First Edition,
James Stewart, Daniel Clegg

Publisher: Richard Stratton

Sr. Development Editor: Laura Wheel

Development Editor: Nicole Mollica

Sr. Editorial Assistant: Haeree Chang

Associate Media Editor : Andrew Coppola

Sr. Marketing Manager: Barb Bartoszek

Marketing Coordinator: Michael Ledesma

Marketing Communications Manager:
    Mary Anne Payumo

Content Project Manager: Cheryll Linthicum

Sr. Art Director: Vernon Boes

Print Buyer: Becky Cross

Rights Acquisitions Specialist: Roberta Broyer

Production Service: TECH·arts

Text Designer: Lisa Henry

Photo and Text Researcher: Terri Wright

Copy Editor: Kathi Townes, TECH·arts

Illustrator: TECH·arts

Cover Designer: Lisa Henry

Cover Image: Globe Encounters
    © MIT Senseable City Lab

Compositor: Stephanie Kuhns, TECH·arts

Unless otherwise noted, all art in this textbook
is © Cengage Learning.

Library of Congress Control Number: 2011932384

Student Edition:
ISBN-13: 978-0-534-42382-7
ISBN-10: 0-534-42382-5

Brooks/Cole
20 Davis Drive
Belmont, CA 94002-3098
USA

Cengage Learning is a leading provider of customized learning solutions with office locations around the globe, including Singapore, the United Kingdom, Australia, Mexico, Brazil, and Japan. Locate your local office at **www.cengage.com/global**.

Cengage Learning products are represented in Canada by Nelson Education, Ltd.

To learn more about Brooks/Cole, visit
www.cengage.com/brookscole.

Purchase any of our products at your local college store or at our preferred online store **www.cengagebrain.com**.

K06T11

Printed in Canada
1 2 3 4 5 6 7 15 14 13 12 11

Contents

7 Functions of Several Variables

Appendixes

Index

Brief Applied Calculus is designed as an introduction to calculus for students in business, economics, and the social and life sciences. We take an informal, intuitive approach to the subject, without sacrificing mathematical integrity. And we aim to balance conceptual understanding with technical skills.

There is a wide range of students' abilities in the applied calculus course and we have taken steps to address this phenomenon: On the one hand, we know that there are many students with weak algebraic skills and so we provide careful, patient explanations in the exposition and plenty of algebraic steps in the solutions of the examples. And before some of the exercise sets we include warm-up problems called "Prepare Yourself" to practice the skills needed for the exercises that follow. On the other hand, in such a course there are usually some students who are quite mathematically talented and accordingly, not wanting to neglect them, we have placed a collection of problems labeled "Challenge Yourself" at the end of most exercise sets.

Features

Modeling and Real-World Data

It is important for students to understand what a mathematical model is. We discuss the meaning of mathematical modeling in the first section of the book (pages 10–11) and we consistently use data to construct models throughout the book. We spent a great deal of time looking in libraries, contacting companies and government agencies, and searching the Internet for interesting real-world data to introduce, motivate, and illustrate the concepts of calculus. As a result, many of the examples and exercises deal with functions defined by such numerical data or graphs. See, for instance, Exercise 1.5.47 (computing power of microchips), Example 3 in Section 2.1 (how driving speed affects gas mileage), Example 10 in Section 2.3 (US national debt), Exercise 2.3.52 (life expectancy), Exercise 2.3.54 (Starbucks locations), Exercise 2.3.56 (swimming speed of salmon), Exercise 2.4.13 (unemployment rate), and Example 3 in Section 5.3 (San Francisco power consumption).

Conceptual Exercises

The most effective way to foster conceptual understanding is through the problems that we assign. To that end we have devised various types of problems. Some exercises request explanations of the meanings of the basic concepts of the section. (See, for instance, Exercises 2.2.1 and 2.2.45). Similarly, all the review sections begin with a Concept Check. Other exercises test conceptual understanding through graphs or tables (Exercises 2.3.21–24, 2.3.52–57, 4.3.31–34, 7.2.5–8) or interpretations of mathematical statements (Exercises 2.3.39–46, 2.4.10–14, 4.3.39–41). Another type of exercise uses verbal description to test conceptual understanding (Exercises 2.4.49–50).

Prepare Yourself

Some exercise sets are preceded by problems designed to prepare students by reviving their algebraic skills or reminding them of techniques studied earlier in the book. In Section 2.2, for instance, students are asked to factor or simplify algebraic expressions in preparation for calculating limits.

Challenge Yourself

Exercise sets conclude with problems designed to challenge students to think a little more deeply about the ideas of the section.

Projects

One way of involving students and making them active learners is to have them work (perhaps in groups) on extended projects that give a feeling of substantial accomplishment when completed. For instance, the project called *The Gini Index* (page 340) uses actual census data and explores how to measure income distribution among inhabitants of a given country. It is a nice application of areas between curves.

Animations

We have devised animations that we call *Tools for Enriching Calculus* (TEC). TEC uses a discovery and exploratory approach that brings the subject alive. In sections of the book where technology is particularly appropriate, marginal icons direct students to TEC animations that provide a laboratory environment in which they can explore the topic in different ways and at different levels. **Visuals are animations of figures in text; Modules are more elaborate activities and include exercises**. Instructors can choose to become involved at several different levels, ranging from simply encouraging students to use the Visuals and Modules for independent exploration, to assigning specific exercises from those included with each Module, or to creating additional exercises, labs, and projects that make use of the Visuals and Modules.

Concept Check / Chapter Review

At the beginning of each chapter review section is what we call Concept Check, which is a collection of questions designed to make sure that students know the main concepts of the chapter. The answers to these Concept Check questions are all collected on the back endpapers, thus constituting a review of the entire course.

www.stewartcalculus.com

This site includes the following:

- Lies My Calculator and Computer Told Me
- History of Mathematics, with links to the better historical websites
- Links, for particular topics, to outside web resources
- Tools for Enriching Calculus (TEC) Modules and Visuals

Content

1 Functions and Models From the beginning, multiple representations of functions are stressed: verbal, numerical, visual, and algebraic. A discussion of mathematical models leads to a review of the standard functions, including exponential and logarithmic functions, from these four points of view. We have two reasons for including the exponential and logarithmic functions in Chapter 1: They give us a greater variety of functions to illustrate the differentiation rules, particularly the Product Rule. And they provide a wealth of models of real-world situations to use in the early chapters.

2 The Derivative The material on limits is motivated by a prior discussion of shrinking the interval in average rates of change. Limits are treated from descriptive, graphical, numerical, and algebraic points of view. Sections 2.3 and 2.4 deal with derivatives (especially with functions defined graphically and numerically) before the differentiation rules are covered in Chapter 3. Here the examples and exercises explore the meanings of derivatives in various contexts. Section 2.4 foreshadows, in an intuitive way, the material on shapes of curves that is studied in greater depth in Chapter 4. This enables a better discussion of cost and revenue functions in marginal analysis in Section 3.2.

3 Techniques of Differentiation All the basic functions are differentiated here, including exponential and logarithmic functions. Many examples of derivatives computed in applied settings are given, where students are asked to explain their meanings. In order to study a major application to economics as soon as possible, marginal analysis is introduced just after shortcuts for polynomial derivatives are presented. (The topic is revisited in more detail in Section 4.7.) Exponential growth and decay are covered, as well as logistic models.

4 Applications of Differentiation With the differentiation rules in hand, we pursue the classic applications of derivatives: related rates, maximum and minimum values, curve sketching, and optimization problems. Applications to business and economics include maximizing profit, elasticity, and managing inventory.

5 Integrals The definite integral is motivated by the problem of using marginal cost to find total cost and the associated area problem. Emphasis is placed on explaining the meanings of integrals in various contexts and on estimating their values from graphs and tables. The Midpoint Rule, discussed here, is adequate for most purposes and is generally much more accurate than the Trapezoidal Rule. (But the Trapezoidal Rule and Simpson's Rule are in Appendix C for those instructors who wish to cover them.) The Fundamental Theorem of Calculus is reformulated as the Net Change Theorem and is applied to rates of change in the social and natural sciences. The main techniques of integration, the Substitution Rule and Integration by Parts, conclude the chapter.

6 Applications of Integration Here we present applications to economics and biology, as well as areas between curves, enabling the project on the Gini Index. Separable differential equations lead to a discussion of logistic growth and improper integrals are used to treat basic ideas of probability.

7 Functions of Several Variables Functions of two or more variables are studied from verbal, numerical, visual, and algebraic points of view. In particular, we introduce partial derivatives by looking at a specific column in a table of values of the heat index (perceived air temperature) as a function of the actual temperature and the relative humidity. The treatment of maximum and minimum values includes Lagrange multipliers and their interpretation in economics. We have not included double integrals in this chapter because their applications are largely to the natural sciences. But for instructors who wish to cover them, a treatment of double integrals is included in Appendix D.

Ancillaries

Brief Applied Calculus is supported by a complete set of ancillaries developed under our direction. Each piece has been designed to enhance student understanding and to facilitate creative instruction. Included are new media and technologies that have been developed to help students to visualize calculus and instructors to customize content to better align with the way they teach their course. The list on pages xv–xvii describes each of these ancillaries.

Acknowledgments

We are grateful to the following reviewers and focus group participants for sharing their knowledge and judgment with us. We have learned something from each of them.

Reviewers

Jurg Bolli, *University of New Mexico*

Sarah Clark, *South Dakota State University*

Wendi Dass, *Germanna Community College*

Karabi Datta, *Northern Illinois University*

Perry Gillespie, *Fayetteville State University*

Jean Harper, *State University of New York at Fredonia*

Frederick Hoffman, *Florida Atlantic University*

Lucyna Kabza, *Southeastern Louisiana University*

Kristine King, *University of South Alabama*

Solange Kouemou, *Florida International University*

Mark MacLean, *University of British Columbia*

Augustine Mascuilli, *Pace University*

Scott McDaniel, *Middle Tennessee State University*
Ian Musson, *University of Wisconsin–Milwaukee*
James Peirce, *University of Wisconsin–La Crosse*
Georgia Pyrros, *University of Delaware*
John Racquet, *State University of New York at Albany*
Marlene Sims, *Kennesaw State University*
William Slough, *Eastern Illinois University*
Yanbo Ye, *University of Iowa*

Focus Group Participants

Mary Brown, *Harrisburg Area Community College*
Anna Butler, *Polk State College*
Rebecca Cajucom, *Pierce College*
Joe Castillo, *Broward College*
Jerry Chen, *Suffolk County Community College*
Debbie Garrison, *Valencia College*
Don Hancock, *Pepperdine University*
Aminul Huq, *University of Minnesota–Rochester*
Raja Khoury, *Collin College*
Marko Kranjc, *Western Illinois University*
Richard Leedy, *Polk State College*
Timothy Lucas, *Pepperdine University*
Phillip Miller, *Indiana University Southeast*
Lyra Neville, *Harrisburg Area Community College*
Luis Ortiz-Franco, *Chapman University*
Altay Ozgener, *State College of Florida*
Stephen Proietti, *Northern Essex Community College*
Yuri Yatsenko, *Houston Baptist University*

In addition, we would like to thank Kathi Townes and Stephanie Kuhns of TECH-arts for their production services; Marv Riedesel and Mary Johnson for their accuracy checking and proofing of the pages; Gina Sanders for checking the answers; and the following Brooks/Cole staff: Laura Wheel, Haeree Chang, Cheryl Linthicum, Vernon Boes, Barb Bartoszek, Andrew Coppola, Ashley Pickering, and Nicole Mollica.

Special thanks go to our editors, Liz Covello and Richard Stratton, for patiently and skillfully guiding this project to its completion.

JAMES STEWART
DANIEL CLEGG

Ancillaries for Instructors

PowerLecture (978-1-111-42654-5)
This comprehensive DVD contains all art from the text in both jpeg and Power-Point formats, key equations and tables from the text, complete pre-built Power-Point lectures, the Instructor's Guide, Solution Builder, and ExamView testing software.

Instructor's Guide for *Brief Applied Calculus*
Douglas Shaw
Each section of the text is discussed from several viewpoints. The Instructor's Guide contains suggested time to allot, points to stress, text discussion topics, core materials for lecture, workshop/discussion suggestions, group work exercises in a form suitable for handout, and suggested homework assignments. The Instructor's Guide is available on the PowerLecture DVD.

Complete Solutions Manual for *Brief Applied Calculus* (978-1-133-36433-7)
Daniel Clegg
The Complete Solutions Manual provides worked-out solutions to all exercises in the text.

Solution Builder (www.cengage.com/solutionbuilder)
This online instructor database offers complete worked-out solutions to all exercises in the text. Solution Builder allows you to create customized, secure solutions printouts (in PDF format) matched exactly to the problems you assign in class.

ExamView Testing
Create, deliver, and customize tests in print and online formats with ExamView, an easy-to-use assessment and tutorial software. ExamView contains hundreds of multiple-choice and free response test items. ExamView testing is available on the PowerLecture DVD.

Ancillaries for Instructors and Students

TEC Tools for Enriching™ Calculus
James Stewart, Harvey Keynes, Daniel Clegg, and developer *Hu Hohn*
Tools for Enriching Calculus (TEC) functions as both a powerful tool for instructors, as well as a tutorial environment in which students can explore and review selected topics. The Flash simulation modules in TEC include instructions, written and audio explanations of the concepts, and exercises. TEC is available on **www.stewartcalculus.com** as well as in CourseMate, Enhanced WebAssign, and the instructors' PowerLecture DVD.

Cengage Customizable YouBook
YouBook is a Flash-based eBook that is interactive and customizable! Containing all the content from Stewart/Clegg's *Brief Applied Calculus*, YouBook features a text edit tool that allows instructors to modify the textbook narrative as needed. With YouBook, instructors can quickly re-order entire sections and chapters or hide any content they don't teach to create an eBook that perfectly matches their syllabus. Instructors can further customize the text by publishing web links. Additional media assets include: animated figures, video clips, highlighting, notes, and more! YouBook is available in Enhanced WebAssign.

CourseMate (www.cengagebrain.com)
Interested in a simple way to complement your text and course content with study and practice materials? CourseMate brings course concepts to life with interactive learning, study, and exam preparation tools that support the printed textbook. CourseMate for Stewart/Clegg's *Brief Applied Calculus* includes: an interactive eBook, quizzes, flashcards, TEC animations and modules, videos, and more! For instructors, CourseMate includes Engagement Tracker, a first-of-its-kind tool that monitors student engagement.

WebAssign Enhanced WebAssign (www.webassign.net)
WebAssign's homework delivery system lets instructors deliver, collect, grade, and record assignments via the web. Enhanced WebAssign for Stewart/Clegg's *Brief Applied Calculus* includes: Cengage YouBook interactive eBook, Personal Study Plans, a Show My Work feature, Answer Evaluator, Visualizing Calculus animations and modules, quizzes, videos, and more!

Ancillaries for Students

Student Solutions Manual for *Brief Applied Calculus* (978-0-534-42387-2)
Daniel Clegg
Provides complete worked-out solutions to all odd-numbered exercises in the text, giving students a chance to check their answers and ensure they took the correct steps to arrive at an answer.

Succeeding in Applied Calculus: Algebra Success, Second Edition
(978-0-495-11153-5)
Warren B. Gordon

Written for students in need of an algebra refresher, this supplemental book is a convenient source of reference and review for students enrolled in an applied calculus course. It is a quick, easy-to-use resource that provides "just-in-time" algebra review for only those algebra topics that are most essential to the study of applied calculus.

Reading a calculus textbook is different from reading a newspaper or a novel, or even another textbook. Don't be discouraged if you have to read a passage more than once in order to understand it. You should have pencil and paper and calculator at hand to sketch a diagram or make a calculation.

Some students start by trying their homework problems and read the text only if they get stuck on an exercise. We suggest that a far better plan is to read and understand a section of the text before attempting the exercises. In particular, you should look at the definitions to see the exact meanings of the terms. And before you read each example, we suggest that you cover up the solution and try solving the problem yourself. You'll get a lot more from looking at the solution if you do so.

Part of the aim of this course is to train you to think logically. Learn to write the solutions of the exercises in a connected, step-by-step fashion with explanatory sentences—not just a string of disconnected equations or formulas. We also want you to be able to interpret your results in context: many of the exercises ask you to explain the meaning of the answers you get.

The answers to the odd-numbered exercises appear at the back of the book, in Appendix E. Some exercises ask for a verbal explanation or interpretation or description. In such cases there is no single correct way of expressing the answer, so don't worry that you haven't found the definitive answer. In addition, there are often several different forms in which to express a numerical or algebraic answer, so if your answer differs from the book's, don't immediately assume you're wrong. For example, if the answer you obtain is $\sqrt{3}/3$ and the answer given in the back of the book is $1/\sqrt{3}$, then you're right and rationalizing the denominator will show that the answers are equivalent.

We want to draw your attention to the website **www.stewartcalculus.com**. There you will find *Graphing Calculators and Computers* (with a discussion of these graphing devices and some of the pitfalls you may encounter), *Lies My Calculator and Computer Told Me* (explaining why calculators sometimes give the wrong answer), *History of Mathematics*, and links to outside resources.

Also on the website is TEC: *Tools for Enriching Calculus*, which is referred to by means of the symbol **TEC**. It directs you to *Visuals* and *Modules* in which you can explore aspects of calculus for which the computer is particularly useful.

The icon ⊞ indicates an exercise that definitely requires the use of either a graphing calculator or a computer with graphing software. But that doesn't mean that graphing devices can't be used to check your work on the other exercises as well. You will also encounter the symbol ⊘, which warns you against committing an error. We have placed this symbol in the margin in situations where we have observed that a large proportion of our students tend to make the same mistake.

Calculus is an exciting subject, justly considered to be one of the greatest achievements of the human intellect. We hope you will discover that it is not only useful but also intrinsically beautiful.

JAMES STEWART

DANIEL CLEGG

Diagnostic Test

Success in calculus depends to a large extent on knowledge of the mathematics that precedes calculus. The following test is intended to diagnose weaknesses that you might have. After taking the test you can check your answers against the given answers and, if necessary, refresh your skills by referring to the review materials that are provided.

1. Evaluate each expression without using a calculator.

(a) $(-3)^4$ (b) -3^4 (c) 3^{-4}

(d) $\dfrac{5^{23}}{5^{21}}$ (e) $\left(\dfrac{2}{3}\right)^{-2}$ (f) $16^{-3/4}$

2. Simplify each expression. Write your answer without negative exponents.

(a) $\sqrt{200} - \sqrt{32}$

(b) $(3a^3b^3)(4ab^2)^2$

(c) $\left(\dfrac{3x^{3/2}y^3}{x^2y^{-1/2}}\right)^{-2}$

3. Expand and simplify.

(a) $3(x + 6) + 4(2x - 5)$ (b) $(x + 3)(4x - 5)$

(c) $(\sqrt{a} + \sqrt{b})(\sqrt{a} - \sqrt{b})$ (d) $(2x + 3)^2$

(e) $(x + 2)^3$

4. Factor each expression.

(a) $4x^2 - 25$ (b) $2x^2 + 5x - 12$

(c) $x^3 - 3x^2 - 4x + 12$ (d) $x^4 + 27x$

(e) $x^3y - 4xy$

5. Simplify the rational expression.

(a) $\dfrac{x^2 + 3x + 2}{x^2 - x - 2}$ (b) $\dfrac{2x^2 - x - 1}{x^2 - 9} \cdot \dfrac{x + 3}{2x + 1}$

(c) $\dfrac{x^2}{x^2 - 4} - \dfrac{x + 1}{x + 2}$ (d) $\dfrac{\dfrac{y}{x} - \dfrac{x}{y}}{\dfrac{1}{y} - \dfrac{1}{x}}$

6. Rationalize the expression and simplify.

(a) $\dfrac{\sqrt{10}}{\sqrt{5} - 2}$ (b) $\dfrac{\sqrt{4 + h} - 2}{h}$

7. Solve the equation. (Find only the real solutions.)

(a) $x + 5 = 14 - \frac{1}{2}x$ (b) $\dfrac{2x}{x + 1} = \dfrac{2x - 1}{x}$

(c) $x^2 - x - 12 = 0$ (d) $2x^2 + 4x + 1 = 0$

(e) $x^4 - 3x^2 + 2 = 0$ (f) $2x(4 - x)^{-1/2} - 3\sqrt{4 - x} = 0$

8. Solve each inequality. Write your answer using interval notation.

     (a) $-4 < 5 - 3x \leqslant 17$ (b) $x^2 < 2x + 8$

     (c) $x(x - 1)(x + 2) > 0$

9. State whether each equation is true or false.

     (a) $(p + q)^2 = p^2 + q^2$ (b) $\sqrt{ab} = \sqrt{a}\,\sqrt{b}$

     (c) $\sqrt{a^2 + b^2} = a + b$ (d) $\dfrac{1 + TC}{C} = 1 + T$

     (e) $\dfrac{1}{x - y} = \dfrac{1}{x} - \dfrac{1}{y}$ (f) $\dfrac{1/x}{a/x - b/x} = \dfrac{1}{a - b}$

10. Find an equation for the line that passes through the point $(2, -5)$ and

     (a) has slope -3

     (b) is parallel to the x-axis

     (c) is parallel to the y-axis

     (d) is parallel to the line $2x - 4y = 3$

11. Find an equation for the circle that has center $(-1, 4)$ and passes through the point $(3, -2)$.

12. Let $A(-7, 4)$ and $B(5, -12)$ be points in the plane.

     (a) Find the slope of the line that contains A and B.

     (b) Find an equation of the line that passes through A and B. What are the intercepts?

     (c) Find the length of the segment AB.

13. Sketch the region in the xy-plane defined by the equation or inequalities.

     (a) $y = 1 - \frac{1}{2}x$ (b) $y < 1 - \frac{1}{2}x$

     (c) $y = x^2 - 1$ (d) $x^2 + y^2 = 4$

     (e) $-1 \leqslant y \leqslant 3$

▪ Answers to Diagnostic Test

1. (a) 81   (b) -81 (c) $\frac{1}{81}$

    (d) 25   (e) $\frac{9}{4}$ (f) $\frac{1}{8}$

2. (a) $6\sqrt{2}$ (b) $48a^5b^7$ (c) $\dfrac{x}{9y^7}$

3. (a) $11x - 2$ (b) $4x^2 + 7x - 15$

    (c) $a - b$ (d) $4x^2 + 12x + 9$

    (e) $x^3 + 6x^2 + 12x + 8$

4. (a) $(2x - 5)(2x + 5)$ (b) $(2x - 3)(x + 4)$

    (c) $(x - 3)(x - 2)(x + 2)$

    (d) $x(x + 3)(x^2 - 3x + 9)$

    (e) $xy(x - 2)(x + 2)$

5. (a) $\dfrac{x + 2}{x - 2}$ (b) $\dfrac{x - 1}{x - 3}$

    (c) $\dfrac{1}{x - 2}$ (d) $-(x + y)$

6. (a) $5\sqrt{2} + 2\sqrt{10}$ (b) $\dfrac{1}{\sqrt{4 + h} + 2}$

7. (a) 6   (b) 1   (c) $-3, 4$

    (d) $-1 \pm \frac{1}{2}\sqrt{2}$ (e) $\pm 1, \pm\sqrt{2}$ (f) $\frac{12}{5}$

8. (a) $[-4, 3)$ (b) $(-2, 4)$

    (c) $(-2, 0) \cup (1, \infty)$

9. (a) False   **(b)** True   **(c)** False
    (d) False   **(e)** False   **(f)** True

10. (a) $y = -3x + 1$ **(b)** $y = -5$
    (c) $x = 2$ **(d)** $y = \frac{1}{2}x - 6$

11. $(x + 1)^2 + (y - 4)^2 = 52$

12. (a) $-\frac{4}{3}$
    (b) $4x + 3y + 16 = 0$; x-intercept -4, y-intercept $-\frac{16}{3}$
    (c) 20

13. (a)   **(b)**   **(c)**

    (d)   **(e)**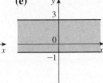

If you have had difficulty with these problems, you may wish to consult the review of algebra in Appendix A or the review of analytic geometry and lines in Appendix B.

Brief Applied Calculus

After learning a core group of basic functions, we will be armed with the tools to create formulas that describe scenarios as diverse as trends in the stock market, world population, historic Olympic wins, the growth of computing power, and the popularity of a new product.   © Michael Nagle/Bloomberg via Getty Images

1

Functions and Models

The fundamental objects that we deal with in calculus are functions. This chapter prepares the way for calculus by discussing the basic ideas concerning functions, their graphs, and ways of transforming and combining them. We stress that a function can be represented in different ways: by an equation, in a table, by a graph, or in words. We look at the main types of functions that will be needed in our study of calculus and describe the process of using these functions as mathematical models of real-world phenomena.

1.1 Functions and Their Representations

■ Introduction to Functions

Mathematical relationships can be observed in virtually every aspect of our environment and daily lives. Populations, financial markets, the spread of diseases, setting the price of a new product, and the effects of pollution on an ecosystem can all be analyzed using mathematics.

Many mathematical relationships can be considered as *functions*. A function is a correspondence in which one quantity is determined by another. For instance, each day that the US stock market is open corresponds to a closing price of Google stock. We say that the daily closing price of the stock is a function of the date.

For additional illustrations, consider the following four situations.

A. The area A of a square plot of land depends on the length s of one side of the plot. The rule that connects s and A is given by the equation $A = s^2$. With each positive number s there is associated one value of A, and we can say that A is a *function* of s.

B. The human population of the world P depends on the time t. The table gives estimates of the world population for certain years. For instance, when $t = 1950$, $P \approx 2,560,000,000$. But for each value of the time t there is a corresponding value of P, and we say that P is a function of t.

C. The cost C of mailing an envelope depends on its weight w. Although there is no simple formula that connects w and C, the post office has a rule for determining C when w is known.

D. The vertical acceleration a of the ground as measured by a seismograph during an earthquake is a function of the elapsed time t. Figure 1 shows a graph generated by seismic activity during the Northridge earthquake that shook Los Angeles in 1994. For a given value of t, the graph provides a corresponding value of a.

Year	Population (millions)
1900	1650
1910	1750
1920	1860
1930	2070
1940	2300
1950	2560
1960	3040
1970	3710
1980	4450
1990	5280
2000	6080
2010	6870

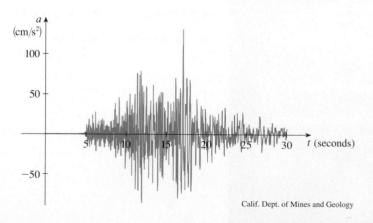

FIGURE 1
Vertical ground acceleration during the Northridge earthquake

Calif. Dept. of Mines and Geology

Each of these examples describes a rule whereby, given a number (s, t, w, or t), another number (A, P, C, or a) is assigned. In each case we say that the second number is a function of the first number. You can think of a function in terms of an input/output relationship, where the function assigns an output value to each input value it accepts.

> ▪ A **function** is a rule that assigns to each input exactly one output.

Notice that while a function can assign only one output to each input, it is perfectly acceptable for two different inputs to share the same output. Although a function can be defined for any sort of input or output, we usually consider functions for which the inputs and outputs are real numbers.

Notation and Terminology for Functions

We typically refer to a function by a single letter such as f. If x represents an input to the function f, the corresponding output is $f(x)$, read "f of x."

The set of all allowable inputs is called the **domain** of the function.

The **range** of f is the set of all possible output values, $f(x)$, as x varies throughout the domain.

A symbol that represents an arbitrary number in the *domain* of a function f is called an **independent variable**.

A symbol that represents a number in the *range* of f is called a **dependent variable**.

In Example A, for instance, s is the independent variable and A is the dependent variable. (We can choose the value of s independently, but A *depends* on the value of s.) Using function notation we can write $A = f(s)$, where f represents the area function.

It's helpful to think of a function as a **machine** (see Figure 2). If x is in the domain of the function f, then when x enters the machine, it's accepted as an input and the machine produces an output $f(x)$ according to the rule of the function.

For example, many cash registers used in retail stores have a button that, when pressed, automatically computes the sales tax to be added to the total. This button can be thought of as a function: An amount of money is entered as an input, and the machine outputs an amount of tax. Both the domain and range of this function are sets of positive numbers that represent amounts of money.

x (input) f $f(x)$ (output)

FIGURE 2
Machine diagram for a function f

▪ EXAMPLE 1 A Price Function

A cafe sells its basic coffee in three different cup sizes: 8, 10, and 14 ounces. They charge $0.22 per ounce for the drinks.

(a) If the function p is defined so that $p(v)$ is the price of v ounces of coffee, find and interpret the value of $p(10)$.

(b) What are the domain and range of p?

SOLUTION

(a) The function value $p(10)$ represents the output (price) of the function when the input is 10 ounces of coffee. Thus $p(10) = \$0.22 \times 10 = \$2.20$.

(b) If we assume that the cafe sells only 8-, 10-, and 14-ounce coffee drinks, then the only allowable inputs to the price function are the three numbers 8, 10, and 14, so the domain of p is the set $\{8, 10, 14\}$. The range is the set of outputs that correspond to the inputs in the domain: $\{1.76, 2.20, 3.08\}$. ▪

We use braces { } to list the elements of a set.

Although the rule defining a function may be clear, or you may have a list of inputs and outputs for a function, it is often easiest to analyze a function if you can visualize the relationship between the inputs and outputs. The most common method for visualizing a function is to view its graph. If f is a function, then its **graph** is the set of input-output pairs $(x, f(x))$ plotted as points for all x in the domain of f. In other words, the graph of f consists of all points (x, y) in the coordinate plane such that $y = f(x)$ and x is in the domain of f.

If the domain consists of isolated values, as in Example 1, the data are *discrete* and the graph is a collection of individual points, called a *scatter plot*. On the other hand, if the input variable represents a quantity that can vary *continuously* through an interval of values, the graph is a curve or line (see Figure 3). We will define a continuous function more formally in Chapter 2. For now, you can think of a continuous function as one for which you can sketch its graph without lifting your pencil from the paper.

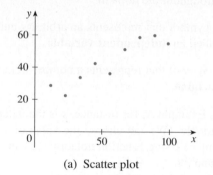
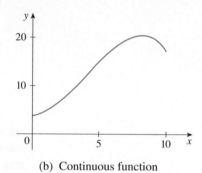

FIGURE 3
Graphs of functions

(a) Scatter plot   (b) Continuous function

The graph of a function f gives us a useful picture of the behavior or "life history" of a function. Since the y-coordinate of any point (x, y) on the graph is $y = f(x)$, we can read the value of $f(x)$ from the graph as being the height of the graph above the point x. (See Figure 4.) The graph of f also allows us to picture the domain of f on the x-axis and its range on the y-axis as in Figure 5.

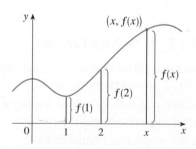

FIGURE 4

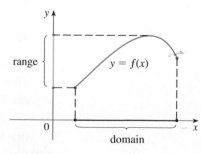

FIGURE 5

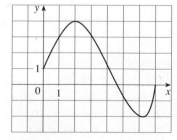

FIGURE 6

▪ **EXAMPLE 2   Reading Information from a Graph**

The graph of a function f is shown in Figure 6.

(a) Find the values of $f(1)$ and $f(5)$.

(b) What are the domain and range of f?

Recall that a closed bracket is used with interval notation to indicate an included value, while an open parenthesis indicates that the endpoint of the interval is not included. For instance, the interval $[2, 5)$ is equivalent to $\{x \mid 2 \leqslant x < 5\}$. An interval is called *closed* if it includes both endpoints; an *open* interval includes neither endpoint. A detailed review is included in Appendix A.

SOLUTION

(a) We see from Figure 6 that the point $(1, 3)$ lies on the graph of f, so the value of f at 1 is $f(1) = 3$. (In other words, the point on the graph that lies above $x = 1$ is 3 units above the x-axis.)

When $x = 5$, the graph lies about 0.7 unit below the x-axis, so we estimate that $f(5) \approx -0.7$.

(b) We see that $f(x)$ is defined when $0 \leqslant x \leqslant 7$, so the domain of f is the closed interval $[0, 7]$. Notice that f takes on all values from -2 to 4, so the range of f is

$$\{y \mid -2 \leqslant y \leqslant 4\} = [-2, 4] \qquad \blacksquare$$

▪ Representations of Functions

We have seen four possible ways to represent a function:

- verbally   (by a description in words)
- numerically   (by a table of values)
- visually   (by a graph)
- algebraically   (by an explicit formula)

If a single function can be represented in several ways, it is often useful to go from one representation to another to gain additional insight into the function. But certain functions are described more naturally by one method than by another. With this in mind, let's reexamine the four situations that we considered at the beginning of this section.

A. The most useful representation of the area of a square plot of land as a function of its side length is probably the algebraic formula $A(s) = s^2$, though it is possible to compile a table of values or sketch a graph (half a parabola). Because a square has to have a positive side length, the domain is $\{s \mid s > 0\} = (0, \infty)$, and the range is also $(0, \infty)$.

B. We are given a description of the function in words: $P(t)$ is the human population of the world at time t. For convenience, we can measure $P(t)$ in millions and let $t = 0$ represent the year 1900. Then the table of values of world population at the left provides a convenient representation of this function. If we plot these values, we get the scatter plot in Figure 7.

t	$P(t)$ (millions)
0	1650
10	1750
20	1860
30	2070
40	2300
50	2560
60	3040
70	3710
80	4450
90	5280
100	6080
110	6870

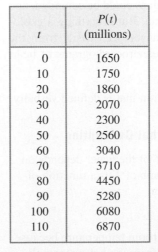

FIGURE 7

This scatter plot is a useful representation; the graph allows us to absorb all the data at once. What about a formula? Of course, it's impossible to devise an explicit formula that gives the exact human population $P(t)$ at any time t. But it is possible to find an expression for a function that *approximates* $P(t)$. In fact, using methods explained in Section 1.5, we obtain the approximation

$$P(t) \approx (1436.53) \cdot (1.01395)^t$$

Figure 8 shows that this function is a reasonably good "fit." Notice that here we have graphed a continuous curve as an approximation to discrete data. We will soon see that the ideas of calculus can be applied to discrete data as well as explicit formulas.

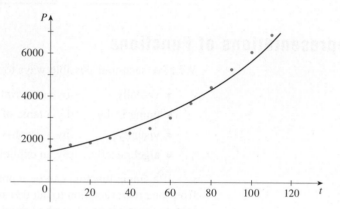

FIGURE 8

A function defined by a table of values is called a *tabular* function.

w (ounces)	$C(w)$ (dollars)
$0 < w \leqslant 1$	0.88
$1 < w \leqslant 2$	1.05
$2 < w \leqslant 3$	1.22
$3 < w \leqslant 4$	1.39
$4 < w \leqslant 5$	1.56
.	.
.	.
.	.

C. Again the function is described in words: $C(w)$ is the cost of mailing a large envelope with weight w. The rule that the US Postal Service used as of 2011 is as follows: The cost is 88 cents for up to 1 oz, plus 17 cents for each additional ounce (or less), up to 13 oz. The table of values at the left is the most convenient representation for this function, though it is possible to sketch a graph (see Example 10).

D. The graph shown in Figure 1 is the most natural representation of the vertical acceleration function $a(t)$. It's true that a table of values could be compiled, and it is even possible to devise an approximate formula. But everything a geologist needs to know—amplitudes and patterns—can be seen easily from the graph. (The same is true for the patterns seen in electrocardiograms of heart patients and polygraphs for lie-detection.)

In the next example we sketch the graph of a function that is defined verbally.

▪ EXAMPLE 3   Drawing a Graph from a Verbal Description

When you turn on a hot-water faucet, the temperature T of the water depends on how long the water has been running. Draw a rough graph of T as a function of the time t that has elapsed since the faucet was turned on.

SOLUTION

The initial temperature of the running water is close to room temperature because the water has been sitting in the pipes. When the water from the hot-water tank starts flowing from the faucet, T increases quickly. In the next phase, T is constant at the temperature of the heated water in the tank. When the tank is drained, T

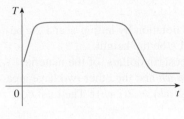

FIGURE 9

t	$N(t)$
1	41.4
3	25.1
5	15.5
7	10.2
9	6.0

decreases to the temperature of the water supply. This enables us to make the rough sketch of T as a function of t in Figure 9. ■

A more accurate graph of the function in Example 3 could be obtained by using a thermometer to measure the temperature of the water at 10-second intervals. In general, researchers collect experimental data and use them to sketch the graphs of functions, as the next example illustrates.

■ EXAMPLE 4 A Numerically Defined Function

The data shown in the margin give weekly sales figures for a video game shortly after its release. Let $N(t)$ be the number of copies sold, in thousands, during the week ending t weeks after the game's release. Sketch a scatter plot of these data, and use the scatter plot to draw a continuous approximation to the graph of $N(t)$. Then use the graph to estimate the number of copies sold during the sixth week.

SOLUTION

We plot the five points corresponding to the data from the table in Figure 10. The data points in Figure 10 look quite well behaved, so we simply draw a smooth curve through them by hand as in Figure 11. (Later in this chapter you will see how to find an algebraic formula that approximates the data.)

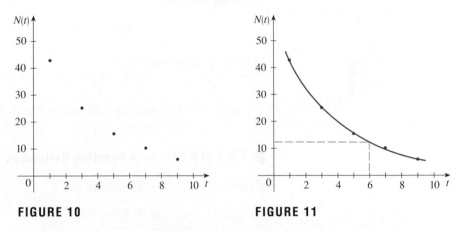

FIGURE 10 **FIGURE 11**

From the graph, it appears that $N(6) \approx 12.5$, so we estimate that 12,500 units were sold during the sixth week. ■

In the following example we start with a verbal description of a function in a physical situation and obtain an explicit algebraic formula. The ability to do this is a useful skill in solving optimization problems such as maximizing the profit of a company.

■ EXAMPLE 5 Expressing a Cost as a Function

A rectangular storage container with an open top has a volume of 10 m³. The length of its base is twice its width. Material for the base costs $10 per square meter; material for the sides costs $6 per square meter. Express the cost of materials as a function of the width of the base.

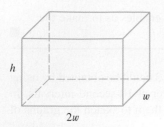

h

w

$2w$

FIGURE 12

SOLUTION

We draw a diagram as in Figure 12 and introduce notation by letting w and $2w$ be the width and length of the base, respectively, and h be the height.

The area of the base is $(2w)w = 2w^2$, so the cost, in dollars, of the material for the base is $10(2w^2)$. Two of the sides have area wh and the other two have area $2wh$, so the cost of the material for the sides is $6[2(wh) + 2(2wh)]$. The total cost is therefore

$$C = 10(2w^2) + 6[2(wh) + 2(2wh)] = 20w^2 + 36wh$$

To express C as a function of w alone, we need to eliminate h and we do so by using the fact that the volume is 10 m³. Thus

$$\text{volume} = \text{width} \cdot \text{length} \cdot \text{height} = w(2w)h = 10$$

which gives

$$h = \frac{10}{2w^2} = \frac{5}{w^2}$$

Substituting this into the expression for C, we have

$$C = 20w^2 + 36w\left(\frac{5}{w^2}\right) = 20w^2 + \frac{180}{w}$$

Therefore the equation

$$C(w) = 20w^2 + \frac{180}{w} \qquad w > 0$$

expresses C as a function of w.   ▪

In the next two examples we look at functions given by algebraic formulas.

▪ **EXAMPLE 6**   **A Function Defined by a Formula**

If $f(x) = 2x^2 - 5x + 1$, evaluate

(a) $f(-3)$ **(b)** $f(4) - f(2)$ **(c)** $\dfrac{f(1 + h) - f(1)}{h}$ $(h \neq 0)$

SOLUTION

(a) Replace x by -3 in the expression for $f(x)$:

$$f(-3) = 2(-3)^2 - 5(-3) + 1 = 2 \cdot 9 + 15 + 1 = 18 + 15 + 1 = 34$$

(b) $f(4) - f(2) = [2(4)^2 - 5(4) + 1] - [2(2)^2 - 5(2) + 1] = 13 - (-1) = 14$

(c) We first evaluate $f(1 + h)$ by replacing x by $1 + h$ in the expression for $f(x)$:

$$f(1 + h) = 2(1 + h)^2 - 5(1 + h) + 1$$

$$= 2(1 + 2h + h^2) - 5(1 + h) + 1$$

$$= 2 + 4h + 2h^2 - 5 - 5h + 1 = 2h^2 - h - 2$$

The expression

$$\frac{f(1 + h) - f(1)}{h}$$

in Example 6 is called a **difference quotient** and occurs frequently in calculus. We will will begin making use of it in Chapter 2.

Then we substitute into the given expression and simplify:

$$\frac{f(1 + h) - f(1)}{h} = \frac{(2h^2 - h - 2) - (2 - 5 + 1)}{h}$$

$$= \frac{2h^2 - h - 2 - (-2)}{h}$$

$$= \frac{2h^2 - h}{h} = \frac{h(2h - 1)}{h} = 2h - 1 \quad ▪$$

▪ EXAMPLE 7

Determining the Domain of a Function Defined by a Formula

Find the domain of each function.

(a) $B(r) = \sqrt{r + 2}$ **(b)** $g(x) = \dfrac{1}{x^2 - x}$

SOLUTION

If a function is given by a formula and the domain is not stated explicitly, the convention is that the domain is the set of all numbers for which the formula makes sense and defines a real number.

(a) Because the square root of a negative number is not defined (as a real number), the domain of B consists of all values of r such that $r + 2 \geqslant 0$. This is equivalent to $r \geqslant -2$, so the domain is the interval $[-2, \infty)$.

(b) Since

$$g(x) = \frac{1}{x^2 - x} = \frac{1}{x(x - 1)}$$

and division by 0 is not allowed, we see that $g(x)$ is not defined when $x = 0$ or $x = 1$. Thus the domain of g is $\{x \mid x \neq 0, x \neq 1\}$. ▪

The graph of a function is a curve or scatter plot in the xy-plane. But the question arises: Which graphs in the xy-plane represent functions and which do not? This is answered by the following test.

> ▪ **The Vertical Line Test** A curve or scatter plot in the xy-plane is the graph of a function of x if and only if no vertical line intersects the graph more than once.

The reason for the truth of the Vertical Line Test can be seen in Figure 13.

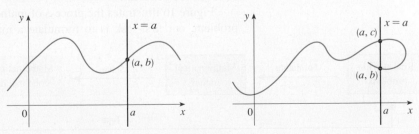

FIGURE 13

If each vertical line $x = a$ intersects a curve only once, at (a, b), then exactly one functional value is defined by $f(a) = b$. But if a line $x = a$ intersects the curve twice, at (a, b) and (a, c), then the curve can't represent a function because a function can't assign two different output values to an input a.

■ **EXAMPLE 8** **Using the Vertical Line Test**

Determine whether the graph represents a function.

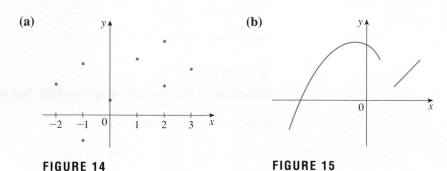

(a)   (b)

FIGURE 14   **FIGURE 15**

SOLUTION

(a) Notice that if we draw a vertical line on the scatter plot in Figure 14 at $x = -1$ or at $x = 2$, the line will intersect two of the points. Therefore the scatter plot does not represent a function.

(b) No matter where we draw a vertical line on the graph in Figure 15, the line will intersect the graph at most once, so this is the graph of a function. Notice that the "gap" in the graph does not pose any trouble; it is acceptable for a vertical line not to intersect the graph at all. ■

■ Mathematical Modeling

In Example B on page 5, we drew a scatter plot of the world population data and then found an explicit equation that approximated the behavior of the population data. The function P we used is called a *mathematical model* for the population. A **mathematical model** is a mathematical description (usually by means of a function or an equation) of a real-world scenario, such as the demand for a company's product or the life expectancy of a person at birth. Although a function used as a model may not exactly match observed data, it should be a close enough approximation to allow us to understand and analyze the situation, and perhaps to make predictions about future behavior.

Figure 16 illustrates the process of mathematical modeling. Given a real-world problem, our first task is to formulate a mathematical model by identifying and

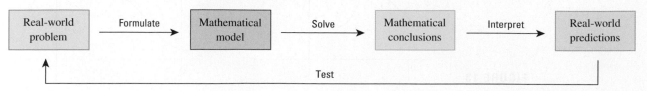

FIGURE 16 The modeling process

naming the independent and dependent variables and making assumptions that simplify the situation enough to make it mathematically tractable. We use our knowledge of the physical situation and our mathematical skills to develop equations that relate the variables. In situations where there is no physical law to guide us, we may need to collect data (either from the Internet or a library or by conducting our own experiments) and examine the data in the form of a table or a graph. In the next few sections, we will see a variety of different types of algebraic equations that are often used as mathematical models.

The second stage is to apply the mathematics that we know (such as the calculus that will be developed throughout this book) to the mathematical model that we have formulated in order to derive mathematical conclusions. Then, in the third stage, we take those mathematical conclusions and interpret them as information about the original real-world situation by way of offering explanations or making predictions. The final step is to test our predictions by checking against new real data. If the predictions don't compare well with reality, we need to refine our model or formulate a new model and start the cycle again.

Keep in mind that a mathematical model is rarely a completely accurate representation of a physical situation—it is an *idealization*. A good model simplifies reality enough to permit mathematical calculations but is accurate enough to provide valuable conclusions. It is important to realize the limitations of the model. In the end, Mother Nature and financial markets have not always been predictable!

■ Piecewise Defined Functions

In some instances, no single formula adequately describes the behavior of a quantity. A population may exhibit one growth pattern for 20 years but then change to a different trend. In such cases we can use a function with different formulas in different parts of the domain. We call such functions *piecewise defined functions*, and the next two examples illustrate the concept.

■ EXAMPLE 9 Graphing a Piecewise Defined Function

A function f is defined by

$$f(x) = \begin{cases} 1 - x & \text{if } x \leq -1 \\ x^2 & \text{if } x > -1 \end{cases}$$

Evaluate $f(-2), f(-1),$ and $f(1)$ and sketch the graph.

SOLUTION

Remember that a function is a rule. For this particular function the rule is the following: First look at the value of the input x. If it happens that $x \leq -1$, then the value of $f(x)$ is $1 - x$. On the other hand, if $x > -1$, then the value of $f(x)$ is x^2.

Since $-2 \leq -1$, we have $f(-2) = 1 - (-2) = 3$.

Since $-1 \leq -1$, we have $f(-1) = 1 - (-1) = 2$.

Since $1 > -1$, we have $f(1) = 1^2 = 1$.

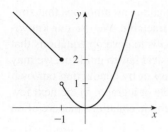

FIGURE 17

How do we draw the graph of f? We observe that if $x \le -1$, then $f(x) = 1 - x$, so the part of the graph of f that lies to the left of $x = -1$ must coincide with the line $y = 1 - x$, which has slope -1 and y-intercept 1. (Linear equations are reviewed in Section 1.3.) If $x > -1$, then $f(x) = x^2$, so the part of the graph of f that lies to the right of the line $x = -1$ must coincide with the graph of $y = x^2$, which is a parabola. This enables us to sketch the graph in Figure 17. The solid dot indicates that the point $(-1, 2)$ is included on the graph; the open dot indicates that the point $(-1, 1)$ is excluded from the graph. ■

■ **EXAMPLE 10   A Step Function**

In Example C at the beginning of this section we considered the cost $C(w)$ of mailing a large envelope with weight w. In effect, this is a piecewise defined function because, from the table of values on page 6, we have

$$C(w) = \begin{cases} 0.88 & \text{if } 0 < w \le 1 \\ 1.05 & \text{if } 1 < w \le 2 \\ 1.22 & \text{if } 2 < w \le 3 \\ 1.39 & \text{if } 3 < w \le 4 \\ \vdots \end{cases}$$

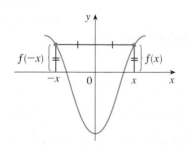

FIGURE 18

The graph is shown in Figure 18. You can see why functions similar to this one are called *step functions*—they jump from one value to the next. ■

■ | **Symmetry**

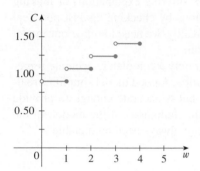

FIGURE 19 An even function

If a function f satisfies $f(-x) = f(x)$ for every number x in its domain, then f is called an **even function**. For instance, the function $f(x) = x^2$ is even because

$$f(-x) = (-x)^2 = x^2 = f(x)$$

The geometric significance of an even function is that its graph is symmetric with respect to the y-axis (see Figure 19). This means that if we have plotted the graph of f for $x \ge 0$, we obtain the entire graph simply by reflecting this portion about the y-axis.

If f satisfies $f(-x) = -f(x)$ for every number x in its domain, then f is called an **odd function**. For example, the function $f(x) = x^3$ is odd because

$$f(-x) = (-x)^3 = -x^3 = -f(x)$$

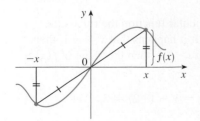

FIGURE 20 An odd function

The graph of an odd function is symmetric about the origin (see Figure 20). If we already have the graph of f for $x \ge 0$, we can obtain the entire graph by rotating this portion through $180°$ about the origin. Note that a function does not have to be either even or odd; many are neither.

■ **EXAMPLE 11   Testing for Symmetry**

Determine whether each of the following functions is even, odd, or neither even nor odd.

(a) $f(x) = x^5 + x$ **(b)** $g(x) = 1 - x^4$ **(c)** $h(x) = 2x - x^2$

SOLUTION

(a)
$$f(-x) = (-x)^5 + (-x) = (-1)^5 x^5 + (-x)$$
$$= -x^5 - x = -(x^5 + x)$$
$$= -f(x)$$

Therefore f is an odd function.

(b)
$$g(-x) = 1 - (-x)^4 = 1 - x^4 = g(x)$$

So g is even.

(c)
$$h(-x) = 2(-x) - (-x)^2 = -2x - x^2$$

Since $h(-x) \neq h(x)$ and $h(-x) \neq -h(x)$, we conclude that h is neither even nor odd. ■

The graphs of the functions in Example 11 are shown in Figure 21. Notice that the graph of h is symmetric neither about the y-axis nor about the origin.

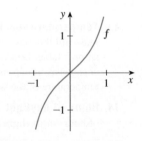
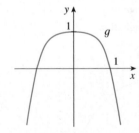
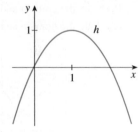

FIGURE 21   (a) Odd function   (b) Even function   (c) Neither even nor odd

---

Exercises 1.1

1. **Price function**   A nursery sells potting soil for $0.40 per pound, and the soil is available in 4-lb, 10-lb, and 50-lb bags. If $f(x)$ is the price of a bag of potting soil that weighs x pounds,
   (a) find and interpret the value of $f(10)$.
   (b) determine the domain and range of f.

2. **Price function**   An Internet retailer charges $4.99 to ship an order that totals less than $25 and $5.99 for an order up to $75, and offers free shipping for an order over $75. If $g(p)$ is the shipping cost for an order totaling p dollars, state the domain and range of g.

3. **Population function**   Let $P(t)$ be the population, in thousands, of a city t years after January 1, 2000. Interpret the equation $P(8) = 64.3$. What does $P(4.5)$ represent?

4. **Blood alcohol content**   Let $B(t)$ be the blood alcohol content (measured as the percentage by volume of alcohol in the blood) of a dinner guest t hours after her arrival. Interpret the equation $B(1.25) = 0.06$ in this context.

5. **Fuel economy**   Let $F(s)$ be the average fuel economy of a particular car, measured in miles per gallon, when the car is being driven at s mi/h. What does the equation $F(65) = 24.7$ say in this context?

6. **Loan payments** Let $N(r)$ be the number of $300 monthly payments required to repay an $18,000 auto loan when the interest rate is r percent. What does the equation $N(6.5) = 73$ say in this context?

7. The graph of a function f is given.
    (a) State the value of $f(-1)$.
    (b) Estimate the value of $f(2)$.
    (c) For what values of x is $f(x) = 2$?
    (d) Estimate the values of x such that $f(x) = 0$.
    (e) State the domain and range of f.

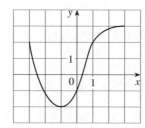

8. The graphs of f and g are given.
    (a) State the values of $f(-4)$ and $g(3)$.
    (b) For what values of x is $f(x) = g(x)$?
    (c) Estimate the solutions of the equation $f(x) = -1$.
    (d) State the domain and range of f.
    (e) State the domain and range of g.

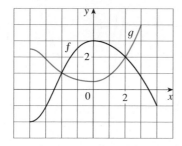

9. **Earthquakes** Figure 1 was recorded by an instrument operated by the California Department of Mines and Geology at the University Hospital of the University of Southern California in Los Angeles. Use it to estimate the range of the vertical ground acceleration function at USC during the Northridge earthquake.

10. In this section we discussed examples of ordinary, every-day functions: Population is a function of time, postage cost is a function of weight, water temperature is a function of time. Give three other examples of functions from everyday life that are described verbally. What can you say about the domain and range of each of your functions? If possible, sketch a rough graph of each function.

11. **Weight function** The graph gives the weight of a certain person as a function of age. Describe in words how

this person's weight varies over time. What do you think happened when this person was 30 years old?

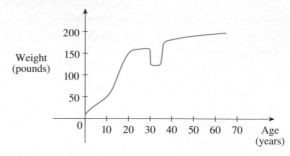

12. **Distance function** The graph gives a salesman's distance from his home as a function of time on a certain day. Describe in words what the graph indicates about his travels on this day.

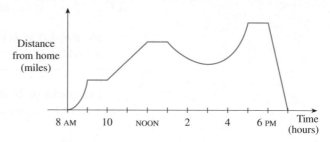

13. **Temperature function** You put some ice cubes in a glass, fill the glass with cold water, and then let the glass sit on a table. Describe how the temperature of the water changes as time passes. Then sketch a rough graph of the temperature of the water as a function of the elapsed time.

14. **Hours of daylight** Sketch a rough graph of the number of hours of daylight as a function of the time of year.

15. **Temperature function** Sketch a rough graph of the outdoor temperature as a function of time during a typical spring day.

16. **Market value** Sketch a rough graph of the market value of a new car as a function of time for a period of 20 years. Assume the car is well maintained.

17. **Retail sales** Sketch a rough graph of the average daily amount of a particular type of coffee bean (measured in pounds) sold by a store as a function of the price of the beans.

18. **Temperature function** You place a frozen pie in an oven and bake it for an hour. Then you take it out and let it cool before eating it. Describe how the temperature of the pie changes as time passes. Then sketch a rough graph of the temperature of the pie as a function of time.

19. **Lawn height** A homeowner mows the lawn every Wednesday afternoon. Sketch a rough graph of the height of the grass as a function of time over the course of a four-week period.

20. Air travel   An airplane flies from an airport and lands an hour later at another airport, 400 miles away. If t represents the time in minutes since the plane has left the terminal building, let $x(t)$ be the horizontal distance traveled and $y(t)$ be the altitude of the plane.

(a) Sketch a possible graph of $x(t)$.

(b) Sketch a possible graph of $y(t)$.

(c) Sketch a possible graph of the ground speed.

21. Phone subscribers   The number N (in millions) of US cellular phone subscribers is shown in the table. (End of year estimates are given.)

t	1996	1998	2000	2002	2004	2006
N	44	69	109	141	182	233

(a) Use the data to sketch a rough graph of N as a function of t.

(b) Use your graph to estimate the number of cell-phone subscribers at the end of 2001 and 2005.

22. Temperature   Temperature readings T (in °F) were recorded every two hours from midnight to 2:00 PM in Baltimore on September 26, 2007. The time t was measured in hours from midnight.

t	0	2	4	6	8	10	12	14
T	68	65	63	63	65	76	85	91

(a) Use the readings to sketch a rough graph of T as a function of t.

(b) Use your graph to estimate the temperature at 11:00 AM.

23. If $f(x) = 3x^2 - x + 2$, find $f(2), f(-2), f(a), f(-a)$, $f(a + 1), 2f(a), f(2a), f(a^2), [f(a)]^2$, and $f(a + h)$.

24. If $g(t) = 4t - t^2$, find $g(3), g(-1), g(x), g(x - 2)$, and $g(x + h)$.

25–30 ▪ Evaluate the difference quotient for the given function. Simplify your answer.

25. $f(x) = x^2 + 1$, $\dfrac{f(4 + h) - f(4)}{h}$

26. $f(x) = 2x^2 - x$, $\dfrac{f(t + h) - f(t)}{h}$

27. $f(x) = 4 + 3x - x^2$, $\dfrac{f(3 + h) - f(3)}{h}$

28. $f(x) = x^3$, $\dfrac{f(a + h) - f(a)}{h}$

29. $f(x) = \dfrac{1}{x}$, $\dfrac{f(x) - f(a)}{x - a}$

30. $f(x) = \dfrac{x + 3}{x + 1}$, $\dfrac{f(x) - f(1)}{x - 1}$

31–34 ▪ Find the domain of the function.

31. $f(x) = \dfrac{x}{3x - 1}$

32. $f(x) = \dfrac{3x + 4}{x^2 - x}$

33. $f(t) = \sqrt{2t + 6}$

34. $g(u) = \sqrt{u - 4} + 1.5u$

35–36 ▪ Determine whether the scatter plot is the graph of a function of x. Explain how you reached your conclusion.

35.

36.

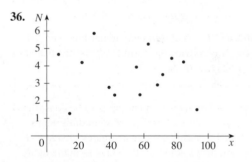

37–40 ▪ Determine whether the curve is the graph of a function of x. If it is, state the domain and range of the function.

37.

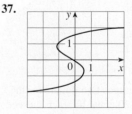

38.

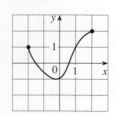

39.

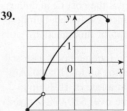

40.

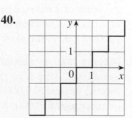

41–44 ▪ Evaluate $f(-3)$, $f(0)$, and $f(2)$ for the piecewise defined function. Then sketch the graph of the function.

41. $f(x) = \begin{cases} x + 2 & \text{if } x < 0 \\ 1 - x & \text{if } x \geq 0 \end{cases}$

42. $f(x) = \begin{cases} 3 - \frac{1}{2}x & \text{if } x < 2 \\ 2x - 5 & \text{if } x \geq 2 \end{cases}$

43. $f(x) = \begin{cases} x + 1 & \text{if } x \leq -1 \\ x^2 & \text{if } x > -1 \end{cases}$

44. $f(x) = \begin{cases} -1 & \text{if } x \leq 1 \\ 7 - 2x & \text{if } x > 1 \end{cases}$

45–48 ▪ Find a formula for the described function and state its domain.

45. A rectangle has perimeter 20 m. Express the area of the rectangle as a function of the length of one of its sides.

46. A rectangle has area 16 m². Express the perimeter of the rectangle as a function of the length of one of its sides.

47. Surface area   An open rectangular box with volume 2 m³ has a square base. Express the surface area of the box as a function of the length of a side of the base.

48. Height and width   A closed rectangular box with volume 8 ft³ has length twice the width. Express the height of the box as a function of the width.

49. Box design   A box with an open top is to be constructed from a rectangular piece of cardboard with dimensions 12 in. by 20 in. by cutting out equal squares of side x at each corner and then folding up the sides as in the figure. Express the volume V of the box as a function of x.

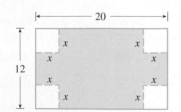

50. Taxi fares   A taxi company charges two dollars for the first mile (or part of a mile) and 20 cents for each succeeding tenth of a mile (or part). Express the cost C, in dollars, of a ride as a function of the distance x traveled, in miles, for $0 < x < 2$, and sketch the graph of this function.

51. Income tax   In a certain country, income tax is assessed as follows. There is no tax on income up to \$10,000. Any income beyond \$10,000 is taxed at a rate of 10%, up to an income of \$20,000. Any income over \$20,000 is taxed at 15%.

(a) Sketch the graph of the tax rate R as a function of the income I.

(b) How much tax is assessed on an income of \$14,000? On \$26,000?

(c) Sketch the graph of the total assessed tax T as a function of the income I.

52. The functions in Example 10 and Exercises 50 and 51(a) are called *step functions* because their graphs look like stairs. Give two other examples of step functions that arise in everyday life.

53–54 ▪ Graphs of f and g are shown. Decide whether each function is even, odd, or neither. Explain your reasoning.

53.

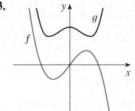

54.

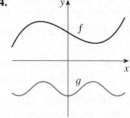

55. (a) If the point $(5, 3)$ is on the graph of an even function, what other point must also be on the graph?

    (b) If the point $(5, 3)$ is on the graph of an odd function, what other point must also be on the graph?

56. A function f has domain $[-5, 5]$ and a portion of its graph is shown.

    (a) Complete the graph of f if it is known that f is even.

    (b) Complete the graph of f if it is known that f is odd.

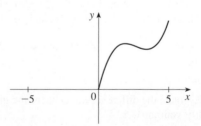

57–62 ▪ Determine whether f is even, odd, or neither. If you have a graphing calculator, use it to check your answer visually.

57. $f(x) = \dfrac{x}{x^2 + 1}$ **58.** $f(x) = \dfrac{x^2}{x^4 + 1}$

59. $f(x) = \dfrac{x}{x + 1}$ **60.** $f(x) = x|x|$

61. $f(x) = 1 + 3x^2 - x^4$ **62.** $f(x) = 1 + 3x^3 - x^5$

■ Challenge Yourself

63. If f and g are both even functions and $h(x) = f(x) + g(x)$, is h even? If f and g are both odd functions, is h odd? What if f is even and g is odd? Justify your answers.

64. Window area   A Norman window has the shape of a rectangle surmounted by a semicircle. If the perimeter of the window is 30 ft, express the area A of the window as a function of the width x of the window.

© Brooks Cole / Cengage Learning

$\longmapsto x \longrightarrow$

1.2 Combining and Transforming Functions

In this section we form new functions by combining existing functions in various ways. We also learn how to transform functions by shifting, stretching, or reflecting their graphs. These skills will enable you to use a basic set of functions, studied in the sections ahead, to design specific functions that model a wide variety of applications.

■ Combinations of Functions

Two functions f and g can be combined to form new functions using the operations of addition, subtraction, multiplication, and division in a manner similar to the way we add, subtract, multiply, and divide real numbers. For instance, we can define a new function h that is the sum of f and g by the equation $h(x) = f(x) + g(x)$. This means that the output of the new function h is the sum of the outputs of the individual functions f and g. This definition makes sense if both $f(x)$ and $g(x)$ are defined. Thus the domain of the function h consists of only those values that belong to both the domain of f and the domain of g.

Suppose a company has two different shipping centers, one on the West Coast and the other on the East Coast. If $W(t)$ is the number of packages shipped from the western facility t weeks after the start of the year, and $E(t)$ is the number of packages shipped from the eastern facility t weeks after the start of the year, then we can define a new function $N(t)$ by

$$N(t) = W(t) + E(t)$$

Thus $N(t)$ measures the combined number of packages sent from both shipping centers t weeks after the start of the year. Notice that the input for each function is the same; if the inputs of two functions are not measuring the same quantities, the sum of the functions is not meaningful.

We can subtract, multiply, or divide functions in a similar way. For instance, $k(x) = f(x) \, g(x)$ means that the output of the function k is the product of the outputs of the functions f and g. The domain of each of these new functions consists of all the numbers that appear in both the domain of f and the domain of g, with the exception that if we divide f by g, we must ensure that no division by 0 will occur. So the domain of $q(x) = f(x)/g(x)$ is all values shared by the domains of f and g where $g(x) \neq 0$.

■ **EXAMPLE 1** **Combining Two Functions**

If $N(v) = \sqrt{v}$ and $T(v) = 3 - v$, find equations and the domains for the functions $A(v) = N(v) \, T(v)$ and $B(v) = N(v)/T(v)$.

SOLUTION

The domain of $N(v) = \sqrt{v}$ is $[0, \infty)$, all the real numbers greater than or equal to 0. The domain of $T(v) = 3 - v$ is $\mathbb{R}$, all real numbers. The domain of $A(v) = N(v) \, T(v)$ consists of those values that are shared by both these domains, namely $[0, \infty)$. The formula for the product function is

$$A(v) = N(v) \, T(v) = \sqrt{v} \, (3 - v)$$

Similarly,

$$B(v) = \frac{N(v)}{T(v)} = \frac{\sqrt{v}}{3 - v}$$

Notice that $T(v) = 0$ when $v = 3$, so 3 must be excluded from the domain of B. Thus the domain of B is all real numbers greater than or equal to 0, except 3. In set-builder notation, we write $\{v \mid v \geq 0, v \neq 3\}$. ■

■ **EXAMPLE 2** **Combining Revenue and Cost Functions**

Suppose the annual revenue, in millions of dollars, of a company is $R(t) = 0.2t^2 + 3t + 5$, where t is measured in years and $t = 0$ corresponds to the year 2000. The annual cost, in millions of dollars, for the company is $C(t) = 4t + 9$.

(a) Find a formula for the function $P(t) = R(t) - C(t)$.

(b) Compute and interpret $P(7)$.

SOLUTION

(a)
$$P(t) = R(t) - C(t) = (0.2t^2 + 3t + 5) - (4t + 9)$$
$$= 0.2t^2 + 3t + 5 - 4t - 9$$
$$= 0.2t^2 - t - 4$$

(b) We can find $P(7)$ by subtracting the output values of the functions R and C, or we can use the formula from part (a) directly:

$$P(7) = 0.2(7^2) - 7 - 4 = -1.2$$

Notice that $P(t)$ is the annual revenue minus the annual cost, so it represents the annual profit for the company. Since $t = 7$ corresponds to 2007, and the output is negative, we know that during 2007 the company lost 1.2 million dollars.   ■

■ Composition of Functions

There is another way of combining two functions to form a new function. As a simple illustration, suppose that a company's annual profit for year t is given by $P(t)$ and the total amount of tax the company pays, $f(P)$, is determined by its profit P. Since the tax paid is a function of profit and profit is, in turn, a function of t, it follows that the amount of tax paid is ultimately a function of t. In effect, the output of the profit function P can be used as the input for the tax function f, and $f(P(t))$ is the amount of tax the company paid during year t. This new function is called the *composition* of the functions P and f.

If we have equations for two functions, we can write a formula for their composition. For example, suppose $y = f(t) = \sqrt{t}$ and $t = g(x) = x^2 + 1$. Now y is a function of t and t is a function of x, so y can be considered as a function of x. We compute this by substitution:

$$y = f(t) = f(g(x)) = f(x^2 + 1) = \sqrt{x^2 + 1}$$

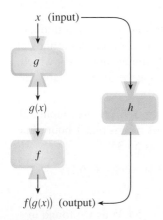

x (input)

g

$g(x)$

h

f

$f(g(x))$ (output)

FIGURE 1
The h machine is composed of the g machine (first) and then the f machine.

> ■ **Definition**   Given two functions f and g, the **composition** of f and g is defined by
>
> $$h(x) = f(g(x))$$

The domain of $h(x) = f(g(x))$ is the set of all values x in the domain of g such that $g(x)$ is in the domain of f. In other words, $f(g(x))$ is defined whenever both $g(x)$ and $f(g(x))$ are defined. It is probably easier to picture the composition of f and g with a machine diagram (see Figure 1).

■ EXAMPLE 3   Composing Two Functions

Let $f(x) = x^2$ and $g(x) = x - 3$. If $h(x) = f(g(x))$ and $k(x) = g(f(x))$, compute $h(5)$ and $k(5)$.

SOLUTION

First let's trace the path the input 5 takes under the function h. Since $h(5) = f(g(5))$, we first input 5 into the inner function g, where $g(5) = 2$. The output 2 is then used as an input into the outer function f, which gives an output of $f(2) = 2^2 = 4$. Thus $h(5) = f(g(5)) = f(2) = 4$. Similarly, $k(5) = g(f(5)) = g(25) = 22$. Notice that the original input always goes through the inner function first, and the resulting output is used as an input into the outer function.

We can also write formulas for h and k:

$$h(x) = f(g(x)) = f(x - 3) = (x - 3)^2$$
$$k(x) = g(f(x)) = g(x^2) = x^2 - 3$$

Then it is easy to compute

$$h(5) = (5 - 3)^2 = 2^2 = 4 \qquad \text{and} \qquad k(5) = 5^2 - 3 = 25 - 3 = 22 \quad ■$$

NOTE: You can see from Example 3 that, in general, $f(g(x)) \neq g(f(x))$. Remember, the notation $f(g(x))$ means that the function g is applied first and then f is applied second. In Example 3, $f(g(x))$ is the function that *first* subtracts 3 and *then* squares; $g(f(x))$ is the function that *first* squares and *then* subtracts 3.

■ EXAMPLE 4 Interpreting a Composition of Functions

The altitude of a small airplane t hours after taking off is given by $A(t) = -2.8t^2 + 6.7t$ thousand feet, where $0 \leq t \leq 2$. The air temperature in the area at an altitude of x thousand feet is $f(x) = 68 - 3.5x$ degrees Fahrenheit.

(a) What does the composition $h(t) = f(A(t))$ measure?

(b) Compute $h(1)$ and interpret your result in this context.

(c) Find a formula for $h(t)$.

(d) Does $A(f(x))$ give a meaningful result in this context?

SOLUTION

(a) The hours t that the airplane has been flying is first used as an input into the inner function A, which outputs the altitude of the plane $A(t)$ in thousands of feet. This altitude in turn is used as an input into the outer function f, which outputs a temperature in degrees Fahrenheit. Thus h is the air temperature at the airplane's location t hours after take-off.

(b) The input 1 first enters the function A, giving $A(1) = 3.9$. We then input 3.9 into the function f, which gives $f(3.9) = 54.35$. This means that 1 hour after take-off, the air temperature at the plane's location is $54.35°$F.

(c)
$$h(t) = f(A(t)) = f(-2.8t^2 + 6.7t) = 68 - 3.5(-2.8t^2 + 6.7t)$$
$$= 9.8t^2 - 23.45t + 68$$

Using this direct formula, you can verify that $h(1) = 54.35$ as we found in part (b).

(d) Although we could compute a formula for $A(f(x))$, it wouldn't be a meaningful quantity here. The inner function f outputs a temperature in °F, but this is not an appropriate value to pass to the outer function A as an input, because A is a function of t, a number of hours. ■

So far we have used composition to build complicated functions from simpler ones. But we will see in later chapters that in calculus, it is often useful to be able to *decompose* a complicated function into simpler ones, as in the following example.

■ EXAMPLE 5 Decomposing a Function

If $L(t) = (2t - 1)^3$, find functions f and g such that $L(t) = f(g(t))$.

SOLUTION

The formula for L says: First double t and subtract 1, then cube the result. One option is to think of $2t - 1$ as the inner function and call it g. Then

It is not important what letter we use to represent the variable in the outer function f. The function $f(x) = x^3$ is the same function as $f(a) = a^3$ or $f(q) = q^3$.

$g(t) = 2t - 1$ and $L(t) = (g(t))^3$. The outer function is the cubing function, so if we let $f(x) = x^3$, then

$$L(t) = f(g(t)) = f(2t - 1) = (2t - 1)^3$$

Note that there are other choices we could have made, such as $g(t) = 2t$ and $f(x) = (x - 1)^3$, but the first solution is probably the most useful one.   ▪

▪ Transformations of Functions

Next we discuss how to modify a function to change the shape or location of its graph. Armed with these techniques, we can use familiar graphs to design functions that will fit a wide variety of applications. The first of these *transformations* we will consider are called **translations**. If you compare the graphs of $y = f(x)$ and $y = f(x) + 3$ in Figure 2, you will notice that the shapes are identical, but the second graph is located 3 units higher on the coordinate plane. The second function increases each output of the first function by 3, so each point on its graph moves 3 units higher. In effect, we have shifted the entire graph upward 3 units. Similarly, $y = f(x) - 3$ shifts the graph of $y = f(x)$ downward 3 units.

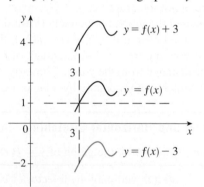

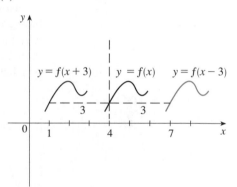

FIGURE 2
Vertical translations of the graph of f

FIGURE 3
Horizontal translations of the graph of f

Next compare the graph of $y = f(x)$ with the graph of $y = f(x + 3)$ in Figure 3. The graph of $y = f(x + 3)$ is the same as the graph of $y = f(x)$ but shifted 3 units to the *left*. To see why this is the case, note that if $g(x) = f(x + 3)$, then $g(1) = f(1 + 3) = f(4)$, so the output corresponding to $x = 4$ in the graph of f is plotted with $x = 1$ in the graph of g, 3 units to the left. Similarly, $y = f(x - 3)$ shifts the graph of f to the right 3 units.

▪ Vertical and Horizontal Shifts   Suppose c is a positive number.

translation of the graph of $y = f(x)$	equation
shift c units upward	$y = f(x) + c$
shift c units downward	$y = f(x) - c$
shift c units to the right	$y = f(x - c)$
shift c units to the left	$y = f(x + c)$

We can also **stretch** (or compress) graphs. For instance, compare the graphs of $y = f(x)$ and $y = 2f(x)$ in Figure 4. The second graph has a shape similar to the first, but it has been stretched vertically by a factor of 2. Each output of the original function is doubled, so the vertical distance between each point of the graph and the x-axis is doubled. If we graph $y = \frac{1}{2}f(x)$, each output is halved, so the graph appears to be compressed vertically (toward the x-axis).

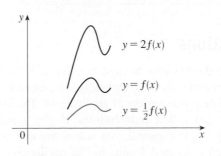

FIGURE 4
Stretching the graph of f vertically

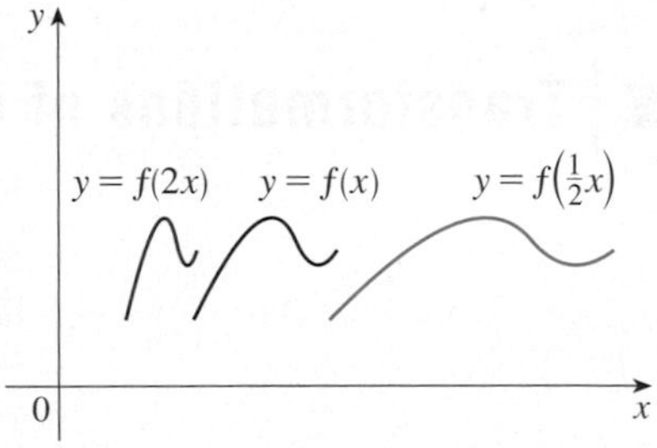

FIGURE 5
Stretching the graph of f horizontally

Now compare the graphs of $y = f(x)$ and $y = f(2x)$ in Figure 5. This time we have compressed the graph horizontally (toward the y-axis) by a factor of 2. To see why this occurs, observe that if $g(x) = f(2x)$, then $g(1) = f(2 \cdot 1) = f(2)$, so the output corresponding to $x = 2$ in the graph of f is plotted with $x = 1$ in the graph of g, half the distance from the y-axis. Similarly, the graph of $y = f(\frac{1}{2}x)$ is the graph of $y = f(x)$ stretched horizontally by a factor of 2.

▪ **Vertical and Horizontal Stretching**   Suppose $c > 1$.

transformation of the graph of $y = f(x)$	equation
stretch vertically by a factor of c	$y = cf(x)$
compress vertically by a factor of c	$y = \frac{1}{c}f(x)$
compress horizontally by a factor of c	$y = f(cx)$
stretch horizontally by a factor of c	$y = f(\frac{1}{c}x)$

Finally, we can **reflect** graphs in either a vertical or horizontal direction. If we compare the graphs of $y = f(x)$ and $y = -f(x)$ in Figure 6, the graph of $y = -f(x)$ is the graph of $y = f(x)$ but flipped upside down. Each point (x, y) on the original graph is replaced by the point $(x, -y)$, so the graph appears to be reflected about the

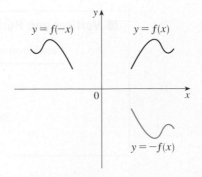

FIGURE 6
Reflecting the graph of f

x-axis. If you compare the graph of $y = f(x)$ with the graph of $y = f(-x)$ in Figure 6, you'll notice that this time the x-values are made opposite, so the graph appears reflected about the y-axis.

▪ Vertical and Horizontal Reflections

reflection of the graph of $y = f(x)$	equation
reflect about the x-axis	$y = -f(x)$
reflect about the y-axis	$y = f(-x)$

Figure 7 illustrates several combinations of various transformations.

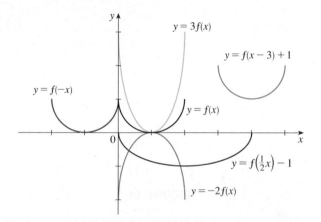

FIGURE 7

▪ EXAMPLE 6   Sketching Transformations of a Function

Given the graph of $y = \sqrt{x}$, use transformations to graph $y = \sqrt{x} - 2$, $y = \sqrt{x - 2}$, $y = -\sqrt{x}$, $y = 2\sqrt{x}$, and $y = \sqrt{-x}$.

SOLUTION

The graph of the square root function $y = \sqrt{x}$ is shown in Figure 8(a). If we let $f(x) = \sqrt{x}$, then $y = \sqrt{x} - 2 = f(x) - 2$, so the graph is shifted 2 units downward. Similarly, $y = \sqrt{x - 2} = f(x - 2)$ shifts the graph 2 units to the right, $y = -\sqrt{x} = -f(x)$ reflects the graph about the x-axis, $y = 2\sqrt{x} = 2f(x)$ stretches the graph vertically by a factor of 2, and $y = \sqrt{-x} = f(-x)$ reflects the graph about the y-axis.

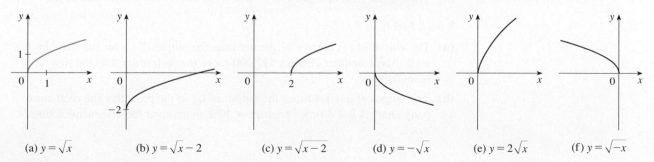

(a) $y = \sqrt{x}$ (b) $y = \sqrt{x} - 2$ (c) $y = \sqrt{x - 2}$ (d) $y = -\sqrt{x}$ (e) $y = 2\sqrt{x}$ (f) $y = \sqrt{-x}$

FIGURE 8

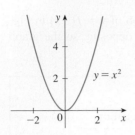

FIGURE 9

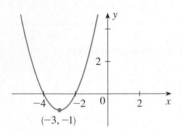

FIGURE 10

▪ EXAMPLE 7 Sketching Multiple Transformations

Given the graph of the function $y = x^2$ shown in Figure 9, sketch the graphs of
(a) $f(x) = (x + 3)^2 - 1$ and **(b)** $g(x) = -\frac{1}{3}x^2 + 2$.

SOLUTION

(a) The graph of $y = (x + 3)^2$ is the graph of $y = x^2$ shifted 3 units to the left.
If we then shift the graph down 1 unit, we have the graph of
$f(x) = (x + 3)^2 - 1$ shown in Figure 10.

(b) The graph of $y = -\frac{1}{3}x^2$ is the graph of $y = x^2$ compressed vertically by a
factor of 3 and reflected across the x-axis. [See Figure 11(a).] Shift the resulting graph up 2 units to arrive at the graph of $g(x) = -\frac{1}{3}x^2 + 2$ as shown in
Figure 11(b).

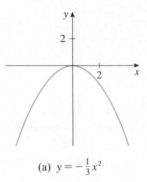

(a) $y = -\frac{1}{3}x^2$

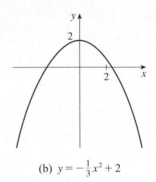

(b) $y = -\frac{1}{3}x^2 + 2$

FIGURE 11

▪ EXAMPLE 8 Interpreting Transformations of Functions

Let $C(x)$ be the amount, in thousands of dollars, that a manufacturer charges for
an order of x thousand computer memory chips.

(a) The price (in thousands of dollars) that a rival manufacturer charges to
provide x thousand chips is given by $f(x) = C(x) + 12$. How does the rival
company's price compare to that of the first company?

(b) What if the amount that the rival company charges for an order of x thousand
chips is given by $g(x) = 1.4C(x)$?

(c) What if the rival charges $h(x) = C(x - 2)$ to provide x thousand chips?

SOLUTION

(a) The output of f is always 12 greater than the output of C (for the same input),
so the rival supplier charges $12,000 more for each order than the first
manufacturer.

(b) The output of g is 1.4 times the output of C, so the price that the rival company charges is 1.4 times greater, or 40% more, than the first manufacturer's
price.

(c) The graph of $h(x)$ is the graph of $C(x)$ shifted two units to the right. This means that for the same price, the rival manufacturer will supply 2000 more chips than the first manufacturer. For instance, if $x = 10$ then $h(10) = C(8)$. ▪

▪ Exercises 1.2

1. Class attendance Let $M(t)$ be the number of male students and $F(t)$ the number of female students that attended a math class at a local university on day t of this year. If we define a function g where $g(t) = M(t) + F(t)$, describe what g measures.

2. Price of gas Let $A(x)$ be the total amount charged to a consumer for x gallons of premium gasoline at a particular gas station, and let $T(x)$ be the total amount of tax the station pays for x gallons of the gasoline. What does the function $f(x) = A(x) - T(x)$ measure?

3. Bank holdings Let $g(n)$ be the amount of gold, in ounces, that a bank has in its vault at the end of the nth day of this year, and let $v(n)$ be the value, in dollars, of one ounce of gold at the end of the nth day of this year. What does the function $f(n) = g(n) v(n)$ measure?

4. Investments Let $P(t)$ be the daily closing price of one share of General Electric stock t days after January 1, 2010, and let $Q(t)$ be the number of shares owned by a pension fund at the end of that same day. What does the function $g(t) = P(t) Q(t)$ measure?

5. Crops A farm devotes $A(x)$ acres of its land to growing corn during year x. If $B(x)$ is the number of bushels of corn the farm yielded during year x, what does the function $C(x) = B(x)/A(x)$ represent?

6. Phone usage Let $M(n)$ be the total number of minutes Kathi talked on her cellular phone during the nth month of last year, and let $C(n)$ be the amount she paid for her phone service during that month. What does the function $h(n) = C(n)/M(n)$ represent?

7. Salary An employee's annual salary, in thousands of dollars, is given by $S(t) = 42 + 1.8t$, where t is the year with $t = 0$ corresponding to 2000, and $C(t) = 16.4 + 0.6t$ is the total amount of commissions, in thousands of dollars, the employee earned that year.

(a) Find a formula for the function $f(t) = S(t) + C(t)$.

(b) Compute $f(4)$ and interpret your result in this context.

8. Revenue and profit The annual revenue of a small store, in thousands of dollars, is given by $R(t) = 645 + 21t$, where t is the year, with $t = 0$ corre-

sponding to 2000. Similarly, the store's annual profit is given by $P(t) = 175 + 16t - 0.3t^2$.

(a) Write a formula for the annual cost function $C(t)$ for the store.

(b) Compute $C(3)$ and interpret your result in this context.

9. If $f(x) = x^2 - 5x$ and $g(x) = 3x + 12$, write a formula for each of the following functions.

(a) $A(x) = f(x) + g(x)$ **(b)** $B(x) = f(x) - g(x)$
(c) $C(x) = f(x) g(x)$ **(d)** $D(x) = f(x)/g(x)$

10. If $p(x) = \sqrt{x + 1}$ and $q(x) = 2x - 4$, write a formula for each of the following functions. What is the domain?

(a) $A(x) = p(x) + q(x)$ **(b)** $B(x) = p(x) - q(x)$
(c) $C(x) = p(x) q(x)$ **(d)** $D(x) = p(x)/q(x)$

11. If $f(x) = x^2 + 1$, $g(t) = 4t - 2$, $A(t) = f(g(t))$, and $B(x) = g(f(x))$, compute $A(3)$ and $B(3)$.

12. If $h(n) = 2 - 5n$, $p(n) = n^2 - 3$, $u(n) = h(p(n))$, and $v(n) = p(h(n))$, compute $u(2)$ and $v(2)$.

13. If $M(t) = t + \sqrt{t}$, $N(t) = 3t + 7$, $C(t) = M(N(t))$, and $D(t) = N(M(t))$, compute $C(3)$ and $D(4)$.

14. If $f(t) = t^3 + 2$, $g(x) = 2x + 3$, $p(x) = f(g(x))$, and $r(t) = g(f(t))$, compute $p(-1)$ and $r(-2)$.

15–20 ▪ Find the functions $p(x) = f(g(x))$ and $q(x) = g(f(x))$.

15. $f(x) = x^2 - 1$, $g(x) = 2x + 1$

16. $f(x) = 1 - x^3$, $g(x) = 1/x$

17. $f(x) = x^3 + 2x$, $g(x) = 1 - \sqrt{x}$

18. $f(x) = 1 - 3x$, $g(x) = 5x^2 + 3x + 2$

19. $f(x) = x + \dfrac{1}{x}$, $g(x) = x + 2$

20. $f(x) = \sqrt{2x + 3}$, $g(x) = x^2 + 1$

21. Surfboard production Let $N(t)$ be the number of surfboards a manufacturer produces during year t. If $P(x)$ is the profit, in thousands of dollars, the manufacturer earns by selling x surfboards, what does the function $f(t) = P(N(t))$ represent?

22. Car maintenance If $C(m)$ is the average annual cost for maintaining a Honda Civic that has been driven m thousand miles and $f(t)$ is the number of miles on Sean's Honda Civic t years after he purchased it, what does the function $g(t) = C(f(t))$ represent?

23. Carpooling As fuel prices increase, more drivers carpool. The function $f(p)$ gives the average percentage of commuters who carpool when the cost of gasoline is p dollars per gallon. If $g(t)$ is the average monthly price (per gallon) of gasoline, where t is the time in months beginning January 1, 2011, which composition gives a meaningful result, $f(g(t))$ or $g(f(p))$? Describe what the resulting function measures.

24. Home prices People are moving into a small community and driving the home prices higher. Suppose $p(t)$ is the population of the community t years after January 1, 2000, and $f(n)$ is the median home price when the population of the area is n people. Which function gives a meaningful result, $p(f(n))$ or $f(p(t))$? What does it represent in this context?

25. Scuba diving The pressure a scuba diver experiences at a depth of d feet is approximately $P(d) = 14.7 + 0.433d$ PSI (pounds per square inch). Suppose that for the first portion of Paul's dive, his depth after m minutes is $f(m) = 0.5m + 3\sqrt{m}$ feet.

(a) Write a formula for the function $A(m) = P(f(m))$. What does A measure?

(b) Compute $A(25)$ and interpret your result in this context.

26. Electric power A town produces a portion of its electricity using windmills. Suppose that with winds that average s mi/h, the windmills generate $p(s) = \sqrt{1400s}$ kilowatts of power. The town estimates that $f(x) = 0.34x$ is the number of people that can be supported by a power level of x kilowatts.

(a) Write a formula for the function $r(s) = f(p(s))$. What does r measure?

(b) Compute $r(18)$ and interpret your result in this context.

27. Use the given graphs of f and g to evaluate each expression.

(a) $f(g(2))$ (b) $g(f(0))$

(c) $f(g(0))$ (d) $f(f(4))$

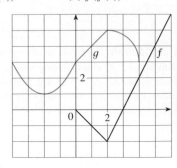

28. Use the table to evaluate each expression.

(a) $f(g(1))$ (b) $g(f(1))$

(c) $g(f(3))$ (d) $f(g(6))$

x	1	2	3	4	5	6
$f(x)$	3	1	4	2	2	5
$g(x)$	6	3	2	1	2	3

29–32 ▪ Find functions f and g so that $h(x) = f(g(x))$.

29. $h(x) = (x^2 + 1)^{10}$ **30.** $h(x) = \sqrt{x^3 - 1}$

31. $h(x) = \sqrt{2x^2 + 5}$ **32.** $h(x) = \dfrac{1}{x^2 - 5}$

33. Suppose the graph of f is given. Write equations (in terms of f) for the graphs that are obtained from the graph of f as follows.

(a) Shift 4 units upward.

(b) Shift 4 units downward.

(c) Shift 4 units to the right.

(d) Shift 4 units to the left.

(e) Reflect about the x-axis.

(f) Reflect about the y-axis.

(g) Stretch vertically by a factor of 3.

(h) Shrink vertically by a factor of 3.

34. Explain how the following graphs are obtained from the graph of $y = f(x)$.

(a) $y = 5f(x)$ (b) $y = f(x - 5)$

(c) $y = -f(x)$ (d) $y = -5f(x)$

(e) $y = f(5x)$ (f) $y = 5f(x) - 3$

35. The graph of $y = f(x)$ is given. Match each equation with its graph and give reasons for your choices.

(a) $y = f(x - 4)$ (b) $y = f(x) + 3$

(c) $y = \frac{1}{3}f(x)$ (d) $y = -f(x + 4)$

(e) $y = 2f(x + 6)$

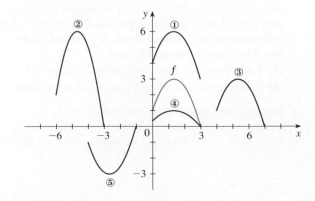

36. The graph of f is given. Draw the graph of each of the following functions.

(a) $y = f(x + 4)$ (b) $y = f(x) + 4$

(c) $y = 2f(x)$ (d) $y = -\frac{1}{2}f(x) + 3$

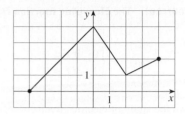

37. The graph of f is given. Use it to graph the following functions.

(a) $y = f(2x)$ (b) $y = f\left(\frac{1}{2}x\right)$

(c) $y = f(-x)$ (d) $y = -f(-x)$

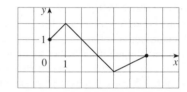

38–42 ▪ The graph of $y = \sqrt{x}$ is shown in Figure 8(a). Use transformations to graph each of the following functions.

38. $y = \sqrt{x} + 3$

39. $y = \sqrt{x + 3}$ **40.** $y = -\frac{1}{2}\sqrt{x}$

41. $y = -\sqrt{x - 1}$ **42.** $y = \sqrt{-x} + 2$

43–46 ▪ The graph of $y = x^2$ is shown in Figure 9. Use transformations to graph each of the following functions.

43. $y = -x^2 + 2$ **44.** $y = (x - 1)^2 - 4$

45. $f(x) = \frac{1}{4}x^2 - 3$

46. $g(x) = -(x + 5)^2 + 3$

---

47. Given the graph of $y = \sqrt{x}$ as shown in Figure 8(a), use transformations to create a function whose graph is as shown.

(a)   (b)

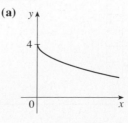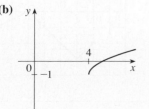

48. Given the graph of $y = x^2$ as shown in Figure 9, use transformations to create a function whose graph is as shown.

(a)   (b)

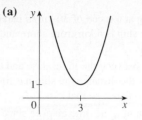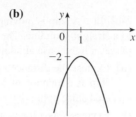

49. Water depth The depth, in feet, of water in a reservoir is given by $f(t)$, where t is the time in months beginning January 1, 2000.

(a) If a second reservoir's water depth is given by $g(t) = f(t) - 15$, how do the water levels of the two reservoirs compare?

(b) What if the second reservoir's depth is $g(t) = f(t - 2)$?

(c) What if the second reservoir's depth is $g(t) = f(t + 2)$?

(d) What if the second reservoir's depth is $g(t) = 0.8f(t)$?

50. Temperature The temperature, in degrees Fahrenheit, at Bob Hope Airport in California x days after the start of the year is given by $T(x)$.

(a) If the temperature x days after the start of the year at Los Angeles International Airport (LAX) is given by $h(x) = T(x) - 8$, how does the temperature at LAX compare to the temperature at Bob Hope Airport?

(b) What if the temperature at LAX is $h(x) = 0.9T(x)$?

51. Music sales The number of songs sold, in thousands, during the nth month of last year by an Internet music service is $A(n)$.

(a) If a rival service sold $B(n) = 1.3A(n)$ songs, how does the number of songs sold by the rival service compare to that of the first service?

(b) What if the rival service sold $B(n) = A(n) + 23$ songs?

(c) What if the rival service sold $B(n) = A(n - 1) + 5$ songs?

52. Bear population An ecologist has been observing the populations of brown bears and black bears in a region of Alaska. Let $R(t)$ represent the estimated number of brown bears, and $L(t)$ the estimated number of black bears, t years after January 1, 1990.

(a) If there are always 500 more black bears than brown bears, write a formula [in terms of $R(t)$] for $L(t)$.

(b) If there are always 15% fewer black bears than brown bears, write a formula [in terms of $R(t)$] for $L(t)$.

(c) If the number of black bears at any point in time matches the number of brown bears two years prior, write a formula [in terms of $R(t)$] for $L(t)$.

53. Motion   A ship is moving at a speed of 30 km/h parallel to a straight shoreline. The ship is 6 km from shore and it passes a lighthouse at noon.

(a) Express the distance s between the lighthouse and the ship as a function of d, the distance the ship has traveled since noon; that is, find f so that $s = f(d)$.

(b) Express d as a function of t, the time elapsed since noon; that is, find g so that $d = g(t)$.

(c) Find $f(g(t))$. What does this function represent?

54. Motion   An airplane is flying at a speed of 350 mi/h at an altitude of one mile and passes directly over a radar station at time $t = 0$.

(a) Express the horizontal distance d (in miles) that the plane has flown as a function of t.

(b) Express the distance s between the plane and the radar station as a function of d.

(c) Use composition to express s as a function of t.

55. Water ripple   A stone is dropped into a lake, creating a circular ripple that travels outward at a speed of 60 cm/s.

(a) Express the radius r of this circle as a function of the time t in seconds.

(b) If A is the area of this circle as a function of the radius, find $A(r(t))$ and interpret it.

56. Electric current   The **Heaviside function** H is defined by

$$H(t) = \begin{cases} 0 & \text{if } t < 0 \\ 1 & \text{if } t \geq 0 \end{cases}$$

It is used in the study of electric circuits to represent the sudden surge of electric current, or voltage, when a switch is instantaneously turned on.

(a) Sketch the graph of the Heaviside function.

(b) Sketch the graph of the voltage $V(t)$ in a circuit if the switch is turned on at time $t = 0$ and 120 volts are applied instantaneously to the circuit. Write a formula for $V(t)$ in terms of $H(t)$.

(c) Sketch the graph of the voltage $V(t)$ in a circuit if the switch is turned on at time $t = 5$ seconds and 240 volts are applied instantaneously to the circuit. Write a formula for $V(t)$ in terms of $H(t)$. (Note that starting at $t = 5$ corresponds to a translation.)

▪ Challenge Yourself

57–58 ▪ Find a formula for $p(x) = f(g(h(x)))$.

57. $f(x) = \sqrt{x - 1}, \quad g(x) = x^2 + 2, \quad h(x) = x + 3$

58. $f(x) = 2x - 1, \quad g(x) = x^2, \quad h(x) = 1 - x$

59. The graph of a function $y = f(x)$ is given.

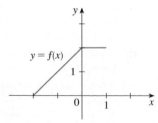

Write an equation (in terms of f) for the function whose graph is as shown.

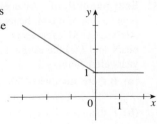

60–61 ▪ If f is the graph given in Exercise 37, write a formula (in terms of f) for the function whose graph is shown.

60.

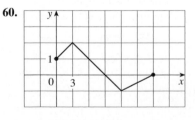

61.

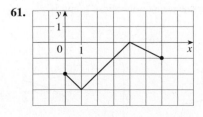

62–63 ▪ If f is the graph given in Exercise 36, write a formula (in terms of f) for the function whose graph is shown.

62.

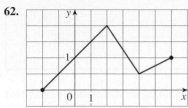

63.

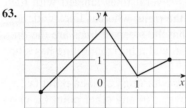

64. Use the given graphs of f and g to estimate the value of $h(x) = f(g(x))$ for $x = -5, -4, -3, \ldots, 5$. Use these estimates to sketch a rough graph of h.

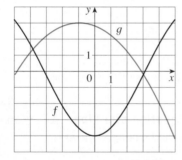

65–66 ▪ The graph of $y = \sqrt{3x - x^2}$ is given. Use transformations to create a function whose graph is as shown.

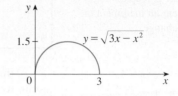

65.   **66.**

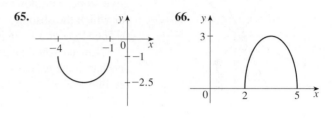

67. Let f and g be linear functions with equations $f(x) = m_1x + b_1$ and $g(x) = m_2x + b_2$. If $h(x) = f(g(x))$, is h also a linear function? If so, what is the slope of its graph?

68. If you invest x dollars at 4% interest compounded annually, then the amount $A(x)$ of the investment after one year is $A(x) = 1.04x$. Find formulas for $A(A(x))$, $A(A(A(x)))$, and $A(A(A(A(x))))$. What do these compositions represent? Find a formula for the composition of n copies of A.

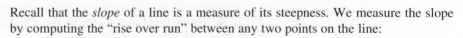

1.3   Linear Models and Rates of Change

Of the many different types of functions that can be used to model relationships observed in the real world, one of the most common is the *linear function*. When we say that one quantity is a linear function of another, we mean that the graph of the function is a line.

▪ Review of Lines

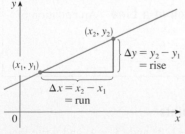

FIGURE 1

Recall that the *slope* of a line is a measure of its steepness. We measure the slope by computing the "rise over run" between any two points on the line:

$$\text{slope} = \frac{\text{rise}}{\text{run}}$$

As we can see in Figure 1, the rise is simply the difference or change in y-values between the two points and the run is the difference in x-values. Thus we can think of the slope as the "change in y over the change in x."

The Greek letter Δ (capital delta) is used to represent an increment or an amount of change.

(1) ■ Definition   The **slope** of the line that passes through the points (x_1, y_1) and (x_2, y_2) is

$$m = \frac{\text{change in } y}{\text{change in } x} = \frac{\Delta y}{\Delta x} = \frac{y_2 - y_1}{x_2 - x_1}$$

A line has the same slope everywhere, so it makes no difference which two points we use to compute slope. Figure 2 shows several lines labeled with their slopes. Lines with positive slope slant upward to the right, whereas lines with negative slope slant downward to the right. Notice that the steepest lines are the ones for which the absolute value of the slope is largest, and a horizontal line has slope 0. The slope of a vertical line is not defined.

Now let's find an equation of the line that passes through a given point (x_1, y_1) and has slope m. If we compute the slope from (x_1, y_1) to any other point (x, y) on the line, we get

$$m = \frac{y - y_1}{x - x_1}$$

which can be written in the form

$$y - y_1 = m(x - x_1)$$

This equation is satisfied by all points on the line, including (x_1, y_1), and *only* by points on the line. Therefore it is an equation of the given line.

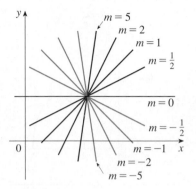

FIGURE 2

(2) ■ Point-Slope Form of the Equation of a Line   An equation of the line passing through the point (x_1, y_1) and having slope m is

$$y - y_1 = m(x - x_1)$$

Equation 2 becomes even simpler if we use the point at which a (nonvertical) line intersects the y-axis. The x-coordinate there is 0 and the y-value, called the **y-intercept**, is traditionally denoted by b. (See Figure 3.) Thus the line passes through the point $(0, b)$ and Equation 2 becomes

$$y - b = m(x - 0)$$

which simplifies to the following:

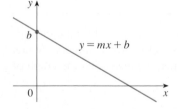

FIGURE 3

(3) ■ Slope-Intercept Form of the Equation of a Line   An equation of the line with slope m and y-intercept b is

$$y = mx + b$$

■ EXAMPLE 1   A Line through Two Points

Find an equation of the line through the points $(-1, 2)$ and $(3, -4)$ and write the equation in slope-intercept form.

SOLUTION

By Definition 1 the slope of the line is

$$m = \frac{-4 - 2}{3 - (-1)} = -\frac{3}{2}$$

Using Equation 2 with $x_1 = -1$ and $y_1 = 2$, we obtain

$$y - 2 = -\frac{3}{2}(x + 1)$$

which can be written as

$$y - 2 = -\frac{3}{2}x - \frac{3}{2} \quad \text{or} \quad y = -\frac{3}{2}x + \frac{1}{2} \qquad ■$$

See Appendix B for a more detailed review of slope and lines, along with additional examples and exercises.

The equation of a nonvertical line can always be written in slope-intercept form, which reveals the slope and y-intercept at a glance.

■ **EXAMPLE 2   Graphing a Linear Equation**

Sketch the graph of the equation $y = -\frac{3}{4}x + 5$.

SOLUTION

The equation is in slope-intercept form, which allows us to identify the graph as a line with slope $m = -\frac{3}{4}$ and y-intercept 5. To sketch the line we start at the point $(0, 5)$. The slope is negative, so we move through a "rise" of -3 (actually a downward movement) and a run of 4 (to the right) to arrive at the point $(4, 2)$. The graph is the line through these two points as shown in Figure 4.

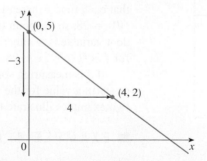

FIGURE 4

Rate of Change and Linear Functions

We defined the slope of a line as the ratio of the change in y, Δy, to the change in x, Δx. Thus we can interpret the slope as the **rate of change** of y with respect to x. If f is a linear function, then its graph is a line and we can think of the slope as the ratio of the change in output to the change in input. In this context, the slope measures the rate of change of the function.

$$\text{rate of change} = \frac{\text{change in output}}{\text{change in input}} = \frac{\Delta y}{\Delta x} = \text{slope of line}$$

For instance, a slope of 4 means that a change in input will cause a change in output that is four times larger. The slope of a given line is the same at all points, so a characteristic feature of linear functions is that *the rate of change is constant*:

Linear functions grow at a constant rate.

A rate of change is always measured by a ratio of units: output units per input unit.

■ EXAMPLE 3   **Slope of a Linear Function**

A company that produces snowboards has seen its annual sales increase linearly. In 2005, it sold 31,300 snowboards, and it sold 38,200 snowboards in 2011. Compute the slope of the linear function that gives annual sales as a function of the year. What does the slope represent in this context?

SOLUTION
The slope is

$$m = \frac{\text{change in output}}{\text{change in input}} = \frac{38{,}200 - 31{,}300}{2011 - 2005} = \frac{6900}{6} = 1150$$

and the units are number of snowboards per year. Thus the number of snowboards the company produces is increasing at a rate of 1150 per year.   ■

Because the graph of a linear function is a line, we can write an equation for a linear function using the slope-intercept form given by Equation 3:

$$f(x) = mx + b$$

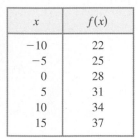

x	$f(x)$
-10	22
-5	25
0	28
5	31
10	34
15	37

For instance, consider a linear function with values as given in the table. Observe that each time x increases by 5, $y = f(x)$ increases by 3, so the slope is $\frac{3}{5}$. We have $f(0) = 28$, so the point $(0, 28)$ is on the line and 28 is the y-intercept. (If the dependent variable is A rather than y, we would call it the A-intercept.) Thus an equation for f is $f(x) = \frac{3}{5}x + 28$.

If f is measuring some quantity, we can think of 28 (when $x = 0$) as the initial or starting value for the function. Values change from there at a rate of $\frac{3}{5}$. The next two examples illustrate this point.

■ EXAMPLE 4   **A Linear Cost Function**

The owner of a car-wash business estimates that it costs $C(x) = 4.5x + 340$ dollars to wash x cars in one day.

(a) What is the rate of change? What does it mean in this context?

(b) What is the C-intercept? What does it represent here?

SOLUTION

(a) The rate of change is 4.5, the coefficient of x, and the units are dollars per car. Thus each additional car washed adds $4.50 to the cost for that day.

(b) The C-intercept is 340. This value is the initial output corresponding to $x = 0$, so it represents the fixed cost, $340, of operating the car wash for a day, whether or not any cars are washed.   ■

■ **EXAMPLE 5** **Writing an Equation for a Linear Function**

(a) As dry air moves upward, it expands and cools. If the ground temperature is 20°C and the temperature at a height of 1 km is 10°C, express the temperature T (in °C) as a function of the height h (in kilometers), assuming that a linear model is appropriate.

(b) Draw the graph of the function in part (a). What does the slope represent in this context?

(c) What is the temperature at a height of 2.5 km?

SOLUTION

(a) Because we are assuming that T is a linear function of h, we can write

$$T = mh + b$$

We are given two function values, $T(0) = 20$ and $T(1) = 10$, so the slope of the graph is

$$m = \frac{\Delta T}{\Delta h} = \frac{T(1) - T(0)}{1 - 0} = \frac{10 - 20}{1 - 0} = -10$$

The initial temperature value is $T(0) = 20$, so the T-intercept is $b = 20$ and the linear function is

$$T(h) = -10h + 20$$

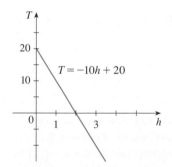

(b) The graph is sketched in Figure 5. The slope is $m = -10$ and it represents the rate of change of T with respect to h; the units are degrees Celsius per kilometer (°C/km). The rate of change is negative, so the temperature decreases by 10°C for each rise in elevation of 1 km.

(c) At a height of $h = 2.5$ km, the temperature is

$$T(2.5) = -10(2.5) + 20 = -5°C \qquad ■$$

FIGURE 5

We don't need to know the starting value for a linear function to write an equation. We can simply use the point-slope formula given in Equation 2.

■ **EXAMPLE 6** **Writing a Linear Model Using the Point-Slope Form**

A pump has been pouring water into a swimming pool. The data in the table show the water volume of the pool every two hours after the pump was activated.

(a) Explain why a linear model is appropriate.

(b) Write an equation for a linear function to model the data.

(c) Use your model to predict the volume of water in the pool after 17.5 hours.

(d) When will the amount of water in the pool reach 6000 gallons?

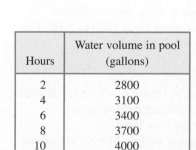

Hours	Water volume in pool (gallons)
2	2800
4	3100
6	3400
8	3700
10	4000

SOLUTION

(a) The volume of water increases 300 gallons during each two-hour interval. Thus the rate of change is constant, indicating a linear relationship between the input and output.

(b) The rate of change is 300 gallons every two hours, or 150 gallons per hour, so the slope is $m = 150$. If we let $V(t)$ be the water volume in gallons t hours after the pump is activated, then $V(2) = 2800$, so the point $(2, 2800)$ is on the graph. The point-slope formula gives

$$V - 2800 = 150(t - 2) = 150t - 300$$

$$V = 150t + 2500$$

Thus the water volume in gallons t hours after the pump is turned on is given by $V(t) = 150t + 2500$.

(c) The volume of water in the pool after 17.5 hours is

$$V(17.5) = 150(17.5) + 2500 = 5125 \text{ gallons}$$

(d) We solve $V(t) = 6000$:

$$150t + 2500 = 6000$$

$$150t = 3500$$

$$t = \frac{3500}{150} = \frac{70}{3}$$

Thus the amount of water in the pool reaches 6000 gallons after $23\frac{1}{3}$ hours, or 23 hours 20 minutes.   ■

Fitting a Model to Data

Although linear functions are often used as models, few functional relationships in the world around us are perfectly linear. Nevertheless, many observed data are *approximately* linear, and a linear function is still an appropriate model to use. The goal is to write an equation that "fits" the data accurately enough to capture the basic trend and allow further analysis. But how do we know if an equation fits the data well enough? The next example illustrates one approach we can take.

■ EXAMPLE 7   Modeling with a Linear Function

Table 1 on page 35 lists the average carbon dioxide level in the atmosphere, measured in parts per million at Mauna Loa Observatory from 1980 to 2008. Use the data in Table 1 to find a model for the carbon dioxide level.

SOLUTION

We use the data in Table 1 to make the scatter plot in Figure 6, where t represents time, in years, and C represents the CO_2 level, in parts per million (ppm). To simplify the input values, let $t = 0$ correspond to the year 1980.

Notice that the data points appear to lie close to a straight line, so it's natural to choose a linear model in this case. But there are many possible lines that approximate these data points, so which one should we use? One strategy is to use two of the points from the scatter plot to write an equation. Different pairs of points will generate different results, so the points should be chosen wisely. You may wish to first draw a line with a ruler on the scatter plot to help select two points.

TABLE 1

Year	CO$_2$ level (in ppm)
1980	338.7
1982	341.2
1984	344.4
1986	347.2
1988	351.5
1990	354.2
1992	356.3
1994	358.6
1996	362.4
1998	366.5
2000	369.4
2002	373.2
2004	377.5
2006	381.9
2008	385.6

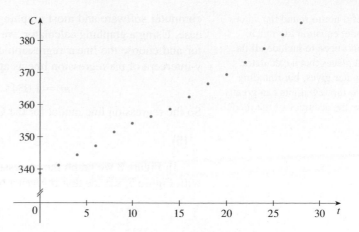

FIGURE 6   Scatter plot for the average CO$_2$ level

From the scatter plot, it appears that a line passing through the points (4, 344.4) and (20, 369.4) will fit the data reasonably well. The slope of this line is

$$\frac{369.4 - 344.4}{20 - 4} = \frac{25.0}{16} = 1.5625$$

and its equation is

$$C - 344.4 = 1.5625(t - 4)$$

or

(4) $$C = 1.5625t + 338.15$$

Equation 4 gives one possible linear model for the carbon dioxide level; it is graphed in Figure 7.

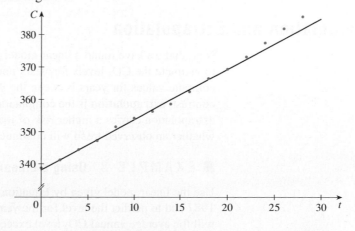

FIGURE 7
Linear model through two data points

▪ Regression Lines

A more sophisticated procedure for finding a linear model to fit data is called the *method of least squares*. In this method, the vertical distance from each data point to a line is measured, and the squares of the distances are added together. Of all possible lines, the line that gives the smallest of such sums, called the **regression line**, is chosen as the best-fitting line. This process is tedious to carry out by hand, but

Be careful not to round the values in a model equation too much. You don't need to include all the decimal places that a calculator or computer gives, but rounding values to too few digits can greatly decrease the accuracy of the model.

computer software and most graphing calculators can determine the equation with ease. Using a graphing calculator, we enter the data from Table 1 into the data editor and choose the linear regression command. The calculator gives the slope and y-intercept of the regression line as approximately

$$m = 1.6543 \qquad b = 337.41$$

So the regression line model for the CO_2 level is

(5) $$C = 1.6543t + 337.41$$

In Figure 8 we graph the regression line as well as the data points. Comparing with Figure 7, we see that it gives a better fit than our first linear model.

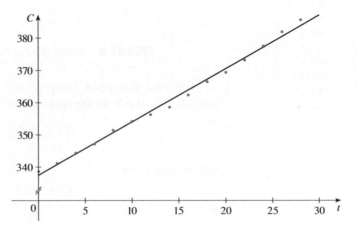

FIGURE 8
The regression line

Notice that here the regression line does not include *any* of the original data points. In many cases, the best-fitting line is not one that passes through data points.

Interpolation and Extrapolation

Now that we have found a linear model for the carbon dioxide levels, we can use it to estimate the CO_2 levels for years not listed in Table 1. If we use the model to compute values for years between the years in the table, we are **interpolating**. In contrast, **extrapolation** is the computation of values outside given data. In general, extrapolation carries a higher risk of inaccuracy because it is often hard to predict whether an observed trend will continue.

▪ EXAMPLE 8   Using a Linear Model to Interpolate and Extrapolate

Use the linear model given by Equation 5 to estimate the average CO_2 level for 1987 and to predict the level for the year 2016. According to this model, when will the average annual CO_2 level exceed 410 parts per million?

SOLUTION

The year 1987 corresponds to $t = 7$. Using Equation 5, we estimate that the average CO_2 level in 1987 was

$$C(7) = (1.6543)(7) + 337.41 \approx 348.99 \text{ ppm}$$

This is an example of interpolation because we have estimated a value *between* observed values. (In fact, the Mauna Loa Observatory reported that the average CO_2 level in 1987 was 348.93 ppm, so our estimate is quite accurate.)

To predict the level for the year 2016, we use :

$$C(36) = (1.6543)(36) + 337.41 \approx 396.96 \text{ ppm}$$

So we predict that the average CO_2 level in the year 2016 will be 396.96 ppm. This is an example of extrapolation because we have predicted a value *outside* the observed years 1980–2008. Consequently, we are far less certain about the accuracy of our prediction.

To determine when our model predicts a CO_2 level of 410 ppm, we solve $C(t) = 410$:

$$1.6543t + 337.41 = 410$$

$$1.6543t = 72.59$$

$$t = \frac{72.59}{1.6543} \approx 43.88$$

The solution corresponds to a value between the years 2023 and 2024, and since the CO_2 levels are annual averages, only integer inputs give meaningful function values. Comparing $C(43) \approx 408.54$ and $C(44) \approx 410.20$, we see that the CO_2 level will first exceed an annual average of 410 ppm in 2024. Note that this prediction is somewhat risky because it involves a time quite a few years beyond the observed values, and no one can say whether the same patterns will continue. ■

In the preceding example, we used a continuous function to model discrete data. We will often find this to be a useful technique. However, we must use caution in interpreting results found using the continuous model, as the example illustrates.

■ Exercises 1.3

1–4 ■ Find the slope of the line through the given points.

1. $(3, 7)$, $(5, 10)$

2. $(1, 8)$, $(4, -6)$

3. $(45, 1860)$, $(26, 2240)$

4. $(185, 5600)$, $(210, 8150)$

5–14 ■ Find an equation of the line that satisfies the given conditions. Express the equation in slope-intercept form.

5. Slope 3, y-intercept -2

6. Slope $\frac{2}{5}$, y-intercept 4

7. Through $(2, -3)$, slope 6

8. Through $(-3, -5)$, slope $-\frac{7}{2}$

9. Through $(2, 1)$ and $(1, 6)$

10. Through $(-1, -2)$ and $(4, 3)$

11. Through $(4, 84)$ and $(13, -312)$

12. Through $(6, 70)$ and $(16, 300)$

13. x-intercept 1, y-intercept -3

14. x-intercept -8, y-intercept 6

15. Sketch a line through the point $(-2, 6)$ with slope $-\frac{1}{5}$.

16. Sketch a line with slope $\frac{4}{7}$ and y-intercept -3.

17–20 ■ Find the slope and y-intercept of the line. Then draw its graph.

17. $2x + 5y = 15$

18. $3x - 8y = -10$

19. $-5x + 6y = 42$

20. $8y - 3x = 48$

21–24 ■ Find the slope and intercepts of the linear function. Then sketch a graph.

21. $f(x) = -2x + 14$

22. $g(t) = -0.5t + 5$

23. $A(t) = 0.2t - 4$

24. $P(v) = 3v - 1$

25. Write an equation for a linear function h where $h(7) = 329$ and $h(11) = 553$.

26. Write an equation for a linear function w where $w(42) = 230.8$ and $w(74) = 819.6$.

27. Television ratings The weekly ratings, in millions of viewers, of a recent television program are given by $L(w)$, where w is the number of weeks since the show premiered. If L is a linear function where $L(8) = 5.32$ and $L(12) = 8.36$, compute the slope of L and explain what it represents in this context.

28. Consumer demand A movie studio is releasing a new DVD, and the studio estimates that if the DVD is priced at $19.99, it will sell 6.68 million copies, whereas if it is priced at $15.99, it will sell 11.27 million copies. If f is a linear function that gives the number of copies sold (in millions) at a given price, what is the slope of f? What does it represent in this context?

29. Depreciation A small company purchased a new copy machine for $16,500 and the company's accountant plans to depreciate (for tax purposes) the machine to a value of $0 over five years. If $V(t)$ is the value of the machine after t years, and V is a linear function, what is the slope of V? What does the slope represent in this context?

30. Matric suction Matric suction is the pressure that causes water to flow from wetter soil to dryer soil and often decreases linearly with depth. A researcher has taken samples at a location at various depths. Let $g(d)$ be the matric suction, measured in kilopascals (kPa), at a depth of d cm. Assuming that g is a linear function, if $g(35) = 78$ and $g(104) = 43$, what is the slope of g? What does the slope represent in this context?

31. Taxes Suppose the taxes a company pays are approximately $T(p) = 0.26p + 15.4$ thousand dollars, where p is the company's annual profit in thousands of dollars. What is the rate of change, and what does it measure in this context?

32. Earth's surface temperature Some scientists believe that the average surface temperature of the earth has been rising steadily. They have modeled the temperature by the linear function $T = 0.02t + 8.50$, where T is temperature in °C and t represents years since 1900.

(a) What are the slope and T-intercept? What do they represent in this context?

(b) Use the equation to predict the average global surface temperature in 2100.

33. Drug dosage If the recommended adult dosage for a drug is D, in mg, then to determine the appropriate dosage c for a child of age a, pharmacists use the equation $c = 0.0417D(a + 1)$. Suppose the dosage for an adult is 200 mg.

(a) Find the slope of the graph of c (as a function of a). What does it represent?

(b) What is the dosage for a newborn?

34. Consumer demand The manager of a weekend flea market knows from past experience that if he charges x dollars for a rental space at the flea market, then the

number y of spaces he can rent is given by the equation $y = 200 - 4x$.

(a) Sketch a graph of this linear function. (Remember that the rental charge per space and the number of spaces rented can't be negative quantities.)

(b) What do the slope, the y-intercept, and the x-intercept of the graph represent?

35. Temperature scales The relationship between the Fahrenheit (F) and Celsius (C) temperature scales is given by the linear function $F = \frac{9}{5}C + 32$.

(a) Sketch a graph of this function.

(b) What is the slope of the graph and what does it represent? What is the F-intercept and what does it represent?

36. Driving distance Jason leaves Detroit at 2:00 PM and drives at a constant speed west along I-96. He passes Ann Arbor, 40 mi from Detroit, at 2:50 PM.

(a) Express the distance traveled in terms of the time elapsed.

(b) Draw the graph of the equation in part (a).

(c) What is the slope of this line? What does it represent?

37. Cricket chirping rate Biologists have noticed that the chirping rate of crickets of a certain species is related to temperature, and the relationship appears to be very nearly linear. A cricket produces 113 chirps per minute at 70°F and 173 chirps per minute at 80°F.

(a) Find a linear equation that models the temperature T as a function of the number of chirps per minute N.

(b) What is the slope of the graph? What does it represent?

(c) If the crickets are chirping at 150 chirps per minute, estimate the temperature.

38. Manufacturing cost The manager of a furniture factory finds that it costs $2200 to manufacture 100 chairs in one day and $4800 to produce 300 chairs in one day.

(a) Express the cost as a function of the number of chairs produced, assuming that it is linear. Then sketch the graph.

(b) What is the slope of the graph and what does it represent?

(c) What is the y-intercept of the graph and what does it represent?

39. Ocean water pressure At the surface of the ocean, the water pressure is the same as the air pressure above the water, 15 lb/in². Below the surface, the water pressure increases by 4.34 lb/in² for every 10 ft of descent.

(a) Express the water pressure as a function of the depth below the ocean surface.

(b) At what depth is the pressure 100 lb/in²?

40. Car expense The monthly cost of driving a car depends on the number of miles driven. Lynn found that in May it cost her $380 to drive 480 mi and in June it cost her $460 to drive 800 mi.

(a) Express the monthly cost C as a function of the distance driven d, assuming that a linear relationship gives a suitable model.

(b) Use part (a) to predict the cost of driving 1500 miles per month.

(c) Draw the graph of the linear function. What does the slope represent?

(d) What does the C-intercept represent?

(e) Why does a linear function give a suitable model in this situation?

41. Ulcer rates   The table shows (lifetime) peptic ulcer rates (per 100 population) for various family incomes as reported by the National Health Interview Survey.

Income	Ulcer rate (per 100 population)
$4,000	14.1
$6,000	13.0
$8,000	13.4
$12,000	12.5
$16,000	12.0
$20,000	12.4
$30,000	10.5
$45,000	9.4
$60,000	8.2

(a) Make a scatter plot of these data and decide whether a linear model is appropriate.

(b) Find and graph a linear model using the third and last data points.

(c) According to the model, how likely is someone with an income of $90,000 to suffer from peptic ulcers? Is this an example of interpolation, or extrapolation?

(d) Do you think it would be reasonable to apply the model to someone with an income of $200,000?

42. US public debt   The table gives the amount of debt held by the public in the United States (at the end of the year) as estimated by the Congressional Budget Office of the US federal government.

Year	Public debt (billions of dollars)
2004	4296
2005	4665
2006	4971
2007	5246
2008	5494
2009	5716
2010	5919

(a) Make a scatter plot of these data and decide whether a linear model is appropriate.

(b) Find and graph a linear model using the second and sixth data points.

(c) According to the model, when will the public debt reach $5.1 trillion? Is this an example of interpolation, or extrapolation?

(d) Use the model to predict the projected public debt in 2014.

43. Vehicle value   In general, a used car is worth more if it has low mileage. The table shows how the value of a particular vehicle is affected by different mileage figures.

Mileage	Value
20,000	$14,245
30,000	$13,520
40,000	$12,520
50,000	$11,645
60,000	$10,970

(a) Make a scatter plot of these data and use two of the data points to write a linear model for the data.

(b) Use the model to estimate the value of the same car if it has been driven only 12,000 miles.

(c) For how many miles driven does the model predict a value of $0? Is this realistic?

44. Hospital visits   The Center for Disease Control compiles the average annual hospital emergency room visits due to falls for children of various ages. Annual averages during recent years are listed in the table.

Age	Number of visits per 10,000 population
9	295.4
11	276.1
13	268.1
15	215.8
17	186.0
19	176.5

(a) Make a scatter plot of these data. Is a linear model appropriate?

(b) Use two of the data points to write a linear model for the data.

(c) Use the model to estimate the average number of emergency room visits per 10,000 population for 18-year-olds. How does your estimated value compare with the published value of 182.4?

45. Ulcer rates   Exercise 41 lists data for peptic ulcer rates for various family incomes.

   (a) Find the least squares regression line for these data.

   (b) Use the linear model in part (a) to estimate the ulcer rate for an income of $25,000.

   (c) According to the model, how likely is someone with an income of $80,000 to suffer from peptic ulcers?

46. Cricket chirping rates   Biologists have observed that the chirping rate of crickets of a certain species appears to be related to temperature. The table shows the chirping rates for various temperatures.

Temperature (°F)	Chirping rate (chirps/min)	Temperature (°F)	Chirping rate (chirps/min)
50	20	75	140
55	46	80	173
60	79	85	198
65	91	90	211
70	113		

   (a) Make a scatter plot of the data.

   (b) Find and graph the regression line.

   (c) Use the linear model in part (b) to estimate the chirping rate at 100°F.

47. Olympics   The table gives the winning heights for the men's Olympic pole vault competitions up to the year 2004.

Year	Height (m)	Year	Height (m)
1896	3.30	1960	4.70
1900	3.30	1964	5.10
1904	3.50	1968	5.40
1908	3.71	1972	5.64
1912	3.95	1976	5.64
1920	4.09	1980	5.78
1924	3.95	1984	5.75
1928	4.20	1988	5.90
1932	4.31	1992	5.87
1936	4.35	1996	5.92
1948	4.30	2000	5.90
1952	4.55	2004	5.95
1956	4.56		

   (a) Make a scatter plot and decide whether a linear model is appropriate.

   (b) Find and graph the regression line.

   (c) Use the linear model to predict the height of the winning pole vault at the 2008 Olympics and compare with the actual winning height of 5.96 meters.

   (d) Is it reasonable to use the model to predict the winning height at the 2100 Olympics?

48. US public debt   The table gives the projected amount of debt held by the public in the United States as estimated by the Congressional Budget Office.

Year	Public debt (billions of dollars)
2011	6012
2012	5955
2013	5884
2014	5784
2015	5658

   (a) Make a scatter plot of these data and decide whether a linear model is appropriate.

   (b) Find and graph the least squares regression line.

   (c) Use the regression line model to predict the projected public debt in 2020.

   (d) Exercise 42 lists similar data for prior years, and in part (d) the public debt is estimated for 2014. How does your predicted value compare to the value given here? What can you conclude about extrapolation?

49. A *family of functions* is a collection of functions whose equations are related.

   (a) What do all members of the family of linear functions $f(x) = 3x + c$ have in common? Sketch several members of the family.

   (b) What do all members of the family of linear functions $f(x) = ax + 3$ have in common? Sketch several members of the family.

   (c) Which function belongs to both families?

50. What do all members of the family of linear functions $f(x) = c - (x + 3)$ have in common? Sketch several members of the family.

51. Two linear functions are graphed below. Which function has the greater rate of change?

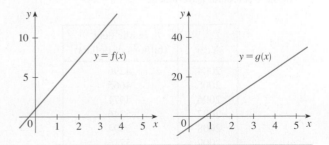

52. Two linear functions $f(x) = ax + b$ and $g(x) = cx + d$ are graphed below.

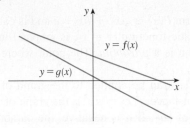

(a) Is $a > c$, or is $a < c$?

(b) Is $b > d$, or is $b < d$?

53. Health spending   The table shows the US national health expenditures, as a percentage of gross domestic product (GDP), for various years.

Year	National health expenditure (percent of GDP)
1995	13.4
1998	13.2
1999	13.2
2000	13.3
2001	14.1
2002	14.9
2003	15.3

Write a piecewise function that models these data.

54. US public debt   Use the data given in Exercises 42 and 48 to write a piecewise function that models the estimated public debt for the years 2004–2015.

1.4 Polynomial Models and Power Functions

While linear functions may be the most basic and commonly occurring models, there are many other kinds of functions that are often used in modeling. In this section we look at polynomials, power functions, and rational functions. We will see that there is some overlap between the categories; some functions qualify as all three types.

A function P is called a **polynomial** if it can be written as

$$P(x) = a_n x^n + a_{n-1} x^{n-1} + \cdots + a_2 x^2 + a_1 x + a_0$$

where n is a nonnegative integer and the numbers $a_0, a_1, a_2, \ldots, a_n$ are constants called the **coefficients** of the polynomial. Examples of polynomials are

$$f(x) = x^5 - 3x^8 + 4x^2$$

$$g(t) = 1.737t^3 - 2.49t^2 + 8.51t + 4.12$$

$$P(v) = 2v^6 - v^4 + \tfrac{2}{5}v^3 + \sqrt{2}$$

The domain of any polynomial is $\mathbb{R} = (-\infty, \infty)$. The largest exponent that appears on the input variable is called the **degree** of the polynomial. The degrees of the polynomials above are 8, 3, and 6, respectively. The linear functions $f(x) = mx + b$ studied in the previous section are polynomials of degree 1. (Recall that $x^1 = x$.)

Polynomials are commonly used to model various quantities that occur in the natural and social sciences. For instance, we will soon see why economists often use a polynomial $P(x)$ to represent the cost of producing x units of a commodity.

■ Quadratic Functions

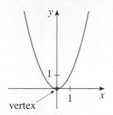

FIGURE 1
Graph of the quadratic
function $y = x^2$

A polynomial of degree 2 is of the form $P(x) = ax^2 + bx + c$ and is called a **quadratic function**. The simplest quadratic function, $y = x^2$, is graphed in Figure 1. The graph of every quadratic function is a *parabola*. The point where the graph changes direction is called the *vertex*.

The transformations of Section 1.2 can be applied to the graph of $y = x^2$ to graph other quadratic functions. For instance, $f(x) = ax^2$ is the graph of $y = x^2$ but stretched or compressed by a factor of $|a|$. If a is negative, the parabola will be reflected across the x-axis so that it opens downward. In this case, the vertex is the highest point on the graph. When a is positive, the vertex is the lowest point on the graph. By shifting the graph of $f(x) = ax^2$ horizontally and vertically, we can obtain the graph of any quadratic function. The result is an equation in the form $y = a(x - h)^2 + k$ that can always be expanded to the form $y = ax^2 + bx + c$.

■ **Standard Form of a Quadratic Function**   A quadratic function $f(x) = ax^2 + bx + c$ can be expressed in the standard form

$$f(x) = a(x - h)^2 + k$$

The graph is a parabola with vertex (h, k); the parabola opens upward if $a > 0$ or downward if $a < 0$.

■ **EXAMPLE 1   Graphing a Quadratic Function**

Graph the quadratic function $g(x) = -2(x - 1)^2 + 4$.

SOLUTION

The graph of $y = -2x^2$ is the graph of $y = x^2$ but stretched vertically by a factor of 2 and reflected across the x-axis as shown in Figure 2. Then we shift the graph 1 unit to the right and 4 units upward to obtain the graph of the function $g(x) = -2(x - 1)^2 + 4$. (See Figure 3.)

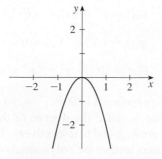

FIGURE 2
Graph of $y = -2x^2$

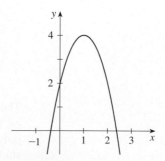

FIGURE 3
Graph of $y = -2(x - 1)^2 + 4$

The transformations shift the vertex to $(1, 4)$, the highest point on the graph.   ■

Year	Population
1970	4250
1975	3200
1980	2400
1985	2200
1990	2450
1995	3100
2000	4350
2005	5800

▪ **EXAMPLE 2** **Writing a Quadratic Model**

Protection efforts have reversed the declining population of an endangered species. The table lists the estimated population of the species for various years.

(a) Draw a scatter plot of these data and explain why a quadratic model is appropriate.

(b) Write the equation of a quadratic function that models the data.

(c) Use the model to predict the animal population in the year 2018.

SOLUTION

(a)

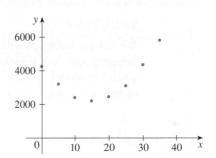

FIGURE 4

For simplicity, we let $x = 0$ correspond to the year 1970. The points of the scatter plot, shown in Figure 4, appear to coincide with a parabola so a quadratic model is a good choice.

(b) From the scatter plot, it appears that the point $(15, 2200)$ is approximately the location of the vertex of the parabola. Then a quadratic function with this vertex is $y = a(x - 15)^2 + 2200$. To determine a, we select a point from the scatter plot that the parabola should pass through, such as $(30, 4350)$. Substituting these values for x and y gives

$$4350 = a(30 - 15)^2 + 2200 = 225a + 2200$$

$$2150 = 225a$$

$$a = \frac{2150}{225} = \frac{86}{9} \approx 9.56$$

Thus the equation for the model is $f(x) = 9.56(x - 15)^2 + 2200$, where $f(x)$ is the number of animals x years after 1970. The equation simplifies to $f(x) = 9.56x^2 - 286.8x + 4351$. As you can see from the graph in Figure 5, the model fits the data quite well.

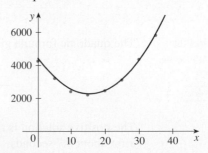

FIGURE 5

(c) The year 2018 corresponds to $x = 48$, and $f(48) \approx 12{,}611$. So we estimate that the animal population will be approximately 12,600 in 2018. ▪

Many graphing calculators and computer programs can fit a quadratic function to data using least squares regression. In the following example we use a graphing calculator to find a quadratic function to model the fall of a ball.

▪ EXAMPLE 3   Modeling Using Quadratic Regression

A ball is dropped from the upper observation deck of the CN Tower in Toronto, 450 m above the ground, and its height h above the ground is recorded at 1-second intervals in Table 1. Find a model to fit the data and use the model to predict the time at which the ball hits the ground.

SOLUTION

We draw a scatter plot of the data in Figure 6 and observe that a linear model is inappropriate. But it looks as if the data points might lie on a parabola, so we try a quadratic model instead. Using a graphing calculator (which uses the least squares method), we obtain the following quadratic model:

(1) $$h = 449.36 + 0.96t - 4.90t^2$$

TABLE 1

Time (seconds)	Height (meters)
0	450
1	445
2	431
3	408
4	375
5	332
6	279
7	216
8	143
9	61

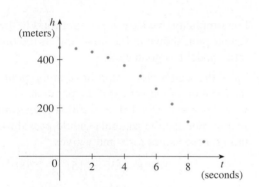

FIGURE 6
Scatter plot for a falling ball

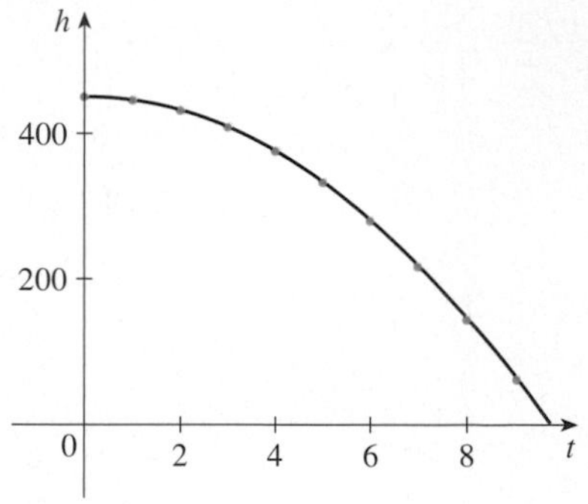

FIGURE 7
Quadratic model for a falling ball

In Figure 7 we plot the graph of Equation 1 together with the data points and see that the quadratic model gives a very good fit.

The ball hits the ground when $h = 0$, so we solve the quadratic equation

$$-4.90t^2 + 0.96t + 449.36 = 0$$

The quadratic formula states that the solutions of the quadratic equation $ax^2 + bx + c = 0$ are

$$x = \frac{-b \pm \sqrt{b^2 - 4ac}}{2a}$$

The quadratic formula gives

$$t = \frac{-0.96 \pm \sqrt{(0.96)^2 - 4(-4.90)(449.36)}}{2(-4.90)}$$

The positive solution is $t \approx 9.67$, so we predict that the ball will hit the ground after about 9.7 seconds. ▪

▪ Polynomial Functions of Higher Degree

A polynomial of degree 3 is of the form

$$P(x) = ax^3 + bx^2 + cx + d \qquad (a \neq 0)$$

and is called a **cubic function**. Figure 8 shows the graph of a cubic function in part (a) and graphs of polynomials of degrees 4 and 5 in parts (b) and (c). We will see later why the graphs have these shapes. In general, the higher the degree, the more changes in direction the graph can have. To describe the behavior of the graph, we say the function is **increasing** if the output values increase as the input values increase. This occurs where the graph rises (as we look from left to right). A function is **decreasing** if the output values decrease as the input values increase, and the graph falls (looking left to right). Polynomial functions of degree 2 or greater can exhibit both increasing and decreasing behavior in different portions of their domains.

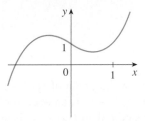
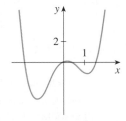
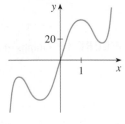

FIGURE 8   (a) $y = x^3 - x + 1$ (b) $y = x^4 - 3x^2 + x$ (c) $y = 3x^5 - 25x^3 + 60x$

▪ EXAMPLE 4   Observing Intervals of Increase and Decrease

The cubic function $g(t) = 2t^3 - 3t^2 - 12t + 24$ is graphed in Figure 9. On what intervals is g increasing? Decreasing?

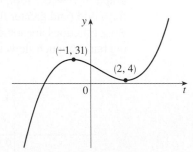

FIGURE 9

SOLUTION

Starting from the left, the graph rises until it reaches the point $(-1, 31)$. The behavior of the graph is described in terms of the input values, so we say that g is increasing for $t < -1$. In interval notation, g is increasing on $(-\infty, -1)$. The graph then falls until it reaches the point $(2, 4)$, after which the graph changes direction and rises. Thus g is decreasing on $(-1, 2)$ and increasing again on $(2, \infty)$. ▪

■ Power Functions

A function of the form $f(x) = x^a$, where a is a constant, is called a **power function**. If a is a positive integer, then the function is simply a polynomial with only one term. The graphs of $f(x) = x^n$ for $n = 1, 2, 3, 4,$ and 5 are shown in Figure 10. We already know the shape of the graphs of $y = x$ (a line through the origin with slope 1) and $y = x^2$ (a parabola, see Figure 1).

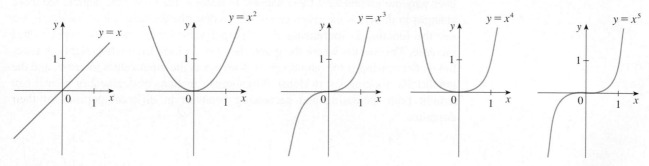

FIGURE 10   Graphs of $f(x) = x^n$ for $n = 1, 2, 3, 4, 5$

The general shape of the graph of $f(x) = x^n$ depends on whether n is even or odd. If n is even, then $f(x) = x^n$ is an even function and its graph is similar to the parabola $y = x^2$. If n is odd, then $f(x) = x^n$ is an odd function and its graph is similar to that of $y = x^3$.

The function $f(x) = x^{1/2}$ is equivalent to $\sqrt{x}$, the **square root function**. Its domain is $[0, \infty)$ (because the square root of a negative number is not defined as a real number) and its graph is the upper half of a parabola. [See Figure 11(a).] Similarly, $g(x) = x^{1/3} = \sqrt[3]{x}$ is the **cube root function**, graphed in Figure 11(b). Notice that the domain is $\mathbb{R}$ because every real number has a cube root. In general, the graph of $y = x^{1/n}$ for n even is similar to the graph of the square root function, and for n odd (and greater than 3) the graph is similar to that of the cube root function. Root functions are sometimes used to model situations where a function is increasing but at a much slower rate than a polynomial function. (See Exercises 52 and 53.)

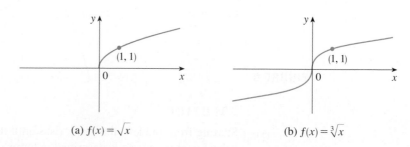

FIGURE 11

Graphs of root functions

(a) $f(x) = \sqrt{x}$ (b) $f(x) = \sqrt[3]{x}$

If the exponent a is a fraction such as $\frac{2}{3}$, the function can be written as

$$x^{2/3} = (x^{1/3})^2 = (\sqrt[3]{x})^2$$

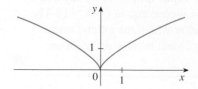

FIGURE 12   The graph of $y = x^{2/3}$ is shown in Figure 12.

■ **EXAMPLE 5** **The Cobb-Douglas Production Function**

In 1928 Charles Cobb and Paul Douglas published a study in which they modeled the growth of the American economy during the period 1899–1922. Although they simplified many of the factors that affect economic performance, their model proved to be remarkably accurate. Their general result has been used in many settings, ranging from individual firms to global economic questions, and has become known as the **Cobb-Douglas production function**. A particular case of their model (corresponding to the year 1910) is $P(x) = 42.6x^{1/4}$, where x represents the amount of capital invested in the economy (the monetary worth of all machinery, equipment, and buildings), and $P(x)$ is the total production (the monetary value of all goods produced in a year). For simplicity, Cobb and Douglas measured both quantities in reference to their values in 1899, which were assigned to be 100. Thus $P(x) = 200$ is twice the production of 1899.

(a) If the capital invested is increased from a level of 100 to 150, what is the effect on production?

(b) In 1910, the total capital was estimated to be at a level of 208. What was the production level?

SOLUTION

(a) We compute

$$P(100) = 42.6(100^{1/4}) \approx 134.7$$

$$P(150) = 42.6(150^{1/4}) \approx 149.1$$

The second value is $149.1/134.7 \approx 1.107$ times larger than the first, so by increasing the capital invested by 50%, a 10.7% increase in production could be expected.

(b) The model predicts that the production level was about

$$P(208) = 42.6(208^{1/4}) \approx 161.8 \qquad ■$$

If the exponent a in a power function x^a is negative, recall that the function can be rewritten using the property $x^{-n} = 1/x^n$. In particular, $f(x) = x^{-1} = 1/x$ is the **reciprocal function**. The graph, shown in Figure 13, has both the x-axis and the y-axis as asymptotes. In other words, the curve approaches the x-axis at the left and right, and it approaches the y-axis as the x-values approach 0. (We will study asymptotes in more detail in Chapter 4.) The right half of the graph can be used to model situations where the output continues to shrink as the input increases.

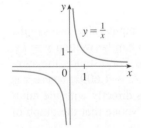

FIGURE 13

The reciprocal function

■ | Rational Functions

A **rational function** f is a function formed by dividing one polynomial by another:

$$f(x) = \frac{P(x)}{Q(x)}$$

where P and Q are polynomials. The domain consists of all values of x such that $Q(x) \neq 0$ (to prevent division by zero). A simple example of a rational function is

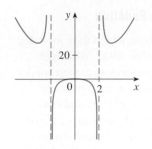

FIGURE 14
$f(x) = \dfrac{2x^4 - x^2 + 1}{x^2 - 4}$

the function $f(x) = 1/x$, whose domain is $\{x \mid x \neq 0\}$; this is the reciprocal function graphed in Figure 13. Graphs of rational functions can be considerably more complex than the graphs of polynomial or power functions. For instance, they often have different components separated by asymptotes. The function

$$f(x) = \frac{2x^4 - x^2 + 1}{x^2 - 4}$$

is a typical rational function. Its domain is $\{x \mid x \neq \pm 2\}$, and its graph is shown in Figure 14.

▪ **EXAMPLE 6   An Average Cost Function**

Suppose the cost for a company to produce x toys during one day is $C(x) = 0.2x^2 + 6x + 850$ dollars. Then the *average cost* per toy is

$$a(x) = \frac{\text{total cost}}{\text{number of toys}} = \frac{C(x)}{x} = \frac{0.2x^2 + 6x + 850}{x}$$

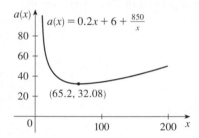

FIGURE 15

or, equivalently, $a(x) = 0.2x + 6 + 850/x$ dollars per toy. The graph of the average cost function is shown in Figure 15. (We will investigate average cost functions in more detail in Section 3.2.) Using a graphing calculator, we estimate that the coordinates of the lowest point on the graph are $(65.2, 32.08)$, so the company will achieve the minimum average cost per toy, about \$32, if it produces 65 toys per day.   ▪

▪ Direct and Inverse Variation

When the output value of a function is *proportional* to the input value, we say the output y **varies directly** with the input x. The relationship can be written $y = kx$, where k is a constant called the **constant of proportionality**. For instance, the number N of kilometers in a mile is given (approximately) by $N = 1.609x$, where x is the number of miles. Thus the number of kilometers varies directly with the number of miles, and the constant of proportionality is 1.609. Notice that the graph of $N = 1.609x$ is a line through the origin with slope 1.609.

We say that an output y **varies inversely** with an input x if the output is proportional to the *reciprocal* of the input. The relationship is $y = k/x$, where k is a constant. For example, such a relationship arises in physics and chemistry in connection with Boyle's Law, which says that, when the temperature is constant, the volume V of a gas is inversely proportional to the pressure P:

$$V = \frac{C}{P}$$

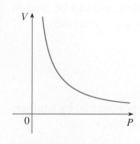

FIGURE 16
Volume as a function of pressure at constant temperature

where C is a constant. Thus the graph of V as a function of P (see Figure 16) has the same general shape as the right half of the reciprocal function graphed in Figure 13.

■ **EXAMPLE 7** **Direct and Inverse Variation**

(a) If A varies directly with t, and $A = 66$ when $t = 4$, write an equation for A in terms of t.

(b) Suppose $f(x)$ varies inversely with the square of x. If $f(5) = 240$, find the value of $f(8)$.

SOLUTION

(a) Because A varies directly with t, we have $A = kt$ for some constant k. We know $A = 66$ when $t = 4$, so $66 = k \cdot 4$ and $k = 66/4 = 16.5$. Thus $A = 16.5t$.

(b) First we write an equation for f. We know $f(x) = k/x^2$, and $f(5) = 240$, so $k/5^2 = 240$ and

$$k = 240 \cdot 5^2 = 240 \cdot 25 = 6000$$

Thus $f(x) = 6000/x^2$ and

$$f(8) = \frac{6000}{8^2} = \frac{375}{4} = 93.75$$

■

■ Exercises 1.4

1–2 ■ State whether each function is (i) a polynomial (state its degree), (ii) a power function, or (iii) a rational function.

1. (a) $g(w) = w^4$ (b) $f(x) = \sqrt[5]{x}$

(c) $A(t) = -2t^7 + 3t - 1$

(d) $r(x) = \dfrac{x^2 + 1}{x^3 + x}$

2. (a) $y = \dfrac{x - 6}{x + 6}$ (b) $y = 1.3t^3 - 8.52t^5$

(c) $y = x^{-1}$ (d) $y = x^{10}$

3–4 ■ Match each equation with its graph. Explain your choices. (Don't use a computer or graphing calculator.)

3. (a) $y = x^2$ (b) $y = x^5$ (c) $y = \sqrt{x}$

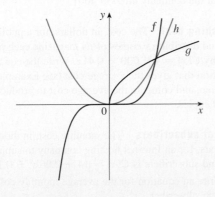

4. (a) $y = 3x$ (b) $y = x^3$ (c) $y = \sqrt[3]{x}$

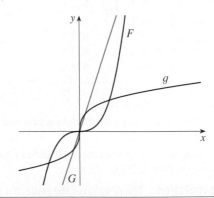

5–12 ■ Use the transformations of Section 1.2 to graph the quadratic function. State the coordinates of the vertex.

5. $f(x) = (x + 2)^2 + 5$

6. $g(x) = (x - 3)^2 + 2$

7. $K(t) = -t^2 + 3$

8. $L(r) = -2r^2 - 1$

9. $A(p) = \frac{1}{2}(p - 1)^2 - 2$

10. $N(t) = 3(t + 4)^2 - 5$

11. $C(x) = -2(x - 3)^2 - 4$

12. $J(u) = -\frac{1}{3}(u + 1)^2 + 2$

13–18 ▪ Write an equation for the given quadratic function. Give your equation in the form $f(x) = ax^2 + bx + c$.

13.
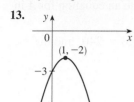

14.
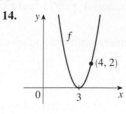

15. Vertex $(0, 22)$, graph passes through the point $(8, 6)$

16. Vertex $(8, 15)$, graph passes through the point $(11, -12)$

17. Vertex $(55, 1840)$, graph has y-intercept $(0, 2203)$

18. Vertex $(6.4, 0.36)$, graph passes through the point $(1.4, 3.86)$

---

19. State population   The table gives the population of North Dakota, as of July 1 of each year, as estimated by the US Census Bureau.

Year	Population (thousands)
2000	641.1
2001	636.3
2002	633.8
2003	633.4
2004	634.4

(a) Draw a scatter plot of these data.

(b) Write the equation of a quadratic function that models the data.

(c) Use your model to predict the population of North Dakota on July 1, 2011.

20. State population   The table gives the projected population of Ohio, as of July 1 of each year, as estimated by the US Census Bureau.

Year	Population (millions)
2005	11.478
2010	11.576
2015	11.635
2020	11.644
2025	11.606
2030	11.551

(a) Draw a scatter plot of these data.

(b) Write the equation of a quadratic function that models the data.

(c) Use your model to predict the population of Ohio on July 1, 2018.

21. Health insurance   The table gives the percentage of persons in the United States under the age of 65 whose health insurance is provided by Medicaid.

Year	Percentage
1995	11.5
1997	9.7
1999	9.1
2001	10.4
2003	12.5

(a) Draw a scatter plot of these data.

(b) Write the equation of a quadratic function that models the data.

(c) Use your model to estimate the percentage of persons under the age of 65 covered by Medicaid in 2002.

22. Hiring consultants   The number of consultants a large corporation has hired during various years is listed in the table.

Year	Number of consultants
1988	151
1992	433
1996	608
2000	735
2004	778
2008	740

(a) Draw a scatter plot of these data.

(b) Write the equation of a quadratic function that models the data.

(c) Use your model to estimate the number of consultants that the company hired in 2001.

23. Publishing cost   The cost, in dollars, for a publisher to print and distribute x copies of its magazine each month is given by $C(x) = 131,000 + 0.41x$. Write the equation for a function that gives the average cost (see Example 6) per magazine, and compute the average cost to produce 65,000 copies.

24. Internet subscribers   The monthly cost, in thousands of dollars, for an Internet hosting company to support x thousand subscribers is $C(x) = 14 + 0.06x^2 - 0.18x$.

(a) Write an equation for the average monthly cost of each subscriber.

(b) Use a graphing calculator to estimate the number of subscribers for which the average cost is a minimum.

25. Online music sales   After some market research, an online music service has estimated that the number of music files, in thousands, that they can expect to sell annually if the files are priced at p dollars each is approximately $f(p) = 142 - 91.4\sqrt{p}$.

  (a) Write an equation for the annual revenue the service can expect when the files are p dollars each.

  (b) Use a graphing calculator to estimate the price at which the annual revenue is maximized. How many music files will be sold at this price?

26. Fundraising   A charity is holding a fundraising dinner, and estimates that $g(p) = 150 - 0.008p^2$ people will attend when they charge p dollars per plate.

  (a) Write an equation for the function that gives the total funds raised when each attendee is charged p dollars.

  (b) Use a graphing calculator to estimate the price at which the benefit dinner will raise the most funds. How much money will the dinner raise at this price?

27–30 ▪ The graph of a polynomial function f is shown. State the interval(s) on which f is increasing and the interval(s) on which f is decreasing.

27.

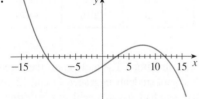

28.

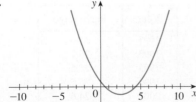

29.

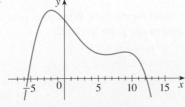

30.

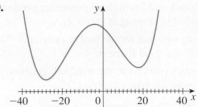

31. Use a graphing calculator to estimate the intervals on which $f(x) = 2x^3 - 5x^{4/3}$ is increasing and the intervals where f is decreasing.

32. Use a graphing calculator to estimate the intervals on which

$$g(x) = \frac{6x^2 - 4x + 5}{x^2 + 1}$$

is increasing and the intervals where g is decreasing.

33. Suppose A varies directly with t. If $A = 80$ when $t = 25$, write an equation for A in terms of t. Then find the value of A when $t = 36$.

34. Suppose C varies directly with the square of x. If $C = 12$ when $x = 4$, write an equation for C in terms of x. Then find the value of C when $x = 9$.

35. Suppose $f(x)$ is proportional to the cube of x. If $f(2) = 14.4$, find the value of $f(5)$.

36. Suppose $g(x)$ is proportional to the square root of x. If $g(4) = 465$, find the value of $g(9)$.

37. If $T(p)$ is inversely proportional to p and $T(8) = 2$, what is the value of $T(30)$?

38. If $Q(v)$ varies inversely with the square of v and $Q(5) = 0.0168$, what is the value of $Q(2)$?

39. Thunder and lightning   During a thunderstorm you see the lightning before you hear the thunder because light travels much faster than sound. The distance between you and the storm varies directly as the time elapsed between the lightning and the thunder. Suppose that the thunder from a storm 1.5 miles away from you takes 8 seconds to reach you.

  (a) Write an equation for the distance D between you and the storm in terms of the time t elapsed between the lightning and thunder.

  (b) If you later observe a 14-second interval between the lightning and thunder, how far away is the storm?

40. Electrical power   A ranch uses a large windmill to supply electrical power. Suppose the power generated by the windmill is directly proportional to the cube of the wind speed, and 96 watts of power are produced when the wind is blowing at 20 mi/h.

(a) Write an equation for the power generated P in terms of the wind speed w.

(b) How much power will the windmill generate if the wind speed increases to 30 mi/h?

41. Sound level   The loudness L of a sound, measured in decibels (dB), is inversely proportional to the square of the distance d from the source of the sound. A person 10 ft from a lawn mower experiences a sound level of 70 dB; how loud is the lawn mower when the person is 100 ft away?

42. Stopping distance   The stopping distance D of Gilbert's car after the brakes have been applied varies directly as the square of the speed s. When traveling at 50 mi/h, Gilbert can stop his car in 240 ft. What is the maximum speed the car can be traveling to stop in 160 ft?

43. Sound frequency   The frequency of vibration of a violin string is inversely proportional to the length of the string. (The constant of proportionality is positive and depends on the properties of the string.) What effect does doubling the length of the string have on the frequency of its vibration?

44. Pendulum period   The period of a pendulum (the time it takes for the pendulum to swing back and forth) varies directly with the square root of the length of the pendulum. How would the length need to be changed in order to double the period?

45. House building   The table gives the number of new (privately owned) housing units completed in the United States during each year as reported by the US Census Bureau.

Year	Housing units completed (in thousands)
1999	1605
2000	1574
2001	1571
2002	1648
2003	1679
2004	1842
2005	1931

Find a quadratic model for the data. Then use the model to estimate the number of housing units completed in 2008.

46. Automobile emissions   A study by the US Office of Science and Technology in 1972 estimated the cost (in 1972 dollars) to reduce automobile emissions by certain percentages:

Reduction in emissions (%)	Cost per car (in $)	Reduction in emissions (%)	Cost per car (in $)
50	45	75	90
55	55	80	100
60	62	85	200
65	70	90	375
70	80	95	600

Find a cubic model that captures the "diminishing returns" trend of these data.

47. World population   Use the data in the table to model the population of the world in the 20th century by a cubic function. Then use your model to estimate the population in the year 1925.

Year	Population (millions)	Year	Population (millions)
1900	1650	1960	3040
1910	1750	1970	3710
1920	1860	1980	4450
1930	2070	1990	5280
1940	2300	2000	6080
1950	2560	2010	6870

48. Drinking and driving   The table gives the percentage of US high school students in grades 11 and 12 who reported that they had driven a vehicle after drinking alcohol.

Year	Percentage
1991	30.6
1993	25.7
1995	25.5
1999	25.5
2001	24.6
2003	22.4

Find a cubic model for the data. What does your model estimate the percentage to be for 2002?

49. Cobb-Douglas production function The Cobb-Douglas production function given in Example 5 corresponding to the year 1920 is approximately

$$P(x) = 52.5x^{1/4}$$

where x measures the amount of capital invested as detailed in the example. The level of capital for 1920 was estimated to be 407. If this level were 20% higher that year, what would have been the effect on production?

50. Planetary distances The table shows the mean (average) distances d of the planets from the Sun (taking the unit of measurement to be the distance from planet Earth to the Sun) and their periods T (time of revolution in years).

Planet	d	T
Mercury	0.387	0.241
Venus	0.723	0.615
Earth	1.000	1.000
Mars	1.523	1.881
Jupiter	5.203	11.861
Saturn	9.541	29.457
Uranus	19.190	84.008
Neptune	30.086	164.784

(a) Fit a power model to the data.

(b) Kepler's Third Law of Planetary Motion states that

"The square of the period of revolution of a planet is proportional to the cube of its mean distance from the Sun."

Does your model corroborate Kepler's Third Law?

Challenge Yourself

51. Illumination Many physical quantities are connected by *inverse square laws*, that is, by power functions of the form $f(x) = kx^{-2}$. In particular, the illumination of an object by a light source is inversely proportional to the square of the distance from the source. Suppose that after dark you are in a room with just one lamp lit, and you are trying to read a book. The light is too dim and so you move halfway to the lamp. How much brighter is the light?

52. Species vs. area It makes sense that the larger the area of a region, the larger the number of species that inhabit the region. Many ecologists have modeled the species-area relation with a power function and, in particular, the number of species S of bats living in caves in central Mexico has been related to the surface area A of the caves by the equation $S = 0.7A^{0.3}$.

(a) The cave called *Misión Imposible* near Puebla, Mexico, has a surface area of $A = 60$ m². How many species of bats would you expect to find in that cave?

(b) If you discover that four species of bats live in a cave, estimate the area of the cave.

53. Species vs. area The table shows the number N of species of reptiles and amphibians inhabiting Caribbean islands and the area A of the island in square miles.

Island	A	N
Saba	4	5
Monserrat	40	9
Puerto Rico	3,459	40
Jamaica	4,411	39
Hispaniola	29,418	84
Cuba	44,218	76

(a) Use a power function to model N as a function of A.

(b) The Caribbean island of Dominica has area 291 mi². How many species of reptiles and amphibians would you expect to find on Dominica?

1.5 Exponential Models

You may have heard a newscaster say that a company or industry was "growing exponentially." Populations and financial markets can also grow exponentially. What does this mean? Suppose a bacteria population (cultured in ideal conditions) is doubling every hour. If we begin with 1000 bacteria, then after one hour we have $1000 \times 2 = 2000$ bacteria, after two hours we have $2000 \times 2 = 4000$ bacteria, after three hours we have 8000 bacteria, and so on. The population is not growing in a linear fashion: If we had a constant rate of change, the population would consistently increase by 1000 every hour. Instead, we have a constant *percentage* rate of growth; the population increases by 100% every hour.

To model the growth of the bacteria population, let $p(t)$ be the number of bacteria after t hours. Then

$$p(0) = 1000$$

$$p(1) = 2 \times p(0) = 2 \times 1000$$

$$p(2) = 2 \times p(1) = 2 \times (2 \times 1000) = 2^2 \times 1000$$

$$p(3) = 2 \times p(2) = 2 \times (2^2 \times 1000) = 2^3 \times 1000$$

It seems from this pattern that, in general,

$$p(t) = 2^t \times 1000 = 1000(2^t)$$

Our result is a constant multiple of the *exponential function* $y = 2^t$. It is called an exponential function because the variable, t, is the exponent. It should not be confused with the power function $y = t^2$.

Introduction to Exponential Functions

In general, an **exponential function** is a function of the form

$$f(x) = a^x$$

where a is a positive constant called the **base**. Every exponential function has domain $\mathbb{R} = (-\infty, \infty)$, although this may not be immediately apparent. For instance, if we input a positive integer such as 8, then we simply have

$$f(8) = a^8 = \underbrace{a \cdot a \cdot \cdots \cdot a}_{8 \text{ factors}}$$

If we input a negative integer such as -3, then recall that

$$f(-3) = a^{-3} = \frac{1}{a^3}$$

We can input 0 as well: $f(0) = a^0 = 1$. Reciprocal inputs such as $1/3$ become roots:

$$f\left(\tfrac{1}{3}\right) = a^{1/3} = \sqrt[3]{a}$$

In fact, any rational number input x can be expressed as a fraction p/q (where p and

q are integers), in which case

$$f(x) = f(p/q) = a^{p/q} = \sqrt[q]{a^p} = \left(\sqrt[q]{a}\right)^p$$

What if x is an irrational number, like $\sqrt{2}$? We can define $a^{\sqrt{2}}$ by a limiting process using rational approximations to $\sqrt{2}$. Because $\sqrt{2}$ can be approximated by 1.4, 1.41, 1.414, 1.4142, ... with increasing accuracy, so $a^{\sqrt{2}}$ is approximated by $a^{1.4}$, $a^{1.41}$, $a^{1.414}$, $a^{1.4142}$, ... For our purposes, it is enough to know that a calculator can generate an (approximate) value. Thus a^x is defined for any real number input.

The range of all exponential functions (except $1^x = 1$) is $(0, \infty)$. (An exponential function can never output 0 or a negative number.)

The graph of $y = 2^x$ is shown in Figure 1 and the graphs of members of the family of functions $y = a^x$ are shown in Figure 2 for various values of the base a. Notice that all of these graphs pass through the same intercept point $(0, 1)$ because $a^0 = 1$ for all positive values of a. Notice also that as the base a gets larger, the exponential function grows more rapidly (for $x > 0$).

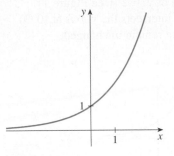

FIGURE 1

$y = 2^x$

If $0 < a < 1$, then a^x approaches 0 as x becomes large. If $a > 1$, then a^x approaches 0 as x decreases through negative values. In both cases the x-axis is a horizontal asymptote. These matters are discussed in Section 4.4.

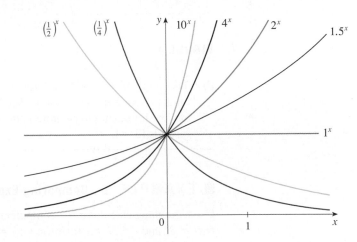

FIGURE 2

You can see from Figure 2 that there are basically two kinds of exponential functions $y = a^x$ (assuming $a \neq 1$). If $0 < a < 1$, the exponential function decreases; if $a > 1$, it increases. These cases are illustrated in Figure 3. Notice that, since $(1/a)^x = 1/a^x = a^{-x}$, the graph of $y = (1/a)^x$ is just the reflection of the graph of $y = a^x$ about the y-axis.

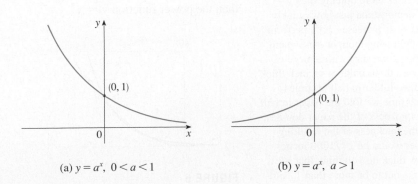

FIGURE 3   (a) $y = a^x$, $0 < a < 1$ (b) $y = a^x$, $a > 1$

■ EXAMPLE 1 Sketching Graphs of Exponential Functions

Sketch the graphs of the functions **(a)** $f(x) = 3 \cdot 2^x$ and **(b)** $g(x) = \left(\frac{1}{2}\right)^x + 3$. What are the domain and range?

SOLUTION

For a review of reflecting and shifting graphs, see Section 1.2.

(a) The graph of f (shown in Figure 4) is the graph of $y = 2^x$ (see Figure 1) stretched vertically by a factor of 3. The graph intersects the y-axis at $(0, 3)$ but the domain, range, and horizontal asymptote remain unchanged.

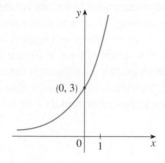

FIGURE 4

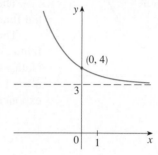

FIGURE 5

(b) The graph of $y = (1/2)^x$ is shown in Figure 2. We shift the graph upward 3 units to obtain the graph of g. (See Figure 5.) The y-intercept is shifted to 4 and the horizontal asymptote is the line $y = 3$. The domain is $\mathbb{R}$ and the range is $(3, \infty)$. ∎

■ EXAMPLE 2 Comparing Exponential and Power Functions

Use a graphing calculator (or computer) to compare the exponential function $f(x) = 2^x$ and the power function $g(x) = x^2$. Which function grows more quickly when x is large?

SOLUTION

Figure 6 shows both functions graphed in the viewing rectangle $[-2, 6]$ by $[0, 40]$. We see that the graphs intersect three times, but for $x > 4$ the graph of $f(x) = 2^x$ stays above the graph of $g(x) = x^2$. Figure 7 gives a more global view and shows that, for large values of x, the exponential function $y = 2^x$ grows far more rapidly than the power function $y = x^2$.

Example 2 shows that $y = 2^x$ increases more quickly than $y = x^2$. To demonstrate just how quickly $f(x) = 2^x$ increases, let's perform the following thought experiment. Suppose we start with a piece of paper a thousandth of an inch thick and we fold it in half 50 times. Each time we fold the paper in half, the thickness of the paper doubles, so the thickness of the resulting paper would be $2^{50}/1000$ inches. How thick do you think that is? It works out to be more than 17 million miles!

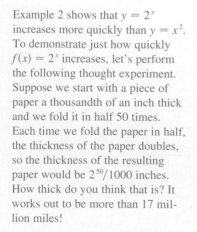

FIGURE 6

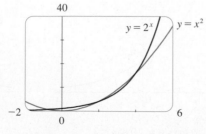

FIGURE 7 ∎

Properties of Exponential Functions

One reason for the importance of the exponential function lies in the following properties. If x and y are integers or rational numbers, then these laws are well known from elementary algebra. It can be proved that they remain true for all real numbers x and y.

■ **Laws of Exponents**   If a and b are positive numbers and x and y are any real numbers, then

1. $a^x \cdot a^y = a^{x+y}$ **2.** $\dfrac{a^x}{a^y} = a^{x-y}$ **3.** $(a^x)^y = a^{xy}$ **4.** $(ab)^x = a^x b^x$

■ **EXAMPLE 3   Using Properties of Exponential Functions**

For more review and practice using the Laws of Exponents, see Appendix A.

Show that each of the following is true.

(a) $8 \cdot (1.6)^{2x} = 8 \cdot (2.56)^x$

(b) $5 \cdot 4^{x/2} = 5 \cdot 2^x$

(c) $\dfrac{10}{5^{x/3}} = 10 \cdot (5^{-1/3})^x$

(d) $3^{4+2t} = 81 \cdot 9^t$

SOLUTION

(a) $8 \cdot (1.6)^{2x} = 8 \cdot ((1.6)^2)^x = 8 \cdot (2.56)^x$

(b) $5 \cdot 4^{x/2} = 5 \cdot (4^{1/2})^x = 5 \cdot (\sqrt{4})^x = 5 \cdot 2^x$

(c) $\dfrac{10}{5^{x/3}} = 10 \cdot (5^{-x/3}) = 10 \cdot (5^{-1/3})^x$

(d) $3^{4+2t} = 3^4 \cdot 3^{2t} = 81 \cdot (3^2)^t = 81 \cdot 9^t$ ■

Applications of Exponential Functions

The exponential function occurs very frequently in mathematical models of nature and society. Any situation where a quantity is growing or shrinking at a constant percentage rate exhibits *exponential growth* or *exponential decay* and can be modeled with a transformed exponential function. Here we give an example where such a model is appropriate to describe population growth. In Section 3.6 we will study many additional applications.

Many graphing calculators (and computer software) have exponential regression capabilities that can fit an exponential model to data. They typically use a least squares technique similar to the linear regression method we used in Section 1.3. The following example uses this technology to model the world's human population over the last century.

▪ EXAMPLE 4 Modeling Population with Exponential Regression

Table 1 shows data for the population of the world in the 20th century and Figure 8 shows the corresponding scatter plot. For simplicity, we have used $t = 0$ to represent 1900.

TABLE 1

Year	Population (millions)
1900	1650
1910	1750
1920	1860
1930	2070
1940	2300
1950	2560
1960	3040
1970	3710
1980	4450
1990	5280
2000	6080
2010	6870

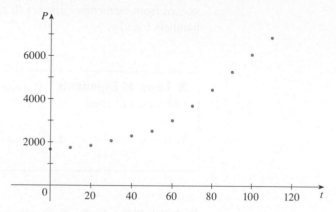

FIGURE 8 Scatter plot for world population growth

The pattern of the data points in Figure 8 suggests exponential growth, so we use a graphing calculator to obtain the exponential model

$$P(t) = (1436.53) \cdot (1.01395)^t$$

Figure 9 shows the graph of this exponential function together with the original data points. We see that the exponential curve fits the data reasonably well. The period of relatively slow population growth is explained by the two world wars and the Great Depression of the 1930s.

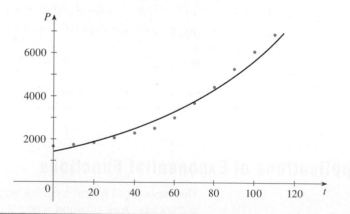

FIGURE 9
Exponential model for
population growth

▪ The Number e

Of all possible bases for an exponential function, there is one that is most convenient for the purposes of calculus. The choice of a base a is influenced by the way the graph of $y = a^x$ crosses the y-axis. Figures 10 and 11 show the *tangent lines* to the graphs of $y = 2^x$ and $y = 3^x$ at the point (0, 1). (Tangent lines will be defined precisely in Section 2.3. For present purposes, you can think of the tangent line to an exponential graph at a point as the line that touches the graph only at that point. It has the same direction as the exponential graph at that point.) If we measure the slopes of these tangent lines at $(0, 1)$, we find that $m \approx 0.7$ for $y = 2^x$ and $m \approx 1.1$ for $y = 3^x$.

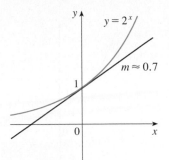

FIGURE 10

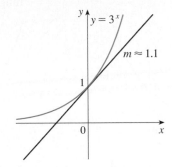

FIGURE 11

It turns out, as we will see in Chapter 3, that some of the formulas of calculus will be greatly simplified if we choose the base a so that the slope of the tangent line to $y = a^x$ at $(0, 1)$ is *exactly* 1. (See Figure 12.) In fact, there *is* such a number; it is an irrational number (it has an infinite nonrepeating decimal representation) and is denoted by the letter e. (This notation was chosen by the Swiss mathematician Leonhard Euler in 1727, probably because it is the first letter of the word *exponential*.) This value also arises naturally in the analysis of compounded interest on a bank account, as one example. In view of Figures 10 and 11, it comes as no surprise that the number e lies between 2 and 3 and the graph of $y = e^x$ lies between the graphs of $y = 2^x$ and $y = 3^x$. (See Figure 13.) We call $y = e^x$ the *natural exponential function*. In Chapter 3 we will see that the value of e, correct to five decimal places, is

$$e \approx 2.71828$$

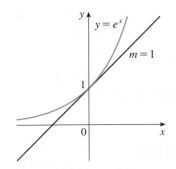

FIGURE 12

The natural exponential function crosses the y-axis with a slope of 1.

TEC Module 1.5 enables you to graph exponential functions with various bases and their tangent lines in order to estimate more closely the value of a for which the tangent has slope 1.

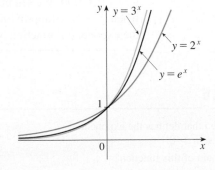

FIGURE 13

■ EXAMPLE 5
Graphing a Transformed Natural Exponential Function

Graph the function $y = \frac{1}{2}e^{-x} - 1$ and state the domain and range.

SOLUTION

We start with the graph of $y = e^x$ from Figures 12 and 14(a) and reflect about the y-axis to get the graph of $y = e^{-x}$ in Figure 14(b). (Notice that the graph crosses the y-axis with a slope of -1). Then we compress the graph vertically by a factor of 2 to obtain the graph of $y = \frac{1}{2}e^{-x}$ in Figure 14(c). Finally, we shift the graph downward one unit to get the desired graph in Figure 14(d). The y-intercept is $-\frac{1}{2}$ and the horizontal asymptote has shifted to $y = -1$. The domain is $\mathbb{R}$ and the range is $(-1, \infty)$.

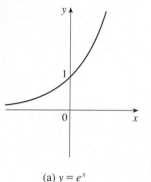

(a) $y = e^x$

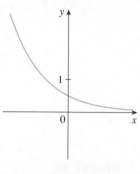

(b) $y = e^{-x}$ (c) $y = \frac{1}{2}e^{-x}$

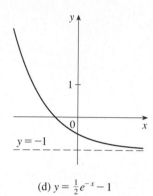

(d) $y = \frac{1}{2}e^{-x} - 1$

FIGURE 14

How far to the right do you think we would have to go for the height of the graph of $y = e^x$ to exceed a million? The next example demonstrates the rapid growth of this function by providing an answer that might surprise you.

▪ EXAMPLE 6

The Rapid Growth of the Natural Exponential Function

Use a graphing device to find the values of x for which $e^x > 1,000,000$.

SOLUTION

In Figure 15 we graph both the function $y = e^x$ and the horizontal line $y = 1,000,000$. We see that these curves intersect when $x \approx 13.8$. Thus, $e^x > 10^6$ when $x > 13.8$ (approximately). Most people would not guess that the values of the exponential function have already surpassed a million when x is only 14! ▪

FIGURE 15

▪ Exercises 1.5

1. (a) Write an equation that defines the exponential function with base $a > 0$.
   (b) What is the domain of this function?
   (c) If $a \neq 1$, what is the range of this function?
   (d) Sketch the general shape of the graph of the exponential function for each of the following cases.
      (i) $a > 1$ **(ii)** $0 < a < 1$

2. (a) How is the number e defined?
   (b) What is an approximate value for e?
   (c) What is the natural exponential function?

3–6 ▪ Graph the given functions on a common screen. How are these graphs related?

3. $y = 2^x$, $y = e^x$, $y = 5^x$, $y = 20^x$

4. $y = e^x$, $y = e^{-x}$, $y = 8^x$, $y = 8^{-x}$

5. $y = 3^x$, $y = 10^x$, $y = \left(\frac{1}{3}\right)^x$, $y = \left(\frac{1}{10}\right)^x$

6. $y = 0.9^x$, $y = 0.6^x$, $y = 0.3^x$, $y = 0.1^x$

7–12 ▪ Make a rough sketch of the graph of the function. Do not use a calculator. Just use the graphs given in Figures 2 and

13 and, if necessary, the transformations of Section 1.2. Indicate the location of the horizontal asymptote.

7. $y = 4^x - 3$

8. $y = 4^{x-3}$

9. $y = -2^{-x}$

10. $y = 2e^x + 1$

11. $f(x) = 3e^{-x}$

12. $g(x) = 2\left(\frac{1}{2}\right)^x - 1$

13. Starting with the graph of $y = e^x$, write the equation of the graph that results from

(a) shifting 2 units downward

(b) shifting 2 units to the right

(c) reflecting about the x-axis

(d) reflecting about the y-axis

(e) reflecting about the x-axis and then about the y-axis

14. Starting with the graph of $y = e^x$, find the equation of the graph that results from

(a) reflecting about the x-axis and then shifting 4 units to the left

(b) reflecting about the y-axis and then shifting 3 units upward

15–20 ■ Simplify each of the following expressions.

15. $x^3 x^5$

16. b^9/b^3

17. $(u^4)^2$

18. $(m^2 n)^4$

19. $\left(\dfrac{p^3}{2}\right)^3$

20. $(2xy^2)^3$

21–24 ■ Write each of the following as an expression using radicals.

21. $4^{2/3}$

22. $7^{5/2}$

23. $e^{1/4}$

24. $w^{3/4}$

25–30 ■ Show that each of the following statements is true.

25. $P \cdot 3^{3x} = P \cdot 27^x$

26. $8^{t/3} = 2^t$

27. $500 \cdot (1.025)^{4t} \approx 500 \cdot (1.1038)^t$

28. $\dfrac{1}{e^{x/2}} = \left(\dfrac{1}{\sqrt{e}}\right)^x$

29. $4^{x+3} = 64 \cdot 4^x$

30. $12e^{0.2t} \approx 12 \cdot (1.2214)^t$

31–34 ■ The table lists some function values. Decide whether the function could be linear, exponential, or neither.

If the function could be linear or exponential, write a possible equation for the function.

31.

x	$f(x)$
0	5
1	10
2	20
3	40
4	80

32.

x	$g(x)$
0	5
1	10
2	15
3	20
4	25

33.

t	$A(t)$
1	12
2	11
3	9
4	6
5	2

34.

n	$P(n)$
1	18
2	6
3	2
4	2/3
5	2/9

35. Bacteria population Under ideal conditions a certain bacteria population is known to double every three hours. Suppose that there are initially 100 bacteria.

(a) What is the size of the population after 15 hours?

(b) What is the size of the population after t hours?

(c) Estimate the size of the population after 20 hours.

(d) Graph the population function and estimate the time for the population to reach 50,000.

36. Bacteria population A bacteria culture starts with 500 bacteria and doubles in size every half hour.

(a) How many bacteria are there after 3 hours?

(b) How many bacteria are there after t hours?

(c) How many bacteria are there after 40 minutes?

(d) Graph the population function and estimate the time for the population to reach 100,000.

37–38 ■ Find the exponential function $f(x) = C \cdot a^x$ whose graph is given.

37.

38.

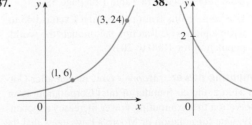

39. If $f(x) = 5^x$, show that

$$\frac{f(x+h) - f(x)}{h} = 5^x \left(\frac{5^h - 1}{h} \right)$$

40. Compensation   Suppose you are offered a job that lasts one month. Which of the following methods of payment do you prefer?

I. One million dollars at the end of the month.

II. One cent on the first day of the month, two cents on the second day, four cents on the third day, and, in general, 2^{n-1} cents on the nth day.

41. Suppose the graphs of $f(x) = x^2$ and $g(x) = 2^x$ are drawn on a coordinate grid where the unit of measurement is 1 inch. Show that, at a distance 2 ft to the right of the origin, the height of the graph of f is 48 ft but the height of the graph of g is about 265 mi.

42. Compare the functions $f(x) = x^5$ and $g(x) = 5^x$ by graphing both functions in several viewing rectangles. Find all points of intersection of the graphs correct to one decimal place. Which function grows more rapidly when x is large?

43. Compare the functions $f(x) = x^{10}$ and $g(x) = e^x$ by graphing both f and g in several viewing rectangles. When does the graph of g finally surpass the graph of f?

44. Use a graph to estimate the values of x such that $e^x > 1{,}000{,}000{,}000$.

45. World population   Use a graphing calculator with exponential regression capability to model the population of the world with the data from 1950 to 2010 in Table 1 on page 58. Use the model to estimate the population in 1993 and to predict the population in the year 2020.

46. World population

(a) Use a graphing calculator to find an exponential model for the population of the world with the data from 1900 to 1950 in Table 1 on page 58.

(b) Use your results from part (a) and Exercise 45 to write a piecewise function that models the world population for 1900 to 2010.

47. Computing power   *Moore's Law*, named after Gordon Moore, the co-founder of Intel Corporation, is an observation that computing power increases exponentially. One formulation of Moore's Law states that the number of transistors on integrated circuits doubles every 18 months. The table lists the number of transistors on various Intel processors for selected years.

Year	Processor	Transistors (in millions)
1982	80286	0.134
1985	386	0.275
1989	486	1.2
1993	Pentium	3.1
1995	Pentium Pro	5.5
1997	Pentium II	7.5
1999	Pentium III	28
2001	Pentium 4	42

(a) Use a graphing calculator to find an exponential model for these data. (Use $t = 0$ to represent 1980.)

(b) Use the model to estimate how long it takes for the number of transistors to double. How close is this to Moore's Law?

(c) In 2004, Intel introduced the Itanium 2 processor carrying 592 million transistors. How does this compare with the number predicted by the model in part (a)?

48. US population   The table gives the population of the United States, in millions, for the years 1900–2010.

Year	Population	Year	Population
1900	76	1960	181
1910	92	1970	205
1920	106	1980	227
1930	123	1990	249
1940	132	2000	281
1950	152	2010	309

Use a graphing calculator with exponential regression capability to model the US population since 1900. Use the model to estimate the population in 1925 and to predict the population in the year 2020.

49. Animal population   Some populations at first increase with exponential growth but eventually slow down and stabilize at a particular level, called the *carrying capacity*. Such quantities can be modeled by functions of the form

$$P(t) = \frac{M}{1 + Ae^{-kt}}$$

called *logistic functions*. The value M is the carrying capacity. Suppose an animal population, in thousands, is modeled by

$$P(t) = \frac{23.7}{1 + 4.8e^{-0.2t}}$$

where t is the number of years after January 1, 2000.

(a) According to the model, what is the animal population on January 1, 2007?

(b) What is the carrying capacity of the population?

(c) Sketch a graph of the function. Then use the graph to estimate when the number of animals reaches reaches half the carrying capacity.

50. Market penetration   Consumer ownership of a particular product (such as a refrigerator or microwave oven) over time can sometimes follow a logistic model; ownership increases swiftly at first, but eventually market saturation occurs and virtually everyone who is capable of and interested in owning the product has purchased it. Suppose the percentage of households owning a certain product is given by

$$g(t) = \frac{0.94}{1 + 2.5e^{-0.3t}}$$

where t is the number of years after 1980.

(a) The carrying capacity is 0.94. What does this represent in this context? Why can't the carrying capacity be larger than 1?

(b) What percentage of households owned the product in 1990?

(c) Use a graphing calculator to estimate when 90% of households owned the product.

■ Challenge Yourself

51. Starting with the graph of $y = 2^x$, find the equation of the graph that results from
   (a) reflecting about the line $y = 3$
   (b) reflecting about the line $x = -4$

52. Starting with the graph of $y = e^x$, find the equation of the graph that results from
   (a) reflecting about the line $y = 4$
   (b) reflecting about the line $x = 2$

53. If you graph the function $f(x) = \dfrac{1 - e^{1/x}}{1 + e^{1/x}}$ you'll see that f appears to be an odd function. Prove it.

54. Graph several members of the family of functions

$$f(x) = \frac{1}{1 + ae^{bx}}$$

where $a > 0$. How does the graph change when b changes? How does it change when a changes?

1.6 Logarithmic Functions

■ Introduction to Logarithms

In Section 1.5 we looked at a bacteria population that started with 1000 bacteria and doubled every hour. If t is the time in hours and N is the population in thousands, then we can say N is a function of t: $N = f(t)$. Several values are listed in Table 1. Suppose, however, that we change our point of view and become interested in the time required for the population to reach various levels. In other words, we are thinking of the function in reverse: We would like to input the population N and receive the number of hours t as the output, so $t = g(N)$. This function g is called the *inverse function* of f. Its values are shown in Table 2; they are simply the

TABLE 1 N as a function of t

t (hours)	$N = f(t)$ = population at time t (in thousands)
0	1
1	2
2	4
3	8
4	16
5	32

TABLE 2 t as a function of N

N	$t = g(N)$ = time (in hours) to reach N thousand bacteria
1	0
2	1
4	2
8	3
16	4
32	5

values from Table 1 with the columns reversed. The inputs of f become the outputs of g, and vice versa.

The inverse of the exponential function $f(x) = a^x$ (assuming that $a > 0$ and $a \neq 1$) is called the **logarithmic function with base a** and is denoted by $\log_a$. The population of the bacteria in Table 1 is given by $f(t) = 2^t$, so its inverse (in Table 2) is $g(N) = \log_2 N$. In words, the value of $\log_2 N$ is the exponent to which the base 2 must be raised to give N. Since $f(3) = 2^3 = 8$, we have $g(8) = \log_2 8 = 3$. In general,

$$\log_a b = c \iff a^c = b$$

■ EXAMPLE 1   Evaluating a Logarithm

The value of $\log_5 125$ is 3, because $5^3 = 125$.   ■

■ EXAMPLE 2
Converting between Logarithmic and Exponential Forms

Write the logarithmic expression $\log_4 w = r$ in an equivalent exponential form.

SOLUTION

In the logarithmic expression, r is the exponent to which 4 is raised to get w:

$$4^r = w$$ ■

The most commonly used bases for logarithms are 10 and e. In fact, these are normally the only bases for which calculators have logarithm keys. When the base is 10, the subscript 10 is often omitted. Thus $\log x$ is assumed to be the logarithmic function with base 10, called the **common logarithm**. It is the inverse of the exponential function $y = 10^x$.

■ The Natural Logarithmic Function

For the purposes of calculus, we will soon see that the most convenient choice of a base for logarithms is the number e, which was defined in Section 1.5. The logarithm with base e is called the **natural logarithm** and has a special notation:

Notation for Logarithms
Most textbooks in calculus and the sciences, as well as calculators, use the notation $\ln x$ for the natural logarithm and $\log x$ for the common logarithm. In the more advanced mathematical and scientific literature and in computer languages, however, the notation $\log x$ often denotes the natural logarithm.

$$\log_e x = \ln x$$

The natural logarithmic function is the inverse of the natural exponential function e^x. Thus

(1) $$\ln b = c \iff e^c = b$$

In particular,

$$\ln e = 1$$

because the exponent to which e must be raised to return e is 1, and

$$\ln 1 = 0$$

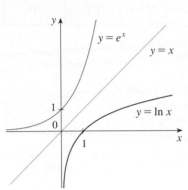

FIGURE 1

because $e^0 = 1$.

The natural logarithmic function $\ln x$ has domain $(0, \infty)$ and range $\mathbb{R}$. (Because $\ln x$ is the inverse of e^x, its domain is the range of e^x, and its range is the domain of e^x.) The graph of $y = \ln x$, shown in Figure 1, is the reflection of the graph of $y = e^x$ about the line $y = x$. The logarithmic function has a vertical asymptote along the y-axis and x-intercept 1, whereas the exponential function has a horizontal asymptote along the x-axis and y-intercept 1. The fact that $y = e^x$ is a very rapidly increasing function for $x > 0$ is reflected in the fact that $y = \ln x$ is a very slowly increasing function for $x > 1$. Notice that the values of $\ln x$ become very large negative as x approaches 0.

■ **EXAMPLE 3**   **Sketching the Graph of a Logarithmic Function**

Sketch the graph of the function $y = \ln(x - 2) - 1$.

SOLUTION

We start with the graph of $y = \ln x$ as given in Figure 1. Using the transformations of Section 1.2, we shift it 2 units to the right to get the graph of $y = \ln(x - 2)$ and then we shift it 1 unit downward to get the graph of $y = \ln(x - 2) - 1$. (See Figure 2.)

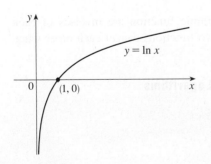 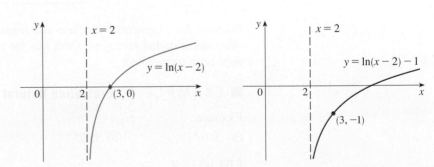

FIGURE 2   ■

Although $\ln x$ is an increasing function, it grows *very* slowly when $x > 1$. In fact, $\ln x$ grows more slowly than any positive power of x. (Compare this to the fact that e^x grows more rapidly than any power of x.) To illustrate this fact, we compare approximate values of the functions $y = \ln x$ and $y = x^{1/2} = \sqrt{x}$ in the following table and we graph them in Figures 3 and 4. You can see that initially the graphs of $y = \sqrt{x}$ and $y = \ln x$ grow at comparable rates, but eventually the root function far surpasses the logarithm.

x	1	2	5	10	50	100	500	1000	10,000	100,000
$\ln x$	0	0.69	1.61	2.30	3.91	4.6	6.2	6.9	9.2	11.5
$\sqrt{x}$	1	1.41	2.24	3.16	7.07	10.0	22.4	31.6	100	316

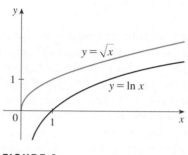

FIGURE 3   **FIGURE 4**

▪ Properties of Logarithms

The following **cancellation equations** say that if we form the composition of the natural logarithmic and exponential functions, in either order, the output is simply the original input.

(2)

$$\ln(e^x) = x$$
$$e^{\ln x} = x \quad (x > 0)$$

Because the exponential function and logarithmic function are inverses of each other, these equations say in effect that the two functions cancel each other when applied in succession.

▪ EXAMPLE 4 Evaluating Natural Logarithms

Evaluate:

(a) $\ln(e^4)$ (b) $\ln 25$

SOLUTION

(a) From the first cancellation equation in (2), $\ln(e^4) = 4$. Another way to look at it: $\ln(e^4) = 4$ is the exponent to which e must be raised to get e^4, namely 4.

(b) The value of $\ln 25$ is the exponent to which e must be raised to get 25, but this is not a number we can determine by hand. Using a calculator, the value is approximately 3.2189. Thus $e^{3.2189} \approx 25$. ▪

■ **EXAMPLE 5** **Solving a Basic Logarithmic Equation**

Find x if $\ln x = 5$.

SOLUTION 1

From (1) we see that

$$\ln x = 5 \qquad \text{means} \qquad e^5 = x$$

Therefore $x = e^5$.

(If you have trouble working with the "ln" notation, just replace it by $\log_e$. Then the equation becomes $\log_e x = 5$; so, by the definition of logarithm, $e^5 = x$.)

SOLUTION 2

Start with the equation

$$\ln x = 5$$

and apply the exponential function to both sides of the equation:

$$e^{\ln x} = e^5$$

But the second cancellation equation in (2) says that $e^{\ln x} = x$. Therefore

$$x = e^5 \qquad\qquad ■$$

The following properties of logarithmic functions follow from the corresponding properties of exponential functions given in Section 1.5.

■ **Laws of Logarithms**   If x and y are positive numbers, then

1. $\ln(xy) = \ln x + \ln y$

2. $\ln\left(\dfrac{x}{y}\right) = \ln x - \ln y$

3. $\ln(x^r) = r \ln x$ (where r is any real number)

■ **EXAMPLE 6** **Simplifying a Logarithmic Function**

Show that $f(t) = \ln(5e^{3t})$ is a linear function.

SOLUTION

Using the first law of logarithms, $f(t)$ can be written as $\ln 5 + \ln(e^{3t})$. But the first cancellation equation in (2) says that $\ln(e^{3t}) = 3t$, so we have $f(t) = 3t + \ln 5$, a linear function with slope 3 and y-intercept $\ln 5 \approx 1.6094$.   ■

▪ **EXAMPLE 7** **Using Properties of Logarithms**

Express $\ln a + \frac{1}{2} \ln b$ as a single logarithm.

SOLUTION

Using Laws 3 and 1 of logarithms, we have

$$\ln a + \tfrac{1}{2} \ln b = \ln a + \ln b^{1/2}$$
$$= \ln a + \ln \sqrt{b}$$
$$= \ln\left(a\sqrt{b}\right) \qquad \blacksquare$$

▪ | # Solving Exponential Equations

Logarithms can be used to solve exponential equations. By taking the natural logarithm of each side of an equation, we can use the properties of logarithms to solve for a variable in an exponent regardless of the base of the exponential expression.

▪ **EXAMPLE 8** **Solving an Exponential Equation**

Solve the equation $e^{5-3x} = 10$.

SOLUTION

We take natural logarithms of both sides of the equation and use (2):

$$\ln(e^{5-3x}) = \ln 10$$
$$5 - 3x = \ln 10$$
$$3x = 5 - \ln 10$$
$$x = \tfrac{1}{3}(5 - \ln 10)$$

Using a calculator, we can approximate the solution: to four decimal places, $x \approx 0.8991$. ▪

▪ **EXAMPLE 9** **Solving an Exponential Equation**

Solve the equation $3^x = 18$. Give a decimal number solution, rounded to four decimal places.

SOLUTION

Take natural logarithms of both sides of the equation:

$$\ln(3^x) = \ln 18$$

Using Law 3 of logarithms, we can write

$$x \cdot \ln 3 = \ln 18$$

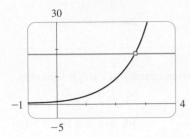

FIGURE 5

We can use a graphing calculator to check our work in Example 9. Figure 5 shows that the graph of $y = 3^x$ intersects the horizontal line $y = 18$ at $x \approx 2.63$.

Dividing both sides by ln 3 gives

$$x = \frac{\ln 18}{\ln 3} \approx 2.6309$$

A quick check confirms that $3^{2.6309} \approx 18$. ▪

▪ **EXAMPLE 10** **An Exponential Model for Light Intensity**

Light decreases in intensity exponentially as it passes through a substance. Suppose the intensity of a beam of light passing through the murky water in a pond can be modeled by $I(x) = I_0 \cdot 2^{-x/18}$, where I_0 is the initial intensity of the light and x is the distance in feet that the beam has traveled through the water. How far has the beam traveled when its intensity is reduced to 10% of its original intensity?

SOLUTION

Because 10% of the original intensity is $0.10I_0$, we need to solve the equation $I_0 \cdot 2^{-x/18} = 0.1I_0$. We start by isolating the exponential expression (divide both sides by I_0), and then we take natural logarithms of both sides:

$$2^{-x/18} = 0.1$$

$$\ln(2^{-x/18}) = \ln(0.1)$$

$$-\frac{x}{18} \cdot \ln 2 = \ln(0.1)$$

$$x = -\frac{18}{\ln 2} \cdot \ln(0.1) \approx 59.795$$

Thus the intensity is reduced to 10% of the original intensity after the light has passed through about 59.8 feet of the water. ▪

▪ **Exercises 1.6**

1. (a) How is the logarithmic function $y = \log_a x$ defined?
   (b) What is the domain of this function?
   (c) What is the range of this function?

2. (a) What is the natural logarithm?
   (b) What is the common logarithm?
   (c) Sketch the graphs of the natural logarithm function and the natural exponential function with a common set of axes.

3–6 ▪ Find the exact value (without using a calculator) of each expression.

3. (a) $\log_2 64$ (b) $\log_6 \frac{1}{36}$

4. (a) $\log_2 8$ (b) $\log_8 2$

5. (a) $\ln e^3$ (b) $e^{\ln 7}$

6. (a) $\ln e^{\sqrt{2}}$ (b) $e^{3 \ln 2}$

7–10 ▪ Use a calculator to evaluate the quantity correct to four decimal places.

7. $\ln 100$

8. $3 \ln(e + 2)$

9. $\dfrac{\ln 28}{\ln 4}$

10. $\dfrac{\ln 6}{5 \ln 3}$

11–12 ▪ Write the logarithmic expression in an equivalent exponential form.

11. (a) $\log_8 4 = \frac{2}{3}$ (b) $\log_6 u = v$

12. (a) $\ln 12 \approx 2.4849$ (b) $C = \ln A$

13–14 ▪ Write the exponential expression in an equivalent logarithmic form.

13. (a) $10^3 = 1000$ (b) $y = 4^x$

14. (a) $e^x = 2$ (b) $R = e^{3t}$

15–18 ▪ Make a rough sketch of the graph of each function. Do not use a calculator. Just use the graph given in Figure 1 and the transformations of Section 1.2.

15. $y = -\ln x$ **16.** $y = \ln(-x)$

17. $y = \ln(x + 1) + 3$ **18.** $y = \ln(x - 4) - 2$

19. Starting with the graph of $y = \ln x$, find the equation of the graph that results from
    (a) shifting 3 units upward
    (b) shifting 3 units to the left
    (c) reflecting about the x-axis
    (d) reflecting about the y-axis

20. If we start with the graph of $y = \ln x$, reflect the graph about the x-axis, and then shift the graph down 4 units, what is the equation of the resulting graph?

21. Suppose that the graph of $y = \ln x$ is drawn on a coordinate grid where the unit of measurement is an inch. How many miles to the right of the origin do we have to move before the height of the curve reaches 3 ft?

22. Compare the functions $f(x) = x^{0.1}$ and $g(x) = \ln x$ by graphing both f and g in several viewing rectangles. When does the graph of f finally surpass the graph of g?

23–26 ▪ State whether each of the following is true or false.

23. $\ln(c + d) = \ln c + \ln d$ **24.** $\ln(cd) = \ln c + \ln d$

25. $\ln(u/3) = \dfrac{\ln u}{\ln 3}$ **26.** $(\ln x)^2 = 2 \ln x$

27–30 ▪ Express the given quantity as a single logarithm.

27. $2 \ln 4 - \ln 2$ **28.** $\ln 3 + 2 \ln x$

29. $3 \ln u - 2 \ln 5$ **30.** $\ln x + a \ln y - b \ln z$

31. (a) Explain why $y = \ln(x^3)$ and $y = 3 \ln x$ have the same graph.
    (b) Explain why $y = \ln(x^2)$ and $y = 2 \ln x$ don't have the same graph.

32. The graph of the function $f(t) = \ln(3^t)$ is a line through the origin. Explain why this is a linear function. What is the slope?

33–34 ▪ Solve each equation for x. Give both an exact solution and a decimal approximation, rounded to four decimal places.

33. (a) $2 \ln x = 1$ (b) $e^{-x} = 5$

34. (a) $e^{2x+3} - 7 = 0$ (b) $\ln(5 - 2x) = -3$

35–42 ▪ Solve each equation. Give a decimal approximation, rounded to four decimal places.

35. $5^t = 20$ **36.** $1.13^x = 7.65$

37. $2^{x-5} = 3$ **38.** $10^{3-2x} = 42$

39. $8e^{3x} = 31$ **40.** $450e^{0.15t} = 1200$

41. $6 \cdot (2^{x/7}) = 11.4$ **42.** $100 \cdot (4^{-p/5}) = 8.8$

43. County population Suppose the function $P(t) = 437.2(1.036)^t$ is used to model the population, measured in thousands of people, of a county t years after the end of 1995. When will the population reach one million people?

44. Vehicle value The value of Tracy's car is given by $V(t) = 18500(0.78)^t$, where t is the number of years she has owned the vehicle. When will the car be worth only $2000?

45. Water transparency Environmental scientists measure the intensity of light at various depths in a lake to find the "transparency" of the water. Certain levels of transparency are required for the biodiversity of the submerged macrophyte population. In a certain lake the intensity of light at a depth of x feet is given by

$$I = 10e^{-0.008x}$$

where I is measured in lumens. At what depth has the light intensity dropped to 5 lumens?

46. **Engine temperature**   Suppose you're driving a car on a cold winter day (20°F outside) and the engine overheats (at about 220°F). When you park, the engine begins to cool down. The temperature T of the engine x minutes after you park satisfies the equation

$$\ln\left(\frac{T - 20}{200}\right) = -0.11x$$

Find the temperature of the engine after 20 minutes.

47. **Bacteria population**   If a bacteria population starts with 100 bacteria and doubles every three hours, then the number of bacteria after t hours is

$$n = f(t) = 100 \cdot 2^{t/3}$$

(see Exercise 35 in Section 1.5). When will the population reach 50,000?

48. **Electric charge**   When a camera flash goes off, the batteries immediately begin to recharge the flash's capacitor, which stores electric charge given by

$$Q(t) = Q_0(1 - e^{-t/a})$$

(The maximum charge capacity is Q_0 and t is measured in seconds.) How long does it take to recharge the capacitor to 90% of capacity if $a = 2$?

49. **Investment**   Many graphing calculators can fit a logarithmic function $f(x) = a + b \ln x$ to data. The

table shows the time required for a $10,000 investment to reach different values in a particular bank account.

Value	Years
$11,000	2.1
$12,000	4.0
$13,000	5.8
$14,000	7.4
$15,000	9.0
$16,000	10.4
$17,000	11.7
$18,000	13.0

(a) Use a graphing calculator to find a logarithmic model for the data.

(b) Use the model to estimate how long it will take for the account to reach $25,000 in value.

50. **Kiln temperature**   A pottery kiln heated to 2400°F is turned off and allowed to cool. An alert sounds whenever the temperature drops 200°F . The elapsed times, in hours, when the alerts sounded are recorded in the table.

Temperature (°F)	Time (hours)
2200	0.52
2000	1.12
1800	1.80
1600	2.56
1400	3.45
1200	4.48
1000	5.76
800	7.39

(a) Use a graphing calculator to find a logarithmic model for the data.

(b) Use the model to estimate how long it will take for the kiln to cool to 300°F.

Challenge Yourself

51. **Television viewership**   Market researchers estimate that the percentage of households that have viewed a particular television program is given by the logistic function

$$f(t) = \frac{0.41}{1 + 0.52e^{-0.4t}}$$

where t is time in years and $t = 0$ corresponds to January 1, 2005. When will 30% of households have seen the program?

▪ CONCEPT CHECK

1. (a) What is a function? What are its domain and range?

  (b) What is the graph of a function? What is a scatter plot?

  (c) How can you tell whether a given curve is the graph of a function?

2. Discuss four ways of representing a function. Illustrate your discussion with examples.

3. What is a mathematical model?

4. What is a piecewise defined function? Give an example.

5. (a) What is an even function? How can you tell if a function is even by looking at its graph?

  (b) What is an odd function? How can you tell if a function is odd by looking at its graph?

6. Suppose that a function f has domain $(-5, 5)$ and a function g has domain $[0, \infty)$.

  (a) What is the domain of $f + g$?

  (b) What is the domain of fg?

  (c) What is the domain of f/g?

7. How is the composition of functions f and g defined? What is its domain?

8. Suppose the graph of f is given. Write an equation for each of the graphs that are obtained from the graph of f as follows.

  (a) Shift 2 units upward.

  (b) Shift 2 units downward.

  (c) Shift 2 units to the right.

  (d) Shift 2 units to the left.

  (e) Reflect about the x-axis.

  (f) Reflect about the y-axis.

  (g) Stretch vertically by a factor of 2.

  (h) Shrink vertically by a factor of of 2.

  (i) Stretch horizontally by a factor of 2.

  (j) Shrink horizontally by a factor of of 2.

9. Give an example of each type of function.

  (a) Linear function

  (b) Quadratic function

  (c) Polynomial of degree 5

  (d) Power function

  (e) Rational function

  (f) Exponential function

10. What is the slope of a line? How do you compute it? What is the rate of change of a linear function?

11. How do you write an equation for a linear function if you know the slope and a point on the line?

12. What is a regression line?

13. What is the difference between interpolation and extrapolation?

14. What is the shape of the graph of a quadratic function? What is the vertex?

15. Sketch by hand, on the same axes, the graphs of the following functions.

  (a) $f(x) = x$ **(b)** $g(x) = x^2$

  (c) $h(x) = x^3$

16. Draw, by hand, a rough sketch of the graph of each function.

  (a) $y = \sqrt{x}$ **(b)** $y = 1/x$

17. (a) Write an equation for a function whose output varies directly with x.

  (b) Write an equation for a function whose output varies inversely with x.

18. Draw, by hand, a rough sketch of the graph of each function.

  (a) $y = e^x$ **(b)** $y = \ln x$

19. (a) What is an inverse function?

  (b) What is the inverse function of $f(x) = 3^x$?

Answers to the Concept Check can be found on the back endpapers.

■ EXERCISES

1. Let f be the function whose graph is given.

  (a) Estimate the value of $f(2)$.

  (b) Estimate the values of x such that $f(x) = 3$.

  (c) State the domain of f.

  (d) State the range of f.

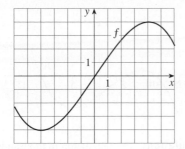

2. Health costs   Let $f(x)$ be the cost, in millions of dollars, to a particular large company for health insurance when the company has x thousand employees. What does the equation $f(8.6) = 14.1$ say in this context?

3. Driving distance   The distance traveled by a car is given by the values in the table.

t (seconds)	0	1	2	3	4	5
d (feet)	0	10	32	70	119	178

  (a) Use the data to sketch the graph of d as a function of t.

  (b) Use the graph to estimate the distance traveled after 4.5 seconds.

4. Crop yield   Sketch a rough graph of the yield of a crop as a function of the amount of fertilizer used.

5–8 ■ Find the domain of the function.

5. $f(x) = \sqrt{4 - 10x}$ **6.** $g(x) = 1/(x + 1)$

7. $y = 2^x + 1$ **8.** $A(x) = 6.3 + \ln(x - 1)$

9. If $p(x) = x^2 - 3x$, find and simplify $p(-2)$, $p(x - 5)$, $\dfrac{p(a) - p(4)}{a - 4}$, and $\dfrac{p(x + h) - p(x)}{h}$.

10. Let f be the piecewise function defined by

$$f(x) = \begin{cases} \frac{1}{2}x + 2 & \text{if } x < 0 \\ 3^x & \text{if } x \geq 0 \end{cases}$$

  (a) Evaluate $f(-4)$, $f(0)$, and $f(2)$.

  (b) Sketch a graph of f.

11. Determine whether f is even, odd, or neither even nor odd. If you have a graphing calculator, use it to check your answer visually.

  (a) $f(x) = 2x^5 - 3x^2 + 2$ (b) $f(x) = x^3 - x^7$

  (c) $f(x) = e^{-x^2}$

12. Functions f and g are graphed below. Decide whether each function is even, odd, or neither. Explain your reasoning.

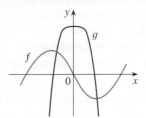

13. Retirement savings   Let $S(t)$ be David's annual salary (in dollars) for year t, and let $P(t)$ be the percentage (given as a decimal number) of David's salary that he contributed to a retirement account during year t. What does the function $h(t) = S(t) \cdot P(t)$ measure?

14. Inventory costs   Let $N(t)$ be the number of computer monitors a manufacturer has stored in its warehouses t weeks after the start of the year. If $C(x)$ is the weekly cost the company incurs to store x monitors, what does the function $g(t) = C(N(t))$ represent?

15. If $f(x) = 3x^2 + 4$ and $g(x) = 2^x - 5$, find each of the following functions.

  (a) $A(x) = f(x) - g(x)$ (b) $B(x) = f(g(x))$

  (c) $C(x) = g(f(x))$

16. If $h(x) = e^{x^2 - 3x}$, find functions f and g so that $h(x) = f(g(x))$.

17. Suppose that the graph of f is given. Describe how the graphs of the following functions can be obtained from the graph of f.

  (a) $y = f(x) + 8$ (b) $y = f(x + 8)$

  (c) $y = 1 + 2f(x)$ (d) $y = f(x - 2) - 2$

  (e) $y = -f(x)$

18. The graph of f is given. Draw the graphs of the following functions.

  (a) $y = f(x - 8)$ (b) $y = -f(x)$

  (c) $y = 2 - f(x)$ (d) $y = \frac{1}{2}f(x) - 1$

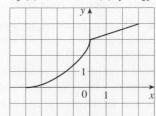

19–22 ▪ Use transformations to sketch the graph of the function.

19. $y = (x - 2)^2 - 3$

20. $y = -\ln(x - 2)$

21. $y = 2e^x + 3$

22. $y = 2 - \sqrt{x}$

---

23. Consumer demand   A company has developed a new product. Let $L(x)$ be the estimated number of units, in thousands, that will be purchased if the price of the product is x dollars. The company's market research shows that $L(15) = 281$ and $L(18) = 245$.

(a) If L is a linear function, what is the slope of L? What does it represent in this context?

(b) Write a formula for L, assuming that L is a linear function.

24. Graph the linear function $p(t) = 2.5t + 4.5$. What is the slope?

25. Life expectancy   Life expectancy improved dramatically in the 20th century. The table gives the life expectancy at birth (in years) of males born in the United States.

Birth year	Life expectancy	Birth year	Life expectancy
1900	48.3	1960	66.6
1910	51.1	1970	67.1
1920	55.2	1980	70.0
1930	57.4	1990	71.8
1940	62.5	2000	73.0
1950	65.6		

Use a scatter plot to write a linear model for the data using two of the data points. Then use your model to predict the life span of a male born in the year 2015.

26. Cost function   A small-appliance manufacturer finds that it costs $9000 to produce 1000 toaster ovens a week and $12,000 to produce 1500 toaster ovens a week.

(a) Express the cost as a function of the number of toaster ovens produced, assuming that it is linear. Then sketch the graph.

(b) What is the slope of the graph and what does it represent?

(c) What is the y-intercept of the graph and what does it represent?

27. Write an equation for a quadratic function with vertex $(-8, 2)$ that passes through the point $(2, 4)$.

28. Cost function   The cost, in dollars, for a furniture manufacturer to produce x units of a particular chair is $C(x) = 0.02x^2 + 1.6x + 4200$.

(a) If the manufacturer increases production from 600 units to 800 units, what is the increase in cost?

(b) Write a formula for the average cost of each chair when x chairs are produced.

29. If A varies inversely with x and $A = 28$ when $x = 112$, write an equation for A in terms of x.

30. Suppose $f(n)$ is proportional to the square of n. If $f(5) = 35$, find the value of $f(10)$.

31. The graph of a function f is shown. State the interval(s) on which f is increasing and the interval(s) on which f is decreasing.

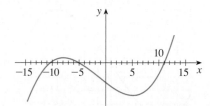

32. City population   The population in thousands of a city is given by $P(t)$, where t is the year, with $t = 0$ corresponding to 2000. In 2000, the population of the city was 45,200. Write a formula for P that satisfies the given description.

(a) The population is increasing by 1650 people per year.

(b) The population is doubling every 20 years.

33. Simplify each of the following expressions.

(a) $(3xy^4)^2$

(b) $e^{2\ln 3}$

(c) $\log_4 16$

34. Show that each of the following statements is true.

(a) $9^{t/2} = 3^t$

(b) $2\ln x + \ln 3 = \ln(3x^2)$

35. Find the exponential function $f(x) = C \cdot a^x$, where $f(0) = 8.3$ and $f(4) = 20.9$.

36. Solve each equation for x.

(a) $e^x = 5$

(b) $\ln x = 2$

(c) $5e^{2x} = 18$

37. Insect population   Suppose an insect population that currently numbers 4000 will triple every five years.

(a) What will the insect population be after 20 years?

(b) Write a formula for the population $P(t)$ after t years.

(c) How long until the population reaches one million?

38. Animal population   The population of a certain species in a limited environment with initial population 100 and carrying capacity 1000 is

$$P(t) = \frac{100{,}000}{100 + 900e^{-t}}$$

where t is measured in years.

(a) Compute $P(5)$.

(b) Graph this function and estimate how long it takes for the population to reach 900.

39. Bacteria population A population of bacteria is doubling every hour. How long does it take for the population to triple?

40. Graph the three functions $y = x^a$, $y = a^x$, and $y = \log_a x$ on the same screen for two or three values of $a > 1$. For large values of x, which of these functions has the largest values and which has the smallest values? *Note:* To enter $\log_a x$ into a graphing calculator, use the *change of base formula*

$$\log_a x = \frac{\ln x}{\ln a}$$

41. Hourly income The table gives the average hourly rate, in dollars, earned by registered nurses in Columbus, Ohio, for various years as estimated by the US Department of Labor.

Year	Hourly rate
1999	$20.90
2001	$23.03
2003	$24.91
2004	$26.63
2005	$27.06

Find the regression line for these data. Then use the regression model to predict the average hourly rate a registered nurse in the area will earn in 2012.

42. Oil imports The table shows the daily crude oil imports into the United States, measured in thousands of barrels per day, as recorded by the US Energy Information Administration during the first week of May for various years.

Year	Crude oil imports (thousand barrels per day)
1990	6286
1993	6495
1996	7223
1999	8439
2002	9160
2005	9992

Find a cubic model for these data. Then use your model to estimate the crude oil imports into the United States during the first week of May in 2001.

2

The concept of a derivative that you will learn in this chapter allows us to compute or estimate how quickly a quantity is changing at a particular moment in time. In a problem on page 131, for instance, you are asked to estimate the rate at which the amount of US currency in circulation was changing in the year 2000.
© Loskutnikov/ Shutterstock

The Derivative

We begin our study of calculus by investigating rates of change for a variety of functions and contexts. Specifically, we see how an average rate of change becomes an instantaneous rate of change through a limiting process. The concept of a limit serves as the foundation for all of calculus; here we explore limits and their properties and see that the special type of limit that is used to find velocities and other rates of change gives rise to the central idea in differential calculus, the derivative. We see how derivatives can be interpreted as rates of change in various situations as well as the slope of a curve, and we learn how the derivative of a function gives information about the original function.

2.1 Measuring Change

One of the major themes of calculus is analyzing how quantities change; most importantly, we are interested in the *rates* at which quantities change. We saw in Chapter 1 that the output values of linear functions change at a constant rate that is equal to the value of slope. But for a nonlinear function, the function values change at varying rates and measuring those rates is not as straightforward.

■ Average Rate of Change

We start our study of change by looking at the *average rate of change* of a function over a particular interval of input values. Suppose y is a function of x, in which case we can write $y = f(x)$. If x changes from x_1 to x_2, then the change in x (also called the **increment** of x) is

$$\Delta x = x_2 - x_1$$

and the corresponding change in y is

$$\Delta y = f(x_2) - f(x_1)$$

We then compute the ratio of change in output to change in input.

■ The **average rate of change of y with respect to x** over the interval $[x_1, x_2]$ is the difference quotient

$$\frac{\Delta y}{\Delta x} = \frac{f(x_2) - f(x_1)}{x_2 - x_1}$$

The units of the average rate of change are always output units per input unit.

■ **EXAMPLE 1** **Finding an Average Rate of Change of a Function**

Compute the average rate of change of the function $g(t) = 4t^2 - 3t$ over the interval $[2, 5]$.

SOLUTION

We have $g(2) = 10$ and $g(5) = 85$, so the average rate of change is

$$\frac{\Delta g}{\Delta t} = \frac{g(t_2) - g(t_1)}{t_2 - t_1} = \frac{g(5) - g(2)}{5 - 2} = \frac{85 - 10}{5 - 2} = \frac{75}{3} = 25 \qquad ■$$

Year	Population (thousands)
1960	57.2
1965	59.0
1970	59.0
1975	59.2
1980	59.9
1985	61.5
1990	63.0
1995	80.0
2000	90.0
2005	97.0
2010	104.6

■ **EXAMPLE 2   Average Rate of Change of a Population**

The population (in thousands) of Aruba for various years is given in the table.

(a) Compute the average rate of change of the population from 1975 to 2010 and interpret your result.

(b) What is the average rate of change of the population from 1965 to 1970?

SOLUTION

Let $P(t)$ represent the population in thousands at year t.

(a) The average rate of change is

$$\frac{\Delta P}{\Delta t} = \frac{P(2010) - P(1975)}{2010 - 1975} = \frac{104.6 - 59.2}{2010 - 1975} = \frac{45.4}{35} \approx 1.30$$

The units are output units per input unit: thousands of people per year. Thus the population of Aruba was increasing at an *average* rate of approximately 1300 people per year from 1975 to 2010.

(b) From 1965 to 1970, the average rate of change is

$$\frac{P(1970) - P(1965)}{1970 - 1965} = \frac{59.0 - 59.0}{5} = \frac{0}{5} = 0 \text{ people/year}$$ ■

In Example 2(b) the average rate of change is 0. Note that this does not mean that the population of Aruba did not change during the five-year interval. It simply means that the population was the same at the start and end of the time interval. In fact, from the information given, we don't know anything about the population between 1965 and 1970.

■ Geometric Interpretation of Average Rate of Change

The difference quotient that appears in the definition of average rate of change should look familiar to you; it is the slope of a line from Section 1.3. Let's consider the graph of a function f. A **secant line** is a line passing through two points on a curve. Figure 1 shows the secant line through the points $(x_1, f(x_1))$ and $(x_2, f(x_2))$.

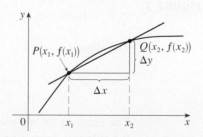

FIGURE 1   average rate of change = slope of secant line through P, Q

The slope of this line is

$$\frac{\Delta y}{\Delta x} = \frac{f(x_2) - f(x_1)}{x_2 - x_1}$$

which is precisely the average rate of change on the interval $[x_1, x_2]$.

> ▪ The average rate of change of a function f over the interval $[x_1, x_2]$ is the slope of the secant line through the points $(x_1, f(x_1))$ and $(x_2, f(x_2))$.

▪ **EXAMPLE 3**   **Estimating an Average Rate of Change from a Graph**

The graph in Figure 2 shows an example of how driving speed affects gas mileage, as published by the US Department of Energy.

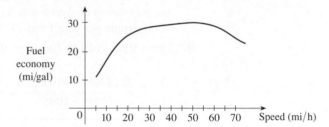

FIGURE 2

Estimate the average rate of change of fuel economy when driving speed increases from 45 mi/h to 70 mi/h.

SOLUTION

We mark points on the graph corresponding to speeds of 45 and 70 mi/h. The average rate of change is the slope of the secant line through these points. (See Figure 3.)

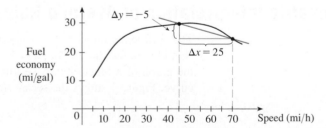

FIGURE 3

The slope is

$$\frac{\Delta y}{\Delta x} \approx \frac{-5}{25} = -0.2$$

The units are mi/gal per mi/h, and because the slope is negative, we know that the output is decreasing (as the input increases). Thus the fuel economy decreases at an average rate of 0.2 mi/gal for each mi/h increase in driving speed between 45 and 70 mi/h.   ▪

■ Shrinking the Interval

The average rate of change can be computed over any finite interval, large or small. However, smaller intervals give a better idea of how a quantity is changing near a particular input value. In the next example, we look at the distance a dropped ball has fallen after t seconds. By comparing average rates of change as the time interval shrinks, we can predict the exact speed at which the ball is traveling at a particular moment in time.

■ EXAMPLE 4   Estimating a Speed at One Moment in Time

Suppose that a ball is dropped from the upper observation deck of the CN Tower in Toronto, 450 m above the ground. Estimate the speed of the ball after 5 seconds.

SOLUTION

Through experiments carried out four centuries ago, Galileo discovered that the distance fallen by any freely falling body is proportional to the square of the time it has been falling. (This model for free fall neglects air resistance.) If the distance fallen after t seconds is denoted by $s(t)$ and measured in meters, then Galileo's law is expressed by the equation

$$s(t) = 4.9t^2$$

If we compute an average rate of change, we are finding an average speed (the units are meters per second). The difficulty in finding the speed at precisely 5 seconds is that we are dealing with a single instant of time ($t = 5$), so no time interval is involved. However, we can approximate the desired speed by computing the average speed over the brief time interval of a tenth of a second from $t = 5$ to $t = 5.1$:

$$\text{average speed} = \frac{\text{distance traveled}}{\text{time elapsed}}$$

$$= \frac{s(5.1) - s(5)}{5.1 - 5}$$

$$= \frac{4.9(5.1)^2 - 4.9(5)^2}{0.1} = 49.49 \text{ m/s}$$

We could make our approximation more accurate by looking at even smaller time intervals. The following table shows the results of similar calculations of the average rate of change over successively smaller time periods.

Time interval	Average speed (m/s)
$5 \leqslant t \leqslant 6$	53.9
$5 \leqslant t \leqslant 5.1$	49.49
$5 \leqslant t \leqslant 5.05$	49.245
$5 \leqslant t \leqslant 5.01$	49.049
$5 \leqslant t \leqslant 5.001$	49.0049

The CN Tower in Toronto was the tallest freestanding building in the world for 32 years.

© 2003 Brand X Pictures/Jupiter Images/Fotosearch

It appears that as we shorten the time period, the average speed is becoming closer to 49 m/s. It would be reasonable to say that the speed of the ball after 5 seconds is 49 m/s.   ▪

In Example 4, we estimated that the speed of the falling ball was 49 m/s because the average speeds appeared to be approaching 49 m/s as we shortened the time interval. But how do we know 49 is the correct value? We could continue to shrink the time interval smaller and smaller, but this is a never-ending process. If there is in fact a value that the average speeds are approaching as the time interval shrinks, it is called the *limit* of the average speeds. We will explore the concept of a limit in the next section, after which we will be equipped to determine that the speed of the ball after 5 seconds, according to our model, is indeed exactly 49 m/s.

▪ Exercises 2.1

1–4 ▪ Find the average rate of change of the function over the given interval.

1. $f(x) = x^2 + 5x$, $[1, 3]$

2. $g(t) = 2t^3 - 4t + 1$, $[0, 2]$

3. $A(v) = \sqrt{v + 3}$, $[6, 13]$

4. $C(x) = 4x/(x + 2)$, $[4, 8]$

5–8 ▪ Find the average rate of change of the function over the given interval. Give a decimal approximation rounded to three decimal places.

5. $P(t) = 4.7 \ln t + 1.8$, $[16, 84]$

6. $h(t) = 1.85(3^t)$, $[2.9, 4.1]$

7. $N(w) = 5e^{0.2w}$, $[16, 22]$

8. $q(u) = 34.1/(1 + 2e^{-0.4u})$, $[150, 225]$

9. Gold prices   The table shows London's closing price, in dollars, of one ounce of gold for various days in 2006.

Date	Closing price
January 11	$544.40
February 8	$548.75
March 15	$556.50
April 19	$624.75
May 17	$699.50

(a) Compute the average rate of change of the closing price from March 15 to May 17 and interpret your result. What are the units?

(b) Find the average rate of change from January 11 to February 8.

10. H5N1 flu cases   The table shows the cumulative number of confirmed cases of H5N1 avian influenza (bird flu) worldwide on various dates in 2006, according to the World Health Organization.

Date	Cases
January 5	144
February 2	161
March 1	174
April 3	190
May 4	206
June 6	225

(a) Compute the average rate of change of the number of cases from March 1 to June 6 and interpret your result.

(b) Find the average rate of change from April 3 to May 4.

11. Advertising   Let $f(x)$ be the number, in thousands, of vehicles a manufacturer estimates will be sold when x million dollars are spent on advertising. If $f(1.8) = 240$ and $f(2.5) = 325$, compute the average rate of change for $1.8 \leq x \leq 2.5$. What does your result mean in this context?

12. Storage costs   Let $c(x)$ be the annual storage cost, in thousands of dollars, that a microchip company incurs when x thousand chips are held in inventory. If $c(5.1) = 3.75$ and $c(6.3) = 4.12$, find the average rate of change for $5.1 \leq x \leq 6.3$ and interpret your result.

13. Investment   The balance in an investment account t years after the account is opened is given by $9500(1.064^t)$. Compute the average rate of change for $2.5 \leq t \leq 4.5$ and interpret your result in this context.

14. Mold population The mass in grams of a mold colony growing in a lab experiment is given by $5.4e^{0.28t}$ where t is the number of hours after the start of the experiment. What is the average rate of change in the mass of the colony from the end of the second hour to the end of the sixth hour?

15. Battery charge A rechargeable battery is plugged into a charger. The graph shows the percentage of full charge that the battery reaches as a function of the time elapsed in hours. Estimate the average rate of change from 3 to 6 hours and interpret your result.

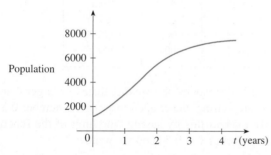

16. Fish population The graph shows the fish population of a lake as a function of t, where t is the time in years after the lake is stocked with fish. Estimate the average rate of change from 1.5 to 3 years and interpret your result.

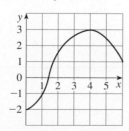

17. The graph of a function f is shown.

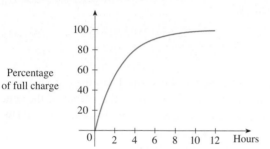

(a) Sketch on the graph a secant line whose slope represents the average rate of change over $[2, 6]$. Then compute this average rate of change.

(b) Is the average rate of change of f over $[0, 3]$ positive or negative?

(c) Which interval gives a larger average rate of change, $[1, 2]$ or $[3, 4]$?

18. The graph of a function g is shown.

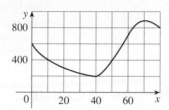

(a) Find the average rate of change of g over $[20, 60]$.

(b) Identify an interval on which the average rate of change of g is 0.

(c) Which interval gives a larger average rate of change, $[40, 60]$ or $[40, 70]$?

19. Projectile motion The height in feet of a baseball, t seconds after being thrown straight upward, is given by $h(t) = 36t - 16t^2$.

(a) Find the average speed of the ball for

    (i) $0 \le t \le 1$ (ii) $0.5 \le t \le 1$

    (iii) $0.9 \le t \le 1$ (iv) $0.99 \le t \le 1$

(b) Estimate the speed of the ball after 1 second.

20. Motion The position of a particle moving along a straight path is given by $f(x) = 3x^2 - 0.8x^3$ where x is the time elapsed in minutes and $f(x)$ is measured in feet.

(a) Find the average speed of the particle for

    (i) $2 \le x \le 2.1$ (ii) $2 \le x \le 2.01$

    (iii) $2 \le x \le 2.001$

(b) Estimate the speed of the particle after 2 minutes.

21. Projectile motion If a ball is thrown into the air with a speed of 40 ft/s, its height in feet t seconds later is given by $y = 40t - 16t^2$.

(a) Find the average speed for the time period beginning when $t = 2$ and lasting

    (i) 0.5 second   (ii) 0.1 second

    (iii) 0.05 second   (iv) 0.01 second

(b) Estimate the speed when $t = 2$.

22. Projectile motion If an arrow is shot upward on the moon with a speed of 58 m/s, its height in meters t seconds later is given by $h = 58t - 0.83t^2$.

(a) Find the average speed over the given time intervals:

    (i) $[1, 2]$ (ii) $[1, 1.5]$

    (iii) $[1, 1.1]$ (iv) $[1, 1.01]$

    (v) $[1, 1.001]$

(b) Estimate the speed when $t = 1$.

2.2 Limits

In Example 4 of Section 2.1, we saw how the concept of a limit arises when we want to find the speed of a falling ball at a given instant. We now turn our attention to limits in general and methods for computing them. This knowledge will allow us to compute precise speeds and other rates of change in the next section.

▪ Introduction to Limits

Consider the function f defined by

$$f(x) = \frac{x - 1}{x^2 - 1}$$

If we use inputs that are closer and closer to the value 1, do the outputs approach any particular value? Note that we are not asking about the output of the function *at* $x = 1$, just *near* $x = 1$. (In fact, this function is not defined at $x = 1$.) The following table gives values of $f(x)$ for values of x close to 1.

x	$f(x)$	x	$f(x)$
0.8	0.55556	1.2	0.45455
0.9	0.52632	1.1	0.47619
0.95	0.51282	1.05	0.48780
0.98	0.50505	1.02	0.49505
0.99	0.50251	1.01	0.49751
0.995	0.50125	1.005	0.49875
0.999	0.50025	1.001	0.49975

Regardless of whether we start with values of x smaller than 1 or larger than 1, it appears that the values of $f(x)$ are getting closer and closer to the number 0.5 as x gets closer and closer to 1. We express this by saying "the limit of the function $f(x) = (x - 1)/(x^2 - 1)$ as x approaches 1 is 0.5" and we write

$$\lim_{x \to 1} \frac{x - 1}{x^2 - 1} = 0.5$$

In general, we use the following notation.

▪ **Definition**   We write

$$\lim_{x \to a} f(x) = L$$

and say   "the limit of $f(x)$, as x approaches a, equals L"

if the values of $f(x)$ approach L as the values of x approach a (but are not equal to a).

The idea is that if we write

$$\lim_{t \to 5} A(t) = 12$$

we mean that the values of $A(t)$ tend to get closer and closer to 12 as t gets closer and closer to 5. In fact, we can make the ouput values $A(t)$ arbitrarily close to 12 (as close to 12 as we like) by choosing values of t that are sufficiently close to 5 (but not equal to 5).

An alternative notation for

$$\lim_{x \to a} f(x) = L$$

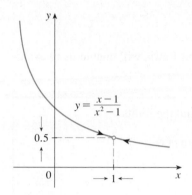

$$y = \frac{x - 1}{x^2 - 1}$$

FIGURE 1

is $f(x) \to L$ as $x \to a$

which is usually read "$f(x)$ approaches L as x approaches a."

Figure 1 shows the graph of the function $f(x) = (x - 1)/(x^2 - 1)$ that we have been studying. Notice that as the x-values get closer to 1, the y-values get closer to 0.5. The fact that there is no point on the graph at $x = 1$ is irrelevant. The only thing that matters is how f is defined *near $x = 1$*.

■ EXAMPLE 1   Estimating a Limit of a Function

Estimate the value of $\lim\limits_{r \to 0} \dfrac{(3 + r)^2 - 9}{r}$.

SOLUTION

The tables list values of the function for several values of r near 0.

r	$\dfrac{(3 + r)^2 - 9}{r}$
0.2	6.2
0.1	6.1
0.01	6.01
0.001	6.001
0.0001	6.0001

r	$\dfrac{(3 + r)^2 - 9}{r}$
-0.2	5.8
-0.1	5.9
-0.01	5.99
-0.001	5.999
-0.0001	5.9999

Both tables suggest that as r approaches 0, the output values approach 6. So we make the guess that

$$\lim_{r \to 0} \frac{(3 + r)^2 - 9}{r} = 6$$

For a further explanation of why calculators sometimes give false values, visit

www.stewartcalculus.com

and click on *Lies My Calculator and Computer Told Me*. In particular, see the section called *The Perils of Subtraction*.

It turns that the value of the limit that we estimated in Example 1 is correct. However, we must be cautious in using a table of values or a graph to guess the value of a limit. If we were to continue choosing values of r even closer to 0 than those listed in the tables of Example 1, eventually a calculator would probably return erroneous values. The cause for the possible inaccuracy lies in how small the numbers involved in the computation can be. At some point, errors in subtracting small numbers and other round-off errors made by the calculator become an issue. (In Exercises 9 and 10 you can see examples where these types of errors occur.) So how do we know if a guess we are making is accurate? We need a foolproof method for calculating limits.

▪ Evaluating Limits

The following properties of limits, called the Limit Laws, will enable us to evaluate many limits with certainty.

▪ **Limit Laws** Suppose that c is a constant and the limits

$$\lim_{x \to a} f(x) \qquad \text{and} \qquad \lim_{x \to a} g(x)$$

exist. Then

1. $\lim_{x \to a} [f(x) + g(x)] = \lim_{x \to a} f(x) + \lim_{x \to a} g(x)$

2. $\lim_{x \to a} [f(x) - g(x)] = \lim_{x \to a} f(x) - \lim_{x \to a} g(x)$

3. $\lim_{x \to a} [cf(x)] = c \lim_{x \to a} f(x)$

4. $\lim_{x \to a} [f(x) g(x)] = \lim_{x \to a} f(x) \cdot \lim_{x \to a} g(x)$

5. $\lim_{x \to a} \dfrac{f(x)}{g(x)} = \dfrac{\lim\limits_{x \to a} f(x)}{\lim\limits_{x \to a} g(x)} \qquad \text{if } \lim_{x \to a} g(x) \neq 0$

These five laws can be stated verbally as follows:

Sum Law
1. The limit of a sum is the sum of the limits.

Difference Law
2. The limit of a difference is the difference of the limits.

Constant Multiple Law
3. The limit of a constant times a function is the constant times the limit of the function.

Product Law
4. The limit of a product is the product of the limits.

Quotient Law
5. The limit of a quotient is the quotient of the limits (provided that the limit of the denominator is not 0).

In addition, if we use the Product Law repeatedly, we obtain the following law.

Power Law
6. $\lim_{x \to a} [f(x)]^n = \left[\lim_{x \to a} f(x) \right]^n \qquad$ where n is a positive integer

---

We will not be formally proving these laws in this text. However, it is not hard to believe that these properties are true. For instance, if $f(x)$ is close to L and $g(x)$ is close to M, it is reasonable to conclude that $f(x) + g(x)$ is close to $L + M$. This gives us an intuitive basis for believing that Law 1 is true.

The next example demonstrates how these laws allow us to evaluate a limit.

■ **EXAMPLE 2**   **Using the Limit Laws to Evaluate a Limit**

Evaluate $\lim_{x \to 5} f(x)$ for $f(x) = 2x^2 - 3x + 4$.

SOLUTION

Let's first look at the limit of each term of f individually. The last term is a constant, and $\lim_{x \to 5} 4 = 4$ because as the inputs approach 5, the outputs are always 4 (and so they approach 4). The middle term is a product of the constant 3 and x. We have $\lim_{x \to 5} x = 5$ because the output is the same as the input, so if x approaches 5, the output also approaches 5. Limit Law 3 says that the limit of a constant times a function is the constant times the limit of the function, so

$$\lim_{x \to 5} 3x = 3 \cdot 5 = 15$$

The first term of f is the product of 2 and x^2. From Limit Law 6 we know that

$$\lim_{x \to 5} x^2 = \left[\lim_{x \to 5} x \right]^2 = 5^2 = 25$$

and then, from Law 3,

$$\lim_{x \to 5} 2x^2 = 2 \cdot 25 = 50$$

Finally, using Limit Laws 1 and 2 we have

$$\lim_{x \to 5} f(x) = \lim_{x \to 5} (2x^2) - \lim_{x \to 5} (3x) + \lim_{x \to 5} 4 = 50 - 15 + 4 = 39$$ ■

Notice that in Example 2, the limit of $f(x)$ as $x \to 5$ is 39 and $f(5) = 39$. In other words, we would have gotten the correct answer for the limit simply by substituting 5 for x. This suggests an easy way to evaluate limits, but when are we justified in computing a limit by direct substitution?

■ | **Continuity**

When the value of a function at $x = a$ is the same as the limit when x approaches a, the function is called *continuous* at a.

> ■ **Definition**   A function f is **continuous at a number a** if
>
> $$\lim_{x \to a} f(x) = f(a)$$

A function is called *continuous on an interval* if it is continuous at every value in the interval. Geometrically, you can think of a function that is continuous on an interval as a function whose graph has no break in it (on the interval). The graph can be drawn without removing your pen from the paper. Most physical phenomena are continuous. For instance, your height varies continuously with time; when a child grows from 3 feet to 4 feet tall, no values between those heights are skipped. If you draw a graph of your height over your lifetime, there will be no gap or jump in the curve. The graph in Figure 2 is continuous everywhere except at $x = -1$ and $x = 2$, where you can observe breaks in the graph.

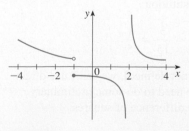

FIGURE 2

It can be proved that all of the functions we studied in Chapter 1 are continuous *on their domains*.

▪ The following types of functions are continuous at every number in their domains:

linear functions	polynomials	rational functions
power functions		root functions
exponential functions		logarithmic functions

When we know a function is continuous, we can evaluate its limit using *direct substitution*:

$$\lim_{x \to a} f(x) = f(a)$$

▪ EXAMPLE 3   Using Continuity to Evaluate a Limit

Let f be the function defined by

$$f(x) = \frac{x - 1}{x^2 - 1}$$

(a) State the domain of f.

(b) Find $\lim_{x \to 4} f(x)$.

(c) Find $\lim_{x \to 1} f(x)$.

SOLUTION

(a) The domain of f consists of all values except those that result in division by 0. This occurs when

$$x^2 - 1 = 0$$
$$x^2 = 1$$
$$x = \pm 1$$

Thus the domain is all real numbers except 1 and -1.

(b) The function f is a rational function, so it is continuous on its domain. The number 4 is in the domain of f, so we know f is continuous there and we are justified in computing the limit by direct substitution:

$$\lim_{x \to 4} f(x) = f(4) = \frac{4 - 1}{4^2 - 1} = \frac{3}{15} = \frac{1}{5}$$

(c) Because 1 is not in the domain of f [$f(1)$ is not defined], we can't use direct substitution to evaluate the limit. Instead, we need to do some preliminary algebra. First we factor the denominator as a difference of squares:

$$\frac{x - 1}{x^2 - 1} = \frac{x - 1}{(x + 1)(x - 1)}$$

The numerator and denominator have a common factor of $x - 1$. We can cancel the common factor as long as it is not equal to 0. When we take the limit as x approaches 1, however, we do not consider $x = 1$ itself. Thus $x \neq 1$ and $x - 1 \neq 0$, so the limit becomes

$$\lim_{x \to 1} \frac{x - 1}{x^2 - 1} = \lim_{x \to 1} \frac{x - 1}{(x + 1)(x - 1)} = \lim_{x \to 1} \frac{1}{x + 1}$$

We are left with a simplified function, and because $1/(x + 1)$ is continuous at $x = 1$ (it is a rational function and 1 is in its domain), we can evaluate the limit by direct substitution:

$$\lim_{x \to 1} \frac{1}{x + 1} = \frac{1}{1 + 1} = \frac{1}{2}$$

Thus $\lim_{x \to 1} f(x) = \frac{1}{2}$.   ■

The limit in part (c) of Example 3 is the same limit we explored at the beginning of the section. We guessed, from the numerical evidence, that the limit is 0.5, but this example proves that our guess was correct.

Notice that when we computed the limit in part (c) of the example, we replaced the given function $f(x) = (x - 1)/(x^2 - 1)$ with the simplified version $1/(x + 1)$. The two functions have the same limit as $x \to 1$ because they have the same outputs except when $x = 1$. Bear in mind that in computing a limit as x approaches 1, we don't consider what happens when x is actually *equal* to 1. This is an important idea to remember as we evaluate limits, and it plays a key role in the next example, where we revisit the limit from Example 1.

■ EXAMPLE 4   Simplifying before Using Direct Subsitution

Find $\lim_{r \to 0} \dfrac{(3 + r)^2 - 9}{r}$.

SOLUTION

We can't use direct substitution to evaluate the limit because the function is not continuous at 0 (0 is not in its domain). So we first simplify the function:

$$\frac{(3 + r)^2 - 9}{r} = \frac{9 + 6r + r^2 - 9}{r} = \frac{6r + r^2}{r}$$

$$= \frac{r(6 + r)}{r} = 6 + r$$

In the last step, we are justified in canceling an r in the numerator and denominator because when we compute the limit as r approaches 0, we assume $r \neq 0$. Then

$$\lim_{r \to 0} \frac{(3 + r)^2 - 9}{r} = \lim_{r \to 0} (6 + r) = 6 + 0 = 6$$

We have evaluated the simplified limit by direct substitution because $6 + r$ is a linear function that is continuous everywhere. This result agrees with our guess in Example 1.   ■

▪ EXAMPLE 5   Rationalizing a Numerator

Find $\lim\limits_{t \to 0} \dfrac{\sqrt{t^2 + 9} - 3}{t^2}$.

SOLUTION

Here 0 is not in the domain of the function, so we can't evaluate the limit by direct substitution. To rewrite the function in a different form, we rationalize the numerator. This procedure results in a fraction without a root in the numerator.

Rationalizing the numerator here is accomplished by multiplying numerator and denominator by the same expression we see in the numerator, but changing the sign from minus to plus (and then simplifying). See Appendix A for additional examples of rationalizing numerators and denominators.

$$\lim_{t \to 0} \frac{\sqrt{t^2 + 9} - 3}{t^2} = \lim_{t \to 0} \left[\frac{\sqrt{t^2 + 9} - 3}{t^2} \cdot \frac{\sqrt{t^2 + 9} + 3}{\sqrt{t^2 + 9} + 3} \right]$$

$$= \lim_{t \to 0} \frac{(t^2 + 9) - 9}{t^2 (\sqrt{t^2 + 9} + 3)} = \lim_{t \to 0} \frac{t^2}{t^2 (\sqrt{t^2 + 9} + 3)}$$

$$= \lim_{t \to 0} \frac{1}{\sqrt{t^2 + 9} + 3}$$

As in Example 4, we are able to cancel the t^2 in the numerator and denominator because we assume $t \neq 0$ when computing the limit as $t \to 0$. The resulting function is continuous at 0 (0 is in its domain), so we evaluate the limit by direct substitution:

$$\lim_{t \to 0} \frac{\sqrt{t^2 + 9} - 3}{t^2} = \lim_{t \to 0} \frac{1}{\sqrt{t^2 + 9} + 3} = \frac{1}{\sqrt{(0^2 + 9)} + 3}$$

$$= \frac{1}{3 + 3} = \frac{1}{6} \qquad ▪$$

▪ EXAMPLE 6   A Limit that Does Not Exist

Find $\lim\limits_{x \to 0} \dfrac{1}{x^2}$ if it exists.

SOLUTION

As x becomes close to 0, x^2 also becomes close to 0, and $1/x^2$ becomes very large. (See the table at the left.) In fact, it appears from the graph of the function $f(x) = 1/x^2$ shown in Figure 3 that as x appoaches 0, the values of $f(x)$ get larger and larger. Thus the values of $f(x)$ do not approach a number, so $\lim_{x \to 0} (1/x^2)$ does not exist.

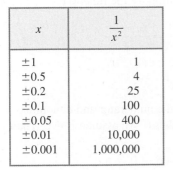

x	$\dfrac{1}{x^2}$
± 1	1
± 0.5	4
± 0.2	25
± 0.1	100
± 0.05	400
± 0.01	10,000
± 0.001	1,000,000

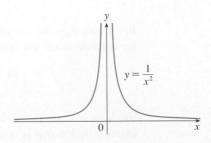

FIGURE 3

■ One-Sided Limits

Sometimes we need to consider the limit of $f(x)$ as x approaches a number a from the right or from the left. The **right-hand limit** of $f(x)$ as x approaches a is written $\lim_{x \to a^+} f(x)$. The symbol "$x \to a^+$" means that we consider only values of x that are larger than a. On a graph, we approach the value a from the right. Similarly, the symbol "$x \to a^-$" indicates a **left-hand limit** and it means that we consider only values of x that are smaller than a; on a graph, we approach a from the left.

■ EXAMPLE 7 Finding One-Sided Limits

The Heaviside function H is defined by

$$H(t) = \begin{cases} 0 & \text{if } t < 0 \\ 1 & \text{if } t \geq 0 \end{cases}$$

[This function is named after the electrical engineer Oliver Heaviside (1850–1925) and can be used to describe an electric current that is switched on at time $t = 0$.] Its graph is shown in Figure 4.

As t approaches 0 from the left, $H(t)$ approaches 0, so

$$\lim_{t \to 0^-} H(t) = 0$$

As t approaches 0 from the right, $H(t)$ approaches 1. Thus

$$\lim_{t \to 0^+} H(t) = 1$$ ■

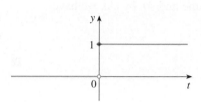

FIGURE 4
The Heaviside function

Notice in Example 7 that there is a jump in the graph at $t = 0$. The function is not continuous at 0, and there is no single number that $H(t)$ approaches as t approaches 0. Therefore, although the one-sided limits exist as t approaches 0, $\lim_{t \to 0} H(t)$ does not exist. In general, if the one-sided limits do not agree, then the limit does not exist:

$$\textbf{(1)} \quad \lim_{x \to a} f(x) = L \quad \text{if and only if} \quad \lim_{x \to a^-} f(x) = L \quad \text{and} \quad \lim_{x \to a^+} f(x) = L$$

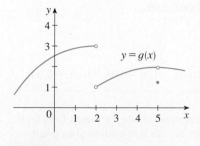

FIGURE 5

■ EXAMPLE 8 Determining One-Sided Limits from a Graph

The graph of a function g is shown in Figure 5. Use it to state the values (if they exist) of the following limits.

(a) $\lim_{x \to 2^-} g(x)$ **(b)** $\lim_{x \to 2^+} g(x)$ **(c)** $\lim_{x \to 2} g(x)$

(d) $\lim_{x \to 5^-} g(x)$ **(e)** $\lim_{x \to 5^+} g(x)$ **(f)** $\lim_{x \to 5} g(x)$

SOLUTION

From the graph we see that the values of $g(x)$ approach 3 as x approaches 2 from the left, but they approach 1 as x approaches 2 from the right. Therefore

$$\textbf{(a)}\ \lim_{x\to 2^-} g(x) = 3 \quad\text{and}\quad \textbf{(b)}\ \lim_{x\to 2^+} g(x) = 1$$

(c) Since the left and right limits are different, we conclude from (1) that $\lim_{x\to 2} g(x)$ does not exist.

The graph also shows that

$$\textbf{(d)}\ \lim_{x\to 5^-} g(x) = 2 \quad\text{and}\quad \textbf{(e)}\ \lim_{x\to 5^+} g(x) = 2$$

(f) This time the left and right limits are the same and so, by (1), we have

$$\lim_{x\to 5} g(x) = 2$$

Despite this fact, notice that $g(5) \neq 2$. ∎

▪ Prepare Yourself

1. If
$$f(x) = \frac{3^x - 2^x}{x}$$
evaluate $f(0.01)$. Round to four decimal places.

2. If
$$g(t) = \frac{\sqrt{t^2 + 1} - 1}{t^2}$$
evaluate $g(0.1)$. Round to four decimal places.

3. Factor the polynomial.
(a) $x^2 - 5x - 24$ (b) $a^2 - 25$
(c) $2w^2 - 7w - 15$ (d) $b^3 + 1$

4. Simplify as much as possible.
(a) $\dfrac{x^2 - 2x - 3}{x^2 - 7x + 12}$ (b) $\dfrac{(c + 2)^2 - 4}{c}$

(c) $\dfrac{\dfrac{1}{q} - \dfrac{1}{3}}{q - 3}$

5. Multiply and then simplify: $\dfrac{\sqrt{x + 1} - 2}{x - 3} \cdot \dfrac{\sqrt{x + 1} + 2}{\sqrt{x + 1} + 2}$

6. Let
$$A(t) = \begin{cases} 1 - t^2 & \text{if } t < 1 \\ 2^t & \text{if } t \geq 1 \end{cases}$$
Evaluate **(a)** $A(3)$, **(b)** $A(-2)$, and **(c)** $A(1)$.

▪ Exercises 2.2

1. Explain in your own words what is meant by the equation
$$\lim_{x\to 2} f(x) = 5$$
Is it possible for this statement to be true and yet $f(2) = 3$? Explain.

2–4 ▪ Guess the value of the limit (if it exists) by evaluating the function at the given numbers (correct to six decimal places).

2. $\lim_{t\to 1} \dfrac{t^3 - 1}{t - 1}$,
$t = 1.1, 1.05, 1.01, 1.001, 0.9, 0.95, 0.99, 0.999$

3. $\lim_{x\to 2} \dfrac{x^2 - x - 2}{x^2 - 2x}$,
$x = 2.1, 2.05, 2.01, 2.005, 2.001,$
$1.9, 1.95, 1.99, 1.995, 1.999$

4. $\lim_{h\to 0} \dfrac{\ln(1 + h)}{h}$,
$h = 0.1, 0.05, 0.01, 0.001, -0.1, -0.05, -0.01, -0.001$

5–8 ▪ Use a table of values to estimate the value of the limit. If you have a graphing device, use it to confirm your result graphically.

5. $\lim_{x\to 0} \dfrac{\sqrt{x + 4} - 2}{x}$ **6.** $\lim_{r\to 0} \dfrac{3^r - 1}{r}$

7. $\lim\limits_{x \to 1} \dfrac{x^6 - 1}{x^{10} - 1}$

8. $\lim\limits_{x \to 0} \dfrac{9^x - 5^x}{x}$

---

9. (a) Evaluate $h(x) = \left(\sqrt{x^4 + 1} - 1\right)/x^4$ for $x = 1, 0.5$, 0.2, and 0.1.

(b) Guess the value of $\lim\limits_{x \to 0} \dfrac{\sqrt{x^4 + 1} - 1}{x^4}$.

(c) Evaluate $h(x)$ for $x = 0.05, 0.01, 0.001$, and 0.0001. Are you still confident that your guess in part (b) is correct?

(d) Explain why you eventually obtained 0 values in part (c). It may help to evaluate just the numerator of $h(x)$ for $x = 0.001$ and 0.0001. Is your calculator giving you the correct values?

 (e) Graph the function h in the viewing rectangle $[-1, 1]$ by $[0, 1]$. Does the graph corroborate your guess from part (b)? Then change the viewing rectangle to $[-0.005, 0.005]$ by $[0.48, 0.52]$. Can you explain the appearance of the graph?

10. (a) Evaluate $g(t) = \left(\sqrt{t^2 + 4} - 2\right)/t^2$ for $t = 1, 0.5$, 0.2, and 0.1.

(b) Guess the value of $\lim\limits_{t \to 0} \dfrac{\sqrt{t^4 + 4} - 2}{t^2}$.

(c) Evaluate $g(t)$ for successively smaller values of t until you finally reach 0 values for $g(t)$. Are you still confident that your guess in part (b) is correct? Explain why you eventually obtained 0 values.

 (d) Graph the function g in the viewing rectangle $[-1, 1]$ by $[0, 1]$. Then zoom in toward the point where the graph appears to cross the y-axis to estimate the limit of $g(t)$ as t approaches 0. Continue to zoom in until you observe distortions in the graph of g. Compare with the results of part (c).

11–14 ■ Evaluate the limit and justify each step by indicating the appropriate Limit Laws.

11. $\lim\limits_{x \to 2} (x^3 + 2x^2 + 1)$

12. $\lim\limits_{t \to -1} (5t^2 - 3t + 2)$

13. $\lim\limits_{v \to 1} \dfrac{v^2 - 5}{v}$

14. $\lim\limits_{x \to 4} (3x - 9)^4$

15–18 ■ Use continuity to evaluate the limit. Round your answer to three decimal places.

15. $\lim\limits_{t \to 1} (3e^t - 4)$

16. $\lim\limits_{x \to 3.5} (2^x + 0.8)$

17. $\lim\limits_{m \to 2} \left(\dfrac{\ln m}{m + 2} \right)$

18. $\lim\limits_{u \to 0.3} \left(\dfrac{u^2 - 4u}{3u + 5} \right)$

19. (a) Determine the domain of $f(x) = \dfrac{x^2 - 4}{x - 2}$.

(b) Find $\lim\limits_{x \to 1} f(x)$.

(c) Find $\lim\limits_{x \to 2} f(x)$.

20. (a) Find the domain of $g(t) = \dfrac{t^2 - 3t - 4}{t + 1}$.

(b) Find $\lim\limits_{t \to 3} g(t)$.

(c) Find $\lim\limits_{t \to -1} g(t)$.

21. (a) Find the domain of $A(z) = \dfrac{2z - 6}{z^2 - 5z + 6}$.

(b) Find $\lim\limits_{z \to 0} A(z)$.

(c) Find $\lim\limits_{z \to 3} A(z)$.

22. (a) Find the domain of $R(x) = \dfrac{x^2 - 2x - 8}{x^2 - 16}$.

(b) Find $\lim\limits_{x \to 2} R(x)$.

(c) Find $\lim\limits_{x \to 4} R(x)$.

23–38 ■ Evaluate the limit.

23. $\lim\limits_{t \to 4} (3t - 7)$

24. $\lim\limits_{x \to -2} (4x^2 + x)$

25. $\lim\limits_{x \to 3} \dfrac{x^2 + 5}{x + 5}$

26. $\lim\limits_{w \to 5} \dfrac{3w^2 + 1}{w}$

27. $\lim\limits_{x \to 2} \dfrac{x^2 + x - 6}{x - 2}$

28. $\lim\limits_{x \to 4} \dfrac{x^2 - 4x}{x^2 - 3x - 4}$

29. $\lim\limits_{t \to -3} \dfrac{t^2 - 9}{2t^2 + 7t + 3}$

30. $\lim\limits_{x \to -4} \dfrac{x^2 + 5x + 4}{x^2 + 3x - 4}$

31. $\lim\limits_{h \to 0} \dfrac{(4 + h)^2 - 16}{h}$

32. $\lim\limits_{x \to -1} \dfrac{x^2 + 2x + 1}{x^4 - 1}$

33. $\lim\limits_{x \to -2} \dfrac{x + 2}{x^3 + 8}$

34. $\lim\limits_{h \to 0} \dfrac{\sqrt{1 + h} - 1}{h}$

35. $\lim\limits_{x \to 7} \dfrac{\sqrt{x + 2} - 3}{x - 7}$

36. $\lim\limits_{x \to 0} \dfrac{\sqrt{x^2 + b^2} - b}{x^2}, \quad b > 0$

37. $\lim\limits_{x \to -4} \dfrac{\dfrac{1}{4} + \dfrac{1}{x}}{4 + x}$

38. $\lim\limits_{t \to 0} \left(\dfrac{1}{t} - \dfrac{1}{t^2 + t} \right)$

39–42 ■ Use a table of values or a graph to explain why the limit does not exist.

39. $\lim\limits_{x \to 0} \dfrac{3}{x^4}$

40. $\lim\limits_{x \to 2} \dfrac{x}{(x - 2)^2}$

41. $\lim\limits_{t \to 0} \dfrac{e^t}{t}$

42. $\lim\limits_{x \to 0^+} \ln x$

---

43. Use the given graph of f to state the value of each quantity, if it exists. If it does not exist, explain why.

  (a) $\lim\limits_{x \to 1^-} f(x)$ (b) $\lim\limits_{x \to 1^+} f(x)$ (c) $\lim\limits_{x \to 1} f(x)$

  (d) $\lim\limits_{x \to 5} f(x)$ (e) $f(5)$

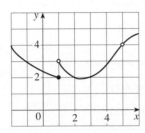

44. For the function whose graph is given, state the value of each quantity, if it exists. If it does not exist, explain why.

  (a) $\lim\limits_{x \to 0} f(x)$ (b) $\lim\limits_{x \to 3^-} f(x)$ (c) $\lim\limits_{x \to 3^+} f(x)$

  (d) $\lim\limits_{x \to 3} f(x)$ (e) $f(3)$

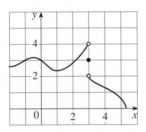

45. Explain what it means to say that

$$\lim_{x \to 1^-} f(x) = 3 \qquad \text{and} \qquad \lim_{x \to 1^+} f(x) = 7$$

In this situation is it possible that $\lim_{x \to 1} f(x)$ exists? Explain.

46. Drug administration A patient receives a 150-mg injection of a drug every four hours. The graph shows the amount $f(t)$ of the drug in the bloodstream after t hours. Find

$$\lim_{t \to 12^-} f(t) \qquad \text{and} \qquad \lim_{t \to 12^+} f(t)$$

and explain the significance of these one-sided limits.

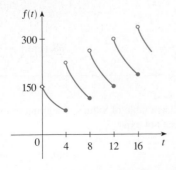

47. Let

$$g(x) = \begin{cases} -x & \text{if } x \leqslant -1 \\ 1 - x^2 & \text{if } -1 < x < 1 \\ x - 1 & \text{if } x > 1 \end{cases}$$

  (a) Evaluate each of the following limits, if it exists.

  (i) $\lim\limits_{x \to 1^+} g(x)$ (ii) $\lim\limits_{x \to 1} g(x)$ (iii) $\lim\limits_{x \to 0} g(x)$

  (iv) $\lim\limits_{x \to -1^-} g(x)$ (v) $\lim\limits_{x \to -1^+} g(x)$ (vi) $\lim\limits_{x \to -1} g(x)$

  (b) Sketch the graph of g.

48. Let

$$f(t) = \begin{cases} t^2 & \text{if } t < 0 \\ e^t & \text{if } t \geqslant 0 \end{cases}$$

  (a) Evaluate each of the following limits, if it exists.

  (i) $\lim\limits_{t \to -1} f(t)$ (ii) $\lim\limits_{t \to 0^-} f(t)$ (iii) $\lim\limits_{t \to 0^+} f(t)$

  (iv) $\lim\limits_{t \to 0} f(t)$ (v) $\lim\limits_{t \to 2} f(t)$

  (b) Explain why f is not continuous at $t = 0$.

  (c) Sketch the graph of f.

49. Recall that

$$|x| = \begin{cases} x & \text{if } x \geqslant 0 \\ -x & \text{if } x < 0 \end{cases}$$

  Let $f(x) = |x|/x$.

  (a) Find $\lim\limits_{x \to 0^+} f(x)$.

  (b) Find $\lim\limits_{x \to 0^-} f(x)$.

  (c) Does $\lim\limits_{x \to 0} f(x)$ exist? Explain.

50. Let $F(x) = \dfrac{x^2 - 1}{|x - 1|}$.

  (a) Find

  (i) $\lim\limits_{x \to 1^+} F(x)$ (ii) $\lim\limits_{x \to 1^-} F(x)$

  (b) Does $\lim_{x \to 1} F(x)$ exist?

  (c) Sketch the graph of F.

51. Parking fees A parking lot charges $3 for the first hour (or part of an hour) and $2 for each succeeding hour (or part), up to a daily maximum of $10.

  (a) Sketch a graph of the cost of parking at this lot as a function of the time parked there.

  (b) Discuss the discontinuities of this function and their significance to someone who parks in the lot.

52. Explain why each function is continuous or discontinuous.

  (a) The temperature at a specific location as a function of time

  (b) The temperature at a specific time as a function of the distance due west from New York City

(c) The altitude above sea level as a function of the distance due west from New York City

(d) The cost of a taxi ride as a function of the distance traveled

53. (a) Estimate the value of

$$\lim_{x \to 0} \frac{x}{\sqrt{1 + 3x} - 1}$$

by graphing the function $f(x) = x/(\sqrt{1 + 3x} - 1)$ and zooming in toward the point where the graph appears to cross the y-axis.

(b) Make a table of values of $f(x)$ for x close to 0 and guess the value of the limit.

(c) Evaluate the limit to prove that your guess is correct.

54. (a) Estimate the value of

$$\lim_{x \to 0} \frac{6^x - 2^x}{x}$$

by graphing the function $f(x) = (6^x - 2^x)/x$ and zooming in toward the point where the graph appears to cross the y-axis. State your answer correct to two decimal places.

(b) Check your answer in part (a) by evaluating $f(x)$ for values of x that approach 0.

55. Use the graph of the function $f(x) = 1/(1 + e^{1/x})$ to state the value of each limit, if it exists. If it does not exist, explain why.

(a) $\displaystyle\lim_{x \to 0^-} f(x)$ **(b)** $\displaystyle\lim_{x \to 0^+} f(x)$ **(c)** $\displaystyle\lim_{x \to 0} f(x)$

56. (a) What is wrong with the following equation?

$$\frac{x^2 + x - 6}{x - 2} = x + 3$$

(b) In view of part (a), explain why the equation

$$\lim_{x \to 2} \frac{x^2 + x - 6}{x - 2} = \lim_{x \to 2} (x + 3)$$

is correct.

■ Challenge Yourself

57. For what value of the constant c is the function f continuous on $(-\infty, \infty)$?

$$f(x) = \begin{cases} cx^2 + 2x & \text{if } x < 2 \\ x^3 - cx & \text{if } x \geq 2 \end{cases}$$

58. Show by means of an example that $\lim_{x \to a} [f(x) + g(x)]$ may exist even though neither $\lim_{x \to a} f(x)$ nor $\lim_{x \to a} g(x)$ exists.

59. If $\displaystyle\lim_{x \to 1} \frac{f(x) - 8}{x - 1} = 10$, find $\displaystyle\lim_{x \to 1} f(x)$.

2.3 Rates of Change and Derivatives

■ Introduction to Instantaneous Rate of Change

In Section 2.1 we studied average rate of change. For instance, if you drive 90 miles over a two-hour period, your average rate of change (which is speed in this case) is 90 miles/2 hours = 45 mi/h. This normally doesn't mean that you were traveling at 45 mi/h for the entire two hours. If you watch the speedometer of a car as you travel in city traffic for instance, you see that the needle doesn't stay still for very long; that is, the speed of the car is not constant. We assume from watching the speedometer that the car has a definite speed at each moment, but how is the "instantaneous" speed defined?

In Example 4 in Section 2.1, we looked at the speed of a ball dropped from the CN Tower in Toronto. We estimated the speed of the ball after 5 seconds to be

49 m/s by computing average speeds over smaller and smaller time intervals, as shown in the following table.

Time interval	Δt	Average speed (m/s)
$5 \leq t \leq 6$	1	53.9
$5 \leq t \leq 5.1$	0.1	49.49
$5 \leq t \leq 5.05$	0.05	49.245
$5 \leq t \leq 5.01$	0.01	49.049
$5 \leq t \leq 5.001$	0.001	49.0049

0	49

We estimated the speed after 5 seconds to be 49 m/s because that is the value that the average speeds appear to be approaching as Δt approaches 0. (We can't actually use $\Delta t = 0$ because then the average speed is undefined.) The instantaneous speed is defined to be the *limit* of the average speeds as the length of the time interval approaches 0. We use this concept in the following example to prove that the instantaneous speed of the falling ball after 5 seconds is in fact 49 m/s.

■ EXAMPLE 1 Instantaneous Speed of a Falling Ball

Suppose that a ball is dropped from the upper observation deck of the CN Tower in Toronto, 450 m above the ground. Find the speed of the ball after 5 seconds.

SOLUTION

We know from Example 4 in Section 2.1 that the distance, in meters, the ball falls after t seconds is

$$s(t) = 4.9t^2$$

The average speed over a time interval $[t_1, t_2]$ is

$$\frac{\Delta s}{\Delta t} = \frac{s(t_2) - s(t_1)}{t_2 - t_1}$$

Now we want to examine average speeds as Δt approaches 0. We use time intervals that start at 5 seconds, so $t_1 = 5$, but the value of t_2 will change and get closer and closer to 5. For simplicity, let's use t in place of t_2; then the average speed over the interval $[5, t]$ is

$$\frac{s(t) - s(5)}{t - 5} = \frac{4.9t^2 - 4.9(5^2)}{t - 5} = \frac{4.9t^2 - 4.9(25)}{t - 5}$$

The instantaneous speed after 5 seconds is the limit of the average speeds as Δt approaches 0, or equivalently, as t approaches 5.

$$\text{instantaneous speed} = \lim_{t \to 5} \frac{4.9t^2 - 4.9(25)}{t - 5}$$

The expression in the limit is not defined at $t = 5$, so we can't evaluate the limit by direct substitution. First we simplify the expression algebraically:

$$\frac{4.9t^2 - 4.9(25)}{t - 5} = \frac{4.9(t^2 - 25)}{t - 5} = \frac{4.9(t + 5)(t - 5)}{t - 5}$$

$$= 4.9(t + 5) \qquad (t \neq 5)$$

Because $4.9(t + 5)$ is a linear function, we know it is continuous and we can evaluate the limit by direct substitution:

$$\text{instantaneous speed} = \lim_{t \to 5} \frac{4.9t^2 - 4.9(25)}{t - 5} = \lim_{t \to 5} 4.9(t + 5)$$

$$= 4.9(5 + 5) = 49 \text{ m/s} \qquad ■$$

The process we used in Example 1 applies not only to speed but to any rate of change. Generally speaking,

$$\text{instantaneous rate of change} = \lim_{\Delta x \to 0} (\text{average rate of change}) = \lim_{\Delta x \to 0} \frac{\Delta y}{\Delta x}$$

To determine the instantaneous rate of change of a function f at an input value x_1, we first compute average rates of change over intervals $[x_1, x_2]$. Then we determine the limit of these average rates of change as the length of the interval approaches 0, or equivalently, as x_2 approaches x_1.

The units for the instantaneous rate of change are the same as those for the average rate of change: output units per input unit.

> **(1)** ▪ **Definition**   The **instantaneous rate of change** of a function f at the input value x_1 is
>
> $$\lim_{\Delta x \to 0} \frac{\Delta y}{\Delta x} = \lim_{x_2 \to x_1} \frac{f(x_2) - f(x_1)}{x_2 - x_1}$$
>
> if this limit exists.

We can express this definition in an alternate form that is sometimes easier to use. Suppose the interval begins at $x_1 = a$ and we use h to represent the length of the interval (so $\Delta x = h$). Then the interval ends at $x_2 = a + h$, and the average rate of change of f over the interval is

$$\frac{\Delta y}{\Delta x} = \frac{f(x_2) - f(x_1)}{x_2 - x_1} = \frac{f(a + h) - f(a)}{(a + h) - a} = \frac{f(a + h) - f(a)}{h}$$

The instantaneous rate of change is the limit as the length of the interval, h, approaches 0.

> **(2)** ▪ The **instantaneous rate of change** of a function f at the input value a is
>
> $$\lim_{h \to 0} \frac{f(a + h) - f(a)}{h}$$
>
> if this limit exists.

▪ EXAMPLE 2   Computing an Instantaneous Rate of Change

Find the instantaneous rate of change of $g(x) = x^2$ at $x = 3$.

SOLUTION 1

Using Definition 1 we have $x_1 = 3$ and, for simplicity, we can use x for x_2. The instantaneous rate of change is

$$\lim_{x_2 \to x_1} \frac{g(x_2) - g(x_1)}{x_2 - x_1} = \lim_{x \to 3} \frac{g(x) - g(3)}{x - 3} = \lim_{x \to 3} \frac{x^2 - 9}{x - 3}$$

To evaluate the limit, we first simplify:

$$\lim_{x \to 3} \frac{x^2 - 9}{x - 3} = \lim_{x \to 3} \frac{(x + 3)(x - 3)}{x - 3} = \lim_{x \to 3} (x + 3)$$

$$= 3 + 3 = 6$$

We evaluated the limit using direct substitution because $x + 3$ is a linear function and therefore continuous at 3.

SOLUTION 2

According to (2), the instantaneous rate of change at $x = 3$ is

$$\lim_{h \to 0} \frac{g(3 + h) - g(3)}{h} = \lim_{h \to 0} \frac{(3 + h)^2 - 9}{h}$$

We need to simplify the expression before we can evaluate the limit.

$$\frac{(3 + h)^2 - 9}{h} = \frac{9 + 6h + h^2 - 9}{h} = \frac{6h + h^2}{h}$$

$$= \frac{h(6 + h)}{h} = 6 + h$$

where $h \neq 0$. Then the instantaneous rate of change is

$$\lim_{h \to 0} \frac{g(3 + h) - g(3)}{h} = \lim_{h \to 0} (6 + h) = 6 + 0 = 6 \qquad ▪$$

▪ Tangent Lines

Let's look at the geometric interpretation of instantaneous rate of change. Figure 1 shows a graph of a function f along with a secant line through a point P on the curve at $x = x_1$ and a nearby point Q on the curve at $x = x_2$. We know from Section 2.1 that the slope of this secant line is the average rate of change of f over the interval $[x_1, x_2]$. But Definition 1 says that the instantaneous rate of change is the limit of these average rates of change as x_2 approaches x_1. This corresponds to the point Q moving closer and closer to P. As Q approaches P, the secant line rotates to a limiting position. The line at this limiting position is called the *tangent line* at $x = x_1$. (See Figure 2.)

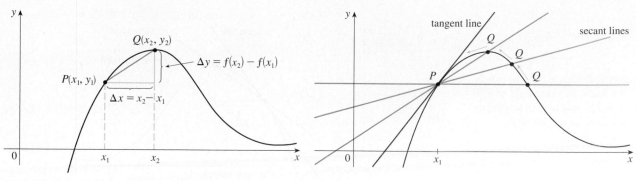

FIGURE 1   **FIGURE 2**

TEC In Visual 2.3A you can see how the process in Figure 2 works for additional functions.

You can think of the tangent line as the unique line that touches the curve at $x = x_1$ and has the same direction as the curve at that location. Because the tangent line occupies the limiting position of the secant lines, its slope is the limit of the slopes of the secant lines.

---

(3) ▪ Definition The **tangent line** to the curve $y = f(x)$ at the point $(x_1, f(x_1))$ is the line through this point with slope

$$m = \lim_{x_2 \to x_1} \frac{f(x_2) - f(x_1)}{x_2 - x_1}$$

provided that this limit exists.

Comparing Definitions 1 and 3, we see that *the slope of the tangent line is the instantaneous rate of change.* Consequently, we can also express the slope of the tangent line in the form given in (2).

---

(4) ▪ The **tangent line** to the curve $y = f(x)$ at the point $(a, f(a))$ is the line through this point with slope

$$m = \lim_{h \to 0} \frac{f(a + h) - f(a)}{h}$$

provided that this limit exists.

This version is illustrated in Figure 3 on page 100. The point P is at $x = a$ and h represents the length of the interval, so the point Q is at $x = a + h$. The slope of the secant line through P and Q is

$$\frac{\Delta y}{\Delta x} = \frac{f(a + h) - f(a)}{h}$$

As the length of the interval h approaches 0, the point Q approaches point P (at $x = a$) and the secant lines approach the tangent line, whose slope is the instantaneous rate of change given in (4).

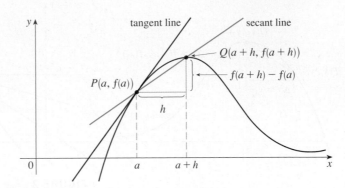

Figure 3 illustrates the case where $h > 0$ and Q is to the right of P. If it happened that $h < 0$, however, Q would be to the left of P.

FIGURE 3

▪ **EXAMPLE 3** **Finding an Equation of a Tangent Line**

(a) Find the slope of the tangent line to the curve $y = 2/x$ at the point where $x = 4$.

(b) Write an equation of the tangent line.

SOLUTION

(a) If we let $f(x) = 2/x$, then from (4) the slope of the tangent line at $x = 4$ is

$$m = \lim_{h \to 0} \frac{f(4 + h) - f(4)}{h} = \lim_{h \to 0} \frac{\dfrac{2}{4 + h} - \dfrac{1}{2}}{h}$$

We need to simplify the expression before we can evaluate the limit:

$$\frac{\dfrac{2}{4 + h} - \dfrac{1}{2}}{h} = \frac{\dfrac{2}{4 + h} \cdot \dfrac{2}{2} - \dfrac{1}{2} \cdot \dfrac{4 + h}{4 + h}}{h}$$

$$= \frac{\dfrac{4 - (4 + h)}{2(4 + h)}}{h} = \frac{\dfrac{4 - 4 - h}{2(4 + h)}}{h}$$

$$= \frac{-h}{2(4 + h) \cdot h} = \frac{-1}{2(4 + h)} \qquad (h \neq 0)$$

The slope of the tangent line at $x = 4$ is

$$\lim_{h \to 0} \frac{f(4 + h) - f(4)}{h} = \lim_{h \to 0} \frac{-1}{2(4 + h)} = \frac{-1}{2(4 + 0)} = -\frac{1}{8}$$

We used direct substitution to evaluate the simplified limit because $-1/[2(4 + h)]$ is a rational function that is continuous at $h = 0$.

(b) The tangent line has slope $-\frac{1}{8}$ and passes through the point $(4, f(4)) = \left(4, \frac{1}{2}\right)$ so its equation is

$$y - y_1 = m(x - x_1)$$

$$y - \tfrac{1}{2} = -\tfrac{1}{8}(x - 4)$$

$$y = -\tfrac{1}{8}x + 1$$

The graphs of f and its tangent line at $x = 4$ are shown in Figure 4. ▪

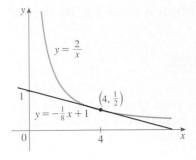

FIGURE 4

We sometimes refer to the slope of the tangent line to a curve at a point as the **slope of the curve** at that point. The idea is that if we zoom in far enough toward the point, the curve looks almost like a straight line. Figure 5 illustrates this procedure for the curve $y = x^2$. The more we zoom in, the more the parabola looks like a line. In other words, the curve becomes almost indistinguishable from its tangent line.

TEC Visual 2.3B shows an animation of Figure 5.

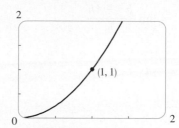
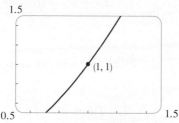
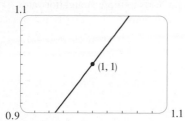

FIGURE 5 Zooming in toward the point $(1, 1)$ on the parabola $y = x^2$

▪ Derivatives

We have seen from Definitions 1–4 that whenever we calculate an instantaneous rate of change or slope of a tangent line, we encounter a limit of the form

$$\lim_{x_2 \to x_1} \frac{f(x_2) - f(x_1)}{x_2 - x_1}$$

or, equivalently,

$$\lim_{h \to 0} \frac{f(a + h) - f(a)}{h}$$

Since this type of limit occurs so widely, it is given a special name and notation.

---

(5) ▪ Definition   The **derivative of a function f at a number a**, denoted by $f'(a)$, is

$$f'(a) = \lim_{h \to 0} \frac{f(a + h) - f(a)}{h}$$

if this limit exists.

$f'(a)$ is read "f prime of a."

We can state the definition of the derivative (5) in an equivalent way that is similar to the limit in Definition 1. If we consider h to be the difference between a value x and the value a, so $h = x - a$, then $x = a + h$ and h approaches 0 if and only if x approaches a. Thus Definition 5 becomes

(6)

$$f'(a) = \lim_{x \to a} \frac{f(x) - f(a)}{x - a}$$

▪ EXAMPLE 4   Calculating a Derivative Value

Find the derivative of the function $f(x) = x^2 - 8x + 9$

(a) at $x = 2$.   **(b)** at the number a.

SOLUTION

(a) From Definition 5 we have

Definitions 5 and 6 are equivalent, so we can use either one to compute a derivative. However, Definition 5 often leads to more straightforward computations.

$$f'(2) = \lim_{h \to 0} \frac{f(2 + h) - f(2)}{h} = \lim_{h \to 0} \frac{[(2 + h)^2 - 8(2 + h) + 9] - (-3)}{h}$$

$$= \lim_{h \to 0} \frac{4 + 4h + h^2 - 16 - 8h + 9 + 3}{h} = \lim_{h \to 0} \frac{h^2 - 4h}{h}$$

$$= \lim_{h \to 0} \frac{h(h - 4)}{h} = \lim_{h \to 0} (h - 4) = 0 - 4 = -4$$

(b)
$$f'(a) = \lim_{h \to 0} \frac{f(a + h) - f(a)}{h}$$

$$= \lim_{h \to 0} \frac{[(a + h)^2 - 8(a + h) + 9] - [a^2 - 8a + 9]}{h}$$

$$= \lim_{h \to 0} \frac{a^2 + 2ah + h^2 - 8a - 8h + 9 - a^2 + 8a - 9}{h}$$

$$= \lim_{h \to 0} \frac{2ah + h^2 - 8h}{h} = \lim_{h \to 0} \frac{h(2a + h - 8)}{h}$$

$$= \lim_{h \to 0} (2a + h - 8) = 2a + 0 - 8 = 2a - 8$$

■

▪ Interpretations of the Derivative

Because the definitions of the derivative in (5) and (6) involve the same types of limits that we saw in Definitions 1 and 2, we have the following observation.

The phrase "with respect to x" indicates that x is the input variable against which we are measuring the rate of change of the function output.

> ▪ The derivative $f'(a)$ is the instantaneous rate of change of $y = f(x)$ with respect to x when $x = a$.

We also saw that if we sketch the curve $y = f(x)$, then the instantaneous rate of change is the slope of the tangent line to this curve at the point where $x = a$.

> ▪ The tangent line to $y = f(x)$ at $(a, f(a))$ is the line through $(a, f(a))$ whose slope is equal to $f'(a)$, the derivative of f at a.

Thus the geometric interpretation of a derivative [as defined by either (5) or (6)] is as shown in Figure 6.

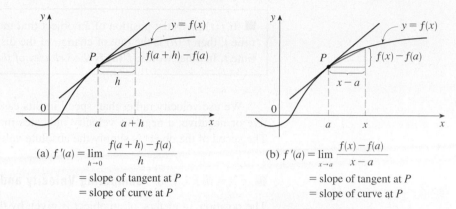

FIGURE 6

Geometric interpretation
of the derivative

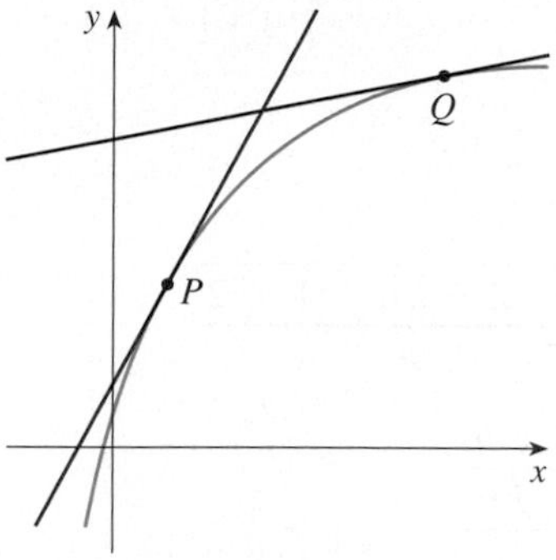

(a) $f'(a) = \lim\limits_{h \to 0} \dfrac{f(a+h) - f(a)}{h}$

= slope of tangent at P

= slope of curve at P

(b) $f'(a) = \lim\limits_{x \to a} \dfrac{f(x) - f(a)}{x - a}$

= slope of tangent at P

= slope of curve at P

This means that when the derivative is large (and therefore the curve is steep, as at the point P in Figure 7), the y-values change rapidly. When the derivative is small, the curve is relatively flat (as at point Q) and the y-values change slowly.

FIGURE 7

The y-values are changing rapidly
at P and slowly at Q.

▪ EXAMPLE 5

Using a Derivative To Find an Equation of a Tangent Line

Find an equation of the tangent line to the parabola $y = x^2 - 8x + 9$ at the point $(3, -6)$.

SOLUTION

From Example 4(b) we know that the derivative of $f(x) = x^2 - 8x + 9$ at the number a is $f'(a) = 2a - 8$. Therefore the slope of the tangent line at $(3, -6)$ is $f'(3) = 2(3) - 8 = -2$. Thus an equation of the tangent line, shown in Figure 8, is

$$y - (-6) = (-2)(x - 3) \qquad \text{or} \qquad y = -2x$$

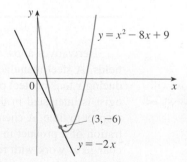

FIGURE 8

▪ | Applications of the Derivative

Remember that the derivative of a function measures the instantaneous rate of change of that function. In Example 1 we computed an instantaneous speed of a falling ball by taking the limit of average speeds. Thus we found the derivative of the distance function s at 5 seconds: $s'(5)$. In general, the following is true.

> ▪ If $f(t)$ gives the position of an object that moves along a straight path at time t, then $f'(a)$ is the rate of change of the displacement with respect to the time t. In other words, $f'(a)$ *is the velocity of the object at time* $t = a$.

We use velocity rather than speed in this case because a velocity can be positive or negative; a negative velocity indicates movement in the reverse direction. The speed of the object is simply the absolute value of the velocity, that is, $|f'(a)|$.

▪ EXAMPLE 6 Computing Velocity and Speed

The position, in meters, of an object is given by the equation $f(t) = 1/(1 + t)$, where t is measured in seconds. Find the velocity and the speed after 2 seconds.

SOLUTION

The derivative of f when $t = 2$ is

$$f'(2) = \lim_{h \to 0} \frac{f(2 + h) - f(2)}{h} = \lim_{h \to 0} \frac{\dfrac{1}{1 + (2 + h)} - \dfrac{1}{1 + 2}}{h}$$

$$= \lim_{h \to 0} \frac{\dfrac{1}{3 + h} - \dfrac{1}{3}}{h} = \lim_{h \to 0} \frac{\dfrac{3 - (3 + h)}{3(3 + h)}}{h}$$

$$= \lim_{h \to 0} \frac{-h}{3(3 + h)h} = \lim_{h \to 0} \frac{-1}{3(3 + h)} = -\frac{1}{9}$$

Thus the velocity after 2 seconds is $f'(2) = -\frac{1}{9}$ m/s, and the speed is

$$|f'(2)| = \left| -\tfrac{1}{9} \right| = \tfrac{1}{9} \text{ m/s} \qquad \blacksquare$$

The concept of *marginal cost* will be explored in detail in Section 3.2.

Derivatives allow us to study many other rates of change in a wide variety of fields. A steel manufacturer is interested in the rate of change of the cost of producing x tons of steel per day with respect to x (called the *marginal cost*). An ecologist is interested in the rate of change of the size of an animal population with respect to time. A chemist might be interested in the rate of change of the concentration of a product in a chemical reaction. Physicists are interested in the rate of change of work with respect to time (called power). You have already seen many instances where rate of change is studied. We look at a few more examples now; further examples will be given in Chapter 4.

▪ EXAMPLE 7 Derivatives in Context

(a) Suppose the number of mold spores, in millions, in a biologist's experiment is given by $N(t)$, where t is the elapsed time in days. Then $N'(t)$ is the (instantaneous) rate of change of the number of mold spores with respect to time after t days. The units are the same as the units for an average rate of change calculation (output inputs per input unit): millions of spores per day.

(b) After market research, a food company estimates that $g(p)$ million boxes of a new cereal can be sold if the price of each box is p dollars. The derivative $g'(p)$ is the rate of change of the number of boxes that will sell with respect to price. For instance, the equation $g'(3.5) = -0.2$ means that at a price of $3.50, the number of boxes sold will decrease (because the derivative is negative) if the price is increased, and the rate at which this will occur is 0.2 million boxes per dollar. This does not necessarily mean that if the price is increased by $1 that 0.2 million fewer boxes will sell. The derivative value -0.2 is the rate of change at the $3.50 price. Once the price shifts, the rate of change may shift as well. (Similarly, if you are driving a car at 45 mi/h right now, it doesn't mean that you will necessarily drive 45 miles after one hour because your speed may fluctuate.) We can say that if the price increases a small amount from $3.50, let's say $0.05, then sales would decrease by about

> The function g described here is an example of a *demand function*. Demand functions will be studied in Section 3.2.

$$\$0.05 \times 0.2 \text{ million boxes/dollar} = 0.01 \text{ million boxes}$$

or 10,000 boxes.

(c) Let $F(s)$ be the average fuel economy of a particular car, measured in miles per gallon, when the car is being driven at s mi/h. The equation $F'(35) = 1.5$ says that when the car is being driven at 35 mi/h, the fuel economy would increase at a rate of 1.5 mi/gal per 1 mi/h speed increase. ■

■ **EXAMPLE 8** **The Derivative of a Cost Function**

A manufacturer produces bolts of a fabric with a fixed width. The cost of producing x yards of this fabric is $C(x)$ dollars.

(a) What is the meaning of the derivative $C'(x)$? What are its units?

(b) In practical terms, what does it mean to say that $C'(1000) = 9$?

(c) Which do you think is greater, $C'(50)$ or $C'(500)$? What about $C'(5000)$?

SOLUTION

(a) The derivative $C'(x)$ is the instantaneous rate of change of $C(x)$ with respect to x; that is, $C'(x)$ means the rate of change of the production cost with respect to the number of yards produced. (This is another example of marginal cost.)

Because

$$C'(x) = \lim_{\Delta x \to 0} \frac{\Delta C}{\Delta x}$$

the units for $C'(x)$ are the same as the units for the average rate of change $\Delta C / \Delta x$. Since ΔC is measured in dollars and Δx in yards, it follows that the units for $C'(x)$ are dollars per yard.

(b) The statement that $C'(1000) = 9$ means that, after 1000 yards of fabric have been manufactured, the rate at which the production cost is increasing is $9/yard. (When $x = 1000$, C is increasing 9 times as fast as x.)

This is an *instantaneous* rate of change, which applies only when exactly 1000 yards have been produced. It is possible that this rate will change (slightly) when the next inch of fabric is manufactured. However, since a change of 1 yard is small compared with 1000 yards already made, we could use the instantaneous rate of change as an approximation to the average rate

> Here we are assuming that the cost function is well behaved; in other words, $C(x)$ doesn't oscillate rapidly near $x = 1000$.

of change for the next yard, and say that the cost of manufacturing the 1001st yard is about $9.

(c) The rate at which the production cost is increasing (per yard) is probably lower when $x = 500$ than when $x = 50$ (the cost of making the 500th yard is less than the cost of the 50th yard) because of economies of scale. (The manufacturer makes more efficient use of the fixed costs of production.) So

$$C'(50) > C'(500)$$

But, as production expands, the resulting large-scale operation might become inefficient and there might be overtime costs. Thus it is possible that the rate of increase of costs will eventually start to rise. If this were the case, we would have

$$C'(5000) > C'(500)$$ ▪

▪ Estimating Derivative Values

If we are unable to calculate a precise derivative value, we can find an approximate value for it either by estimating the value of the limit in Definition 5 or 6 or by estimating the slope of the graph.

▪ EXAMPLE 9   Two Methods for Estimating a Derivative Value

Let $f(x) = 2^x$. Estimate the value of $f'(0)$ in two ways:

(a) By using Definition 5 and taking successively smaller values of h.

(b) By interpreting $f'(0)$ as the slope of a tangent line and using a graphing calculator to zoom in on the graph of $y = 2^x$.

SOLUTION

(a) From Definition 5 we have

$$f'(0) = \lim_{h \to 0} \frac{f(0 + h) - f(0)}{h} = \lim_{h \to 0} \frac{2^h - 1}{h}$$

We are not able to evaluate this limit exactly, so we use a calculator to approximate the values of $(2^h - 1)/h$. From the numerical evidence in the table at the left we see that as h approaches 0, these values appear to approach a number near 0.69. So our estimate is

$$f'(0) \approx 0.69$$

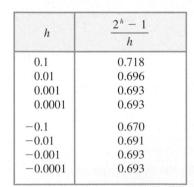

h	$\dfrac{2^h - 1}{h}$
0.1	0.718
0.01	0.696
0.001	0.693
0.0001	0.693
−0.1	0.670
−0.01	0.691
−0.001	0.693
−0.0001	0.693

(b) In Figure 9 we graph the curve $y = 2^x$ and zoom in toward the point $(0, 1)$. We see that the closer we get to $(0, 1)$, the more the curve looks like a straight line. In fact, in Figure 9(c) the curve is practically indistinguishable from its tangent line at $(0, 1)$. Since the x-scale and the y-scale are both 0.01, we estimate that the slope of this line is

$$\frac{0.14}{0.20} = 0.7$$

So our estimate of the derivative is $f'(0) \approx 0.7$. In Section 3.4 we will show that, correct to six decimal places, $f'(0) \approx 0.693147$.

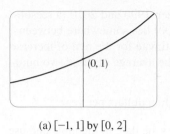

(a) $[-1, 1]$ by $[0, 2]$

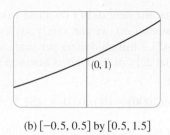

(b) $[-0.5, 0.5]$ by $[0.5, 1.5]$

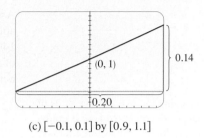

(c) $[-0.1, 0.1]$ by $[0.9, 1.1]$

FIGURE 9   Zooming in on the graph of $y = 2^x$ near $(0, 1)$

In the preceding example, we used an explicit equation to estimate a derivative value. If we have only experimental data or a graph, we can still estimate a derivative and, consequently, an instantaneous rate of change. We demonstrate two techniques for this in the following example.

■ **EXAMPLE 10**   **Estimating the Derivative of a Tabular Function**

Let $D(t)$ be the US national debt at time t. The table in the margin gives approximate values of this function by providing end of year estimates, in billions of dollars, from 1990 to 2010. Interpret and estimate the value of $D'(2000)$.

SOLUTION

The derivative $D'(2000)$ means the instantaneous rate of change of D with respect to t when $t = 2000$, that is, the rate of increase of the national debt in 2000.

According to Equation 5,

t	$D(t)$
1990	3,233.3
1995	4,974.0
2000	5,674.2
2005	7,932.7
2010	13,050.8

$$\textbf{(7)} \qquad D'(2000) = \lim_{h \to 0} \frac{D(2000 + h) - D(2000)}{h}$$

We don't have an equation for $D(t)$, so we won't be able to compute a precise derivative value. But we can compute average rates of change using intervals of equal length on either side of 2000. If we let $h = 5$ in the difference quotient in Equation 7, we have

$$\frac{D(2000 + 5) - D(2000)}{5} = \frac{D(2005) - D(2000)}{5} = \frac{7932.7 - 5674.2}{5}$$

$$= \frac{2258.5}{5} = 451.7$$

which is the average rate of change over the interval $[2000, 2005]$. Similarly, the average rate of change over the interval $[1995, 2000]$ corresponds to putting $h = -5$ into the difference quotient in Equation 7:

$$\frac{D(2000 - 5) - D(2000)}{-5} = \frac{D(1995) - D(2000)}{-5} = \frac{4974.0 - 5674.2}{-5}$$

$$= \frac{-700.2}{-5} = 140.04$$

Assuming that the debt didn't fluctuate wildly between 1990 and 2010 (a reasonably good asumption), we can safely say that $D'(2000)$ lies somewhere between 140.04 and 451.7 billion dollars per year. A good estimate for the rate of increase of the national debt of the United States in 2000 is the average of these two numbers, namely

Averaging the two average rates of change we computed here is equivalent to finding the average rate of change using an interval with 2000 as its midpoint. Such a computation is called a *symmetric difference quotient*; see Exercise 61.

$$D'(2000) \approx \tfrac{1}{2}(140.04 + 451.7) \approx 296 \text{ billion dollars per year}$$

Another way we can estimate $D'(2000)$ is to plot the data in the table and use them to sketch a smooth curve that approximates the graph of D. Then we draw the tangent line at the point where $t = 2000$. The goal is to draw a line that has the same direction as the curve at that point, so that if we were to zoom in very close, the curve and the tangent line would look identical (see Figure 10).

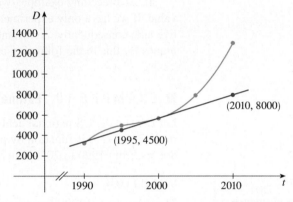

FIGURE 10

Once the tangent line is drawn, we can use two points on the tangent line (not necessarily on the original curve) to determine its slope. The tangent line appears to pass through the points (1995, 4500) and (2010, 8000), so we estimate the slope to be

$$\frac{\Delta y}{\Delta x} = \frac{8000 - 4500}{2010 - 1995} = \frac{3500}{15} \approx 233.33$$

Thus we estimate $D'(2000) \approx 233$ billion dollars per year. ▪

▪ Prepare Yourself

1. If $f(x) = x^2 - 5x$, evaluate and simplify:

(a) $\dfrac{f(a) - f(2)}{a - 2}$

(b) $\dfrac{f(3 + v) - f(3)}{v}$

2. Simplify as much as possible.

(a) $\dfrac{\dfrac{1}{5 + t} - \dfrac{1}{5}}{t}$

(b) $\dfrac{\dfrac{1}{r} - \dfrac{1}{4}}{r - 4}$

3. Evaluate the limit.

(a) $\displaystyle\lim_{x \to 4} \frac{x^2 - 16}{x - 4}$

(b) $\displaystyle\lim_{a \to 4} \frac{a^2 + 2a - 24}{a - 4}$

(c) $\displaystyle\lim_{b \to 0} \frac{(-2 + b)^3 + 8}{b}$

(d) $\displaystyle\lim_{t \to 9} \frac{\sqrt{t} - 3}{t - 9}$

4. Write an equation of the line through the point $(2, -5)$ with slope $\tfrac{3}{4}$.

5. If $f(x) = 2x^2 + 3$, find the average rate of change for $-1 \leqslant x \leqslant 4$.

▪ Exercises 2.3

1–4 ▪ Use Definition 1 or 2 to find the instantaneous rate of change of the function at the given value.

1. $f(t) = t^2$, $t = 5$

2. $g(t) = t^2 + 4t$, $t = 1$

3. $N(x) = 1/x$, $x = 3$

4. $p(r) = 2r^3$, $r = 1$

---

5. Projectile motion The height above the ground of a bullet fired straight upward is $h(t) = 1400t - 16t^2$ feet, where t is the time in seconds after the bullet is fired. Find the instantaneous speed of the bullet 30 seconds after firing.

6. Falling object A rock is dropped from a bridge over a river. The distance, in meters, between the rock and the river t seconds after the rock is dropped is given by $s(t) = 48 - 4.9t^2$. Compute the speed of the rock after 2 seconds.

7. A curve has equation $y = f(x)$.
   (a) Write an expression for the slope of the secant line through the points $(3, f(3))$ and $(x, f(x))$.
   (b) Write an expression for the slope of the tangent line at the point $(3, f(3))$.

8. Motion Suppose an object travels a distance of $f(t)$ miles after t hours.
   (a) Write an expression for the average speed of the object in the time interval from $t = 3$ to $t = 3 + h$.
   (b) Write an expression for the instantaneous speed at $t = 3$.

9–12 ▪ Use Definition 5 to compute the derivative of the function at the given value.

9. $h(x) = 2x^2 + 1$, $x = 3$

10. $f(x) = 3x - x^2$, $x = 1$

11. $g(t) = t^2 + 5t - 2$, $t = 4$

12. $P(t) = t^3 - 4$, $t = -2$

13–16 ▪ Use Definition 6 to compute the derivative of the function at the given value.

13. $f(x) = 4x^2 - x$, $x = 3$

14. $A(u) = 4/u$, $u = -2$

15. $g(t) = t/(t + 1)$, $t = 1$

16. $K(w) = \sqrt{w}$, $w = 4$

---

17. Let $g(t) = 2t^2 + 6$.
   (a) Find $g'(3)$.
   (b) Find $g'(a)$.

18. Let $h(x) = 3x^2 - 4x$.
   (a) Find $h'(1)$.
   (b) Find $h'(a)$.

19. Let $f(x) = 3 - 2x + 4x^2$.
   (a) Find $f'(a)$.
   (b) What is the instantaneous rate of change of f when $x = 5$?
   (c) What is the slope of the tangent line to the graph of f at the point $(-1, 9)$?

20. Let $W(t) = t^2 + 5t - 2$.
   (a) Find $W'(a)$.
   (b) Compute the slope of the graph of W at the point $(2, 12)$.
   (c) Find the instantaneous rate of change of W when $t = 3$.

21. On the given graph of f, mark lengths that represent $f(2)$, $f(2 + h)$, $f(2 + h) - f(2)$, and h. (Choose $h > 0$.) What line has slope $\dfrac{f(2 + h) - f(2)}{h}$?

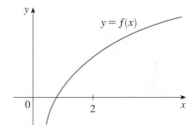

22. For the function f whose graph is shown in Exercise 21, arrange the following numbers in increasing order and explain your reasoning:

$$0 \qquad f'(2) \qquad f(3) - f(2) \qquad \tfrac{1}{2}[f(4) - f(2)]$$

23. For the function g whose graph is given, arrange the following numbers in increasing order and explain your reasoning:

$$0 \qquad g'(-2) \qquad g'(0) \qquad g'(2) \qquad g'(4)$$

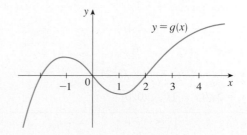

24. Consider the slope of the given curve at each of the five points shown. List these five slopes in decreasing order and explain your reasoning.

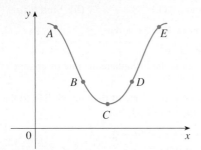

25. Cooling temperature A warm can of soda is placed in a cold refrigerator. Sketch the graph of the temperature of the soda as a function of time. Is the initial rate of change of temperature greater or less than the rate of change after an hour?

26. (a) Find an equation of the tangent line to the graph of $y = g(x)$ at $x = 5$ if $g(5) = -3$ and $g'(5) = 4$.
  (b) If the tangent line to $y = f(x)$ at $(4, 3)$ passes through the point $(0, 2)$, find $f(4)$ and $f'(4)$.

27. If $f(x) = 3x^2 - 5x$, find $f'(2)$ and use it to find an equation of the tangent line to the parabola $y = 3x^2 - 5x$ at the point $(2, 2)$.

28. If $g(x) = 1 - x^3$, find $g'(0)$ and use it to find an equation of the tangent line to the curve $y = 1 - x^3$ at the point $(0, 1)$.

29. (a) If $F(x) = x^2 - 2x$, find $F'(2)$ and use it to find an equation of the tangent line to the curve $y = x^2 - 2x$ at the point $(2, 0)$.
  (b) Illustrate part (a) by graphing the curve and the tangent line on the same screen.

30. (a) If $G(x) = x^2 - 3x + 3$, find $G'(a)$ and use it to find equations of the tangent lines to the curve $y = x^2 - 3x + 3$ at the points $(0, 3)$ and $(2, 1)$.
  (b) Illustrate part (a) by graphing the curve and the tangent lines on the same screen.

31–34 ▪ Find an equation of the tangent line to the curve at the given point.

31. $y = 4.2x - x^2$, $(2, 4.4)$

32. $y = x^3 - 4x$, $(1, -3)$

33. $y = (x - 1)/(x - 2)$, $(3, 2)$

34. $y = 2x/(x + 1)^2$, $(0, 0)$

35–36 ▪ **Motion** A particle moves along a straight line with equation of motion $s = f(t)$, where s is measured in meters and t in seconds. Find the velocity and the speed when $t = 5$.

35. $f(t) = 40t - 6t^2$ **36.** $f(t) = t/(t + 1)$

37. Cost function The cost, in dollars, of producing x units of a certain commodity is

$$C(x) = 5000 + 10x + 0.05x^2$$

  (a) Find the average rate of change of C with respect to x when the production level is changed
    (i) from $x = 100$ to $x = 105$
    (ii) from $x = 100$ to $x = 101$
  (b) Find the instantaneous rate of change of C with respect to x (the marginal cost) when $x = 100$.

38. Revenue function The revenue a company collects after selling x units of its new video camera is $R(x) = 410x - 0.1x^2$.
  (a) Find the average rate of change of R with respect to x when the sales change
    (i) from $x = 300$ to $x = 310$
    (ii) from $x = 300$ to $x = 301$
  (b) Find the instantaneous rate of change of R with respect to x (called the *marginal revenue*) when $x = 300$.

39. Cost function The cost of producing x ounces of gold from a new gold mine is $C = f(x)$ dollars.
  (a) What is the meaning of the derivative $f'(x)$? What are its units?
  (b) What does the statement $f'(800) = 17$ mean?
  (c) Do you think the values of $f'(x)$ will increase or decrease in the short term? What about the long term? Explain.

40. Bacteria population The number of bacteria after t hours in a controlled laboratory experiment is $n = f(t)$.
  (a) What is the meaning of the derivative $f'(5)$? What are its units?
  (b) Suppose there is an unlimited amount of space and nutrients for the bacteria. Which do you think is larger, $f'(5)$ or $f'(10)$? If the supply of nutrients is limited, would that affect your conclusion? Explain.

41. Fuel consumption The fuel consumption, measured in gallons per hour, of a car traveling at a speed of v miles per hour is $c = f(v)$.
  (a) What is the meaning of the derivative $f'(v)$? What are its units?
  (b) Write a sentence (in layman's terms) that explains the meaning of the equation $f'(20) = -0.05$.

42. Consumer demand The quantity, in pounds, of a gourmet ground coffee that is sold by a coffee company at a price of p dollars per pound is $Q = f(p)$.

(a) What is the meaning of the derivative $f'(8)$? What are its units?

(b) Is $f'(8)$ positive or negative? Explain.

43. Heating costs Let $H(t)$ be the daily cost, in dollars, to heat an office building when the outside temperature is t degrees Fahrenheit.

(a) What is the meaning of $H'(58)$? What are its units?

(b) Would you expect $H'(58)$ to be positive or negative? Explain.

44. Loan payments Let $P(r)$ be the amount of the monthly payment for a 30-year mortgage at a fixed interest rate of $r\%$.

(a) What is the meaning of $P'(7)$? What are its units?

(b) Is $P'(7)$ positive or negative? Explain.

45. Profit function Let $P(x)$ be the profit, in dollars, a souvenir shop makes from selling x coffee mugs during a week.

(a) Interpret the statement $P(80) = -125$.

(b) Interpret the statement $P'(80) = 1.5$.

46. Revenue function Let $R(p)$ be the commission a consignment shop takes when they sell an item at p dollars.

(a) Interpret the statement $R(450) = 110$.

(b) Interpret the statement $R'(450) = 0.25$.

47. Let $f(x) = 3^x$. Estimate the value of $f'(1)$ in two ways:

(a) By using Definition 5 and taking successively smaller values of h

(b) By zooming in on the graph of $y = 3^x$ and estimating the slope

48. Let $g(x) = \ln x$. Estimate the value of $g'(3)$ in two ways:

(a) By using Definition 5 and taking successively smaller values of h

(b) By zooming in on the graph of $y = \ln x$ and estimating the slope

49. Temperature Let $T(t)$ be the temperature (in °F) in Phoenix t hours after midnight on September 10, 2008. The table shows values of this function recorded every two hours. What is the meaning of $T'(8)$? Estimate its value.

t	0	2	4	6	8	10	12	14
T	82	75	74	75	84	90	93	94

50. Heart rate A cardiac monitor is used to measure the heart rate of a patient after surgery. It compiles the number of heartbeats after t minutes. When the data in the table are graphed, the slope of the tangent line represents the heart rate in beats per minute.

t (minutes)	Heartbeats
36	2530
38	2661
40	2806
42	2948
44	3080

The monitor estimates this value by calculating the slope of a secant line. Use the data to estimate the patient's heart rate after 42 minutes by averaging the slopes of two secant lines.

51. Motion The table shows the position of a cyclist.

t (seconds)	0	1	2	3	4	5
s (meters)	0	1.4	5.1	10.7	17.7	25.8

(a) Find the average speed for each time period.

(i) $[1, 3]$ (ii) $[2, 3]$

(iii) $[3, 5]$ (iv) $[3, 4]$

(b) Estimate the speed when $t = 3$ by measuring the slope of a tangent to the graph of s.

52. Life expectancy Life expectancy improved dramatically in the 20th century. The table gives values of $E(t)$, the life expectancy at birth, in years, of a male born in the year t in the United States. Estimate and interpret the values of $E'(1910)$ and $E'(1950)$.

t	$E(t)$	t	$E(t)$
1900	48.3	1960	66.6
1910	51.1	1970	67.1
1920	55.2	1980	70.0
1930	57.4	1990	71.8
1940	62.5	2000	74.1
1950	65.6		

53. Mobile-phone subscribers The table shows the estimated percentage P of the population of Brazil that are mobile-phone subscribers. (End of year estimates are given.)

Year	1997	1999	2001	2003	2005	2007
P	2.7	8.8	16.3	25.6	46.3	63.1

(a) Find the average rate of change of P

(i) From 2003 to 2007.

(ii) From 2003 to 2005.

(iii) From 2001 to 2003.

In each case, include the units.

(b) Estimate the instantaneous rate of growth in 2003 by taking the average of two average rates of change. What are its units?

(c) Estimate the instantaneous rate of growth in 2003 by sketching a graph of P and measuring the slope of a tangent.

54. Starbucks locations The number N of locations world-wide of Starbucks coffeehouses is given in the the table. (The number of locations as of October 1 are given.)

Year	N
1998	1886
2000	3501
2002	5886
2004	8569
2006	12,440
2008	16,680

(a) Find the average rate of growth from 2000 to 2008. Include the units.

(b) Estimate the instantaneous rate of growth in 2004 by taking the average of two average rates of change. What are its units?

(c) Estimate the instantaneous rate of growth in 2004 by measuring the slope of a tangent.

55. Oxygen solubility The quantity of oxygen that can dissolve in water depends on the temperature of the water. (So thermal pollution influences the oxygen content of water.) The graph shows how oxygen solubility S varies as a function of the water temperature T.

(a) What is the meaning of the derivative $S'(T)$? What are its units?

(b) Estimate the value of $S'(16)$ and interpret it.

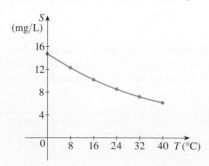

56. Swimming speed of salmon The graph shows the influence of the temperature T on the maximum sustain-able swimming speed S of Coho salmon.

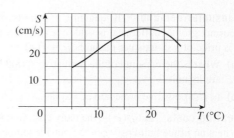

(a) What is the meaning of the derivative $S'(T)$? What are its units?

(b) Estimate the values of $S'(15)$ and $S'(25)$ and interpret them.

57. Draining tank A tank holds 1000 gallons of water, which drains from the bottom of the tank in half an hour. The values in the table show the volume V of water remaining in the tank, in gallons, after t minutes.

t (minutes)	V (gallons)
5	694
10	444
15	250
20	111
25	28
30	0

(a) Use a graphing calculator to find a quadratic model for V.

(b) Use the model to compute $V'(15)$. Then interpret your result in this context.

58. Turkey temperature A roast turkey is taken from an oven when its temperature has reached 185°F and is placed on a table in a room where the temperature is 75°F. The graph shows how the temperature of the turkey decreases and eventually approaches room temperature. By measuring the slope of the tangent, estimate the rate of change of the temperature after an hour.

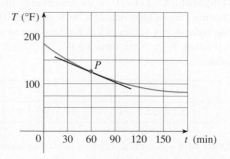

■ Challenge Yourself

59. Draw the graph of a function where the instantaneous rate of change is positive at $x = 1$ and $x = 3$ but the average rate of change on $[1, 3]$ is negative.

60. Draw the graph of a function with domain $[0, 10]$ where the instantaneous rate of change is positive for all values but decreases as the input increases.

61. In Example 10 we averaged two average rates of change to approximate an instantaneous rate of change. An alternative is to use a single average rate of change over an interval centered around the desired input. The result is called a *symmetric difference quotient*.

(a) Compute the average rate of change for the function D in Example 10 on the interval $[1995, 2005]$ and verify that your result agrees with the estimate for $D'(2000)$ computed in the example.

(b) Show that the symmetric difference quotient of a function f at $x = a$ is given by

$$\frac{f(a + d) - f(a - d)}{2d}$$

where d is the distance from a to the endpoints of the interval. Then verify that this is equivalent to averag-ing the average rates of change over the intervals $[a - d, a]$ and $[a, a + d]$.

(c) Use a symmetric difference quotient to estimate $f'(1)$ for $f(x) = x^3 - 2x^2 + 2$ with $d = 0.4$. Draw a graph of f and a secant line whose slope represents the symmetric difference quotient, along with the tangent line to f at $x = 1$. Can you see visually why the symmetric difference quotient is usually a better estimate for a derivative value than an average rate of change over $[1 - d, 1]$ or $[1, 1 + d]$?

62. Here we investigate the derivative of the absolute value function $f(x) = |x|$ at $x = 0$.

(a) Write an expression for $f'(0)$ as a limit. If you calculate this limit as a right-hand limit, what does the value of $f'(0)$ appear to be?

(b) If you calculate this limit as a left-hand limit, what does the value of $f'(0)$ appear to be?

(c) Explain why $f'(0)$ doesn't exist.

(d) Some graphing calculators use symmetric difference quotients to estimate derivatives on a graph. The Texas Instruments TI-84 calculator says that $f'(0) = 0$. Sketch a graph of f and use it to explain why the calculator gave the estimate of 0, and explain why that value is incorrect.

2.4 The Derivative as a Function

In the preceding section we considered the derivative of a function f at a fixed number a:

(1)
$$f'(a) = \lim_{h \to 0} \frac{f(a + h) - f(a)}{h}$$

Here we change our point of view and let the number a vary. If we replace a in Equation 1 by a variable x, we obtain

(2)
$$f'(x) = \lim_{h \to 0} \frac{f(x + h) - f(x)}{h}$$

Given any number x for which this limit exists, we assign to x the number $f'(x)$. So we can regard f' as a new function, called the **derivative of f** and defined by Equation 2. We know that the value of f' at x, $f'(x)$, can be interpreted geometrically as the slope of the tangent line to the graph of f at the point $(x, f(x))$.

A function f is called **differentiable** at a number if the derivative exists there, and we say that f is differentiable on an interval if the derivative exists at every number in that interval. The function f' is called the derivative of f because it has been "derived" from f by the limiting operation in Equation 2. The domain of f' is the set of all numbers where f is differentiable and may be smaller than the domain of f.

The Derivative Function from Graphs, Tables, and Formulas

▪ **EXAMPLE 1**   **Sketching a Graph of f' from a Graph of f**

The graph of a function f is given in Figure 1. Use it to sketch the graph of the derivative f'.

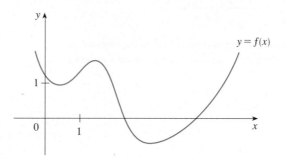

FIGURE 1

SOLUTION

We don't have an equation, but we can estimate the value of the derivative at any value of x by drawing the tangent line at the point $(x, f(x))$ and estimating its slope. For instance, for $x = 5$ we draw the tangent at point P in Figure 2(a) and estimate its slope to be about $\frac{3}{2}$, so $f'(5) \approx 1.5$. This allows us to plot the point P' at $(5, 1.5)$ on the graph of f' directly beneath P. Notice that we are taking the *slope* of the graph of f and using this value as the *y-value* of the graph of f'.

We repeat this procedure at enough points to get a good idea of the shape of the graph, as shown in Figure 2(b). Notice that the tangent lines at points A, B, and C are horizontal, so the derivative is 0 there (horizontal lines have a slope of 0). Because the output of f' is 0 at these points, the graph of f' intersects the x-axis at the points A', B', and C', directly beneath A, B, and C.

Between A and B the tangents have positive slope, so $f'(x)$ is positive there, and the graph of f' appears above the x-axis. But between B and C the tangents have negative slope, so $f'(x)$ is negative there and the graph drops below the x-axis.

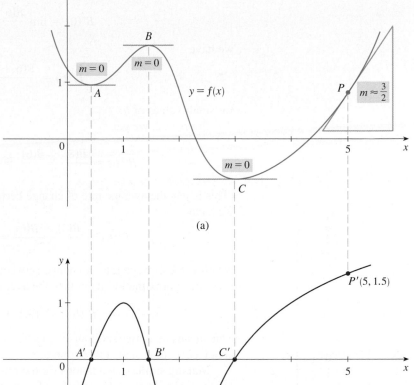

(a)

TEC Visual 2.4 shows an animation of Figure 2 for several functions.

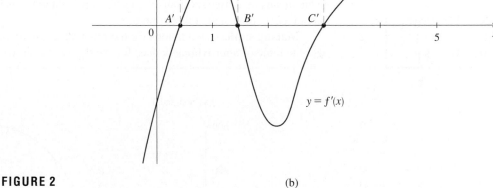

(b)

FIGURE 2

If a function is defined by a table of values, we can construct a table of approximate values of its derivative, as in the next example.

t	$B(t)$
0	1.1
2	3.1
4	8.5
6	22.1
8	60.3
10	138.6
12	278.3
14	429.5
16	541.0
18	589.8
20	609.6

■ **EXAMPLE 2**

Estimating Derivative Values for a Function Given by a Table

Let $B(t)$ be the number of *E. coli* bacteria cells present in a microbiologist's experiment after t hours. The table at the left gives values of $B(t)$, in thousands, for the first 20 hours. Construct a table of values for the derivative of this function.

SOLUTION

We assume that there were no wild fluctuations in the bacteria population between the stated values. Let's start by approximating $B'(6)$, the rate of increase in the number of cells after six hours. Since

$$B'(6) = \lim_{h \to 0} \frac{B(6 + h) - B(6)}{h}$$

we have

$$B'(6) \approx \frac{B(6 + h) - B(6)}{h}$$

for small values of h.

For $h = 2$, we get

$$B'(6) \approx \frac{B(8) - B(6)}{2} = \frac{60.3 - 22.1}{2} = 19.1$$

(This is just the average rate of change between $t = 6$ and $t = 8$.) For $h = -2$, we have

$$B'(6) \approx \frac{B(4) - B(6)}{-2} = \frac{8.5 - 22.1}{-2} = 6.8$$

which is the average rate of change between $t = 4$ and $t = 6$. We get a more accurate approximation if we take the average of these rates of change:

$$B'(6) \approx \tfrac{1}{2}(19.1 + 6.8) = 12.95$$

This means that after six hours the bacteria population was increasing at a rate of about 12.95 thousand cells per hour.

Making similar calculations for the other values (except at the endpoints), we get the table of approximate values for the derivative shown at the left. ▪

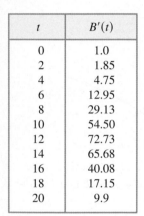

t	$B'(t)$
0	1.0
2	1.85
4	4.75
6	12.95
8	29.13
10	54.50
12	72.73
14	65.68
16	40.08
18	17.15
20	9.9

Figure 3 illustrates Example 2 by showing graphs of the bacteria population function $B(t)$ and its derivative $B'(t)$. Notice how the rate of population growth increases to a maximum after 12 hours and decreases thereafter.

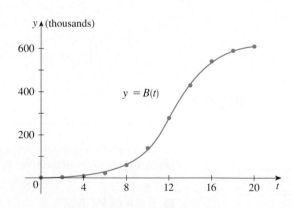

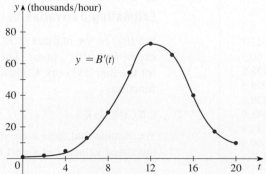

FIGURE 3

Next we find a formula for the derivative of a function.

■ **EXAMPLE 3** **Derivative of a Function Given by a Formula**

(a) If $f(x) = 8x - 2x^2$, find a formula for $f'(x)$.

(b) Illustrate by comparing the graphs of f and f'.

SOLUTION

(a) From Definition 2, we have

$$f'(x) = \lim_{h \to 0} \frac{f(x + h) - f(x)}{h}$$

$$= \lim_{h \to 0} \frac{[8(x + h) - 2(x + h)^2] - [8x - 2x^2]}{h}$$

$$= \lim_{h \to 0} \frac{8x + 8h - 2(x^2 + 2xh + h^2) - 8x + 2x^2}{h}$$

$$= \lim_{h \to 0} \frac{8h - 4xh - 2h^2}{h} = \lim_{h \to 0} \frac{h(8 - 4x - 2h)}{h}$$

$$= \lim_{h \to 0} (8 - 4x - 2h) = 8 - 4x$$

Remember that the limit is determined as h, not x, approaches 0; x should be temporarily regarded as a constant value.

(b) We use a graphing device to graph f and f' in Figure 4. Notice that $f'(x) = 0$ when f has a horizontal tangent line (at $x = 2$), so we see an x-intercept on the graph of f' at $x = 2$. Also, $f'(x)$ is positive when the tangents have positive slope, so the graph of f' is above the x-axis for $x < 2$. For $x > 2$, the slope of f is negative, so the graph of f' appears below the x-axis. The slope of the graph of f grows larger (the graph is steeper) as we look toward the left, so the graph of f' is higher at these x-values. The slope of the graph of f gets larger negative as we look toward the right, so the graph of f' is lower there. When the graph of f is not steep, the values of $f'(x)$ are closer to 0, so the graph of f' is closer to the x-axis. ■

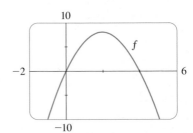

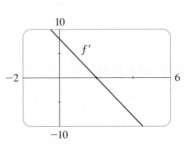

FIGURE 4

■ **EXAMPLE 4** **Graphing the Derivative of a Linear Function**

If $g(t) = 3t - 5$, sketch a graph of g'.

SOLUTION

Since g is a linear function, we know its slope is constant. The slope is 3, so we must have $g'(t) = 3$ for any input t. The graph of g is just the horizontal line $y = 3$, as shown in Figure 5. ■

FIGURE 5

■ How Can Functions Fail To Be Differentiable?

There are situations in which a function is defined at a value but is not differentiable there. For instance, if a function is not continuous at a point, we will see a gap or jump in the graph. We say that the function has a *discontinuity* at that value. A function's graph does not have a tangent line at a discontinuity. Consequently, there is no derivative at that value.

In Section 2.3 we saw that if we zoom in close enough to the graph of a parabola, the curve looks more like a line, and in fact the curve becomes almost indistinguishable from its tangent line. This is always the case for differentiable functions. If a function is continuous but its graph changes direction abruptly (the rate of change shifts suddenly from one value to another), the graph will appear to have a "corner" or "kink" in it. No matter how much we zoom in toward this point, we can't eliminate the corner. Because the graph doesn't appear as a single line after zooming in, no tangent line exists and the function is not differentiable at that point.

Another way a function can be nondifferentiable is if the graph has a vertical tangent line at a point. (Recall that the slope of a vertical line is undefined.) On the graph, this means that the tangent lines become steeper and steeper as we approach that point. Figure 6 illustrates the three possibilities that we have discussed.

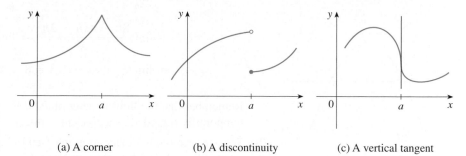

FIGURE 6
Three ways a derivative can fail to exist

(a) A corner   (b) A discontinuity   (c) A vertical tangent

■ **EXAMPLE 5**

A Continuous Function that is Not Differentiable at a Point

If $f(x) = |x|$, sketch a graph of $f'(x)$.

SOLUTION

A graph of f is shown in Figure 7(a).

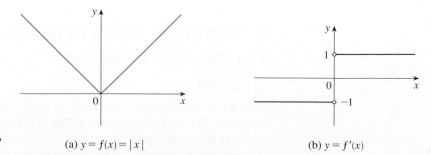

FIGURE 7   (a) $y = f(x) = |x|$ (b) $y = f'(x)$

To the left of $x = 0$, the graph of f is the line $y = -x$ with slope -1, so $f'(x) = -1$ for $x < 0$. To the right of $x = 0$ the graph is the line $y = x$ with slope 1, so $f'(x) = 1$ for $x > 0$. However, at $x = 0$, the graph has a corner and abruptly changes direction. Because there is no tangent line at $x = 0$, $f'(0)$ does not exist. The graph of f' is shown in Figure 7(b). Notice that no point is graphed at $x = 0$. ■

Figures 6 and 7 include several examples of functions that are continuous but not differentiable everywhere. It turns out, however, that if a function is differentiable it must also be continuous.

---

(3) ▪ Theorem   If a function is differentiable at a number, then it is continuous there.

You are asked to prove Theorem 3 in Exercise 54.

▪ Other Notations

So far we have used the notation $f'(x)$ for the derivative of a function f. However, other notations are traditionally used in addition to $f'(x)$, and one notation may be more convenient than another for certain situations. For instance, if we have an equation $y = x^3 + x$, we can refer to the derivative simply as y'.

If we have a function $y = f(x)$, a very useful notation introduced by Leibniz is the symbol dy/dx. (We can also write df/dx.) Although it looks like a fraction, it should not be regarded as a ratio; it is simply a synonym for $f'(x)$. Notice that it looks similar to the expression for average rate of change, $\Delta y/\Delta x$. In fact, since dy/dx is the derivative, it represents the instantaneous rate of change, so we have

$$\frac{dy}{dx} = \lim_{\Delta x \to 0} \frac{\Delta y}{\Delta x}$$

(see Equation 1 in Section 2.3). If we want to indicate the value of a derivative dy/dx in Leibniz notation at a specific number a, we use the notation

$$\left. \frac{dy}{dx} \right|_{x=a}$$

which is a synonym for $f'(a)$.

We can indicate the process of calculating a derivative by writing d/dx. Thus

$$\frac{d}{dx} f(x)$$

represents the derivative of the function f with respect to x and is equivalent to $f'(x)$.

Leibniz

Gottfried Wilhelm Leibniz was born in Leipzig in 1646 and studied law, theology, philosophy, and mathematics at the university there, graduating with a bachelor's degree at age 17. After earning his doctorate in law at age 20, Leibniz entered the diplomatic service and spent most of his life traveling to the capitals of Europe on political missions. In particular, he worked to avert a French military threat against Germany and attempted to reconcile the Catholic and Protestant churches.

His serious study of mathematics did not begin until 1672 while he was on a diplomatic mission in Paris. There he built a calculating machine and met scientists, like Huygens, who directed his attention to the latest developments in mathematics and science. Leibniz sought to develop a symbolic logic and system of notation that would simplify logical reasoning. In particular, the version of calculus that he published in 1684 established the notation and the rules for finding derivatives that we use today.

Unfortunately, a dreadful priority dispute arose in the 1690s between the followers of Newton and those of Leibniz as to who had invented calculus first. Leibniz was even accused of plagiarism by members of the Royal Society in England. The truth is that each man invented calculus independently. Newton arrived at his version of calculus first but, because of his fear of controversy, did not publish it immediately. So Leibniz's 1684 account of calculus was the first to be published.

For instance, the result of Example 3(a) can be expressed as

$$\frac{d}{dx}(8x - 2x^2) = 8 - 4x$$

■ **EXAMPLE 6 Derivative Notations in Context**

(a) Suppose $R = f(t)$ is the amount of rainfall, in inches, t hours after a thunderstorm begins. What does dR/dt represent in this context?

(b) Let $G(x)$ represent the amount, in gallons, of gasoline remaining in the tank of a particular car after the car has been driven x miles. Interpret the expression

$$\left.\frac{dG}{dx}\right|_{x=50}$$

in this context.

SOLUTION

(a) The notation dR/dt represents the derivative of R with respect to t. The derivative is the instantaneous rate of change, so this is the rate at which rain is falling, measured in inches per hour, at time t. If it starts raining harder, you could say that dR/dt is increasing at that moment.

(b) The notation

$$\left.\frac{dG}{dx}\right|_{x=50}$$

represents the rate of change of G with respect to x, evaluated at $x = 50$. So this is the rate at which the amount of gas remaining in the tank is changing when the car has been driven for 50 miles. ▪

▪ Second Derivatives

If f is a differentiable function, then its derivative f' is also a function, so f' may have a derivative of its own, denoted by $(f')' = f''$ (read "f double prime"). This new function f'' is called the **second derivative** of f because it is the derivative of the derivative of f. Using Leibniz notation, we write the second derivative of $y = f(x)$ as

$$\frac{d}{dx}\left(\frac{dy}{dx}\right) = \frac{d^2y}{dx^2}$$

■ **EXAMPLE 7 Computing a Second Derivative Function**

If $f(x) = x^2 - 8x + 9$, find $f''(x)$.

SOLUTION

In Example 4(b) in Section 2.3 we found that the first derivative is $f'(x) = 2x - 8$. This is a linear function with constant slope 2, so its derivative is 2 for every input. Thus $f''(x) = 2$. ▪

We can interpret $f''(x)$ as the slope of the curve $y = f'(x)$ at the point $(x, f'(x))$. In other words, it is the rate of change of the slope of the original curve $y = f(x)$. If, for instance, $f''(2)$ is positive, this means that the slope of the graph of f (not the output of f) is increasing at $x = 2$.

The most familiar example of the second derivative is acceleration. If $s(t)$ is the position of an object that moves in a straight line, we know that its first derivative represents the velocity $v(t)$ of the object as a function of time:

$$v(t) = s'(t)$$

The instantaneous rate of change of velocity with respect to time is called the **acceleration** $a(t)$ of the object. Thus the acceleration function is the derivative of the velocity function and is therefore the second derivative of the position function:

$$a(t) = v'(t) = s''(t)$$

In Leibniz notation, we have

$$v = \frac{ds}{dt} \quad \text{and} \quad a = \frac{dv}{dt} = \frac{d^2s}{dt^2}$$

When you drive a car at a steady speed, there is no change in speed, so you have no acceleration. If you hit the brakes, the speed of the car decreases, so you have introduced negative acceleration. If you step on the gas pedal, the car speeds up, so you have positive acceleration, and you feel your body pressed back into the seat.

■ EXAMPLE 8   Position, Velocity, and Acceleration Functions

A car starts from rest and the graph of its position function is shown in Figure 8, where s is measured in feet and t in seconds. Use it to graph the velocity and acceleration of the car. What is the acceleration at $t = 2$ seconds?

SOLUTION

By measuring the slope of the graph of $s = f(t)$ at $t = 0, 1, 2, 3, 4,$ and 5, and using the method of Example 1, we plot the graph of the velocity function $v = f'(t)$ in Figure 9. The acceleration when $t = 2$ is $a = f''(2)$, the slope of the tangent line to the graph of f' when $t = 2$. We estimate the slope of this tangent line to be

$$a(2) = f''(2) = v'(2) \approx \tfrac{27}{3} = 9 \text{ ft/s}^2$$

Similar measurements enable us to graph the acceleration function in Figure 10.

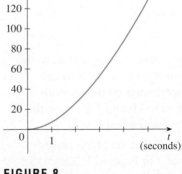

FIGURE 8
Position function of a car

The units for acceleration are feet per second per second, written as ft/s².

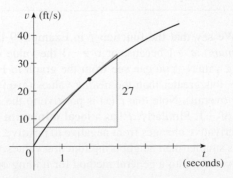

FIGURE 9   Velocity function

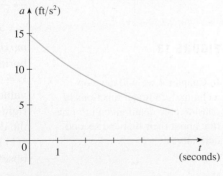

FIGURE 10   Acceleration function   ■

▪ What Do Derivatives Tell Us about a Function?

Many of the applications of calculus depend on our ability to deduce facts about a function f from information concerning its derivative. Because $f'(x)$ represents the slope of the curve $y = f(x)$ at the point $(x, f(x))$, it tells us the direction in which the curve proceeds at each point. So it is reasonable to expect that information about $f'(x)$ will provide us with information about $f(x)$.

In particular, to see how the derivative of f can tell us where a function is increasing or decreasing, look at Figure 11. (Increasing functions and decreasing functions were defined in Section 1.4.) Between A and B and between C and D, the tangent lines have positive slope and so $f'(x) > 0$. Between B and C, the tangent lines have negative slope and so $f'(x) < 0$. Thus it appears that f increases when $f'(x)$ is positive and decreases when $f'(x)$ is negative.

It turns out that what we observed for the function graphed in Figure 11 is always true. We state the general result as follows.

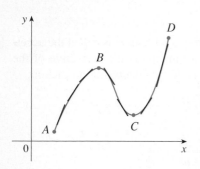

FIGURE 11

> If $f'(x) > 0$ on an interval, then f is increasing on that interval.
>
> If $f'(x) < 0$ on an interval, then f is decreasing on that interval.

▪ EXAMPLE 9

Identifying Information about the Graph of f from the Graph of f'

(a) If it is known that the graph of the derivative f' of a function is as shown in Figure 12, what can we say about f?

(b) If it is known that $f(0) = 0$, sketch a possible graph of f.

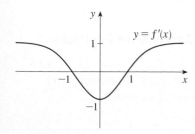

FIGURE 12

SOLUTION

(a) We observe from Figure 12 that $f'(x)$ is negative when $-1 < x < 1$, so the original function f must be decreasing on the interval $(-1, 1)$. Similarly, $f'(x)$ is positive for $x < -1$ and for $x > 1$, so f is increasing on the intervals $(-\infty, -1)$ and $(1, \infty)$. Also note that, since $f'(-1) = 0$ and $f'(1) = 0$, the graph of f has horizontal tangents when $x = -1$ and when $x = 1$.

(b) We use the information from part (a), and the fact that the graph passes through the origin, to sketch a possible graph of f in Figure 13. Notice that $f'(0) = -1$, so we have drawn the curve $y = f(x)$ passing through the origin with a slope of -1. ▪

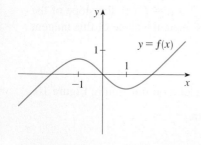

FIGURE 13

In Chapter 4 we will develop techniques to graph functions in considerable detail given information about their first and second derivatives.

We say that the function f in Example 9 has a **local maximum** (or *relative maximum*) at -1 because at $x = -1$ the value of $f(x)$ is larger than at the neighboring values. You can see from the graph in Figure 13 that the output value for $x = -1$ is greater than the nearby values (even though it is not the largest output value overall). Note that $f'(x)$ is positive to the left of -1 and negative just to the right of -1. Similarly, f has a **local minimum** (or *relative minimum*) at 1, where the derivative changes from negative to positive. The graph in Figure 13 shows that $f(1)$ is smaller than neighboring function values. In Chapter 4 we will develop these observations into a general method for finding optimal values of functions.

Next let's see how the sign of $f''(x)$ affects the appearance of the graph of f. Since $f'' = (f')'$, we know that if $f''(x)$ is positive, then f' is an increasing function. This says that the slopes of the tangent lines of the curve $y = f(x)$ increase from left to right. Figure 14 shows the graph of such a function. The slope of this curve becomes progressively larger (regardless of whether the slope is positive or negative) as x increases and we observe that, as a consequence, the curve bends upward. Such a curve is called **concave upward**. In Figure 15, however, $f''(x)$ is negative, which means that f' is decreasing. Thus the slopes of f decrease from left to right and the curve bends downward. This curve is called **concave downward**. We summarize our discussion as follows. (Concavity is discussed in greater detail in Section 4.3.)

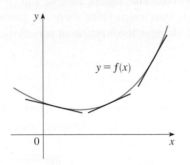

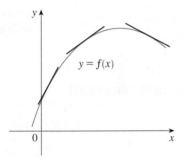

FIGURE 14
Since $f''(x) > 0$, the slopes increase and f is concave upward.

FIGURE 15
Since $f''(x) < 0$, the slopes decrease and f is concave downward.

> If $f''(x) > 0$ on an interval, then f is concave upward on that interval.
>
> If $f''(x) < 0$ on an interval, then f is concave downward on that interval.

■ EXAMPLE 10

Analyzing Rates of Change and Concavity from a Graph

Figure 16 shows a population graph for Cyprian honeybees raised in an apiary. How does the rate of population increase change over time? When is this rate highest? Over what intervals is P concave upward or concave downward?

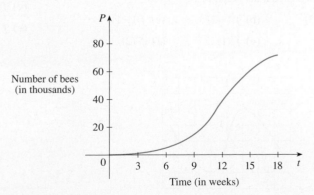

FIGURE 16

SOLUTION

By looking at the slope of the curve as t increases, we see that the rate of increase of the population is initially very small, then gets larger until it reaches a maximum at about $t = 12$ weeks, and decreases as the population begins to level off. As the population approaches its maximum value of about 75,000 (called the *carrying capacity*), the rate of increase, $P'(t)$, approaches 0. The curve appears to be concave upward on $(0, 12)$ and concave downward on $(12, 18)$. ■

In Example 10, the population curve changed from concave upward to concave downward at approximately the point $(12, 38{,}000)$. This point is called an *inflection point* of the curve. The significance of this point is that the rate of population increase has its maximum value there. In general, an **inflection point** is a point where a curve changes its direction of concavity.

■ Prepare Yourself

1. If $g(t) = 6t - 3t^2$, evaluate and simplify

$$\frac{g(t + h) - g(t)}{h}$$

2. If $f(x) = \dfrac{1}{x}$, evaluate and simplify

$$\frac{f(x + c) - f(x)}{c}$$

3. Evaluate the limit:

$$\lim_{h \to 0} \frac{h^2 + 2xh}{h}$$

4. Evaluate the limit:

$$\lim_{h \to 0} \frac{(x + h)^3 + 2(x + h) - x^3 - 2x}{h}$$

5. Simplify as much as possible:

$$\frac{\dfrac{x + a}{x + a + 2} - \dfrac{x}{x + 2}}{a}$$

6. If $f(x) = \sqrt{x}$, compute $f'(4)$.

■ Exercises 2.4

1–2 ■ Use the given graph to estimate the value of each derivative. Then sketch the graph of f'.

1. (a) $f'(-3)$ (b) $f'(-2)$ (c) $f'(-1)$

   (d) $f'(0)$ (e) $f'(1)$ (f) $f'(2)$

   (g) $f'(3)$

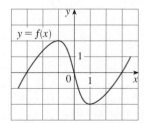

2. (a) $f'(0)$ (b) $f'(1)$

   (c) $f'(2)$ (d) $f'(3)$

   (e) $f'(4)$ (f) $f'(5)$

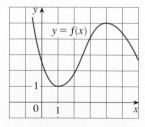

3. Match the graph of each function in (a)–(d) with the graph of its derivative in I–IV. Give reasons for your choices.

(a)
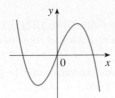

(b)
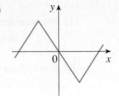

(c)
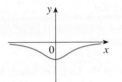

(d)
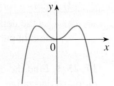

I
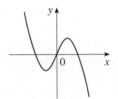

II
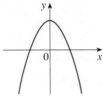

III
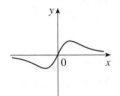

IV

4–7 ■ Trace or copy the graph of the given function f. (Assume that the axes have equal scales.) Then use the method of Example 1 to sketch the graph of f' below it.

4.
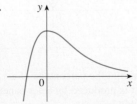

5.
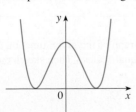

6.
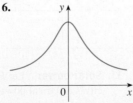

7.
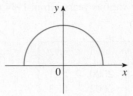

8–9 ■ Make a careful sketch of the graph of f and below it sketch the graph of f' in the same manner as in Exercises 4–7. Can you guess a formula for $f'(x)$ from its graph?

8. $f(x) = \ln x$

9. $f(x) = e^x$

10. Bacteria population Shown is the graph of the population function $P(t)$ for bacteria in a laboratory culture. Use the method of Example 1 to graph the derivative $P'(t)$. What does the graph of P' tell us about the bacteria population?

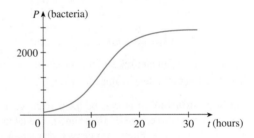

11. Marriage age The graph shows how the average age of first marriage of Japanese men varied in the last half of the 20th century. Sketch the graph of the derivative function $M'(t)$. During which years was the derivative negative?

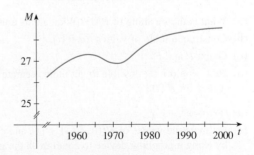

12. Music sales The table (from the Australian Record Industry Association) gives the number N, in millions, of

CDs sold in Australia for various years t from 1994 to 2004.

t	N	t	N
1994	5.6	2000	11.1
1996	7.2	2002	11.3
1998	9.9	2004	9.3

(a) What is the meaning of dN/dt? What are its units?

(b) Construct a table of values for dN/dt.

(c) Sketch a graph of the derivative of N with respect to t.

13. Unemployment rate   The unemployment rate $U(t)$ varies with time. The table (from the Bureau of Labor Statistics) gives the percentage of unemployed in the US labor force from 1999 to 2008.

t	$U(t)$	t	$U(t)$
1999	4.2	2004	5.5
2000	4.0	2005	5.1
2001	4.7	2006	4.6
2002	5.8	2007	4.6
2003	6.0	2008	5.8

(a) What is the meaning of $U'(t)$? What are its units?

(b) Construct a table of values for $U'(t)$.

14. Youth population   Let $P(t)$ be the percentage of Americans under the age of 18 at time t. The table gives values of this function in census years from 1950 to 2000.

t	$P(t)$	t	$P(t)$
1950	31.1	1980	28.0
1960	35.7	1990	25.7
1970	34.0	2000	25.7

(a) What is the meaning of $P'(t)$? What are its units?

(b) Construct a table of values for $P'(t)$.

(c) Graph P and P'.

(d) How would it be possible to get more accurate values for $P'(t)$?

15. Let $f(x) = x^2$.

(a) Estimate the values of $f'(0)$, $f'\left(\frac{1}{2}\right)$, $f'(1)$, and $f'(2)$ by using a graphing device to zoom in on the graph of f.

(b) Use symmetry to deduce the values of $f'\left(-\frac{1}{2}\right)$, $f'(-1)$, and $f'(-2)$.

(c) Use the results from parts (a) and (b) to guess a formula for $f'(x)$.

(d) Use the definition of a derivative to prove that your guess in part (c) is correct.

16. Let $f(x) = x^3$.

(a) Estimate the values of $f'(0)$, $f'\left(\frac{1}{2}\right)$, $f'(1)$, $f'(2)$, and $f'(3)$ by using a graphing device to zoom in on the graph of f.

(b) Use symmetry to deduce the values of $f'\left(-\frac{1}{2}\right)$, $f'(-1)$, $f'(-2)$, and $f'(-3)$.

(c) Use the values from parts (a) and (b) to graph f'.

(d) Guess a formula for $f'(x)$.

(e) Use the definition of a derivative to prove that your guess in part (d) is correct.

17–25 ▪ Find the derivative of the function using the definition of derivative.

17. $f(x) = \frac{1}{2}x - \frac{1}{3}$

18. $g(t) = 3t + 5$

19. $h(v) = 4v^2 - 2$

20. $f(x) = 1.5x^2 - x + 3.7$

21. $f(x) = x^3 - 3x + 5$

22. $y = 2x^3 + 7x^2$

23. $B(p) = 3/p$

24. $N(t) = 1/t^2$

25. $y = \sqrt{x}$

26–30 ▪ Find the derivative of the function using the definition of derivative. State the domain of the function and the domain of its derivative.

26. $f(x) = \dfrac{3 + x}{1 - 3x}$

27. $G(t) = \dfrac{4t}{t + 1}$

28. $f(x) = x + \sqrt{x}$

29. $g(x) = \sqrt{1 + 2x}$

30. $g(t) = \dfrac{1}{\sqrt{t}}$

31–32 ▪ Trace or copy the graph of the given function f. (Assume that the axes have equal scales.) Then sketch the graph of df/dx below it.

31.

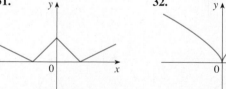

32.

33. Solar power   Let P represent the percentage of a city's electrical power that is produced by solar panels t years after January 1, 2000.

(a) What does dP/dt represent in this context?

(b) Interpret the statement

$$\left.\frac{dP}{dt}\right|_{t=2} = 3.5$$

34. Travel by car   Suppose N is the number of people in the United States who travel by car to another state for a vacation this year when the average price of gasoline is p dollars per gallon. Do you expect dN/dp to be positive or negative? Explain.

35–36 ▪ Use the definition of a derivative to find $f'(x)$ and $f''(x)$. Then graph f, f', and f'' on a common screen and check to see if your answers are reasonable.

35. $f(x) = 1 + 4x - x^2$ **36.** $f(x) = 1/x$

37–38 ▪ The graph of the *derivative* f' of a function f is shown.

(a) On what intervals is f increasing or decreasing?

(b) At what values of x does f have a local maximum or minimum?

(c) If it is known that $f(0) = 0$, sketch a possible graph of f.

37.

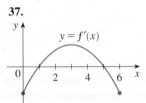

38.

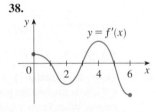

39. Use the given graph of f to estimate the intervals on which the derivative f' is increasing or decreasing.

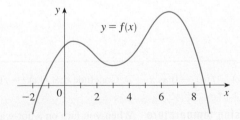

40. (a) Sketch a curve whose slope is always positive and increasing.

(b) Sketch a curve whose slope is always positive and decreasing.

(c) Give possible equations for curves with these properties.

41. Sketch the graph of a function whose first and second derivatives are always negative.

42. Sketch the graph of a function whose first derivative is always negative and whose second derivative is always positive.

43. Federal deficit   The president announces that the federal deficit is increasing, but at a decreasing rate. Interpret this statement in terms of a function and its derivatives.

44. Yeast population   A graph of a population of yeast cells in a new laboratory culture as a function of time is shown.

(a) Describe how the rate of population increase varies.

(b) When is this rate highest?

(c) On what intervals is the population function concave upward or downward?

(d) Estimate the coordinates of the inflection point.

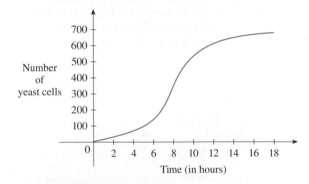

45. Pheasant population   The table gives population densities for ring-necked pheasants (in number of pheasants per acre) on Pelee Island, Ontario.

(a) Describe how the rate of change of population varies.

(b) Estimate the inflection points of the graph. What is the significance of these points?

t	$P(t)$
1927	0.1
1930	0.6
1932	2.5
1934	4.6
1936	4.8
1938	3.5
1940	3.0

46. Motion   A particle is moving along a horizontal straight line. The graph of its position function (the distance to the right of a fixed point as a function of time) is shown.

(a) When is the particle moving toward the right and when is it moving toward the left?

(b) When does the particle have positive acceleration and when does it have negative acceleration?

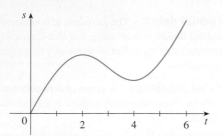

47. Motion The position of an object moving along a straight path is given by $s(t) = 1.7t^2 - 3.1t + 6.8$. Find formulas for the velocity and acceleration functions.

48. Motion The graph of a position function of a car is shown, where s is measured in feet and t in seconds. Use it to graph the velocity and acceleration of the car. What is the acceleration at $t = 10$ seconds?

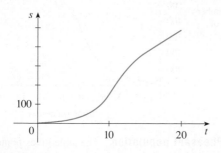

49. Knowledge absorption Let $K(t)$ be a measure of the knowledge you gain by studying for a test for t hours. Which do you think is larger, $K(8) - K(7)$ or

$K(3) - K(2)$? Is the graph of K concave upward or concave downward? Why?

50. Liquid depth Coffee is being poured into the mug shown in the figure at a constant rate (measured in volume per unit time). Sketch a rough graph of the depth of the coffee in the mug as a function of time. Account for the shape of the graph in terms of concavity. What is the significance of the inflection point?

51. If $f(x) = 2x^2 - x^3$, find $f'(x)$ and $f''(x)$. Graph f, f', and f'' on a common screen. Are the graphs consistent with the geometric interpretations of these derivatives?

52. Graph the function using a view where you can identify the inflection point of the curve. Then estimate its coordinates.

    (a) $g(x) = \dfrac{155}{3 + e^{0.02x}}$

    (b) $Q = 3x^3 - 8x^2 - 11x + 21$

53. Graph the function $f(x) = x + \sqrt{|x|}$. Zoom in repeatedly, first toward the point $(-1, 0)$ and then toward the origin. What is different about the behavior of f in the vicinity of these two points? What do you conclude about the differentiability of f?

▪ Challenge Yourself

54. In this exercise we prove Theorem 3. To prove that f is continuous at a value a, we have to show that $\lim_{x \to a} f(x) = f(a)$. We are given that f is differentiable at a. Then we know from Equation 2.3.6 that

$$f'(a) = \lim_{x \to a} \frac{f(x) - f(a)}{x - a}$$

exists. Write $f(x)$ in the form

$$f(x) = f(a) + \frac{f(x) - f(a)}{x - a}(x - a)$$

(Why is this justified?) Now take the limit of both sides as $x \to a$ and use properties of limits to show that

$f(x) - f(a) \to 0$ as $x \to a$. Conclude that $f(x) \to f(a)$ as $x \to a$.

55. Rising temperature When you turn on a hot-water faucet, the temperature T of the water depends on how long the water has been running.

    (a) Sketch a possible graph of T as a function of the time t that has elapsed since the faucet was turned on.

    (b) Describe how the rate of change of T with respect to t varies as t increases.

    (c) Sketch a graph of the derivative of T.

56. Motion The figure shows the graphs of three functions. One is the position function of a car, one is the velocity of the car, and one is its acceleration. Identify each curve, and explain your choices.

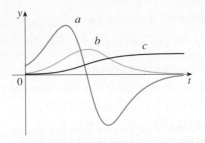

57. The figure shows graphs of f, f', and f''. Identify each curve, and explain your choices.

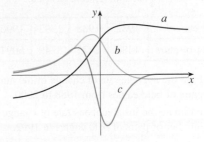

CHAPTER **2** REVIEW

■ CONCEPT CHECK

1. How is the average rate of change of a function on an interval defined? What are the units of the average rate of change? How can you interpret it geometrically?

2. Explain what $\lim_{x \to a} f(x) = L$ means.

3. State the following Limit Laws.
   (a) Sum Law   **(b)** Difference Law
   (c) Constant Multiple Law   **(d)** Product Law
   (e) Quotient Law   **(f)** Power Law

4. (a) What does it mean for f to be continuous at a?
   (b) What does it mean for f to be continuous on the interval $(-\infty, \infty)$? What can you say about the graph of such a function?

5. If we know that a function is continuous at $x = a$, what is the value of $\lim_{x \to a} f(x)$?

6. What is the difference between the limit $\lim_{x \to a} f(x)$ and the one-sided limits $\lim_{x \to a^+} f(x)$ and $\lim_{x \to a^-} f(x)$? What is the connection between them?

7. If $y = f(x)$ and x changes from x_1 to x_2, write expressions for the following.
   (a) The average rate of change of y with respect to x over the interval $[x_1, x_2]$.
   (b) The instantaneous rate of change of y with respect to x at $x = x_1$.

8. What is a tangent line? Write an expression for the slope of the tangent line to the curve $y = f(x)$ at the point $(a, f(a))$.

9. Give two ways we can estimate an instantaneous rate of change if its value cannot be determined precisely.

10. Define the derivative $f'(a)$. Discuss two ways of interpreting this number.

11. If $f(t)$ is the position function of a moving object, how can you interpret the derivative?

12. Given a graph of a function $y = f(x)$, how do you graph the derivative f'?

13. What does it mean for a function to be differentiable at a?

14. Describe several ways in which a function can fail to be differentiable. Illustrate with sketches.

15. Define the second derivative of f. If $f(t)$ is the position function of a particle, how can you interpret the second derivative?

16. (a) What does the sign of $f'(x)$ tell us about f?
   (b) What does the sign of $f''(x)$ tell us about f?

17. What does it mean to say that a curve is concave upward? Concave downward?

> **Answers to the Concept Check can be found on the back endpapers.**

■ Exercises

1–2 ■ Find the average rate of change of the function over the given interval.

1. $h(x) = \sqrt{3x + 5}$, $\quad [1, 4]$

2. $P(t) = 4.1e^{-0.5t}$, $\quad [2, 8]$

3. Revenue Let $R(t)$ be the monthly revenue (in thousands of dollars) of a restaurant t months after the restaurant opened. If $R(6) = 154.2$ and $R(9) = 179.7$, compute the average rate of change for $6 \le t \le 9$. What does your result mean in this context?

4. The table gives the number of pairs of bald eagles in the continental United States for various years, according to the US Fish and Wildlife Service.

Year	1982	1988	1992	1996	2000
Number of pairs	1480	2475	3749	5094	6471

(a) Find the average rate of change of the number of pairs of bald eagles from 1988 to 2000.

(b) Estimate the instantaneous rate of change in the number of pairs of bald eagles in 1992.

5. Use a table of values to estimate the value of $\lim\limits_{t \to 3} \dfrac{2^t - 8}{t - 3}$ accurate to three decimal places.

6. Estimate the value of $\lim\limits_{x \to 1} \dfrac{\sqrt{x} - 1}{x^2 - 1}$ by graphing the function and zooming in toward the point where the graph appears to cross the line $x = 1$.

7–14 ▪ Evaluate the limit.

7. $\lim\limits_{x \to 1} (5x^2 - 4x + 5)$

8. $\lim\limits_{x \to 3} \dfrac{x^2 - 9}{x^2 + 2x - 3}$

9. $\lim\limits_{x \to -3} \dfrac{x^2 - 9}{x^2 + 2x - 3}$

10. $\lim\limits_{t \to 2} \dfrac{t^2 - 4}{t^2 + 3t - 10}$

11. $\lim\limits_{t \to 0} 4e^{-2t}$

12. $\lim\limits_{b \to 1} (\ln b)^2$

13. $\lim\limits_{h \to 0} \dfrac{(h - 1)^3 + 1}{h}$

14. $\lim\limits_{x \to 1} \left(\dfrac{1}{x - 1} + \dfrac{1}{x^2 - 3x + 2} \right)$

---

15. The graph of f is given.

(a) Find each limit, or explain why it does not exist.

    (i) $\lim\limits_{x \to 2^+} f(x)$

    (ii) $\lim\limits_{x \to -3^+} f(x)$

    (iii) $\lim\limits_{x \to -3} f(x)$

    (iv) $\lim\limits_{x \to 4} f(x)$

    (v) $\lim\limits_{x \to 0} f(x)$

(b) At what numbers is f not continuous? Explain.

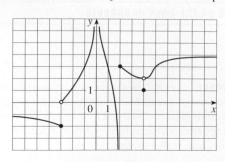

16. Find the instantaneous rate of change of $f(x) = x^2 + 2x$ at $x = 3$ using

(a) Definition 1 in Section 2.3.

(b) Definition 2 in Section 2.3.

17. **Motion**   The position, in meters, of an object moving along a straight path is given by $s = 1 + 2t + t^2/4$, where t is measured in seconds.

(a) Find the average speed over each time period.

    (i) $[1, 3]$ (ii) $[1, 2]$

    (iii) $[1, 1.5]$ (iv) $[1, 1.1]$

(b) Find the instantaneous speed when $t = 1$.

18. **Cost function**   The cost, in dollars, of producing x units of a particular product is $C(x) = 12{,}000 + 31x + 0.08x^2$.

(a) Find the average rate of change of C with respect to x when the production level is changed

    (i) from $x = 75$ to $x = 80$.

    (ii) from $x = 75$ to $x = 76$.

(b) Find the instantaneous rate of change of C with respect to x (the marginal cost) when $x = 75$.

19. Compute the derivative of $g(x) = 0.5x^2 + 4$ at $x = 3$.

20. Find the value of the derivative of $A(t) = 6t - 2t^2$ when $t = 1$.

21. For the function f whose graph is shown, arrange the following numbers in increasing order:

$$0 \qquad 1 \qquad f'(2) \qquad f'(3) \qquad f'(5)$$

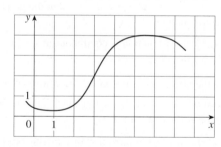

22. (a) Compute $f'(2)$, where $f(x) = x^3 - 2x$.

(b) Find an equation of the tangent line to the curve $y = x^3 - 2x$ at the point $(2, 4)$.

(c) Illustrate part (b) by graphing the curve and the tangent line on the same screen.

23. **Loan cost**   The total cost of repaying a student loan at an interest rate of $r\%$ per year is $C = f(r)$.

(a) What is the meaning of the derivative $f'(r)$? What are its units?

(b) What does the statement $f'(10) = 1200$ mean?

(c) Is $f'(r)$ always positive or does it change sign?

24. **Marketing costs**   Let C represent the cost, in dollars, a marketing company charges to prepare and mail out n thousand promotional brochures.

(a) What does dC/dn represent in this context?

(b) Interpret the statement

$$\left. \frac{dC}{dn} \right|_{n=5} = 250$$

25. (a) If $f(x) = e^{0.5x}$, estimate the value of $f'(1)$ in two ways:

   (i) By using Definition 5 in Section 2.3 and taking successively smaller values of h

   (ii) By zooming in on the graph of f and estimating the slope

(b) Find an approximate equation of the tangent line to the curve $y = e^{0.5x}$ at the point where $x = 1$.

26. Profit function   The table shows the amount of profit P, in thousands of dollars, a company estimates it will earn when x thousand units of its new product are shipped during a one-month period.

x (thousands)	P (thousands of dollars)
10	−121
20	−38
30	71
40	133
50	152
60	144

(a) Estimate the value of $P'(30)$ and interpret your result.

(b) Use a graphing calculator to find a quadratic function f that models the data. Then use the model to compute $f'(30)$. How does your answer compare to the estimate you found in part (a)?

27. Currency circulation   Let $C(t)$ be the total value of US currency (coins and bank notes) in circulation at time t. The table gives values of this function from 1980 to 2005, as of September 30, in billions of dollars. Estimate and interpret the value of $C'(2000)$.

t	1980	1985	1990	1995	2000	2005
$C(t)$	129.9	187.3	271.9	409.3	568.6	758.8

28. College enrollment   The table gives the enrollment $E(t)$ of a liberal arts college for various years t.

t	$E(t)$	t	$E(t)$
1986	5710	1998	7240
1990	6440	2002	6845
1994	6965	2006	7620

(a) What is the meaning of $E'(t)$? What are its units?

(b) Construct a table of values for $E'(t)$.

(c) Graph E and E'.

29. Youth in poverty   The graph shows the percentage P of children under the age of 18 in the United States living below the poverty level for various years.

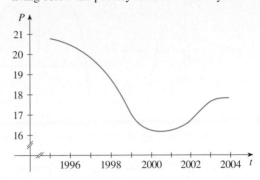

(a) Estimate the average rate of change of P from 1997 to 2001. What are the units?

(b) Estimate and interpret $P'(2002)$.

30. Fertility rate   The *total fertility rate* at time t, denoted by $F(t)$, is an estimate of the average number of children born to each woman (assuming that current birth rates remain constant). The graph of the total fertility rate in the United States shows the fluctuations from 1940 to 1990.

(a) Estimate the values of $F'(1950)$, $F'(1965)$, and $F'(1987)$.

(b) What are the meanings of these derivatives?

(c) Can you suggest reasons for the values of these derivatives?

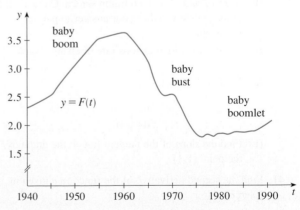

31. The graph of f is shown. State, with reasons, the numbers at which f is not differentiable.

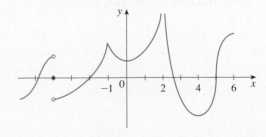

32–34 ▪ Trace or copy the graph of the function. Then sketch a graph of its derivative directly beneath.

32.

33.

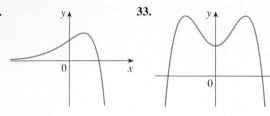

34.

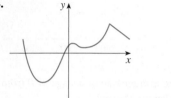

35–39 ▪ Find the derivative of the function (using the definition of derivative).

35. $f(x) = mx + b$ **36.** $P(t) = 5.2t - 3.8$

37. $g(x) = 2x^2 - 3x + 1$ **38.** $r(s) = 5/s$

39. $A(w) = \dfrac{w + 1}{2w - 1}$

40. (a) If $f(x) = 1/(x + 1)$, use the definition of a derivative to find $f'(x)$.

(b) Graph f and f' on a common screen. Compare the graphs to see whether your answer to part (a) is reasonable.

(c) What is the instantaneous rate of change when $x = 2$?

(d) Find the value of

$$\left. \frac{df}{dx} \right|_{x=4}$$

(e) Find the slope of the tangent line to the graph of f at the point $\left(1, \frac{1}{2}\right)$.

41. Find the second derivative of the function g given in Exercise 37.

42. Cost of living The cost of living continues to rise, but at a slower rate. In terms of a function and its derivatives, what does this statement mean?

43. Motion A car starts from rest and its distance traveled is recorded in the table in 2-second intervals.

t (seconds)	0	2	4	6	8	10	12	14
s (feet)	0	8	40	95	180	260	319	373

(a) Estimate the speed after 6 seconds.

(b) Estimate the coordinates of the inflection point of the graph of the position function.

(c) What is the significance of the inflection point?

44. Motion The position function of a toy car traveling along a straight path is $s(t) = 0.05t^3 - 0.8t + 2$, $0 \le t \le 10$, where t is the time in seconds and $s(t)$ is measured in feet.

(a) Find an equation for the velocity of the toy car.

(b) How fast is the car moving after 8 seconds?

(c) Find an equation for the acceleration of the car.

45. The graph of the *derivative* f' of a function f is given.

(a) On what intervals is f increasing or decreasing?

(b) At what values of x does f have a local maximum or minimum?

(c) Where is f concave upward or downward?

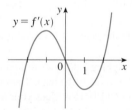

46. The figure shows the graphs of f, f', and f''. Identify each curve, and explain your choices.

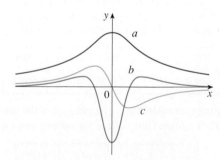

3

For a roller coaster ride to be smooth, the straight stretches of the track need to be connected to the curved segments so that there are no abrupt changes in direction. In the project on page 145 you will see how to design the first ascent and drop of a new coaster for a smooth ride.   © Brett Mulcahy/ Shutterstock

Techniques of Differentiation

In Chapter 2 we learned the importance of derivatives. We learned that the derivative of a function gives us instantaneous rates of change as well as the slope of the function's graph at a point. We learned how to estimate and graph derivatives from tables of values and graphs. When we have a formula for a function, the definition of a derivative allows us to compute a formula for the derivative. But it would be tedious if we always had to use the definition, especially with more complicated functions. So in this chapter we develop rules for finding derivatives without having to use the definition directly. We then use these rules to solve a variety of problems involving rates of change.

3.1   Shortcuts to Finding Derivatives

In this section we learn how to differentiate constant functions, power functions, polynomials, and exponential functions.

Let's start with the simplest of all functions, the constant function $f(x) = c$. The graph of this function is the horizontal line $y = c$, which has slope 0 at every point, so we must have $f'(x) = 0$ at every value of x. (See Figure 1.) This can also be established directly from the definition of a derivative:

$$f'(x) = \lim_{h \to 0} \frac{f(x + h) - f(x)}{h} = \lim_{h \to 0} \frac{c - c}{h}$$

$$= \lim_{h \to 0} 0 = 0$$

In Leibniz notation, we write this rule as follows.

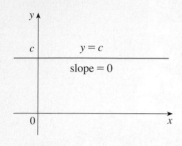

FIGURE 1
The graph of $f(x) = c$ is the line $y = c$, so $f'(x) = 0$.

■ **Derivative of a Constant Function**

$$\frac{d}{dx}(c) = 0$$

■ | Power Functions

We next look at the functions $f(x) = x^n$, where n is a positive integer. If $n = 1$, the graph of $f(x) = x$ is the line $y = x$, which has slope 1 everywhere (see Figure 2). So

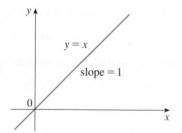

FIGURE 2
The graph of $f(x) = x$ is the line $y = x$, so $f'(x) = 1$.

(1)
$$\frac{d}{dx}(x) = 1$$

(You can also verify Equation 1 from the definition of a derivative.) In Exercise 15 of Section 2.4 we found that

(2)
$$\frac{d}{dx}(x^2) = 2x$$

If $f(x) = x^3$, then

$$f'(x) = \lim_{h \to 0} \frac{f(x + h) - f(x)}{h} = \lim_{h \to 0} \frac{(x + h)^3 - x^3}{h}$$

$$= \lim_{h \to 0} \frac{x^3 + 3x^2 h + 3xh^2 + h^3 - x^3}{h}$$

$$= \lim_{h \to 0} \frac{3x^2 h + 3xh^2 + h^3}{h} = \lim_{h \to 0} (3x^2 + 3xh + h^2) = 3x^2$$

(See Exercise 16 in Section 2.4.) Thus

(3)
$$\frac{d}{dx}(x^3) = 3x^2$$

Similarly, if we use the definition of a derivative to compute the derivative of $f(x) = x^4$, we arrive at

(4)
$$\frac{d}{dx}(x^4) = 4x^3$$

Comparing Equations 1, 2, 3, and 4, we see a pattern emerging. It seems to be a reasonable guess that, when n is a positive integer, $(d/dx)(x^n) = nx^{n-1}$. This turns out to be true. A proof of this fact is included at the end of the section.

▪ **The Power Rule**   If n is a positive integer, then

$$\frac{d}{dx}(x^n) = nx^{n-1}$$

We illustrate the Power Rule using various notations in Example 1.

▪ **EXAMPLE 1   Using the Power Rule to Find Derivative Formulas**

(a) If $f(x) = x^6$, then $f'(x) = 6x^5$.

(b) If $y = x^{1000}$, then $y' = 1000x^{999}$.

(c) If $P = t^4$, then $\dfrac{dP}{dt} = 4t^3$.

(d) $\dfrac{d}{dr}(r^3) = 3r^2$ ▪

What about power functions with negative integer exponents? In Exercise 36 of Section 2.4, we asked you to use the definition of a derivative to establish that

$$\frac{d}{dx}\left(\frac{1}{x}\right) = -\frac{1}{x^2}$$

We can rewrite this equation as

$$\frac{d}{dx}(x^{-1}) = (-1)x^{-2}$$

and so the Power Rule is true when $n = -1$. In fact, it can be shown that it holds for all negative integers.

What if the exponent is a fraction? In Exercise 25 of Section 2.4 you were asked to show, using the definition of derivative, that

$$\frac{d}{dx}\sqrt{x} = \frac{1}{2\sqrt{x}}$$

which can be written as

$$\frac{d}{dx}(x^{1/2}) = \tfrac{1}{2}x^{-1/2}$$

This shows that the Power Rule is true even when $n = \frac{1}{2}$. In fact, it is true for all real numbers n, as we will prove in Section 3.5.

▪ **The Power Rule (General Version)**   If n is any real number, then

$$\frac{d}{dx}(x^n) = nx^{n-1}$$

The Power Rule enables us to compute easily the derivative of any power of x, any root, and any reciprocal of a power or root.

▪ EXAMPLE 2

The Power Rule for Negative and Fractional Exponents

Differentiate: **(a)** $f(x) = \dfrac{1}{x^2}$ **(b)** $y = \sqrt[3]{x}$ **(c)** $A(t) = \dfrac{1}{\sqrt{t}}$

SOLUTION

(a) First rewrite the function as a power of x: $f(x) = x^{-2}$. Then use the Power Rule with $n = -2$:

$$f'(x) = \frac{d}{dx}(x^{-2}) = -2x^{-2-1} = -2x^{-3} = -\frac{2}{x^3}$$

Figure 3 shows the function y in Example 2(b) and its derivative y'. Notice that y has a vertical tangent line at $x = 0$ and so is not differentiable there (y' becomes larger and larger as x approaches 0). Observe that y' is positive everywhere because y is an increasing function.

(b) Because $\sqrt[3]{x}$ can be written as $x^{1/3}$, we have

$$\frac{dy}{dx} = \frac{d}{dx}\left(\sqrt[3]{x}\right) = \frac{d}{dx}(x^{1/3}) = \tfrac{1}{3}x^{(1/3)-1} = \tfrac{1}{3}x^{-2/3}$$

or, equivalently, $1/(3x^{2/3})$.

(c) First we write $A(t)$ as $1/t^{1/2}$ or $t^{-1/2}$. Then

$$A'(t) = \frac{d}{dt}(t^{-1/2}) = -\frac{1}{2}t^{(-1/2)-1} = -\frac{1}{2}t^{-3/2} = -\frac{1}{2t^{3/2}}$$ ▪

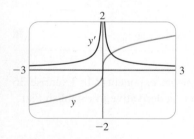

FIGURE 3
$y = \sqrt[3]{x}$

▪ New Derivatives from Old

When new functions are formed from old functions by addition, subtraction, or multiplication by a constant, their derivatives can be calculated in terms of derivatives of the old functions. In particular, the following formula says that *the derivative of a constant times a function is the constant times the derivative of the function.*

▪ **The Constant Multiple Rule**   If c is a constant and f is a differentiable function, then

$$\frac{d}{dx}[cf(x)] = c\frac{d}{dx}f(x)$$

PROOF   Let $g(x) = cf(x)$. Then

$$g'(x) = \lim_{h \to 0} \frac{g(x+h) - g(x)}{h} = \lim_{h \to 0} \frac{cf(x+h) - cf(x)}{h}$$

$$= \lim_{h \to 0} c \left[\frac{f(x+h) - f(x)}{h} \right]$$

$$= c \lim_{h \to 0} \frac{f(x+h) - f(x)}{h} \qquad \text{(by Limit Law 3)}$$

$$= cf'(x) \qquad \blacksquare$$

Geometric Interpretation of the Constant Multiple Rule

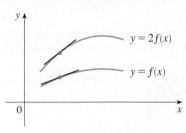

Multiplying by $c = 2$ stretches the graph vertically by a factor of 2. All the rises have been doubled but the runs stay the same. So the slopes are doubled too.

Intuitively, it is not too hard to see why the Constant Multiple Rule is true. For instance, if we double the height of a graph, the slopes of the graph are in turn doubled. If we triple the output values of a function, we also triple the rate of change.

■ **EXAMPLE 3   Using the Constant Multiple Rule**

(a) $\dfrac{d}{dx}(3x^4) = 3\dfrac{d}{dx}(x^4) = 3(4x^3) = 12x^3$

(b) $\dfrac{d}{dx}(-x) = \dfrac{d}{dx}[(-1)x] = (-1)\dfrac{d}{dx}(x) = -1(1) = -1$ ■

The derivative shortcuts we have learned allow us to find the slopes of tangent lines without having to resort to the definition of a derivative, as the next example demonstrates.

■ **EXAMPLE 4   Equation of a Tangent Line**

Find an equation of the tangent line to the curve $y = 0.2x^3$ at the point $(2, 1.6)$. Illustrate by graphing the curve and the tangent line.

SOLUTION
The derivative of $f(x) = 0.2x^3$ is

$$f'(x) = 0.2 \frac{d}{dx}(x^3) = 0.2(3x^2) = 0.6x^2$$

So the slope of the tangent line at $(2, 1.6)$ is $f'(2) = 0.6(2^2) = 2.4$. Therefore an equation of the tangent line is

$$y - 1.6 = 2.4(x - 2) \qquad \text{or} \qquad y = 2.4x - 3.2$$

We graph the curve and its tangent line in Figure 4.   ■

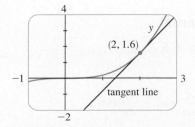

FIGURE 4

The next rule tells us that *the derivative of a sum of functions is the sum of the derivatives.*

Using prime notation, we can write the Sum Rule as

$$(f + g)' = f' + g'$$

▪ **The Sum Rule**   If f and g are both differentiable, then

$$\frac{d}{dx}[f(x) + g(x)] = \frac{d}{dx}f(x) + \frac{d}{dx}g(x)$$

PROOF Let $F(x) = f(x) + g(x)$. Then

$$
\begin{aligned}
F'(x) &= \lim_{h \to 0} \frac{F(x + h) - F(x)}{h} \\[2mm]
&= \lim_{h \to 0} \frac{[f(x + h) + g(x + h)] - [f(x) + g(x)]}{h} \\[2mm]
&= \lim_{h \to 0} \left[\frac{f(x + h) - f(x)}{h} + \frac{g(x + h) - g(x)}{h} \right] \\[2mm]
&= \lim_{h \to 0} \frac{f(x + h) - f(x)}{h} + \lim_{h \to 0} \frac{g(x + h) - g(x)}{h} \qquad \text{(by Limit Law 1)}\\[2mm]
&= f'(x) + g'(x) \hspace{6cm} \blacksquare
\end{aligned}
$$

The Sum Rule can be extended to the sum of any number of functions. For instance, using this rule twice, we get

$$(f + g + h)' = [(f + g) + h]' = (f + g)' + h' = f' + g' + h'$$

By writing $f - g$ as $f + (-1)g$ and applying the Sum Rule and the Constant Multiple Rule, we get the following formula.

▪ **The Difference Rule**   If f and g are both differentiable, then

$$\frac{d}{dx}[f(x) - g(x)] = \frac{d}{dx}f(x) - \frac{d}{dx}g(x)$$

The Constant Multiple Rule, the Sum Rule, and the Difference Rule can be combined with the Power Rule to differentiate any polynomial, as the following examples demonstrate.

▪ **EXAMPLE 5** **Differentiating a Polynomial**

Find the derivative of $g(w) = 12w^5 - 10w^3 - 6w + 5$

SOLUTION

$$
\begin{aligned}
g'(w) &= \frac{d}{dw}(12w^5) - \frac{d}{dw}(10w^3) - \frac{d}{dw}(6w) + \frac{d}{dw}(5) \\[2mm]
&= 12\frac{d}{dw}(w^5) - 10\frac{d}{dw}(w^3) - 6\frac{d}{dw}(w) + \frac{d}{dw}(5) \\[2mm]
&= 12(5w^4) - 10(3w^2) - 6(1) + 0 \\[2mm]
&= 60w^4 - 30w^2 - 6 \hspace{5cm} \blacksquare
\end{aligned}
$$

■ **EXAMPLE 6**   **The Derivative of a Polynomial Model**

The estimated percentage of adults in the state of Arizona who report having no health insurance (according to figures published by the Centers for Disease Control) can be modeled by the function

$$P(t) = -0.197t^3 + 5.54t^2 - 50.2t + 161.4 \qquad 6 \leqslant t \leqslant 12$$

where $t = 0$ corresponds to the year 1990. Use the model to find the instantaneous rate of change of the percentage in 1999.

SOLUTION

First we compute the derivative of P:

$$P'(t) = -0.197(3t^2) + 5.54(2t) - 50.2(1) + 0$$
$$= -0.591t^2 + 11.08t - 50.2$$

The year 1999 corresponds to $t = 9$, so the instantaneous rate of change is

$$P'(9) = -0.591(9^2) + 11.08(9) - 50.2 = 1.649$$

percentage points per year. Thus the percentage of adults in Arizona who report that they do not have health insurance was increasing at a rate of about 1.65 percentage points per year in 1999.   ■

■ **EXAMPLE 7**

Finding Acceleration from a Polynomial Equation of Motion

The equation of motion of a moving object is $s = 2t^3 - 5t^2 + 3t + 4$, where s is measured in centimeters and t in seconds. Find the acceleration as a function of time. What is the acceleration after 2 seconds?

SOLUTION

The velocity and acceleration are

$$v(t) = \frac{ds}{dt} = 2(3t^2) - 5(2t) + 3(1) + 0 = 6t^2 - 10t + 3$$

$$a(t) = \frac{dv}{dt} = 6(2t) - 10(1) + 0 = 12t - 10$$

The acceleration after 2 seconds is $a(2) = 14$ cm/s per second, or cm/s^2.   ■

■ **EXAMPLE 8**

Identifying Points on a Curve with Horizontal Tangent Lines

Find the points on the curve $y = x^4 - 6x^2 + 4$ where the tangent line is horizontal.

SOLUTION

Horizontal tangents occur where the derivative is zero. We have

$$\frac{dy}{dx} = \frac{d}{dx}(x^4) - 6\frac{d}{dx}(x^2) + \frac{d}{dx}(4) = 4x^3 - 12x + 0 = 4x(x^2 - 3)$$

Thus $dy/dx = 0$ if $x = 0$ or $x^2 - 3 = 0$, that is, $x = \pm\sqrt{3}$. So the given curve has horizontal tangents when $x = 0$, $\sqrt{3}$, and $-\sqrt{3}$. The corresponding points are $(0, 4)$, $(\sqrt{3}, -5)$, and $(-\sqrt{3}, -5)$. (See Figure 5.)

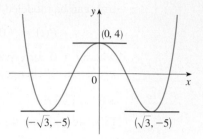

FIGURE 5
The curve $y = x^4 - 6x^2 + 4$ and
its horizontal tangents

▪ Exponential Functions

Let's try to compute the derivative of the exponential function $f(x) = a^x$ using the definition of a derivative:

$$f'(x) = \lim_{h \to 0} \frac{f(x + h) - f(x)}{h} = \lim_{h \to 0} \frac{a^{x+h} - a^x}{h}$$

$$= \lim_{h \to 0} \frac{a^x a^h - a^x}{h} = \lim_{h \to 0} \frac{a^x(a^h - 1)}{h}$$

The factor a^x doesn't depend on h, so we can treat it as a constant and take it in front of the limit:

$$f'(x) = a^x \lim_{h \to 0} \frac{a^h - 1}{h}$$

The resulting limit is not an easy one to evaluate, but notice that the limit is the value of the derivative of f at 0, that is,

$$\lim_{h \to 0} \frac{a^h - 1}{h} = \lim_{h \to 0} \frac{f(0 + h) - f(0)}{h} = f'(0)$$

Therefore we have shown that if the exponential function $f(x) = a^x$ is differentiable at 0, then for any value of x,

(5) $$f'(x) = f'(0)a^x$$

This equation says that *the rate of change of any exponential function is proportional to the function itself.* (The slope is proportional to the height.) You can see in Figure 6 that the slope of the graph of $y = 2^x$ is small to the left of the y-axis, where the function values are also small, and the slope increases along with the function values as we move to the right.

In Example 9 of Section 2.3 we estimated that $f'(0) \approx 0.69$ for $f(x) = 2^x$. Similarly, if $g(x) = 3^x$, then

$$g'(0) = \lim_{h \to 0} \frac{3^h - 1}{h}$$

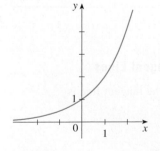

FIGURE 6
$y = 2^x$

h	$\dfrac{3^h - 1}{h}$
0.1	1.1612
0.01	1.1047
0.001	1.0992
0.0001	1.0987

and the table at the left suggests that $g'(0)$ exists and is approximately equal to 1.10. Thus it appears that

$$\text{for } f(x) = 2^x, \qquad f'(0) = \lim_{h \to 0} \frac{2^h - 1}{h} \approx 0.69$$

$$\text{for } f(x) = 3^x, \qquad f'(0) = \lim_{h \to 0} \frac{3^h - 1}{h} \approx 1.10$$

In fact, it can be proved that these limits exist and, correct to six decimal places, the values are

$$\left. \frac{d}{dx}(2^x) \right|_{x=0} \approx 0.693147 \qquad \left. \frac{d}{dx}(3^x) \right|_{x=0} \approx 1.098612$$

Thus from Equation 5 we have

(6) $$\frac{d}{dx}(2^x) \approx (0.69)2^x \qquad \frac{d}{dx}(3^x) \approx (1.10)3^x$$

We will determine precise formulas for these derivatives in Section 3.4. However, we can establish one particular case now: Of all possible choices for the base a in Equation 5, the simplest differentiation formula occurs when $f'(0) = 1$. In view of the estimates of $f'(0)$ for $a = 2$ and $a = 3$, it seems reasonable that there is a number a between 2 and 3 for which $f'(0) = 1$. It is traditional to denote this value by the letter e. (In fact, that is how we introduced e in Section 1.5.) Thus we have the following definition.

▪ **Definition of the Number e**

e is the number such that $\displaystyle \lim_{h \to 0} \frac{e^h - 1}{h} = 1$

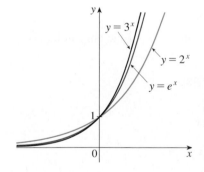

FIGURE 7

We previously saw that e lies between 2 and 3 and the graph of $f(x) = e^x$ lies between the graphs of $y = 2^x$ and $y = 3^x$. (See Figure 7.) Later we will be able to show that, correct to five decimal places,

$$e \approx 2.71828$$

The differentiation formula in Equation 5 achieves its simplest form when the base of the exponential function is e, so that $f'(0) = 1$. This gives the following important differentiation formula.

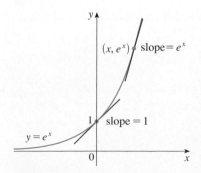

FIGURE 8

▪ **Derivative of the Natural Exponential Function**

$$\frac{d}{dx}(e^x) = e^x$$

TEC Visual 3.1 uses the slope-a-scope to illustrate this formula.

Thus the exponential function $f(x) = e^x$ has the property that *it is its own derivative*. The geometrical significance of this fact is that the slope of a tangent line to the curve $y = e^x$ at a point is equal to the y-coordinate of that point (see Figure 8).

▪ **EXAMPLE 9**

A Derivative Involving the Natural Exponential Function

If $f(x) = 2e^x - x$, find f' and f''.

SOLUTION

Using the Difference and Constant Multiple Rules, we have

$$f'(x) = \frac{d}{dx}(2e^x - x) = 2 \cdot \frac{d}{dx}(e^x) - \frac{d}{dx}(x) = 2e^x - 1$$

In Section 2.4 we defined the second derivative as the derivative of f', so

$$f''(x) = \frac{d}{dx}(2e^x - 1) = 2 \cdot \frac{d}{dx}(e^x) - \frac{d}{dx}(1) = 2e^x$$

The graphs of f and f' are shown in Figure 9.   ▪

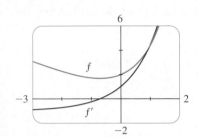

FIGURE 9

▪ **EXAMPLE 10**   **Tangent Lines to the Natural Exponential Function**

What is the slope of the tangent line to the curve $y = e^x$ when $x = 2$? At what point on the curve does the tangent line have slope 2?

SOLUTION

Since $y = e^x$, we have $dy/dx = e^x$. When $x = 2$, the slope of the tangent line is

$$\left.\frac{dy}{dx}\right|_{x=2} = e^2 \approx 7.389$$

The tangent line has slope 2 when $dy/dx = 2$, or

$$e^x = 2 \qquad \Longleftrightarrow \qquad x = \ln 2$$

Therefore the required point is $(\ln 2, 2)$, or approximately $(0.693, 2)$. (See Figure 10.)   ▪

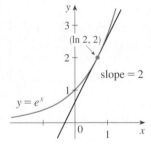

FIGURE 10

▪ | **Proof of the Power Rule**

We must prove that if n is a positive integer, then

$$\frac{d}{dx}(x^n) = nx^{n-1}$$

If $f(x) = x^n$, then

$$f'(x) = \lim_{h \to 0} \frac{f(x + h) - f(x)}{h} = \lim_{h \to 0} \frac{(x + h)^n - x^n}{h}$$

In finding the derivative of x^3 we had to expand $(x + h)^3$. Here we need to expand

The Binomial Theorem is given on Reference Page 1.

$(x + h)^n$ and we use the Binomial Theorem to do so:

$$f'(x) = \lim_{h \to 0} \frac{\left[x^n + nx^{n-1}h + \dfrac{n(n-1)}{2}x^{n-2}h^2 + \cdots + nxh^{n-1} + h^n \right] - x^n}{h}$$

$$= \lim_{h \to 0} \frac{nx^{n-1}h + \dfrac{n(n-1)}{2}x^{n-2}h^2 + \cdots + nxh^{n-1} + h^n}{h}$$

$$= \lim_{h \to 0} \left[nx^{n-1} + \frac{n(n-1)}{2}x^{n-2}h + \cdots + nxh^{n-2} + h^{n-1} \right] = nx^{n-1}$$

because every term except the first has h as a factor and therefore approaches 0. ■

■ Prepare Yourself

1. Write each function in the form of a power function.

(a) $f(x) = \sqrt[3]{x}$

(b) $g(w) = 1/w$

(c) $A(t) = 4/\sqrt{t}$

(d) $B(v) = 8/v^3$

(e) $y = \sqrt[4]{x^3}$

2. If $f(3) = 8$ and $f'(3) = -2$, write an equation of the tangent line to the graph of f at $x = 3$.

■ Exercises 3.1

1–38 ■ Find the derivative of the function.

1. $f(x) = 186.5$

2. $f(x) = \sqrt{30}$

3. $y = x^9$

4. $w(r) = r^{50}$

5. $g(t) = t^{7/2}$

6. $h(v) = v^{-3}$

7. $y = x^{-2/5}$

8. $y = x^{-1/3}$

9. $L(t) = \sqrt[4]{t}$

10. $R(a) = 1/a^4$

11. $y = 8x^6$

12. $y = 2.1x^5$

13. $f(x) = 7/x^2$

14. $R(t) = 5t^{-3/5}$

15. $y = 5x - 3$

16. $y = \frac{1}{2}x + \frac{7}{2}$

17. $f(x) = x^3 - 4x + 6$

18. $f(t) = \frac{1}{2}t^6 - 3t^4 + t$

19. $y = 0.7x^4 - 1.8x^3 + 5.1x$

20. $P = -3.5x^3 + 4.9x^2 + 4.1x - 8.5$

21. $q = e^r + 3.4$

22. $h(t) = 8t^2 + 2t - e^t$

23. $G(x) = x^3 - 4e^x$

24. $y = 5e^x + 3$

25. $f(t) = \frac{1}{4}(t^4 + 8)$

26. $h(x) = (x - 2)(2x + 3)$
[*Hint*: First expand the expression.]

27. $f(q) = \dfrac{6}{q} - \dfrac{3}{q^2}$

28. $A(t) = \dfrac{2}{\sqrt{t}} + \dfrac{3}{t^{2/3}}$

29. $y = x\left(\sqrt{x} + 1/\sqrt{x}\right)$

30. $y = \sqrt{x}\,(x - 1)$

31. $F(x) = \left(\frac{1}{2}x\right)^5$

32. $R(x) = \dfrac{\sqrt{10}}{x^7}$

33. $y = \dfrac{7x^2 - 3x + 5}{x}$ [*Hint*: First rewrite the expression as three separate fractions.]

34. $y = \dfrac{x^2 - 2\sqrt{x}}{x}$

35. $f(y) = \dfrac{A}{y^{10}} + Be^y$

36. $G(v) = ae^v + \dfrac{b}{v} + \dfrac{c}{v^2}$

37. $v = t^2 - \dfrac{1}{\sqrt[4]{t^3}}$

38. $u = \sqrt[3]{t^2} + 2\sqrt{t^3}$

39. Find the slope of the curve $y = 2x^2 + 3/x$ at the point $(3, 19)$.

40. Find the slope of the graph of $g(t) = 7.6\sqrt{t} + 3.9t^2$ when $t = 12$. (Round to two decimal places.)

41–42 ■ Find an equation of the tangent line to the curve at the given point.

41. $y = x^4 + 2e^x$, $(0, 2)$

42. $y = (1 + 2x)^2$, $(1, 9)$

43–44 ▪ Find an equation of the tangent line to the curve at the given point. Illustrate by graphing the curve and the tangent line on the same screen.

43. $y = x + \sqrt{x}$, (1, 2)   **44.** $y = 3x^2 - x^3$, (1, 2)

45. Price of oil   The average worldwide crude oil price, in dollars per barrel, during February, March, and April of 2006 can be modeled by the function $p(t) = 0.12t^2 - 1.6t + 59.5$, where t is the number of weeks after January 1, 2006. Use the model to find the instantaneous rate of change in the crude oil price after the 14th week of the year. What are the units?

46. Birth rate   The number of births per 1000 population in Italy for the years 1970–2010 can be modeled by

$$B(t) = -0.00045t^3 + 0.36t^2 - t + 19$$

where t is the year and $t = 0$ corresponds to 1970. Use the model to find the instantaneous rate of change of Italy's birth rate in 2001.

47. Drag racing   In 1988, John Force set a world record time for a "funny car" in a quarter-mile drag race. (A funny car is a dragster modified to look similar to a standard automobile. The record has since been broken.) The table gives the times to reach various distances during the race, as recorded by the National Hot Rod Association.

Time (seconds)	Distance (feet)	Time (seconds)	Distance (feet)
0	0	4.091	1000
0.885	60	4.646	1254
2.277	330	4.788	1320
3.266	660		

(a) Use a graphing calculator to find a cubic model for the data.

(b) Use the model to find the speed of the car after 4 s.

48. Country population   Midyear estimates of the population of Portugal are given for various years in the table.

Year	Population (millions)	Year	Population (millions)
1980	9.78	2000	10.34
1985	9.90	2005	10.57
1990	9.92	2010	10.74
1995	10.07		

(a) Use a graphing calculator to find a quadratic model for the data.

(b) Use the model to predict the population of Portugal in 2017.

(c) Use the model to find the instantaneous rate of change of the population in 2002.

49–50 ▪ Estimate the value of $f'(a)$ by zooming in on the graph of f. Then differentiate f to find the exact value of $f'(a)$ and compare with your estimate.

49. $f(x) = 3x^2 - x^3$, $a = 1$

50. $f(x) = 1/\sqrt{x}$, $a = 4$

51–54 ▪ Find $f'(x)$. Compare the graphs of f and f' and use them to explain why your answer is reasonable.

51. $f(x) = e^x - 5x$

52. $f(x) = 3x^5 - 20x^3 + 50x$

53. $f(x) = x - 3x^{1/3}$ **54.** $f(x) = x + \dfrac{1}{x}$

55. (a) Use a graphing calculator or computer to graph the function $f(x) = x^4 - 3x^3 - 6x^2 + 7x + 30$ in the viewing rectangle $[-3, 5]$ by $[-10, 50]$.

(b) Using the graph in part (a) to estimate slopes, make a rough sketch, by hand, of the graph of f'. (See Example 1 in Section 2.4.)

(c) Calculate $f'(x)$ and use this expression, with a graphing device, to graph f'. Compare with your sketch in part (b).

56. (a) Use a graphing calculator or computer to graph the function $g(x) = e^x - 3x^2$ in the viewing rectangle $[-1, 4]$ by $[-8, 8]$.

(b) Using the graph in part (a) to estimate slopes, make a rough sketch, by hand, of the graph of g'. (See Example 1 in Section 2.4.)

(c) Calculate $g'(x)$ and use this expression, with a graphing device, to graph g'. Compare with your sketch in part (b).

57–58 ▪ Find the first and second derivatives of the function.

57. $f(x) = x^4 - 3x^3 + 16x$ **58.** $G(r) = \sqrt{r} + \sqrt[3]{r}$

59–60 ▪ Find the first and second derivatives of the function. Check to see that your answers are reasonable by comparing the graphs of f, f', and f''.

59. $f(x) = 0.2x^4 - 1.1x^3 + 0.6x^2 + 2.2x$

60. $f(x) = e^x - x^3$

61. Motion   The equation of motion of a moving object is $s = 4.2t^3 + 3.4t^2 - 6t + 2.5$, where s is measured in feet and t is the time in seconds.

(a) Compute the velocity after 4 seconds.

(b) What is the acceleration after 6 seconds?

62. Motion   The equation of motion of a blue particle is $s = 2t^2 + t^{3/2}$ and the equation of motion of a red particle is $s = 2.4t^2 + 0.2t$, where t is the time in minutes.

(a) Which particle has traveled farther after 3 minutes?

(b) Which particle is moving faster after 3 minutes?

63. Motion   The equation of motion of a particle is $s = t^3 - 3t$, where s is in meters and t is in seconds. Find

(a) the velocity and acceleration as functions of t,

(b) the acceleration after 2 seconds, and

(c) the acceleration when the velocity is 0.

64. Motion   The equation of motion of a particle is $s = 2t^3 - 7t^2 + 4t + 1$, where s is in meters and t is in seconds.

(a) Find the velocity and acceleration as functions of t.

(b) Find the acceleration after 1 second.

(c) Graph the position, velocity, and acceleration functions on the same screen. Comment on the relationships between the graphs.

65. Find the points on the curve $y = 2x^3 + 3x^2 - 12x + 1$ where the tangent is horizontal.

66. For what values of x does the graph of $f(x) = x^3 + 3x^2 + x + 3$ have a horizontal tangent?

■ Challenge Yourself

67. Show that the curve $y = 6x^3 + 5x - 3$ has no tangent line with slope 4.

68. Find the nth derivative of each function by calculating the first few derivatives and observing the pattern that occurs.

(a) $f(x) = x^n$ (b) $f(x) = 1/x$

69. For what values of a and b is the line $2x + y = b$ tangent to the parabola $y = ax^2$ when $x = 2$?

70. Find a parabola with equation $y = ax^2 + bx + c$ that has slope 4 at $x = 1$, slope -8 at $x = -1$, and passes through the point (2, 15).

P R O J E C T   ■ Building a Better Roller Coaster

Suppose you are asked to design the first ascent and drop for a new roller coaster. After studying photographs of your favorite coasters, you decide to make the slope of the ascent 0.8 and the slope of the drop -1.6. You decide to connect these two straight stretches $y = L_1(x)$ and $y = L_2(x)$ with part of a parabola $y = f(x) = ax^2 + bx + c$, where x and $f(x)$ are measured in feet.

For the track to be smooth there can't be abrupt changes in direction, so you want the linear segments L_1 and L_2 to be tangent to the parabola at the transition points P and Q. (See the figure.) To simplify the equations, you decide to place the origin at P.

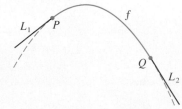

1. Suppose the horizontal distance between P and Q is 100 ft. Write equations in a, b, and c that will ensure that the track is smooth at the transition points.

2. Solve the equations in Problem 1 for a, b, and c to find a formula for $f(x)$.

3. Plot L_1, f, and L_2 to verify graphically that the transitions are smooth.

4. Find the difference in elevation between P and Q.

3.2 Introduction to Marginal Analysis

From an economist's point of view, the goal of a company that produces goods or services is to maximize its profit. To this end, companies are faced with many decisions, such as how many goods to manufacture, how many employees to hire, or how to set the price of goods. Here we examine some of these issues. Specifically, we look at models of cost and revenue for companies. Remember, a model is a simplified representation. There are far too many complex components of a business to reduce thorough analysis to a few concepts and equations. But appropriate models will still be accurate enough to allow us to make valuable conclusions.

■ Average Cost and Marginal Cost

All companies incur costs as they engage in business. Understanding the costs that go into producing a certain number of goods helps companies decide pricing and production levels. There are normally **fixed costs**, which a company pays regardless of whether any goods are produced or not, such as rent on buildings or the cost of machinery. In addition, companies incur **variable costs**, which change with the number of units produced. Variable costs include the cost of raw materials, the salaries of workers, and so on. The sum of these costs gives the **total cost**.

Suppose $C(q)$ is the total cost that a company incurs in producing q units of a particular good. (Think q for *quantity*.) The function C is called a **cost function**. It is often appropriate to represent a total cost function by a quadratic function

$$C(q) = a + bq + cq^2$$

or a cubic function

$$C(q) = a + bq + cq^2 + dq^3$$

Here a represents the fixed costs that don't depend on q and the other terms represent the variable costs. (The cost of raw materials may be proportional to q, but labor costs might depend partly on higher powers of q because of overtime costs and inefficiencies involved in large-scale operations.)

We expect a cost function to be an increasing function, because the cost increases as the number of units produced increases. But how quickly does it increase? Typically, a cost function does not increase linearly, but rather increases more slowly as more units are produced. Increasing the number of units normally means that a company can produce the items more efficiently; workers can specialize and raw materials can be purchased in bulk for better prices, for instance. Economists call this *economies of scale*. However, if the production level is further increased, at some point these benefits are outweighed by other factors such as the inherent inefficiencies of large organizations, additional employees or facilities, or higher costs of resources that are in dwindling supply.

It is helpful for a company to know the average cost of each unit produced. (This information can help a company decide at what price to sell a new product, for example.) The average cost of each unit is easily computed by dividing the total cost by the number of units produced.

▪ **Average Cost**   If $C(q)$ is the total cost of producing q units of a good or service, then the **average cost** per unit is

$$\frac{C(q)}{q}$$

If only a few units are produced, the average cost will be high because the fixed costs are distributed among a small number of items. As production increases, we expect the average cost to decline (the fixed costs are divided among many items). If the average cost declines with additional units, then increasing production will give the company the flexibility to sell the products at a lower price.

If we want to examine how cost changes as production increases without factoring the fixed costs in, then we can look at the rate of change of the cost function. The instantaneous rate of change of cost with respect to q, the number of items produced, is called the *marginal cost*.

▪ **Marginal Cost**   If $C(q)$ is the total cost of producing q units of a good or service, then the **marginal cost** is

$$C'(q) = \frac{dC}{dq}$$

which is the instantaneous rate of change of cost with respect to the number of units produced.

If cost is measured in dollars, the units of marginal cost are dollars per unit. Thus marginal cost tells us the immediate effect on cost if production is increased.

▪ **EXAMPLE 1   Average Cost and Marginal Cost**

Suppose a company has estimated that the cost, in dollars, of producing q items per week is $C(q) = 3000 + 13q - 0.01q^2 + 0.000003q^3$.

(a) What are the fixed costs?

(b) Find a function for the average cost of each unit produced. What is the average cost when 1500 items are produced?

(c) Find the marginal cost function. What is the marginal cost when 1500 units are produced?

(d) What is the actual cost of the 1501st item?

SOLUTION

(a) The fixed costs correspond to the constant term in the cost function. In this case, the fixed weekly costs are $3000.

(b) The average cost is

$$\frac{C(q)}{q} = \frac{3000 + 13q - 0.01q^2 + 0.000003q^3}{q}$$

$$= \frac{3000}{q} + 13 - 0.01q + 0.000003q^2$$

dollars per unit. When 1500 units are produced, the average cost is

$$\frac{C(1500)}{1500} = 6.75$$

dollars per unit.

(c) The marginal cost is

$$C'(q) = 0 + 13(1) - 0.01(2q) + 0.000003(3q^2)$$
$$= 13 - 0.02q + 0.000009q^2$$

When 1500 units are produced, the marginal cost is

$$C'(1500) = 13 - 0.02(1500) + 0.000009(1500^2) = 3.25$$

This means that at this production level, it would cost approximately $3.25 to make one additional (the 1501st) unit.

(d) The actual cost of the 1501st unit is the difference between producing 1501 items and 1500 items:

$$C(1501) - C(1500) \approx 10{,}128.2535 - 10{,}125.00 = \$3.2535 \qquad ▪$$

In Example 1(d), the actual cost of producing an additional unit is very close to the marginal cost. Recall from the definition of the derivative that

$$C'(q) = \lim_{\Delta q \to 0} \frac{\Delta C}{\Delta q} = \lim_{h \to 0} \frac{C(q + h) - C(q)}{h}$$

Since q often takes on only integer values, it may not make literal sense to let h approach 0, but the function used to model the cost is a continuous function that accepts all inputs.

If we assume that the number of items produced is relatively large, then we can take $h = 1$ (so that h is small compared to the value of q) and we can say

$$C'(q) \approx C(q + 1) - C(q)$$

The expression $C(q + 1) - C(q)$ is the cost of producing $q + 1$ units minus the cost of producing q units. In other words, it measures the *actual cost* of producing just the $(q + 1)$st unit. So

the marginal cost is approximately equal
to the cost of producing one additional unit

As in Example 1, $C'(1500)$ is approximately the cost of producing the 1501st item, once 1500 items have been made.

▪ Linear Approximation

The concept of using a derivative value (the instantaneous rate of change) to estimate a change in function values is useful in a variety of settings. For instance, if we know that a bullet leaves a gun barrel traveling at 600 m/s, we can estimate that after half a second the bullet has traveled 300 meters. Even though the bullet slows down after leaving the gun, we expect the change in speed in $\frac{1}{2}$ second to be relatively small, so 300 m should be a reasonable estimate of the distance traveled.

Let's look at this method of estimation graphically. Figure 1 shows the cost function C from Example 1. We can see that the slope of the graph is always varying, so the rate of change is not constant. However, at $q = 1500$ we know that the rate of change is $C'(1500) = 3.25$. If we say that each additional item produced will cost $3.25, it means that we are holding the rate of change fixed as we estimate changes in function values. Now think about graphing these estimated values. If we start at the point $(1500, C(1500))$ and proceed with a constant rate of change of $C'(1500) = 3.25$, it is precisely the tangent line that we are describing. Thus using a derivative to predict nearby function values of a differentiable function corresponds to using values from the tangent line in place of the function values; we call this a **linear approximation**.

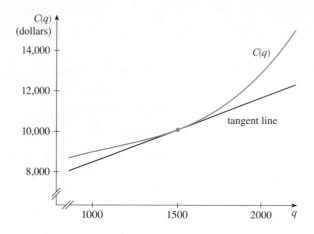

FIGURE 1

We can see from the graph in Figure 1 that the values of C are very close to the tangent line near $q = 1500$, so the linear approximation there predicts a fairly accurate value for the cost of the 1501st item produced. It appears that we could also say that the next 100 items produced would cost about $3.25(100) = \$325$ with a reasonable level of accuracy, because the tangent line is close to the graph of C for $1500 \leq q \leq 1600$. However, if we use values farther away from $q = 1500$, the tangent line does not closely follow the graph of C, so the linear approximation becomes much less accurate. In Figure 2 we see that if we use the tangent line to measure the change in cost of increasing production from 1500 to 2000 items, we get about \$1600 while the actual change in cost is $C(2000) - C(1500) = \$2875$. Thus a linear approximation can be used to predict *nearby* function values, but we cannot wander too far from the point at which we measured the derivative.

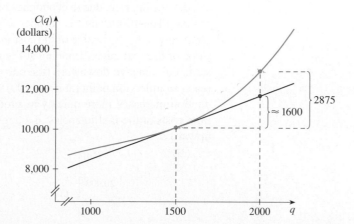

FIGURE 2

We have seen that a linear approximation can simplify computations for quantities such as marginal cost. But linear approximations are especially useful when we have information about a function's derivatives but not enough knowledge of its output values.

In general, if we know a function's value at $x = a$ as well as the derivative value there, we can estimate a nearby function value (at $x = a + \Delta x$, where Δx is small) using the approximation

(1) $$f(a + \Delta x) \approx f(a) + f'(a)\Delta x$$

▪ **EXAMPLE 2**   **Using a Linear Approximation**

Let $E(t)$ be the cost, in dollars, for electricity that a factory incurs when it is in operation t hours during the month. If $E(210) = 1845$ and $E'(210) = 10.12$, use a linear approximation to estimate the monthly cost of electricity if the factory is in operation 218.5 hours.

SOLUTION

We know that the monthly cost when the factory operates for 210 hours is $1845. To estimate the increase in cost when the time increases by 8.5 hours we use the derivative value as the rate of change. Thus the increase in cost is

$$8.5(10.12) = 86.02$$

so we estimate that the adjusted monthly cost will be

$$\$1845 + \$86.02 = \$1931.02$$

In the language of Equation 1, we have $a = 210$ and $\Delta x = 8.5$, so

$$f(218.5) = f(210 + 8.5) \approx f(210) + f'(210) \cdot (8.5)$$
$$= 1845 + 10.12(8.5) = 1845 + 86.02 = 1931.02 \qquad ▪$$

▪ | Minimizing Average Cost

In Figure 3 we show a more complete view of the cost function of Example 1. It has a shape typical of cost functions. For lower production levels, the cost increases but at a decreasing rate, due to economies of scale. Thus marginal cost tends to decrease at first. (The 1000th unit is cheaper to produce than the 100th unit.) Because marginal cost is the derivative of the cost function, we can interpret marginal cost as the slope of the cost curve. Looking at Figure 3, we see that the curve is increasing and starts out concave downward (the rate of change is decreasing). But eventually we reach an inflection point (at $q \approx 1100$) where the trend reverses, and the cost of production increases more quickly as production increases, perhaps because of overtime costs or the inefficiencies of a large-scale operation. Here the curve is concave upward.

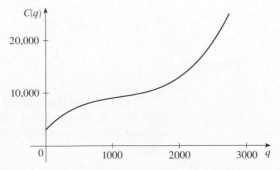

FIGURE 3
The cost function in Example 1

If marginal cost is increasing, does it make sense to increase production? As long as marginal cost is less than average cost, producing an additional unit will lower the average cost. If the goal is to minimize average cost, production should be increased in this case. Conversely, if marginal cost becomes higher than average cost, producing an additional item will raise the average cost. Thus, for a typical cost function,

the minimum average cost per item occurs when

marginal cost is the same as average cost

The graphs of the average cost and marginal cost of the cost function from Example 1 are shown in Figure 4; we can see that the lowest point on the average cost curve is precisely where the two curves intersect.

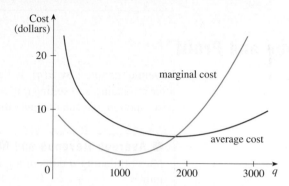

FIGURE 4

Marginal cost and average cost functions in Example 1

▪ EXAMPLE 3   Minimizing Average Cost

A small furniture manufacturer estimates that the cost, in dollars, of producing q units of a particular chair each month is given by

$$C(q) = 10{,}000 + 5q + 0.01q^2$$

How many chairs should be produced in order to minimize the average cost of each chair?

SOLUTION

We obtain the minimum average cost at the production level where average cost and marginal cost are equal. We have

$$\text{average cost} = \frac{C(q)}{q} = \frac{10{,}000 + 5q + 0.01q^2}{q}$$

and

$$\text{marginal cost} = C'(q) = 5 + 0.02q$$

Setting the two functions equal to each other gives

$$\frac{10{,}000 + 5q + 0.01q^2}{q} = 5 + 0.02q$$

$$10{,}000 + 5q + 0.01q^2 = 5q + 0.02q^2$$

$$10{,}000 = 0.01q^2$$

$$1{,}000{,}000 = q^2$$

$$1000 = q$$

Alternatively, we can graph the average cost and marginal cost functions on a calculator (or computer), as shown in Figure 5, and identify the point of intersection. The curves intersect at (1000, 25), so producing 1000 chairs results in the lowest average cost ($25) per chair.

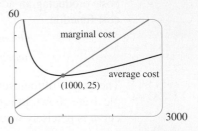

FIGURE 5

▪ Revenue and Profit

Revenue, denoted by $R(q)$, is the total amount of money collected by a company after producing and selling q items, and we call R the **revenue function**. As with the cost function, we can compute the *average revenue* and *marginal revenue* functions.

> ▪ **Average Revenue and Marginal Revenue** If $R(q)$ is the total revenue after producing q units of a good or service, then the **average revenue** per unit is
>
> $$\frac{R(q)}{q}$$
>
> and the **marginal revenue** is
>
> $$R'(q) = \frac{dR}{dq}$$

Marginal revenue can be interpreted as the slope of the graph of the revenue function. We can think of the marginal revenue as approximately the additional income gained by producing and selling one additional unit (assuming that the number of units produced is relatively large).

If a producer always charges the same price for each unit of a product, the marginal revenue is always the same. (And, in fact, the marginal revenue is simply the price of the product.) It is more typical, however, that the price changes. If a significantly larger number of items are produced, the market may become oversaturated, pushing prices down.

In any case, a key question a company needs to answer is how many units to produce in a given time period in order to maximize profit. The **profit function** P is formed by subtracting total cost from total revenue:

$$P(q) = R(q) - C(q)$$

If producing an additional unit adds more revenue than cost, it will increase profit. So to maximize profit, production should be increased as long as marginal revenue is greater than marginal cost. Thus, when $R'(q) > C'(q)$,

to maximize profit, production should be increased to the point

at which marginal revenue and marginal cost are equal

■ EXAMPLE 4 Marginal Revenue and Maximizing Profit

Let's revisit Example 3, where we looked at the cost of manufacturing chairs. Suppose the company estimates that the revenue, in dollars, realized by producing q units of the chair, up to a maximum of 2000 chairs, is given by $R(q) = 48q - 0.012q^2$.

(a) What is the marginal revenue when 1500 chairs are produced?

(b) Determine the number of chairs that the company should produce in order to maximize profit.

SOLUTION

(a) The marginal revenue function is $R'(q) = 48 - 0.024q$, so when 1500 chairs are produced, the marginal revenue is $R'(1500) = 12$ dollars per chair. We can say that the revenue received on the 1501st chair is approximately \$12.

(b) In Example 3 we determined that the marginal cost is $C'(q) = 5 + 0.02q$. Initially we have $R'(q) > C'(q)$, so profit is maximized when marginal revenue equals marginal cost:

$$R'(q) = C'(q)$$

$$48 - 0.024q = 5 + 0.02q$$

$$-0.044q = -43$$

$$q = \frac{43}{0.044} \approx 977.27$$

Thus the furniture manufacturer should produce 977 chairs in order to maximize profit. ■

Figure 6 shows the graphs of the cost and revenue functions in Example 4. Notice that at approximately $q = 977$, tangent lines to the curves have the same slope. Thus the marginal cost and marginal revenue are equal there. To the left of this value, the revenue curve is steeper than the cost curve, indicating that it is advantageous to increase production.

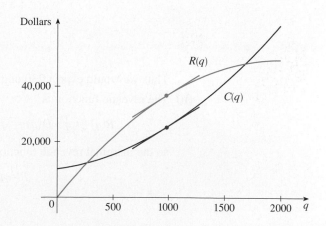

FIGURE 6
The cost and revenue
functions in Example 4

▪ Demand Curves

It is more natural to think of the price of each unit determining the number of units sold, but economists have historically used the number of units sold as the input to the demand function and the price of each unit as the output.

There is normally a relationship between the price of a product or service and the number of units that can be sold. Let $p = D(q)$ be the price per unit that a company can charge if it sells q units. Then D is called the **demand function** (or **price function**) and its graph is called the **demand curve**. We would expect p to be a decreasing function of q (in order to sell more units, a lower price would be required). A typical demand curve is shown in Figure 7.

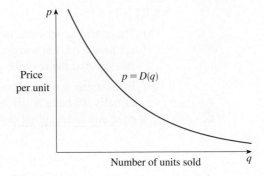

FIGURE 7
The graph of a typical demand function

Because revenue is the number of units sold multiplied by the price of each unit, the revenue function is found by

$$R(q) = q \cdot D(q)$$

▪ EXAMPLE 5   A Demand Function and Maximizing Profit

A company has cost and demand functions

$$C(q) = 84 + 1.26q - 0.01q^2 + 0.00007q^3 \quad \text{and} \quad D(q) = 3.5 - 0.01q$$

(a) If the price of each unit is \$1.20, how many units will be sold?

(b) Determine the production level that will maximize profit for the company.

SOLUTION

(a) The price of each unit is the output of the demand function, so we solve

$$3.5 - 0.01q = 1.20$$
$$0.01q = 2.3$$
$$q = 230$$

Thus we would expect 230 units to be sold at \$1.20 each.

(b) The revenue function is

$$R(q) = q \cdot D(q) = q(3.5 - 0.01q) = 3.5q - 0.01q^2$$

so the marginal revenue function is

$$R'(q) = 3.5 - 0.02q$$

and the marginal cost function is

$$C'(q) = 1.26 - 0.02q + 0.00021q^2$$

Thus marginal revenue is equal to marginal cost when

$$3.5 - 0.02q = 1.26 - 0.02q + 0.00021q^2$$

Solving, we get

$$0.00021q^2 = 2.24$$

$$q^2 = \frac{2.24}{0.00021}$$

$$q = \sqrt{\frac{2.24}{0.00021}} \approx 103$$

Note that initially $R'(q) > C'(q)$, and therefore a production level of 103 units will maximize profit. ▪

Figure 8 shows the graphs of the revenue and cost functions in Example 5. The company makes a profit when $R > C$ and the profit reaches a maximum when $q \approx 103$. Notice that the curves have parallel tangent lines at this production level because marginal revenue equals marginal cost.

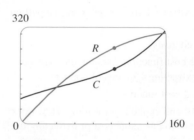

FIGURE 8

▪ Prepare Yourself

1. If $f(x) = 120 + 2.6x + 0.02x^2$, find
    (a) $f'(x)$ **(b)** $f'(50)$

2. If $A(q) = 0.001q^3 + 0.05q^2 + 20q + 350$, compute $A'(400)$.

3. For the function f shown in the graph, estimate the value of $f'(60)$.

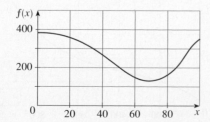

4. Solve for x (round to two decimal places):

$$\frac{250 + 5x + 0.002x^2}{x} = 0.004x + 5 \qquad (x > 0)$$

5. Solve for x (round to the nearest tenth).
    (a) $18e^{-0.5x} = 8$
    (b) $200x^{-1.4} = 75$

▪ Exercises 3.2

1. Cost function   Write an equation for a cost function where the fixed costs are $2000 and the variable costs are $15 per unit.

2. Cost function   If the cost of manufacturing a particular item is $50 per unit and there are no fixed costs, write an equation for the cost function.

3. Candy production   The weekly cost for a small confectioner to produce q chocolate bars is

$$C(q) = 1800 + 0.12q + 0.003q^2$$

(a) Find the average cost function.

(b) Find the marginal cost function.

(c) Compute the average cost and the marginal cost when 500 chocolate bars have been produced. What is the actual cost of the 501st chocolate bar?

4. Mining   A mining company estimates that it will cost

$$C(q) = 73 + 0.6q + 0.004q^2$$

thousand dollars to extract q tons of copper from a particular mine.

(a) What are the fixed costs?

(b) Find the average cost function. What is the average cost per ton of extracting 150 tons of copper?

(c) Find the marginal cost function.

(d) Compute $C'(150)$ and explain its meaning. How does this value compare to the actual cost of extracting the 151st ton of copper?

5. Clothing manufacturing   Suppose that the cost, in dollars, for a company to produce x pairs of a new line of jeans is

$$C(x) = 2000 + 3x + 0.01x^2 + 0.0002x^3$$

(a) Find the marginal cost function.

(b) Find $C'(100)$ and explain its meaning. What does it predict?

(c) Compare $C'(100)$ with the actual cost of manufacturing the 101st pair of jeans.

6. The cost function for a certain commodity is

$$C(q) = 84 + 0.16q - 0.0006q^2 + 0.000003q^3$$

(a) Find and interpret $C'(100)$.

(b) Compare $C'(100)$ with the actual cost of producing the 101st item.

7. Let $C(q)$ be the cost, in millions of dollars, for a company to produce q thousand units of a new product. The table gives several cost values.

(a) Compute the average cost per unit after producing 60,000 units.

(b) Estimate the value of $C'(40)$ and explain its meaning.

(c) Estimate the cost of producing the 40,001st unit.

q	$C(q)$	q	$C(q)$
10	3.1	50	7.3
20	4.7	60	8.8
30	5.6	70	10.6
40	6.3	80	12.4

8. Let $C(q)$ be the cost function in Exercise 7.

(a) Use a graphing calculator to find a cubic model for $C(q)$.

(b) Use the model to write a function for the average cost. What is the average cost after producing 55,000 units?

(c) Use the model to find the marginal cost function. Then compute $C'(40)$ and compare to the estimate found in Exercise 7(b).

9. Let f be a differentiable function. If $f(40) = 378$ and $f'(40) = 6$, use a linear approximation to estimate the value of each of the following.

(a) $f(42)$ (b) $f(38.5)$

10. Let g be a differentiable function. If $g(125) = 5250$ and $g'(125) = -13.8$, use a linear approximation to estimate the value of each of the following.

(a) $g(127.4)$ (b) $g(121)$

11. Paper production   Suppose $C(t)$ is the cost, in thousands of dollars, of producing t tons of white paper. If $C'(10) = 350$, estimate the cost of producing an additional 500 lb of paper once 10 tons have been produced.

12. Water purification   If $C(v)$ is the cost, in dollars, of purifying v gallons of drinking water, and $C'(200,000) = 0.26$, estimate the cost of purifying an additional 3000 gallons of water once 200,000 gallons have been processed.

13. A cost function C is shown in the graph.

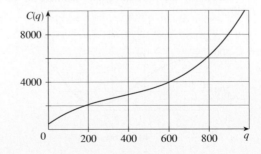

(a) Is the marginal cost higher when 100 units are produced or when 200 units are produced?

(b) Estimate the marginal cost when 600 units are produced.

(c) Estimate the production level that minimizes marginal cost.

14. Graphs of the average cost and the marginal cost of producing an item are shown.

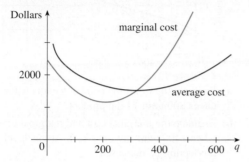

(a) Estimate the production level that minimizes marginal cost.

(b) Estimate the production level that minimizes average cost.

(c) For what production level are marginal cost and average cost equal? What is the significance of this value?

15. Baking   A baker estimates that it costs

$$C(q) = 0.01q^2 + 2q + 250$$

dollars each day to bake q loaves of bread. How many loaves should be baked daily in order to minimize the average cost?

16. Appliance production   A manufacturer's weekly cost, in dollars, for producing q lamps is

$$C(q) = 810 + 3q + 0.002q^2$$

Find the number of lamps that should be produced in order to minimize the average cost.

17. Clothing manufacturing   The cost function

$$C(q) = 6200 + 23q - 0.02q^2 + 0.0001q^3$$

gives the cost a company incurs each month when it produces q pairs of shoes. Graph the average cost and marginal cost functions and use the graph to determine the number of pairs of shoes that should be produced each month in order to minimize the average cost of each pair.

18. Computer production   The monthly cost, in dollars, for a particular factory to produce q laptop computers is given by

$$C(q) = 37,300 + 152q - 0.01q^2 + 0.00014q^3$$

Graph the average cost and marginal cost functions and use the graph to determine the number of laptops that the factory should produce monthly to minimize the average cost of each computer.

19. Appliance production   A manufacturer of power supplies estimates that it will incur a total cost of

$$C(q) = 2500 + 4q + 0.005q^2$$

dollars when producing q power supplies, and it will collect

$$R(q) = 16q - 0.002q^2$$

dollars in revenue.

(a) Write a function for the profit P the manufacturer can expect after producing q power supplies.

(b) Find the marginal cost and marginal revenue functions.

(c) How many power supplies should the manufacturer produce in order to maximize profit?

20. Food production   A food maker estimates that the monthly cost, in dollars, for producing b boxes of its breakfast cereal is

$$C(b) = 1200 + 0.9b + 0.0002b^2$$

The revenue, in dollars, the company earns from selling b boxes of the cereal is given by

$$R(b) = 2.8b - 0.0001b^2$$

(a) Write equations for the average revenue and marginal revenue functions.

(b) Find the number of boxes of cereal that the company should produce monthly in order to maximize profit.

21. The demand function for a company's product is $p = 32e^{-0.6q}$ where q is measured in thousands of units and p is measured in dollars.

(a) What price should the company charge for each unit in order to sell 4500 units?

(b) If the company prices the products at $7.50 each, how many units will sell?

22. Furniture manufacturing   The demand function for a company's new office chair is $p = 43,000q^{-1.2}$.

(a) What price should the company charge for each chair in order to sell 150 chairs?

(b) If the company charges $350 for each chair, how many will sell?

23–24 ▪ For the given cost and demand functions, find the production level that will maximize profit.

23. $C(q) = 680 + 4q + 0.01q^2$, $p = 12 - q/500$

24. $C(q) = 16,000 + 500q - 1.6q^2 + 0.004q^3$,
$p = 1700 - 7q$

25. Jewelry production   A jeweller is considering producing a limited edition diamond bracelet, and she is trying to decide how many bracelets to produce. The table gives her estimated total cost for various production levels as well as the price she would charge for each bracelet.

Number of bracelets	Total cost (thousands)	Price per bracelet
100	$215	$8000
200	$420	$7500
300	$625	$6000
400	$820	$5000
500	$1015	$4200
600	$1205	$3600

(a) Of the production levels listed in the table, which gives the highest profit?

(b) Estimate the marginal cost and marginal revenue when 400 bracelets are made.

(c) According to the estimates in part (b), will increasing the production level higher than 400 bracelets increase profit?

26. Food production   An entrepreneur wants to buy oranges from local growers and produce fresh-squeezed orange juice to sell to restaurants. His estimates for weekly cost and revenue for various production levels are given in the table.

Gallons produced	Total cost	Total revenue
200	$1200	$1000
400	$1960	$1900
600	$2380	$2900
800	$2650	$3700
1000	$3220	$4300
1200	$4100	$4820

(a) Find a cubic model for the cost data and a quadratic model for the revenue data.

(b) Use the models to compute marginal cost and marginal revenue functions. Then determine the production level that maximizes weekly profit.

27. Graphs of the cost function C and revenue function R, in thousands of dollars, for q units of a new product a company is manufacturing are shown.

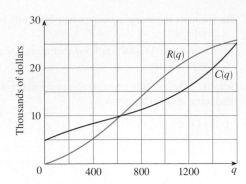

(a) Estimate the profit the company earns when it produces 800 units of the product.

(b) Estimate the marginal cost and marginal revenue when 1400 units are produced. At this level, should production be increased?

(c) Estimate the production level that maximizes profit for the company.

28. In a perfectly competitive market, economists assume that the price p of a product is determined by the market and is the same for all producers. Therefore the revenue after producing q units is pq. Explain why the production level that maximizes profit occurs when the marginal cost equals p.

29. A car-detailing service estimates that its daily cost of waxing q cars is

$$C(q) = 0.08q^2 + 37q + 350$$

If the service collects $65 for each car waxing, find the number of cars the service should wax daily in order to maximize profit.

30. The derivative of the profit function P with respect to q (the number of units produced) is called the *marginal profit function.*

(a) If $R'(q) > C'(q)$ for a particular production level q, is $P'(q)$ positive or negative?

(b) Explain why the maximum profit is reached when $P'(q) = 0$.

(c) Suppose the cost and revenue functions of a commodity are

$$C(q) = 1400 + 4q + 0.003q^2 + 0.0001q^3$$

and

$$R(q) = 26q - 0.001q^2$$

Find the profit function and the marginal profit function.

(d) Compute the marginal profit when $q = 300$ and explain its meaning. What is the significance of the fact that the marginal profit is negative in this case?

3.3 The Product and Quotient Rules

In Section 3.1 we learned the Sum Rule and Difference Rule for derivatives, which state that the derivative of a sum or difference is the sum or difference of the derivatives. Unfortunately, the situation is not so simple when two functions are multiplied or divided. In this section we develop formulas that enable us to differentiate new functions formed from old functions by multiplication or division.

■ The Product Rule

 By analogy with the Sum and Difference Rules, we might be tempted to guess, as Leibniz did more than three centuries ago, that the derivative of a product is the product of the derivatives. We can see, however, that this guess is wrong by looking at a particular example. Let $f(x) = x$ and $g(x) = x^2$. Then the Power Rule gives $f'(x) = 1$ and $g'(x) = 2x$. But if we define $h(x) = f(x) \cdot g(x)$, then $h(x) = x^3$, so $h'(x) = 3x^2$. Thus, $(fg)' \neq f'g'$. The correct formula was discovered by Leibniz (soon after his false start) and is called the Product Rule.

Before stating the Product Rule, let's see how we might discover it. We start by assuming that $u = f(x)$ and $v = g(x)$ are both positive differentiable functions. Then we can interpret the product uv as an area of a rectangle with length u and width v. (See Figure 1.) If x changes by an amount Δx, then the corresponding changes in u and v are

$$\Delta u = f(x + \Delta x) - f(x) \qquad \Delta v = g(x + \Delta x) - g(x)$$

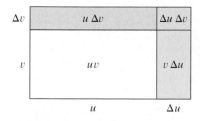

FIGURE 1
The geometry of the Product Rule

and the new value of the product, $(u + \Delta u)(v + \Delta v)$, can be interpreted as the area of the large rectangle in Figure 1 (provided that Δu and Δv happen to be positive).

The change in the area of the rectangle is

(1)
$$\Delta(uv) = (u + \Delta u)(v + \Delta v) - uv = u\,\Delta v + v\,\Delta u + \Delta u\,\Delta v$$

$$= \text{the sum of the three shaded areas}$$

If we divide by Δx, we get the average rate of change of uv.

$$\frac{\Delta(uv)}{\Delta x} = u\,\frac{\Delta v}{\Delta x} + v\,\frac{\Delta u}{\Delta x} + \Delta u\,\frac{\Delta v}{\Delta x}$$

Recall that in Leibniz notation the definition of a derivative can be written as

$$\frac{dy}{dx} = \lim_{\Delta x \to 0} \frac{\Delta y}{\Delta x}$$

If we now let $\Delta x \to 0$, we get the derivative of uv:

$$\frac{d}{dx}(uv) = \lim_{\Delta x \to 0} \frac{\Delta(uv)}{\Delta x} = \lim_{\Delta x \to 0} \left(u\,\frac{\Delta v}{\Delta x} + v\,\frac{\Delta u}{\Delta x} + \Delta u\,\frac{\Delta v}{\Delta x} \right)$$

$$= u \lim_{\Delta x \to 0} \frac{\Delta v}{\Delta x} + v \lim_{\Delta x \to 0} \frac{\Delta u}{\Delta x} + \left(\lim_{\Delta x \to 0} \Delta u \right)\left(\lim_{\Delta x \to 0} \frac{\Delta v}{\Delta x} \right)$$

$$= u\,\frac{dv}{dx} + v\,\frac{du}{dx} + 0 \cdot \frac{dv}{dx}$$

[Note that $\lim_{\Delta x \to 0} \Delta u = 0$ because f is differentiable and therefore continuous by Theorem 3 in Section 2.4, so $\Delta u \to 0$ as $\Delta x \to 0$.] Thus

(2)
$$\frac{d}{dx}(uv) = u\,\frac{dv}{dx} + v\,\frac{du}{dx}$$

Although we started by assuming (for the geometric interpretation) that all the quantities are positive, Equation 1 is always true. (The algebra is valid whether u, v, Δu, and Δv are positive or negative.) So we have proved Equation 2, known as the Product Rule, for all differentiable functions u and v.

In prime notation:

$$(fg)' = fg' + gf'$$

■ **The Product Rule**   If f and g are both differentiable, then

$$\frac{d}{dx}[f(x)g(x)] = f(x)\frac{d}{dx}[g(x)] + g(x)\frac{d}{dx}[f(x)]$$

In words, the Product Rule says that *the derivative of a product of two functions is the first function times the derivative of the second function plus the second function times the derivative of the first function.*

■ **EXAMPLE 1   Using the Product Rule**

If $f(x) = xe^x$, find $f'(x)$ and $f''(x)$.

SOLUTION

By the Product Rule, we have

$$f'(x) = \frac{d}{dx}(xe^x) = x\frac{d}{dx}(e^x) + e^x\frac{d}{dx}(x)$$

$$= \underbrace{x}_{\substack{\text{first} \\ \text{function}}} \cdot \underbrace{e^x}_{\substack{\text{derivative of} \\ \text{the second} \\ \text{function}}} + \underbrace{e^x}_{\substack{\text{second} \\ \text{function}}} \cdot \underbrace{1}_{\substack{\text{derivative of} \\ \text{the first} \\ \text{function}}}$$

$$= (x + 1)e^x$$

Figure 2 shows the graphs of the function f of Example 1 and its derivative f'. Notice that $f'(x)$ is positive when f is increasing and negative when f is decreasing.

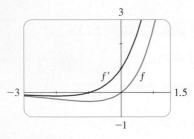

FIGURE 2

Using the Product Rule a second time, we get

$$f''(x) = \frac{d}{dx}[(x + 1)e^x] = (x + 1)\frac{d}{dx}(e^x) + e^x\frac{d}{dx}(x + 1)$$

$$= (x + 1)e^x + e^x \cdot 1 = (x + 2)e^x \qquad ■$$

■ **EXAMPLE 2   An Alternative to the Product Rule**

Differentiate the function $g(t) = 2t^3(3 + 5t)$.

SOLUTION 1

Using the Product Rule, we have

$$g'(t) = 2t^3\frac{d}{dt}(3 + 5t) + (3 + 5t)\frac{d}{dt}(2t^3)$$

$$= 2t^3 \cdot 5 + (3 + 5t) \cdot 6t^2$$

$$= 10t^3 + 18t^2 + 30t^3 = 40t^3 + 18t^2$$

SOLUTION 2

If we first rewrite $g(t)$ by using the distributive property, then we can use the Sum Rule instead of the Product Rule.

$$g(t) = 2t^3(3 + 5t) = 6t^3 + 10t^4$$

$$g'(t) = 18t^2 + 40t^3$$

which is equivalent to the answer given in Solution 1. ■

Example 2 shows that it is sometimes easier to simplify a product of functions than to use the Product Rule. In Example 1, however, the Product Rule is the only possible method.

■ **EXAMPLE 3** **Using the Product Rule without Equations**

If $f(x) = \sqrt{x}\, g(x)$, where $g(4) = 2$ and $g'(4) = 3$, find $f'(4)$.

SOLUTION

Applying the Product Rule, we get

$$f'(x) = \frac{d}{dx}\left[\sqrt{x}\, g(x)\right] = \sqrt{x}\, \frac{d}{dx}\left[g(x)\right] + g(x)\, \frac{d}{dx}\left[\sqrt{x}\,\right]$$

$$= \sqrt{x}\, g'(x) + g(x) \cdot \tfrac{1}{2}x^{-1/2}$$

$$= \sqrt{x}\, g'(x) + \frac{g(x)}{2\sqrt{x}}$$

So

$$f'(4) = \sqrt{4}\, g'(4) + \frac{g(4)}{2\sqrt{4}} = 2 \cdot 3 + \frac{2}{2 \cdot 2} = 6.5 \qquad ■$$

In the next example, we illustrate the Product Rule by estimating the rate of change of a function formed by multiplying two quantities, each of which is increasing.

■ **EXAMPLE 4** **Interpreting the Terms in the Product Rule**

A telephone company wants to estimate the number of new residential phone lines that it will need to install during the upcoming month. At the beginning of January the company had 100,000 subscribers, each of whom had 1.2 phone lines, on average. The company estimated that its subscribership was increasing at the rate of 1000 monthly. By polling its existing subscribers, the company found that each intended to install an average of 0.01 new phone lines by the end of January. Estimate the number of new lines the company will have to install in January by calculating the rate of increase of lines at the beginning of the month.

SOLUTION

Let $s(t)$ be the number of subscribers and let $n(t)$ be the number of phone lines per subscriber at time t, where t is measured in months and $t = 0$ corresponds to the beginning of January. Then the total number of lines is given by

$$L(t) = s(t)n(t)$$

and we want to find $L'(0)$. According to the Product Rule, we have

$$L'(t) = \frac{d}{dt}[s(t)n(t)] = s(t)n'(t) + n(t)s'(t)$$

We are given that $s(0) = 100{,}000$ and $n(0) = 1.2$. The company's estimates concerning rates of increase are that $s'(0) \approx 1000$ and $n'(0) \approx 0.01$. Therefore

$$L'(0) = s(0)n'(0) + n(0)s'(0)$$

$$\approx 100{,}000 \cdot 0.01 + 1.2 \cdot 1000 = 2200$$

The company will need to install approximately 2200 new phone lines in January.

Notice that the two terms arising from the Product Rule come from different sources—old subscribers and new subscribers. One contribution to L' is the number of existing subscribers (100,000) times the rate at which they order new lines (about 0.01 per subscriber monthly). A second contribution is the average number of lines per subscriber (1.2 at the beginning of the month) times the rate of increase of subscribers (1000 monthly).   ▪

▪ The Quotient Rule

Next we wish to find a formula for the derivative of a quotient of two functions. Suppose that f and g are differentiable and let $F(x) = f(x)/g(x)$. If we make the prior assumption that F is differentiable, then we can write $f(x) = F(x) \cdot g(x)$ and apply the Product Rule:

$$f'(x) = F(x) \cdot g'(x) + g(x) \cdot F'(x)$$

Solving this equation for $F'(x)$, we get

$$g(x)F'(x) = f'(x) - F(x)g'(x)$$

$$F'(x) = \frac{f'(x) - F(x)g'(x)}{g(x)} = \frac{f'(x) - \dfrac{f(x)}{g(x)}g'(x)}{g(x)}$$

$$= \frac{f'(x) - \dfrac{f(x)}{g(x)}g'(x)}{g(x)} \cdot \frac{g(x)}{g(x)} = \frac{g(x)f'(x) - f(x)g'(x)}{[g(x)]^2}$$

Thus

$$\left(\frac{f(x)}{g(x)}\right)' = \frac{g(x)f'(x) - f(x)g'(x)}{[g(x)]^2}$$

Although we derived this formula under the assumption that F is differentiable, it can be proved without that assumption in much the same way that we found the Product Rule.

In prime notation:

$$\left(\frac{f}{g}\right)' = \frac{gf' - fg'}{g^2}$$

▪ The Quotient Rule   If f and g are differentiable, then

$$\frac{d}{dx}\left[\frac{f(x)}{g(x)}\right] = \frac{g(x)\dfrac{d}{dx}[f(x)] - f(x)\dfrac{d}{dx}[g(x)]}{[g(x)]^2}$$

In words, the Quotient Rule says that the *derivative of a quotient is the denominator times the derivative of the numerator minus the numerator times the derivative of the denominator, all divided by the square of the denominator.*

The Quotient Rule and the other differentiation formulas enable us to compute the derivative of any rational function, as the next example illustrates.

We can use a graphing device to check that the answer to Example 5 is plausible. Figure 3 shows the graphs of the function of Example 5 and its derivative. Notice that when y grows rapidly (near -2), y' is large. And when y grows slowly, y' is near 0.

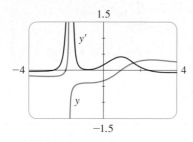

FIGURE 3

■ **EXAMPLE 5**   **Using the Quotient Rule**

Let $y = \dfrac{x^2 + x - 2}{x^3 + 6}$.

Then

$$y' = \frac{(x^3 + 6)\dfrac{d}{dx}(x^2 + x - 2) - (x^2 + x - 2)\dfrac{d}{dx}(x^3 + 6)}{(x^3 + 6)^2}$$

$$= \frac{(x^3 + 6)(2x + 1) - (x^2 + x - 2)(3x^2)}{(x^3 + 6)^2}$$

$$= \frac{(2x^4 + x^3 + 12x + 6) - (3x^4 + 3x^3 - 6x^2)}{(x^3 + 6)^2}$$

$$= \frac{-x^4 - 2x^3 + 6x^2 + 12x + 6}{(x^3 + 6)^2} \qquad ■$$

■ **EXAMPLE 6**   **Equation of a Tangent Line**

Find an equation of the tangent line to the curve $y = e^x/(1 + x^2)$ at the point $(1, e/2)$.

SOLUTION

According to the Quotient Rule, we have

$$\frac{dy}{dx} = \frac{(1 + x^2)\dfrac{d}{dx}(e^x) - e^x\dfrac{d}{dx}(1 + x^2)}{(1 + x^2)^2}$$

$$= \frac{(1 + x^2)e^x - e^x(2x)}{(1 + x^2)^2} = \frac{e^x(1 - x)^2}{(1 + x^2)^2}$$

So the slope of the tangent line at $(1, e/2)$ is

$$\left.\frac{dy}{dx}\right|_{x=1} = \frac{e^1(1 - 1)^2}{(1 + 1^2)^2} = 0$$

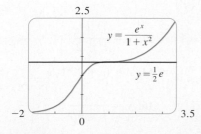

FIGURE 4

This means that the tangent line at $(1, e/2)$ is horizontal and its equation is $y = e/2$ (approximately $y = 1.36$). [See Figure 4. Notice that the function is increasing and crosses its tangent line at $(1, e/2)$.]   ■

NOTE: Don't use the Quotient Rule *every* time you see a quotient. Sometimes it's easier to rewrite a quotient first to put it in a form that is simpler for the purpose of differentiation. For instance, although it is possible to differentiate the function

$$F(x) = \frac{3x^2 + 2\sqrt{x}}{x}$$

using the Quotient Rule, it is much easier to perform the division first and write the function as

$$F(x) = \frac{3x^2}{x} + \frac{2\sqrt{x}}{x} = 3x + 2x^{-1/2}$$

before differentiating.

We summarize the differentiation formulas we have learned so far as follows.

■ **Table of Differentiation Formulas**

$$\frac{d}{dx}(c) = 0 \qquad \frac{d}{dx}(x^n) = nx^{n-1} \qquad \frac{d}{dx}(e^x) = e^x$$

$$(cf)' = cf' \qquad (f+g)' = f'+g' \qquad (f-g)' = f'-g'$$

$$(fg)' = fg' + gf' \qquad \left(\frac{f}{g}\right)' = \frac{gf'-fg'}{g^2}$$

▪ Prepare Yourself

1. Find the derivative of the function.
   (a) $f(x) = 5x^3 + 3x$ (b) $g(x) = 1/x^2$
   (c) $r(x) = \sqrt{x}$ (d) $U(t) = e^t$

2. Write the expression in a form without negative or fractional exponents.
   (a) $5t^{-2}$ (b) $\frac{1}{2}x^{-1/2}$ (c) $4x^{1/3}$

3. If the position, in feet, of a moving object after t minutes is given by $s(t) = 0.2t^3 + 14t + 3$, find each of the following.
   (a) The speed of the object after 2 minutes. What are the units?
   (b) The acceleration of the object after 2 minutes.

▪ Exercises 3.3

1. Find the derivative of $y = (x^2 + 1)(x^3 + 1)$ in two ways: by using the Product Rule and by performing the multiplication first. Do your answers agree?

2. Find the derivative of the function

$$F(x) = \frac{x^2 - 3x^3}{x}$$

in two ways: by using the Quotient Rule and by simplifying first. Show that your answers are equivalent. Which method do you prefer?

3–20 ▪ Differentiate.

3. $f(x) = x^2 e^x$ **4.** $g(x) = \sqrt{x}\,e^x$

5. $y = \dfrac{e^x}{x^2}$ **6.** $y = \dfrac{e^x}{1+x}$

7. $g(x) = \dfrac{3x-1}{2x+1}$ **8.** $f(t) = \dfrac{2t}{4+t^2}$

9. $F(y) = \left(\dfrac{1}{y^2} - \dfrac{3}{y^4}\right)(y + 5y^3)$

10. $R(t) = (t + e^t)(3 - \sqrt{t})$

11. $y = \dfrac{t^2}{3t^2 - 2t + 1}$ **12.** $y = \dfrac{t^3 + t}{t^4 - 2}$

13. $y = (r^2 - 2r)e^r$ **14.** $r = (v + 2e^v)\sqrt{v}$

15. $P = \dfrac{5e^t}{2 + 3t^2}$ **16.** $y = \dfrac{x}{1 - 4/x}$

17. $y = \dfrac{v^3 - 2v\sqrt{v}}{v}$ **18.** $y = \dfrac{x^2}{x - 0.5e^x}$

19. $f(x) = \dfrac{A}{B + Ce^x}$ **20.** $h(v) = \dfrac{1 + av^2}{1 + bv}$

21–22 ▪ Find the slope of the curve at the given x-value. Give a decimal answer rounded to three decimal places.

21. $y = \dfrac{x}{x^2 - 1}$, $\quad x = 2$

22. $y = e^x(0.8x^2 - x)$, $\quad x = 1.5$

23–24 ▪ Find an equation of the tangent line to the given curve at the specified point.

23. $y = 2xe^x$, $\quad (0, 0)$
24. $y = \dfrac{\sqrt{x}}{x + 1}$, $\quad (4, 0.4)$

---

25. Work force productivity If $p(x)$ is the total value of the production when there are x workers in a plant, then the *average productivity* of the work force at the plant is

$$A(x) = \frac{p(x)}{x}$$

(a) Find $A'(x)$. Why does the company want to hire more workers if $A'(x) > 0$?

(b) Show that $A'(x) > 0$ if $p'(x)$ is greater than the average productivity.

26. Motion The position, in inches, of an object moving along a straight path is given by $(t - t^2)e^t$, where t is the time elapsed in seconds. Find the speed of the object after 3.5 seconds.

27. Motion The distance, in feet, a moving object has traveled after t seconds is given by $3t/(2 + t)$. Find the acceleration of the object after 4 seconds.

28. Fabric production A manufacturer produces bolts of a fabric with a fixed width. The quantity q of this fabric (measured in yards) that is sold is a function of the selling price p (in dollars per yard), so we can write $q = f(p)$. (This is the inverse of a demand function, as discussed in Section 3.2.) Then the total revenue earned with selling price p is $R(p) = p \cdot f(p)$.

(a) What does it mean to say that $f(20) = 10{,}000$ and $f'(20) = -350$?

(b) Assuming the values in part (a), find $R'(20)$ and interpret your answer.

29. (a) The curve $y = 1/(1 + x^2)$ is called a **witch of Maria Agnesi**. Find an equation of the tangent line to this curve at the point $\left(-1, \frac{1}{2}\right)$.

(b) Illustrate part (a) by graphing the curve and the tangent line on the same screen.

30. (a) The curve $y = x/(1 + x^2)$ is called a **serpentine**. Find an equation of the tangent line to this curve at the point $(3, 0.3)$.

(b) Illustrate part (a) by graphing the curve and the tangent line on the same screen.

31. (a) If $f(x) = e^x/x^3$, find $f'(x)$.

(b) Check to see that your answer to part (a) is reasonable by comparing the graphs of f and f'.

32. (a) If $f(x) = x/(x^2 - 1)$, find $f'(x)$.

(b) Check to see that your answer to part (a) is reasonable by comparing the graphs of f and f'.

33. (a) If $f(x) = (x - 1)e^x$, find $f'(x)$ and $f''(x)$.

(b) Check to see that your answers to part (a) are reasonable by comparing the graphs of f, f', and f''.

34. (a) If $f(x) = x/(x^2 + 1)$, find $f'(x)$ and $f''(x)$.

(b) Check to see that your answers to part (a) are reasonable by comparing the graphs of f, f', and f''.

35. If $f(x) = x^2/(1 + x)$, find $f''(1)$.

36. If $A(r) = 2r/(r^2 + 1)$, find $A''(1)$.

37. Suppose that $f(5) = 1$, $f'(5) = 6$, $g(5) = -3$, and $g'(5) = 2$. If $A(x) = f(x)\,g(x)$, $B(x) = f(x)/g(x)$, and $C(x) = g(x)/f(x)$, find the following values.

(a) $A'(5)$ (b) $B'(5)$ (c) $C'(5)$

38. Suppose that $f(3) = 4$, $g(3) = 2$, $f'(3) = -6$, and $g'(3) = 5$. If $P(x) = f(x) + g(x)$, $Q(x) = f(x)\,g(x)$, and $R(x) = f(x)/g(x)$, find the following numbers.

(a) $P'(3)$ (b) $Q'(3)$ (c) $R'(3)$

39. If $f(x) = e^x g(x)$, where $g(0) = 2$ and $g'(0) = 5$, find $f'(0)$.

40. If $f(x) = g(x)/e^x$, where $g(1) = -1$ and $g'(1) = 3$, find $f'(1)$.

41. If f and g are the functions whose graphs are shown, let $u(x) = f(x)\,g(x)$ and $v(x) = f(x)/g(x)$.

(a) Find $u'(1)$.   (b) Find $v'(5)$.

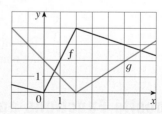

42. Let $P(x) = F(x)\,G(x)$ and $Q(x) = F(x)/G(x)$, where F and G are the functions whose graphs are shown.

(a) Find $P'(2)$.   (b) Find $Q'(7)$.

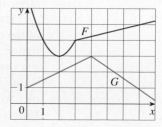

▪ Challenge Yourself

43. Average income   In this exercise we estimate the rate at which the total personal income is rising in the Richmond-Petersburg, Virginia, metropolitan area. In 1999, the population of this area was 961,400, and the population was increasing at roughly 9200 people per year. The average annual income was \$30,593 per capita, and this average was increasing at about \$1400 per year (a little above the national average of about \$1225 yearly). Use the Product Rule and these figures to estimate the rate at which total personal income was rising in the Richmond-Petersburg area in 1999. Explain the meaning of each term in the Product Rule.

44. Charitable donations   A charity organization currently has 42,000 members nationwide and their membership has been increasing by about 800 people each year. The members donate an average of \$75 each per year but this average has been decreasing at a rate of \$10 per year. Use the Product Rule to estimate the rate at which the organization's annual revenue is changing.

45. If g is a differentiable function, find an expression for the derivative of each of the following functions.

(a) $y = xg(x)$ (b) $y = \dfrac{x}{g(x)}$

(c) $y = \dfrac{g(x)}{x}$

46. If f is a differentiable function, find an expression for the derivative of each of the following functions.

(a) $y = x^2 f(x)$ (b) $y = \dfrac{f(x)}{x^2}$

(c) $y = \dfrac{x^2}{f(x)}$ (d) $y = \dfrac{1 + xf(x)}{\sqrt{x}}$

47. If $g(x) = xe^x$, find a formula for the nth derivative of g.

48. (a) If g is differentiable, the **Reciprocal Rule** says that

$$\frac{d}{dx}\left[\frac{1}{g(x)}\right] = -\frac{g'(x)}{[g(x)]^2}$$

Use the Quotient rule to prove the Reciprocal Rule.

(b) Use the Reciprocal Rule to differentiate the function

$$f(x) = \frac{1}{x + 4e^x}$$

(c) Use the Reciprocal Rule to verify that the Power Rule is valid for negative integers, that is,

$$\frac{d}{dx}(x^{-n}) = -nx^{-n-1}$$

for all positive integers n.

3.4 The Chain Rule

Suppose you are asked to differentiate the function

$$F(x) = \sqrt{x^2 + 1}$$

The differentiation formulas you learned in the previous sections of this chapter do not enable you to calculate $F'(x)$.

▪ Composite Functions and the Chain Rule

See Section 1.2 for a review of composite functions.

Observe that F is a composite function. In fact, if we let $f(u) = \sqrt{u}$ and $g(x) = x^2 + 1$, then we can write $F(x) = f(g(x))$. We know how to differentiate both f and g, so it would be useful to have a rule that tells us how to find the derivative of the composition $f(g(x))$ in terms of the derivatives of f and g.

It turns out that the derivative of the composite function $f(g(x))$ is the product of the derivatives of f and g. This fact is one of the most important of the differentiation rules and is called the *Chain Rule*. It seems plausible if we interpret derivatives as rates of change. Regard du/dx as the rate of change of u with respect to x, dy/du as the rate of change of y with respect to u, and dy/dx as the rate of change of y with respect to x. If u changes twice as fast as x and y changes three times as fast as u, then it seems reasonable that y changes six times as fast as x, and so we expect that

$$\frac{dy}{dx} = \frac{dy}{du}\frac{du}{dx}$$

■ **The Chain Rule**   If g is differentiable at x and f is differentiable at $g(x)$, then the composite function F defined by $F(x) = f(g(x))$ is differentiable at x and F' is given by the product

(1) $F'(x) = f'(g(x)) \cdot g'(x)$

In Leibniz notation, if $y = f(u)$ and $u = g(x)$ are both differentiable functions, then

(2) $\dfrac{dy}{dx} = \dfrac{dy}{du}\dfrac{du}{dx}$

James Gregory

The first person to formulate the Chain Rule was the Scottish mathematician James Gregory (1638–1675), who also designed the first practical reflecting telescope. Gregory discovered the basic ideas of calculus at about the same time as Newton. He became the first Professor of Mathematics at the University of St. Andrews and later held the same position at the University of Edinburgh. But one year after accepting that position he died at the age of 36.

Equation 2 is easy to remember because *if dy/du and du/dx were quotients, then we could cancel du.* But du/dx should not be thought of as an actual quotient.

The Chain Rule is difficult to prove and so the formal proof is omitted. However, the following argument shows why the rule makes sense: Let Δu be the change in u corresponding to a change of Δx in x, that is,

$$\Delta u = g(x + \Delta x) - g(x)$$

Then the corresponding change in y is

$$\Delta y = f(u + \Delta u) - f(u)$$

It is tempting to write

$$\frac{dy}{dx} = \lim_{\Delta x \to 0}\frac{\Delta y}{\Delta x}$$

(3) $$= \lim_{\Delta x \to 0}\frac{\Delta y}{\Delta u}\cdot\frac{\Delta u}{\Delta x}$$

$$= \lim_{\Delta x \to 0}\frac{\Delta y}{\Delta u}\cdot\lim_{\Delta x \to 0}\frac{\Delta u}{\Delta x}$$

$$= \lim_{\Delta u \to 0}\frac{\Delta y}{\Delta u}\cdot\lim_{\Delta x \to 0}\frac{\Delta u}{\Delta x} \qquad \text{(Note that } \Delta u \to 0 \text{ as } \Delta x \to 0 \text{ since } g \text{ is continuous.)}$$

$$= \frac{dy}{du}\frac{du}{dx}$$

The only flaw in this reasoning is that in (3) it might happen that $\Delta u = 0$ (even when $\Delta x \neq 0$) and, of course, we can't divide by 0. Nonetheless, this reasoning does at least *suggest* that the Chain Rule is true.

■ **EXAMPLE 1 Using the Chain Rule**

Find $F'(x)$ if $F(x) = \sqrt{x^2 + 1}$.

SOLUTION 1 (using Equation 1)

At the beginning of this section we expressed F as $F(x) = f(g(x))$ where $f(u) = \sqrt{u}$ and $g(x) = x^2 + 1$. Then

$$f'(u) = \tfrac{1}{2}u^{-1/2} = \frac{1}{2\sqrt{u}}$$

so

$$f'(g(x)) = \frac{1}{2\sqrt{g(x)}} = \frac{1}{2\sqrt{x^2 + 1}}$$

Since $g'(x) = 2x$, we have

$$F'(x) = f'(g(x))\, g'(x)$$

$$= \frac{1}{2\sqrt{x^2 + 1}} \cdot 2x = \frac{x}{\sqrt{x^2 + 1}}$$

SOLUTION 2 (using Equation 2)

If we let $u = x^2 + 1$ and $y = \sqrt{u}$, then

$$F'(x) = \frac{dy}{du}\frac{du}{dx} = \frac{1}{2\sqrt{u}}(2x)$$

$$= \frac{1}{2\sqrt{x^2 + 1}}(2x) = \frac{x}{\sqrt{x^2 + 1}} \qquad \blacksquare$$

When using Equation 2 we should bear in mind that dy/dx refers to the derivative of y when y is considered as a function of x (called the *derivative of y with respect to x*), whereas dy/du refers to the derivative of y when considered as a function of u (the derivative of y with respect to u). For instance, in Example 1, y can be considered as a function of x $\left(y = \sqrt{x^2 + 1}\right)$ and also as a function of u $\left(y = \sqrt{u}\right)$. Note that

$$\frac{dy}{dx} = F'(x) = \frac{x}{\sqrt{x^2 + 1}} \qquad \text{whereas} \qquad \frac{dy}{du} = f'(u) = \frac{1}{2\sqrt{u}}$$

NOTE: In using the Chain Rule we work from the outside to the inside. Formula 1 says that *we differentiate the outer function f [evaluated at the inner function $g(x)$] and then we multiply by the derivative of the inner function.*

$$\underbrace{\frac{d}{dx}}_{} \quad \underbrace{f}_{\substack{\text{outer} \\ \text{function}}} \quad \underbrace{(g(x))}_{\substack{\text{evaluated} \\ \text{at inner} \\ \text{function}}} \quad = \quad \underbrace{f'}_{\substack{\text{derivative} \\ \text{of outer} \\ \text{function}}} \quad \underbrace{(g(x))}_{\substack{\text{evaluated} \\ \text{at inner} \\ \text{function}}} \quad \cdot \quad \underbrace{g'(x)}_{\substack{\text{derivative} \\ \text{of inner} \\ \text{function}}}$$

▪ **EXAMPLE 2** **Using the Chain Rule**

Differentiate $y = (x^2 + e^x)^3$.

SOLUTION

If $y = (x^2 + e^x)^3$, then the outer function is the cubing function and the inner function is $x^2 + e^x$, so the Chain Rule gives

$$\frac{dy}{dx} = \frac{d}{dx}\underbrace{(x^2 + e^x)^3}_{\substack{\text{inner} \\ \text{function}}} \quad = \quad \underbrace{3(x^2 + e^x)^2}_{\substack{\text{evaluated} \\ \text{at inner} \\ \text{function}}} \quad \cdot \quad \underbrace{(2x + e^x)}_{\substack{\text{derivative of} \\ \text{inner} \\ \text{function}}}$$

$$\underbrace{\qquad\qquad\qquad\qquad\qquad\qquad}_{\text{derivative of outer function}}$$

$$= 3(x^2 + e^x)^2(2x + e^x) = (6x + 3e^x)(x^2 + e^x)^2 \qquad\qquad ■$$

In Example 2 we combined the Chain Rule with the Power Rule. In a similar fashion, all of the formulas for differentiating functions can be combined with the Chain Rule.

▪ The Power Rule

Let's make explicit the special case of the Chain Rule where the outer function f is a power function. If $y = [g(x)]^n$, then we can write $y = f(u) = u^n$ where $u = g(x)$. By using the Chain Rule and then the Power Rule, we get

$$\frac{dy}{dx} = \frac{dy}{du}\frac{du}{dx} = nu^{n-1}\frac{du}{dx} = n[g(x)]^{n-1}g'(x)$$

---

(4) ▪ The Power Rule Combined with the Chain Rule   If n is any real number and $u = g(x)$ is differentiable, then

$$\frac{d}{dx}(u^n) = nu^{n-1}\frac{du}{dx}$$

Alternatively, $\qquad\qquad \dfrac{d}{dx}[g(x)]^n = n[g(x)]^{n-1} \cdot g'(x)$

Notice that the derivative in Example 1 could be calculated by taking $n = \frac{1}{2}$ in Rule 4.

▪ **EXAMPLE 3   Using the Chain Rule with the Power Rule**

Differentiate $y = (x^3 - 1)^{100}$.

SOLUTION

Taking $u = g(x) = x^3 - 1$ and $n = 100$ in (4), we have

$$\frac{dy}{dx} = \frac{d}{dx}(x^3 - 1)^{100} = 100(x^3 - 1)^{99}\frac{d}{dx}(x^3 - 1)$$

$$= 100(x^3 - 1)^{99} \cdot 3x^2 = 300x^2(x^3 - 1)^{99} \qquad\qquad ■$$

■ **EXAMPLE 4** **Using the Chain Rule with a Root Function**

Find $A'(v)$ if $A(v) = \sqrt[3]{v^2 + v + 1}$.

SOLUTION

First rewrite A: $A(v) = (v^2 + v + 1)^{1/3}$. Then

$$A'(v) = \tfrac{1}{3}(v^2 + v + 1)^{-2/3}\frac{d}{dv}(v^2 + v + 1)$$

$$= \tfrac{1}{3}(v^2 + v + 1)^{-2/3}(2v + 1)$$

$$= \frac{2v + 1}{3(v^2 + v + 1)^{2/3}}$$
■

■ **EXAMPLE 5** **The Chain Rule with the Power Rule**

Differentiate:

$$Q = \frac{4}{(1.9t^2 + 3.2)^2}$$

SOLUTION

Rewrite Q as $Q = 4(1.9t^2 + 3.2)^{-2}$. Then

$$\frac{dQ}{dt} = 4(-2)(1.9t^2 + 3.2)^{-3}\frac{d}{dt}(1.9t^2 + 3.2)$$

$$= -8(1.9t^2 + 3.2)^{-3}(1.9 \cdot 2t)$$

$$= \frac{-30.4t}{(1.9t^2 + 3.2)^3}$$
■

We could also differentiate the function in Example 5 by using the Quotient Rule. Note that the Chain Rule is required when differentiating the denominator.

■ **EXAMPLE 6** **The Chain Rule with the Power and Quotient Rules**

Find the derivative of the function

$$g(t) = \left(\frac{t - 2}{2t + 1}\right)^9$$

SOLUTION

Combining the Power Rule, Chain Rule, and Quotient Rule, we get

$$g'(t) = 9\left(\frac{t - 2}{2t + 1}\right)^8 \frac{d}{dt}\left(\frac{t - 2}{2t + 1}\right)$$

$$= 9\left(\frac{t - 2}{2t + 1}\right)^8 \frac{(2t + 1) \cdot 1 - (t - 2) \cdot 2}{(2t + 1)^2}$$

$$= 9\frac{(t - 2)^8}{(2t + 1)^8} \cdot \frac{2t + 1 - 2t + 4}{(2t + 1)^2}$$

$$= 9\frac{(t - 2)^8}{(2t + 1)^8} \cdot \frac{5}{(2t + 1)^2} = \frac{45(t - 2)^8}{(2t + 1)^{10}}$$
■

■ **EXAMPLE 7** **Using the Chain Rule within the Product Rule**

Differentiate $y = (2x + 1)^5(x^3 - x + 1)^4$.

SOLUTION

In this example we must use the Product Rule before using the Chain Rule:

$$\frac{dy}{dx} = (2x + 1)^5 \frac{d}{dx}(x^3 - x + 1)^4 + (x^3 - x + 1)^4 \frac{d}{dx}(2x + 1)^5$$

$$= (2x + 1)^5 \cdot 4(x^3 - x + 1)^3 \frac{d}{dx}(x^3 - x + 1)$$

$$+ (x^3 - x + 1)^4 \cdot 5(2x + 1)^4 \frac{d}{dx}(2x + 1)$$

$$= 4(2x + 1)^5(x^3 - x + 1)^3(3x^2 - 1) + 5(x^3 - x + 1)^4(2x + 1)^4 \cdot 2$$

The graphs of the functions y and y' in Example 7 are shown in Figure 1. Notice that y' is large when y increases rapidly and $y' = 0$ when y has a horizontal tangent. So our answer appears to be reasonable.

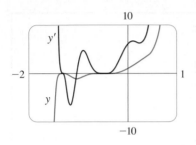

FIGURE 1

Noticing that each term has the common factor $2(2x + 1)^4(x^3 - x + 1)^3$, we could factor it out and write the answer as

$$\frac{dy}{dx} = 2(2x + 1)^4(x^3 - x + 1)^3[2(2x + 1)(3x^2 - 1) + 5(x^3 - x + 1)]$$

$$= 2(2x + 1)^4(x^3 - x + 1)^3(17x^3 + 6x^2 - 9x + 3) \qquad ■$$

■ **EXAMPLE 8** **Differentiating a Composition of Functions**

In Example 4 of Section 1.2, the altitude of a small airplane t hours after taking off was $A(t) = -2.8t^2 + 6.7t$ thousand feet, and the air temperature at an altitude of x thousand feet was $f(x) = 68 - 3.5x$ degrees Fahrenheit. We formed the composition $h(t) = f(A(t))$, which measured the air temperature at the plane's location t hours after take-off. Then $h'(1)$ represents the rate of change of the air temperature with respect to time after one hour.

On the one hand, a formula for h is

$$h(t) = f(A(t)) = f(-2.8t^2 + 6.7t)$$

$$= 68 - 3.5(-2.8t^2 + 6.7t) = 9.8t^2 - 23.45t + 68$$

so

$$h'(t) = 9.8(2t) - 23.45 = 19.6t - 23.45$$

and

$$h'(1) = 19.6(1) - 23.45 = -3.85$$

Thus after one hour, the temperature at the plane's location is decreasing at a rate of 3.85 degrees Fahrenheit per hour.

On the other hand, because the function h is a composition of functions, we can use the Chain Rule to compute $h'(1)$. From Equation 1 we have

$$h'(1) = f'(A(1)) \cdot A'(1) = f'(3.9) \cdot A'(1)$$

$$= (-3.5°\text{F per thousand feet})(1.1 \text{ thousand feet per hour})$$

$$= -3.85°\text{F/h} \qquad ■$$

▪ Exponential Functions

Recall that the natural exponential function is its own derivative:

$$\frac{d}{dx} e^x = e^x$$

When we have a composition of functions where the outer function is the natural exponential function, we must use the Chain Rule. For instance, if $y = e^u$, where u is a differentiable function of x, then

$$\frac{dy}{dx} = \frac{dy}{du} \frac{du}{dx} = e^u \frac{du}{dx}$$

▪ If $u = g(x)$ is a differentiable function, then

(5)
$$\frac{d}{dx}(e^u) = e^u \frac{du}{dx}$$

or, equivalently,

(6)
$$\frac{d}{dx} e^{g(x)} = e^{g(x)} \cdot g'(x)$$

▪ EXAMPLE 9   Using the Chain Rule with an Exponential Function

Differentiate $y = e^{x^2}$.

SOLUTION

Here the inner function is $g(x) = x^2$ and the outer function is the natural exponential function. Using Equation 6 we have

$$\frac{dy}{dx} = \frac{d}{dx}\left(e^{x^2}\right) = e^{x^2} \frac{d}{dx}(x^2) = e^{x^2} \cdot 2x = 2xe^{x^2}$$

▪

▪ EXAMPLE 10

The Rate of Change of an Investment Account Balance

The balance, in dollars, of an investment account after t years is given by $A(t) = 8000e^{0.07t}$. Find $A'(3)$ and interpret your result.

SOLUTION

Using the Constant Multiple Rule and Equation 5, we have

$$A'(t) = 8000e^{0.07t} \frac{d}{dt}(0.07t) = 8000e^{0.07t}(0.07) = 560e^{0.07t}$$

Then $A'(3) = 560e^{0.07(3)} = 560e^{0.21} \approx 690.86$. This means that after three years, the balance of the account is increasing at an instantaneous rate of $690.86 per year.

▪

■ | The Derivative of a^x

We know the derivative of e^x, but now we can use the Chain Rule to differentiate an exponential function a^x with any base $a > 0$. Recall from Equation 2 in Section 1.6 that $a = e^{\ln a}$. So

$$a^x = (e^{\ln a})^x = e^{(\ln a)x}$$

and the Chain Rule (see Equation 5 or 6) gives

$$\frac{d}{dx}(a^x) = \frac{d}{dx}(e^{(\ln a)x}) = e^{(\ln a)x} \frac{d}{dx}(\ln a)x$$

$$= e^{(\ln a)x} \cdot \ln a = a^x \ln a$$

(Note that $\ln a$ is a constant.) So we have the formula

Don't confuse Formula 7 (where x is the *exponent*) with the Power Rule (where x is the *base*):

$$\frac{d}{dx}(x^n) = nx^{n-1}$$

(7)

$$\boxed{\frac{d}{dx}(a^x) = a^x \ln a}$$

In particular, if $a = 2$, we get

(8)

$$\frac{d}{dx}(2^x) = 2^x \ln 2$$

In Section 3.1 we gave the estimate

$$\frac{d}{dx}(2^x) \approx (0.69)2^x$$

This is consistent with the exact formula (8) because $\ln 2 \approx 0.693147$.

■ EXAMPLE 11

Differentiating an Exponential Function with Base other than e

(a) If $P(t) = 1.27^t$, find the value of $P'(8)$.

(b) Find the derivative of $M = 3(5^x) + 4x^2$.

SOLUTION

(a) According to Formula 7, $P'(t) = 1.27^t \ln 1.27$, so

$$P'(8) = 1.27^8 \ln 1.27 \approx 1.618$$

(b) Using the Sum Rule, we have

$$\frac{dM}{dx} = \frac{d}{dx}(3 \cdot 5^x) + \frac{d}{dx}(4x^2)$$

The Constant Multiple Rule then gives

$$\frac{dM}{dx} = 3(5^x \ln 5) + 4 \cdot 2x = 3(5^x \ln 5) + 8x$$

■

■ **EXAMPLE 12**   **Rate of Change of World Population**

In Example 4 of Section 1.5 we used the exponential function

$$P(t) = (1436.53) \cdot (1.01395)^t$$

where $t = 0$ corresponds to the year 1900, to model the world population (measured in millions). Compute and interpret $P'(75)$.

SOLUTION

We use Formula 7 and the Constant Multiple Rule to compute the derivative function:

$$P'(t) = (1436.53) \cdot (1.01395^t \ln 1.01395)$$

Then

$$P'(75) = (1436.53) \cdot (1.01395^{75} \ln 1.01395) \approx 56.25$$

This says that in 1975 the world population was increasing at a rate of about 56.25 million people per year.   ■

We use Formula 7, combined with the Chain Rule, in the next example.

■ **EXAMPLE 13**   **Using the Chain Rule with an Exponential Function**

If $f(x) = 2^{\sqrt{x}}$, find $f'(x)$.

SOLUTION

$$f'(x) = 2^{\sqrt{x}} \ln 2 \frac{d}{dx} \sqrt{x} = 2^{\sqrt{x}} (\ln 2) \cdot \tfrac{1}{2} x^{-1/2}$$

$$= \frac{2^{\sqrt{x}} \ln 2}{2\sqrt{x}}$$

   ■

■ Multiple Chain Rules

The reason for the name "Chain Rule" becomes clear when we make a longer chain by adding another link. Suppose that $y = f(u)$, $u = g(x)$, and $x = h(t)$, where f, g, and h are differentiable functions. Then, to compute the derivative of y with respect to t, we use the Chain Rule twice:

$$\frac{dy}{dt} = \frac{dy}{dx} \frac{dx}{dt} = \frac{dy}{du} \frac{du}{dx} \frac{dx}{dt}$$

■ **EXAMPLE 14**   **A Chain Rule Within a Chain Rule**

Differentiate $y = e^{\sqrt{t^2+2}}$.

SOLUTION

The outer function is the exponential function, the middle function is the square root function, and the inner function is a quadratic function. So we have

$$\frac{dy}{dt} = e^{\sqrt{t^2+2}} \frac{d}{dt}\left(\sqrt{t^2+2}\right)$$

$$= e^{\sqrt{t^2+2}} \cdot \tfrac{1}{2}(t^2+2)^{-1/2}\frac{d}{dt}(t^2+2)$$

$$= e^{\sqrt{t^2+2}} \cdot \tfrac{1}{2}(t^2+2)^{-1/2} \cdot 2t = \frac{te^{\sqrt{t^2+2}}}{\sqrt{t^2+2}}$$ ▪

▪ **EXAMPLE 15   Using the Chain Rule Twice**

If $f(x) = (x + e^{3x})^2$, then

$$f'(x) = 2(x + e^{3x})^1 \frac{d}{dx}(x + e^{3x})$$

$$= 2(x + e^{3x})\left[1 + e^{3x}\frac{d}{dx}(3x)\right]$$

$$= 2(x + e^{3x})[1 + e^{3x} \cdot 3] = 2(x + e^{3x})(1 + 3e^{3x})$$

Notice that we used the Chain Rule twice, first because $(x + e^{3x})$ is the inside function for the squaring function, and later because $3x$ is the inside function for the exponential function.   ▪

▪ Prepare Yourself

1. If $f(g(x)) = \sqrt{5 + 4x}$ and $g(x) = 5 + 4x$, write a formula for $f(x)$.

2. If $f(g(x)) = e^{-3x}$ and $g(x) = -3x$, write a formula for $f(x)$.

3. Find the derivative of the expression.
   (a) $4\sqrt{x}$ **(b)** $5x^6$

(c) $2/\sqrt[3]{t}$

(d) $-8e^t$

(e) $x^2 e^x$

(f) $\dfrac{w}{w^2 + 1}$

4. Find the slope of the graph of $y = 6\sqrt{x} + x^2$ at the point $(4, 28)$.

▪ Exercises 3.4

1–6 ▪ Write the composite function in the form $f(g(x))$. [Identify the inner function $u = g(x)$ and the outer function $y = f(u)$.] Then find the derivative dy/dx.

1. $y = \sqrt{x^2 + 4}$

2. $y = \sqrt{4 + 3x}$

3. $y = (1 - x^2)^{10}$

4. $y = (x^3 + x)^6$

5. $y = e^{\sqrt{x}}$

6. $y = \sqrt{e^x}$

7–38 ▪ Find the derivative of the function.

7. $f(x) = \sqrt{9 - x^2}$

8. $g(t) = \sqrt{t^3 - t}$

9. $F(x) = \sqrt[4]{1 + 2x + x^3}$

10. $F(x) = (x^2 - x + 1)^3$

11. $y = (2x^4 - 8x^2)^7$

12. $y = \sqrt[3]{x + e^x}$

13. $g(t) = \dfrac{1}{(t^4 + 1)^3}$

14. $q(r) = \dfrac{1}{(3r - 1.5r^3)^2}$

15. $A(x) = 5.3e^{0.8x}$

16. $B(t) = 6 + 2e^{-3t}$

17. $y = xe^{-x^2}$

18. $y = \sqrt{3}\,xe^{-5x}$

19. $P(t) = 6^t + 8$

20. $f(x) = 3 - 2 \cdot 5^x$

21. $A = 4500(1.124^t)$

22. $V = 240.2(0.97^x)$

23. $g(x) = (1 + 4x)^5(3 + x - x^2)^8$

24. $h(t) = (t^4 - 1)^3(t^3 + 1)^4$

25. $P(t) = 4^{2+t/3}$

26. $y = 10^{1-x^2}$

27. $L(t) = e^{3 \cdot 2^t}$

28. $r(t) = 0.5e^{t\sqrt{t+1}}$

29. $F(z) = \sqrt{\dfrac{z-1}{z+1}}$

30. $G(y) = \left(\dfrac{y^2}{y+1}\right)^5$

31. $y = \dfrac{r}{\sqrt{r^2+1}}$

32. $y = \dfrac{e^{-x}+1}{e^x+1}$

33. $y = \dfrac{10}{1 + 2e^{-0.3t}}$

34. $P(t) = \dfrac{250}{1 - 0.7e^{0.25t}}$

35. $Q(x) = \sqrt{e^{3x} + x}$

36. $y = (5^{t^2} + 1)^4$

37. $y = e^{\sqrt[3]{x^2+2}}$

38. $y = \sqrt{x + \sqrt{x}}$

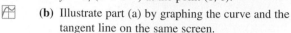 **39. (a)** Graph $y = 4^x$ and its derivative on a common screen. How do the two curves compare?

  (b) Graph $y = (1/4)^x$ and its derivative on a common screen. Do you observe the same relationship as with the curves in part (a)?

40. (a) What types of functions are $f(x) = 3^x$ and $g(x) = x^3$? Compare the differentiation formulas for f and g.

  (b) Which of the two functions in part (b) grows more rapidly when x is large?

41–42 ▪ Find y' and y''.

41. $y = e^{-0.5x}$

42. $y = xe^{-x}$

43–44 ▪ Find $f'(x)$ in two ways, by using the Chain Rule and then by first expanding the expression. Verify that your answers are equivalent.

43. $f(x) = (3x^5 + 1)^2$

44. $f(x) = (x^2 - 5x)^2$

45–46 ▪ Find the slope of the graph of the function at the given input value. Round your answer to three decimal places.

45. $f(x) = 3x - 3^x$, $x = 2$

46. $g(t) = 2(1.5^t)$, $t = 5$

47–48 ▪ Find an equation of the tangent line to the curve at the given point.

47. $y = (1 + 2x)^{10}$, $(0, 1)$ **48.** $y = x^2 - 2^x$, $(3, 1)$

49. (a) Find an equation of the tangent line to the curve $y = 2/(1 + e^{-x})$ at the point $(0, 1)$.

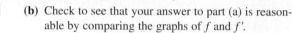 **(b)** Illustrate part (a) by graphing the curve and the tangent line on the same screen.

50. (a) The curve $y = |x|/\sqrt{2 - x^2}$ is called a *bullet-nose curve*. Find an equation of the tangent line to this curve at the point $(1, 1)$.

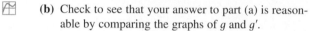 **(b)** Illustrate part (a) by graphing the curve and the tangent line on the same screen.

51. (a) If $f(x) = x\sqrt{2 - x^2}$, find $f'(x)$.

  (b) Check to see that your answer to part (a) is reasonable by comparing the graphs of f and f'.

52. (a) If $g(x) = x \cdot 2^{1-x}$, find $g'(x)$.

  (b) Check to see that your answer to part (a) is reasonable by comparing the graphs of g and g'.

53. Drug concentration   The concentration of a drug in a patient's bloodstream t minutes after an oral dose was administered was approximately $C(t) = 0.6t(0.98^t)$, where $C(t)$ was measured in micrograms per milliliter (μg/mL). Find the rate at which the drug concentration was decreasing two hours after the dose was taken.

54. Consumer demand   A company estimates that it will sell $f(p) = 16(0.92^p)$ thousand units of a particular product when the price is p dollars per unit. The resulting revenue earned is $R(p) = p \cdot f(p)$. Compute $R'(30)$ and interpret your result.

55. Investment account   The value of a particular retirement account is $A(t) = 26{,}800e^{0.07t}$ dollars after t years. Compute and interpret $A'(3.5)$.

56. Investment account   The balance, in dollars, of an investor's savings account is $A(t) = 9500(1.05244^t)$, where t is the number of years since the account was opened. Compute and interpret $A'(4.5)$.

57. Cooling temperature   Suppose a hot cup of coffee is left on a kitchen counter. If the temperature in °F after t hours is given by $F(t) = 75 + 105(0.62^t)$, find the instantaneous rate of change of the coffee's temperature after an hour and a half.

58. US population   The table gives the US population from 1790 to 1860.

  (a) Use a graphing calculator or computer to fit an exponential function to the data. Graph the data points and the exponential model. How good is the fit?

(b) Estimate the rates of population growth in 1800 and 1850 by averaging slopes of secant lines.

(c) Use the exponential model in part (a) to estimate the rates of growth in 1800 and 1850. Compare these estimates with the ones in part (b).

(d) Use the exponential model to predict the population in 1870. Compare with the actual population of 38,558,000. Can you explain the discrepancy?

Year	Population
1790	3,929,000
1800	5,308,000
1810	7,240,000
1820	9,639,000
1830	12,861,000
1840	17,063,000
1850	23,192,000
1860	31,443,000

59. Electrical charge The flash unit on a camera operates by storing charge on a capacitor and releasing it suddenly when the flash is set off. The following data

describe the charge Q remaining on the capacitor, measured in microcoulombs (μC), at time t in seconds.

t	Q
0.00	100.00
0.02	81.87
0.04	67.03
0.06	54.88
0.08	44.93
0.10	36.76

(a) Use a graphing calculator or computer to find an exponential model for the charge. (See Section 1.5.)

(b) The derivative $Q'(t)$ represents the electric current (measured in microamperes, μA) flowing from the capacitor to the flash bulb. Use part (a) to estimate the current when $t = 0.04$ s.

60. If $h(x) = \sqrt{4 + 3f(x)}$, where $f(1) = 7$ and $f'(1) = 4$, find $h'(1)$.

61. Suppose f is differentiable on $\mathbb{R}$. Let $F(x) = f(e^x)$ and $G(x) = e^{f(x)}$. Find expressions for **(a)** $F'(x)$ and **(b)** $G'(x)$.

■ Challenge Yourself

62. A table of values for f, g, f', and g' is given.

x	$f(x)$	$g(x)$	$f'(x)$	$g'(x)$
1	3	2	4	6
2	1	8	5	7
3	7	2	7	9

(a) If $h(x) = f(g(x))$, find $h'(1)$.

(b) If $H(x) = g(f(x))$, find $H'(1)$.

63. If f and g are the functions whose graphs are shown, let $u(x) = f(g(x))$, $v(x) = g(f(x))$, and $w(x) = g(g(x))$. Find each derivative, if it exists. If it does not exist, explain why.

(a) $u'(1)$

(b) $v'(1)$

(c) $w'(1)$

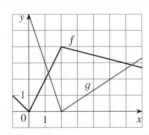

64. Find the 1000th derivative of $f(x) = xe^{-x}$.

65. Find the 30th derivative of $y = e^{2x}$.

66. (a) Write $|x| = \sqrt{x^2}$ and use the Chain Rule to show that

$$\frac{d}{dx}|x| = \frac{x}{|x|}$$

(b) If $f(x) = |x^2 - 4|$, find $f'(x)$ and sketch the graphs of f and f'. Where is f not differentiable?

3.5 Implicit Differentiation and Logarithms

■ Implicit Differentiation

The functions that we have met so far can be described by expressing one variable explicitly in terms of another variable—for example,

$$y = x^3 + 2x \qquad \text{or} \qquad y = 4e^{-x}$$

or, in general, $y = f(x)$. Some functions, however, are defined implicitly by a relation between x and y such as

(1) $$x^2 + y^2 = 25$$

or

(2) $$x^3 + y^3 = 6xy$$

In some cases it is possible to solve such an equation for y as an explicit function (or several functions) of x. For instance, if we solve Equation 1 for y, we obtain $y = \pm\sqrt{25 - x^2}$, so two of the functions determined by the implicit Equation 1 are $f(x) = \sqrt{25 - x^2}$ and $g(x) = -\sqrt{25 - x^2}$. The graphs of f and g are the upper and lower halves of the circle $x^2 + y^2 = 25$. (See Figure 1.)

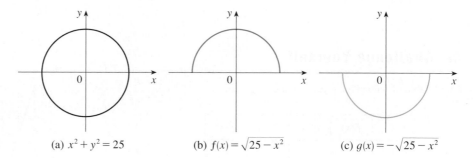

FIGURE 1   (a) $x^2 + y^2 = 25$ (b) $f(x) = \sqrt{25 - x^2}$ (c) $g(x) = -\sqrt{25 - x^2}$

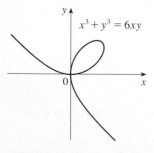

FIGURE 2 The folium of Descartes

It's not easy to solve Equation 2 for y explicitly as a function of x by hand. (It is possible, but the resulting expressions are very complicated.) Nonetheless, (2) is the equation of a curve called the **folium of Descartes** shown in Figure 2 and it implicitly defines y as several functions of x. (It is not possible to solve y explicitly as a single function of x because the graph fails the Vertical Line Test.) When we say that f is a function defined implicitly by Equation 2, we mean that the equation

$$x^3 + [f(x)]^3 = 6xf(x)$$

is true for all values of x in the domain of f.

Fortunately, we don't need to solve an equation for y in terms of x in order to find the derivative of y. Instead we can use the method of **implicit differentiation**. This consists of differentiating both sides of the equation with respect to x and then solving the resulting equation for dy/dx. When differentiating expressions containing y we must remember that y represents a function of x and so the Chain Rule applies. In the examples and exercises of this section it is always assumed that the given equation determines y implicitly as a differentiable function of x so that the method of implicit differentiation can be applied.

■ EXAMPLE 1 The Method of Implicit Differentiation

(a) If $x^2 + y^2 = 25$, find $\dfrac{dy}{dx}$.

(b) Find an equation of the tangent line to the circle $x^2 + y^2 = 25$ at the point $(3, 4)$.

SOLUTION 1

(a) Differentiate both sides of the equation $x^2 + y^2 = 25$ with respect to x:

$$\frac{d}{dx}(x^2 + y^2) = \frac{d}{dx}(25)$$

$$\frac{d}{dx}(x^2) + \frac{d}{dx}(y^2) = 0$$

To compute $\dfrac{d}{dx}(y^2)$ remember that y is a function of x, so we must use the Chain Rule (Equation 2 in Section 3.4), giving

$$\frac{d}{dx}(y^2) = \frac{d}{dy}(y^2)\frac{dy}{dx} = 2y\frac{dy}{dx}$$

If you have trouble seeing why the factor dy/dx is included, write the derivative in function notation:

$$\frac{d}{dx}[f(x)]^2 = 2f(x) \cdot f'(x)$$

Thus we have

$$2x + 2y\frac{dy}{dx} = 0$$

Now we solve this equation for dy/dx:

$$2y\frac{dy}{dx} = -2x \quad \Rightarrow \quad \frac{dy}{dx} = -\frac{2x}{2y} = -\frac{x}{y}$$

(b) At the point $(3, 4)$ we have $x = 3$ and $y = 4$, so

$$\frac{dy}{dx} = -\frac{3}{4}$$

An equation of the tangent line to the circle at $(3, 4)$ is therefore

$$y - 4 = -\tfrac{3}{4}(x - 3) \qquad \text{or} \qquad 3x + 4y = 25$$

SOLUTION 2

(b) Solving the equation $x^2 + y^2 = 25$, we get $y = \pm\sqrt{25 - x^2}$. The point $(3, 4)$ lies on the upper semicircle $y = \sqrt{25 - x^2}$ and so we consider the function $f(x) = \sqrt{25 - x^2}$. Differentiating f using the Chain Rule, we have

Example 1 illustrates that even when it is possible to solve an equation explicitly for y in terms of x, it may be easier to use implicit differentiation.

$$f'(x) = \tfrac{1}{2}(25 - x^2)^{-1/2}\frac{d}{dx}(25 - x^2)$$

$$= \tfrac{1}{2}(25 - x^2)^{-1/2}(-2x) = -\frac{x}{\sqrt{25 - x^2}}$$

So $\qquad f'(3) = -\dfrac{3}{\sqrt{25 - 3^2}} = -\dfrac{3}{4}$

and, as in Solution 1, an equation of the tangent is $3x + 4y = 25$. ▪

▪ **EXAMPLE 2   Finding a Tangent Line Implicitly**

(a) Find dy/dx if $x^3 + y^3 = 6xy$.

(b) Find the tangent line to the folium of Descartes $x^3 + y^3 = 6xy$ at the point $(3, 3)$.

SOLUTION

(a) First we differentiate both sides of $x^3 + y^3 = 6xy$ with respect to x:

$$\frac{d}{dx}(x^3 + y^3) = \frac{d}{dx}(6xy)$$

Regarding y as a function of x, and using the Chain Rule on the y^3 term and the Product Rule on the $6xy$ term, we get

$$3x^2 + 3y^2\frac{dy}{dx} = 6x\frac{dy}{dx} + 6y$$

or

$$x^2 + y^2\frac{dy}{dx} = 2x\frac{dy}{dx} + 2y$$

We now solve for $\dfrac{dy}{dx}$:

$$y^2\frac{dy}{dx} - 2x\frac{dy}{dx} = 2y - x^2$$

Factor out $\dfrac{dy}{dx}$ in the expression on the left.

$$(y^2 - 2x)\frac{dy}{dx} = 2y - x^2$$

$$\frac{dy}{dx} = \frac{2y - x^2}{y^2 - 2x}$$

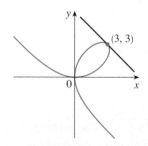

(b) When $x = y = 3$,

$$\frac{dy}{dx} = \frac{2 \cdot 3 - 3^2}{3^2 - 2 \cdot 3} = -1$$

and a glance at Figure 3 confirms that this is a reasonable value for the slope at $(3, 3)$. So an equation of the tangent line to the folium at $(3, 3)$ is

$$y - 3 = -1(x - 3) \qquad \text{or} \qquad x + y = 6 \qquad ▪$$

FIGURE 3

We mentioned that it is possible (but complicated) to solve the equation $x^3 + y^3 = 6xy$ for y in terms of x. Nevertheless, the method of implicit differentiation saves an enormous amount of work in cases such as this. Moreover, implicit differentiation works just as easily for equations such as

$$y^5 + 3x^2y^2 + 5x^4 = 12$$

for which it is *impossible* to find an equivalent expression for y in terms of x.

■ **EXAMPLE 3** **Differentiating a Demand Function Implicitly**

A company estimates that the number of units x of a new product, measured in thousands, that can be sold and the price p, in dollars, of each unit are related by the equation $px^2 + 15px = 30{,}000$. Find dx/dp when the price is $30 per unit.

SOLUTION

We could solve the equation $px^2 + 15px = 30{,}000$ for x and then compute the derivative, but it is easier to differentiate implicitly with respect to p, remembering to treat x as a function of p. (On the left side we use the Product Rule twice.)

$$p \cdot 2x \frac{dx}{dp} + x^2 \cdot 1 + 15\left(p \frac{dx}{dp} + x \cdot 1 \right) = 0$$

$$2px \frac{dx}{dp} + 15p \frac{dx}{dp} = -x^2 - 15x$$

$$(2px + 15p) \frac{dx}{dp} = -x^2 - 15x$$

$$\frac{dx}{dp} = \frac{-x^2 - 15x}{2px + 15p}$$

When $p = 30$ we have $30x^2 + 15(30)x = 30{,}000 \implies x^2 + 15x - 1000 = 0$. The only positive solution is $x = 25$ (you can solve by factoring or use the quadratic formula) so

$$\frac{dx}{dp}\bigg|_{p=30} = \frac{-25^2 - 15(25)}{2(30)(25) + 15(30)} = \frac{-1000}{1950} = -\frac{20}{39} \approx -0.513$$

Thus when the units are priced at $30, an increase in the price will reduce the number of units sold at a rate of about 513 per dollar. ■

■ Derivative of the Natural Logarithmic Function

Armed with the technique of implicit differentiation, we are now prepared to find the derivative of the natural logarithmic function $y = \ln x$.

If we let $y = \ln x$ then $e^y = x$. Differentiating this equation implicitly with respect to x, we get

$$e^y \frac{dy}{dx} = 1$$

Solving for $\dfrac{dy}{dx}$ gives

$$\frac{dy}{dx} = \frac{1}{e^y} = \frac{1}{x}$$

So we have proved the following differentiation formula:

(3)
$$\frac{d}{dx}(\ln x) = \frac{1}{x}$$

▪ **EXAMPLE 4   Differentiating a Logarithm**

If $g(t) = 3t - 4\ln t$, find $g'(t)$. What is the slope of the graph at $t = 1$?

SOLUTION

$$g'(t) = \frac{d}{dt}(3t) - \frac{d}{dt}(4\ln t) = 3 - 4\cdot\frac{1}{t} = 3 - \frac{4}{t}$$

The slope of the graph at $t = 1$ is

$$g'(1) = 3 - \frac{4}{1} = -1$$ ▪

▪ **EXAMPLE 5   The Derivative of a Composite Logarithmic Function**

Differentiate $y = \ln(x^3 + 1)$.

SOLUTION

Here we must use the Chain Rule. Let $u = x^3 + 1$; then $y = \ln u$, so

$$\frac{dy}{dx} = \frac{dy}{du}\frac{du}{dx} = \frac{1}{u}\frac{du}{dx} = \frac{1}{x^3+1}(3x^2) = \frac{3x^2}{x^3+1}$$ ▪

In general, if we combine Formula 3 with the Chain Rule as in Example 5, we get

(4)
$$\frac{d}{dx}(\ln u) = \frac{1}{u}\frac{du}{dx}$$ or $$\frac{d}{dx}[\ln g(x)] = \frac{g'(x)}{g(x)}$$

Figure 4 shows the graph of $f(x) = \ln(x^2 + 3e^x)$ in Example 6 along with its derivative. Notice that f' is negative when f is decreasing.

▪ **EXAMPLE 6   Using the Chain Rule with a Logarithmic Function**

Find $\dfrac{d}{dx}\ln(x^2 + 3e^x)$.

SOLUTION

Using (4), we have

$$\frac{d}{dx}\ln(x^2 + 3e^x) = \frac{1}{x^2+3e^x}\frac{d}{dx}(x^2+3e^x) = \frac{1}{x^2+3e^x}(2x + 3e^x) = \frac{2x + 3e^x}{x^2+3e^x}$$ ▪

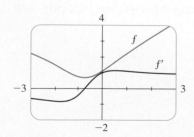

FIGURE 4

■ EXAMPLE 7   **Using the Chain Rule with a Logarithmic Function**

Differentiate $f(x) = \sqrt{\ln x}$.

SOLUTION

First we rewrite the function as $f(x) = (\ln x)^{1/2}$. This time the logarithm is the inner function, so the Chain Rule gives

$$f'(x) = \tfrac{1}{2}(\ln x)^{-1/2}\frac{d}{dx}(\ln x) = \frac{1}{2\sqrt{\ln x}} \cdot \frac{1}{x} = \frac{1}{2x\sqrt{\ln x}}$$ ■

■ EXAMPLE 8

The Natural Logarithm of the Absolute Value Function

Find $f'(x)$ if $f(x) = \ln|x|$.

Figure 5 shows the graph of the function $f(x) = \ln|x|$ in Example 8 and its derivative $f'(x) = 1/x$. Notice that when x is small, the graph of $y = \ln|x|$ is steep and so $f'(x)$ is large (positive or negative).

SOLUTION

Since

$$f(x) = \begin{cases} \ln x & \text{if } x > 0 \\ \ln(-x) & \text{if } x < 0 \end{cases}$$

it follows that

$$f'(x) = \begin{cases} \dfrac{1}{x} & \text{if } x > 0 \\ \dfrac{1}{-x}(-1) = \dfrac{1}{x} & \text{if } x < 0 \end{cases}$$

Thus $f'(x) = 1/x$ for all $x \neq 0$.   ■

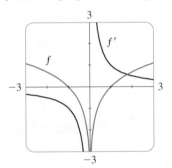

FIGURE 5

The result of Example 8 is worth remembering (and we will revisit it in Chapter 5):

(5)
$$\frac{d}{dx}\ln|x| = \frac{1}{x}$$

■ The Number e as a Limit

As we have seen, one of the main reasons in calculus that the number e is used as a base for both exponential and logarithmic functions is that the differentiation formulas are simplest with this base. We observed in Section 1.5 that e lies somewhere between 2 and 3, but how can we get a more precise value for this mysterious number?

We have shown that if $f(x) = \ln x$, then $f'(x) = 1/x$. Thus $f'(1) = 1$. We now use this fact to express the number e as a limit. If we compute $f'(1)$ from the definition of a derivative, we have

$$f'(1) = \lim_{h \to 0} \frac{f(1 + h) - f(1)}{h} = \lim_{h \to 0} \frac{\ln(1 + h) - \ln 1}{h}$$

$$= \lim_{h \to 0} \frac{1}{h} \ln(1 + h) = \lim_{h \to 0} \ln(1 + h)^{1/h}$$

(We used Law 3 of logarithms from Section 1.6 and the fact that $\ln 1 = 0$.) Because $f'(1) = 1$, we have

$$\lim_{h \to 0} \ln(1 + h)^{1/h} = 1$$

Because exponential functions are continuous, it can be shown that

$$e^{\lim_{h \to 0} \ln(1+h)^{1/h}} = \lim_{h \to 0} e^{\ln(1+h)^{1/h}}$$

But

$$e^{\ln(1+h)^{1/h}} = (1 + h)^{1/h}$$

so we can write

$$\lim_{h \to 0} (1 + h)^{1/h} = \lim_{h \to 0} e^{\ln(1+h)^{1/h}}$$

$$= e^{\lim_{h \to 0} \ln(1+h)^{1/h}} = e^1 = e$$

We can use any variable in place of h, so we have established the following formula.

(6)
$$e = \lim_{x \to 0} (1 + x)^{1/x}$$

The graph of the function $y = (1 + x)^{1/x}$ is shown in Figure 6. Formula 6 says that the graph approaches the value of e at the y-axis. The following table of values shows function values as x approaches 0.

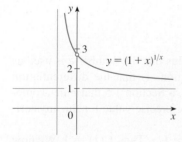

FIGURE 6

x	$(1 + x)^{1/x}$
0.1	2.59374246
0.01	2.70481383
0.001	2.71692393
0.0001	2.71814593
0.00001	2.71826824
0.000001	2.71828047
0.0000001	2.71828169
0.00000001	2.71828181

We can get more and more accurate estimates for e by choosing x-values closer and closer to zero. In fact, correct to seven decimal places,

$$e \approx 2.7182818$$

■ Proof of the General Power Rule

We have used the Power Rule extensively throughout this chapter. In Section 3.1 we proved the rule for positive integer exponents, but now we can prove that it is true for *all* real numbers n.

■ **The Power Rule**   If n is any real number and $f(x) = x^n$, then

$$f'(x) = nx^{n-1}$$

PROOF   Let $y = x^n$. First we assume that $x > 0$ and so $y > 0$. Then take the natural logarithm of both sides:

$$\ln y = \ln x^n$$

Using properties of logarithms, we have

$$\ln y = n \ln x$$

Differentiating both sides with respect to x gives

$$\frac{1}{y} \cdot \frac{dy}{dx} = n \cdot \frac{1}{x}$$

Hence

$$\frac{dy}{dx} = n \frac{y}{x}$$

Finally we substitute $y = x^n$ and obtain

$$\frac{dy}{dx} = n \frac{x^n}{x} = nx^{n-1}$$

If $x = 0$, we can show that $f'(0) = 0$ for $n > 1$ directly from the definition of a derivative.

If $x < 0$ and y is defined, then y could be negative and we take the absolute value of each side before taking the natural logarithm:

$$\ln|y| = \ln|x^n| = \ln|x|^n = n \ln|x|$$

Differentiating (see Equation 5), we again have

$$\frac{1}{y} \cdot \frac{dy}{dx} = n \cdot \frac{1}{x}$$

so we achieve the same result.   ■

■ Table of Differentiation Formulas

We now have a complete set of differentiation formulas. We summarize them here for reference.

	Function	Derivative
Constant function	c	0
Power Rule	x^n	nx^{n-1}
Power Rule with Chain Rule	$[g(x)]^n$	$n[g(x)]^{n-1} g'(x)$
Reciprocal function	$\dfrac{1}{x}$	$-\dfrac{1}{x^2}$
Reciprocal function with Chain Rule	$\dfrac{1}{g(x)}$	$-\dfrac{g'(x)}{[g(x)]^2}$
Natural exponential function	e^x	e^x
Natural exponential function with Chain Rule	$e^{g(x)}$	$e^{g(x)} g'(x)$
Exponential function	a^x	$a^x \ln a$
Natural logarithmic function	$\ln x$	$1/x$
Natural logarithmic function with Chain Rule	$\ln g(x)$	$\dfrac{g'(x)}{g(x)}$
Constant Multiple Rule	$c f(x)$	$c f'(x)$
Sum Rule	$f(x) + g(x)$	$f'(x) + g'(x)$
Difference Rule	$f(x) - g(x)$	$f'(x) - g'(x)$
Product Rule	$f(x) g(x)$	$f(x) g'(x) + g(x) f'(x)$
Quotient Rule	$\dfrac{f(x)}{g(x)}$	$\dfrac{g(x) f'(x) - f(x) g'(x)}{[g(x)]^2}$
Chain Rule	$f(g(x))$	$f'(g(x)) g'(x)$
Chain Rule (Leibniz notation)	$y = f(u), u = g(x)$	$\dfrac{dy}{dx} = \dfrac{dy}{du} \dfrac{du}{dx}$

■ Prepare Yourself

1. If $y = f(x)$, find the derivative with respect to x of the expression.

   (a) $3x + f(x)$ **(b)** $xf(x)$

   (c) $[f(x)]^3$ **(d)** $e^{f(x)}$

2. If $y = f(x)$, evaluate the given derivative.

   (a) $\dfrac{d}{dx}(x + y^4)$ **(b)** $\dfrac{d}{dx}(\sqrt{x} + \sqrt{y})$

   (c) $\dfrac{d}{dx}(x^2y^2 + y)$

   (d) $\dfrac{d}{dx}(e^x - e^y)$

3. If $N = 5a - 2b + a^2\sqrt{b}$, find the indicated derivative.

   (a) dN/da

   (b) dN/db

■ Exercises 3.5

1–2 ■

(a) Find dy/dx by implicit differentiation.

(b) Solve the equation explicitly for y and differentiate to get dy/dx in terms of x.

(c) Check that your solutions to parts (a) and (b) are consistent by substituting the expression for y from part (b) into your solution for part (a).

1. $xy + 2x + 3x^2 = 4$ **2.** $\sqrt{x} + \sqrt{y} = 4$

3–10 ■ Find dy/dx by implicit differentiation.

3. $x^2 + y^2 = 1$ **4.** $x^2 - y^2 = 1$

5. $x^3 + x^2y + 4y^2 = 6$ **6.** $x^2 - 2xy + y^3 = c$

7. $x^2y + xy^2 = 3x$ **8.** $y^5 + x^2y^3 = 1 + ye^{x^2}$

9. $e^{x^2y} = x + y$ **10.** $\sqrt{x + y} = 1 + x^2y^2$

11. If $f(x) + x^2[f(x)]^3 = 10$ and $f(1) = 2$, find $f'(1)$.

12. If $\ln(q + r) = qr$, find dq/dr.

13. If $C + L^3 = e^{2C}$, compute dC/dL when $C = 0$.

14. If $e^{2p} - e^t = p + t$, compute dp/dt when $p = t = 0$.

15–18 ■ Use implicit differentiation to find an equation of the tangent line to the curve at the given point.

15. $x^2 + xy + y^2 = 3$, $(1, 1)$ (ellipse)

16. $x^{2/3} + y^{2/3} = 4$, $\left(-3\sqrt{3}, 1\right)$ (astroid)

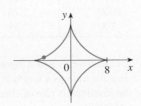

17. $2(x^2 + y^2)^2 = 25(x^2 - y^2)$

   $(3, 1)$

   (leminiscate)

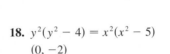

18. $y^2(y^2 - 4) = x^2(x^2 - 5)$

   $(0, -2)$

   (devil's curve)

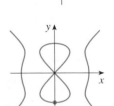

19. Software demand A software company estimates that the number of copies x, measured in thousands, of its new application that will sell when the price is p dollars per copy (up to $100) satisfies the equation $2px^2 + 3px = 58{,}000$. Find dx/dp when the price is $40 per copy and interpret your result.

20. Vehicle demand A company conducted a survey regarding its new electric car and estimates that the selling price p of the vehicle, in thousands of dollars, and the number x of cars that will sell, measured in thousands, are related by the equation $px^2 + 12px = 28{,}000$. Compute and interpret dx/dp and dp/dx when the selling price is $36,000.

21–38 ■ Differentiate the function.

21. $f(x) = 3x - 2\ln x$

22. $g(z) = 9\ln z + 12z + 8.1$

23. $y = 1.5x + \ln x$ **24.** $y = 3\ln t - t^2$

25. $y = (\ln x)^5$ **26.** $f(x) = \ln(x^2 + 10)$

27. $f(x) = \sqrt[5]{\ln x}$ **28.** $f(x) = \ln\sqrt[5]{x}$

29. $y = (\ln x + 1)^2 + (e^x + 1)^2$

30. $f(t) = \dfrac{1 + \ln t}{1 - \ln t}$

31. $F(t) = \ln \dfrac{(2t + 1)^3}{(3t - 1)^4}$

32. $F(y) = y \ln(1 + e^y)$

33. $f(u) = \dfrac{\ln u}{1 + \ln(2u)}$

34. $G(u) = \ln \sqrt{\dfrac{3u + 2}{3u - 2}}$

35. $y = \ln(e^{-x} + xe^{-x})$ **36.** $y = [\ln(3x - 2)]^2$

37. $y = \ln(\ln x)$ **38.** $y = [\ln(1 + e^x)]^2$

39–41 ▪ Find f' and f''.

39. $f(x) = \ln(x^2 - 5)$ **40.** $f(x) = \dfrac{\ln x}{x^2}$

41. $f(x) = e^x \ln x$

42. Find the slope of the graph of

$$M(y) = 2.1y^2 - 11.5 \ln y$$

at $y = 1.7$. Round your answer to three decimal places.

43–44 ▪ Find an equation of the tangent line to the curve at the given point.

43. $y = \ln(x^2 - 3)$, $(2, 0)$

44. $y = x^2 + (\ln x)^2$, $(1, 1)$

45–46 ▪ Find $f'(x)$. Compare the graphs of f and f' and use them to explain why your answer is reasonable.

45. $f(x) = 4 \ln x - x$ **46.** $f(x) = \ln x - 2\sqrt{x}$

47. Use the change of base formula $\log_a x = \dfrac{\ln x}{\ln a}$

to show that $\dfrac{d}{dx} \log_a x = \dfrac{1}{x \ln a}$.

48. Find the derivative of the logarithmic function.
   (a) $f(x) = \log_2 (1 - 3x)$ **(b)** $f(x) = \log_5 (xe^x)$

49. Loudness   The human perception of the loudness of sound with physical intensity I, measured in watts per square meter (W/m^2), is given by $B = 120 + 10 \log_{10} I$, where B is measured in decibels (dB). Find B and dB/dI when the physical intensity of the sound generated by a jet taking off is 100 W/m^2. What are the units?

50. Find equations of the tangent lines to the curve $y = (\ln x)/x$ at the points $(1, 0)$ and $(e, 1/e)$. Illustrate by graphing the curve and its tangent lines.

51. Find dy/dx if $y = \ln(x^2 + y^2)$.

52. Find the slope of the curve with equation $y = \ln(x + 2y)$ at the point $(1, 0)$.

53. (a) The curve with equation $y^2 = 5x^4 - x^2$ is called a *kampyle of Eudoxus*. Find an equation of the tangent line to this curve at the point $(1, 2)$.

   (b) Illustrate part (a) by graphing the curve and the tangent line on a common screen. (If your graphing device will graph implicitly defined curves, then use that capability. If not, you can still graph this curve by graphing its upper and lower halves separately.)

54. (a) The curve with equation $y^2 = x^3 + 3x^2$ is called the *Tschirnhausen cubic*. Find an equation of the tangent line to this curve at the point $(1, -2)$.

   (b) At what points does this curve have a horizontal tangent?

   (c) Illustrate parts (a) and (b) by graphing the curve and the tangent lines on a common screen.

55. (a) Use a graph of $f(x) = \ln(x^2 + e^{-x})$ produced by a graphing calculator to make a rough sketch of the graph of f'.

   (b) Calculate $f'(x)$ and use this expression to graph f'. Compare with your sketch in part (a).

56. Kiln temperature   A pottery kiln heated to 2400°F is turned off and allowed to cool. The table shows the time elapsed as the temperature decreased.

Temperature (°F)	Time (hours)
2200	0.52
2000	1.12
1800	1.80
1600	2.56
1400	3.45
1200	4.48
1000	5.76
800	7.39

Use a graphing calculator to find a logarithmic model $f(x) = a + b \ln x$ for time as a function of temperature and compute the instanteous rate of change when the temperature is 1600°F.

▪ Challenge Yourself

57. Find an equation of the tangent line to the hyperbola

$$\frac{x^2}{a^2} - \frac{y^2}{b^2} = 1$$

at the point (x_0, y_0).

58. Show by implicit differentiation that the tangent to the ellipse

$$\frac{x^2}{a^2} + \frac{y^2}{b^2} = 1$$

at the point (x_0, y_0) is

$$\frac{x_0 x}{a^2} + \frac{y_0 y}{b^2} = 1$$

59. The equation $x^2 - xy + y^2 = 3$ represents a "rotated ellipse," that is, an ellipse whose axes are not parallel to the coordinate axes. Find the points at which this ellipse crosses the x-axis and show that the tangent lines at these points are parallel.

60. Show that the sum of the x- and y-intercepts of any tangent line to the curve $\sqrt{x} + \sqrt{y} = \sqrt{c}$ is equal to c.

61. In proving the general Power Rule we first took the natural logarithm of both sides of the initial equation and used properties of logarithms to simplify the result. Then we found dy/dx by implicit differentiation and finally used the original equation to express any occurences of y in terms of x. This technique is called **logarithmic differentiation**; use it to find the derivative of $y = x^x$.

62. Use logarithmic differentiation to find dy/dx if $x^y = y^x$.

63. The function

$$y = \frac{e^{x^2}\sqrt{x^3 + x}}{(2x + 3)^4}$$

can be differentiated explicitly but the process is messy. Instead, use the technique of logarithmic differentiation to find dy/dx.

64. The Power Rule can be proved using implicit differentiation for the case where n is a rational number, $n = p/q$, and $y = f(x) = x^n$ is assumed beforehand to be a differentiable function. If $y = x^{p/q}$, then $y^q = x^p$. Use implicit differentiation to show that

$$y' = \frac{p}{q} x^{(p/q)-1}$$

3.6 Exponential Growth and Decay

In this chapter we learned that the derivative of an exponential function is a constant multiple of itself. Any quantity that grows or declines at a rate proportional to itself exhibits what is called *exponential growth* and can be described by a transformed exponential function. In this section we will see how this behavior naturally occurs in a wide variety of settings such as population growth, the value of interest-earning investments, and the decay of radioactive material.

▪ Constant Percentage Growth

In Section 1.5 we discussed exponential functions and noted that they model quantities that grow at a constant percentage rate. As an example, we formed the function $p(t) = 1000(2^t)$ to model a bacteria population after t hours that started with 1000 bacteria and doubled every hour.

Similarly, suppose you have a work of art currently worth $5000 that is increasing in value 8% each year. Its value one year from now will be 8% higher:

$$\$5000 + \$5000(0.08) = \$5000(1 + 0.08) = \$5000(1.08)$$

Notice that the original value is multiplied by 1.08, so we can say that the value in one year is 108% of the original value. After the second year, the value will again be multiplied by 1.08:

$$[\$5000(1.08)](1.08) = \$5000(1.08)^2$$

After 3 years the value is $\$5000(1.08)^3$, and so on. This pattern leads to the general expression $\$5000(1.08)^t$ for the value of the artwork after t years. If, instead, the art work is *losing* 8% of its value each year, then it is *maintaining* $100\% - 8\% = 92\%$ of its value year after year, so the value after t years is $\$5000(0.92)^t$.

Any quantity that increases or decreases at a constant percentage rate can be described by an exponential function in a similar way. For instance, in Example 4 of Section 1.5 we modeled the world population, in millions, by the function

$$P(t) = (1436.53) \cdot (1.01395)^t$$

where t represents time measured in years. According to this model, the population each year is 101.395% of the prior year's population, so it is increasing by 1.395% each year.

■ EXAMPLE 1

Modeling a Quantity that Decreases at a Constant Percentage Rate

A used Cessna 172 Skyhawk aircraft is purchased for $56,000. The buyer predicts it will decline in value 3.8% each year.

(a) Write a function to model the value V of the aircraft t years from now.

(b) What is the predicted value of the plane 4.5 years from now?

(c) How long will it take for the value of the plane to decline to $20,000?

SOLUTION

(a) If the value of the aircraft decreases 3.8% each year, then $100\% - 3.8\% = 96.2\%$ of the value is maintained, so we multiply the starting value $56,000 by 0.962 for each year that passes. Thus a model for the value is

$$V(t) = 56,000(0.962)^t$$

dollars after t years.

(b) According to the model, in 4.5 years the plane will be worth

$$V(4.5) = \$56,000(0.962)^{4.5} \approx \$47,041$$

Figure 1 shows the graph of the function from Example 1. As a check on our work in part (c), you can see that the curve intersects the horizontal line $V = 20,000$ when $t \approx 27$.

(c) We solve $V(t) = 20,000$:

$$56,000(0.962)^t = 20,000$$

$$0.962^t = \frac{20,000}{56,000} = \frac{5}{14}$$

$$\ln(0.962^t) = \ln \tfrac{5}{14}$$

$$t \ln 0.962 = \ln \tfrac{5}{14}$$

$$t = \frac{\ln \tfrac{5}{14}}{\ln 0.962} \approx 26.58$$

Thus the plane is worth $20,000 after about 26 and a half years. ■

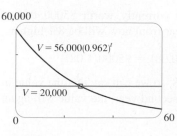

FIGURE 1

▪ Earning Interest

Now we turn our attention to interest-earning investments. If $1000 is invested at 6% annual interest, then the balance is 106% of the original value after one year, and after t years the account is worth $\$1000(1.06)^t$. In general, if an amount P (for principal) is invested at an interest rate r (expressed as a decimal; $r = 0.06$ in this example), then after t years it's worth

(1)
$$A(t) = P(1 + r)^t$$

Usually, however, interest is compounded more frequently, say, n times a year. Then in each compounding period the interest rate is r/n and there are nt compounding periods in t years, so the value of the investment is

(2)
$$A(t) = P\left(1 + \frac{r}{n}\right)^{nt}$$

For instance, after 3 years at 6% interest a $1000 investment will be worth

$$\$1000(1.06)^3 = \$1191.02 \text{ with annual compounding}$$

$$\$1000\left(1 + \tfrac{0.06}{4}\right)^{4\cdot 3} = \$1000(1.015)^{12} = \$1195.62 \text{ with quarterly compounding}$$

$$\$1000\left(1 + \tfrac{0.06}{12}\right)^{12\cdot 3} = \$1000(1.005)^{36} = \$1196.68 \text{ with monthly compounding}$$

$$\$1000\left(1 + \tfrac{0.06}{365}\right)^{365\cdot 3} = \$1197.20 \text{ with daily compounding}$$

You can see that the more often interest is compounded, the more the investor earns.

What happens if we continue to increase n? It would be impractical for a bank to compute compounded interest every hour or every minute, but the concept of a limit enables us to determine the maximum amount of interest that can be earned if the interest is compounded more and more often.

We would like to determine the value of $A(t)$ in Equation 2 as n becomes arbitrarily large. We write $n \to \infty$ and say "n approaches infinity" to indicate that n continues to increase beyond any finite value we could choose. (Note that ∞ is not a number; the notation $n \to \infty$ is used as a convention.) In this case, we say the interest is **compounded continuously**, and the value of the investment is given by

We will investigate limits at infinity in more detail in Section 4.4.

$$A(t) = \lim_{n\to\infty} P\left(1 + \frac{r}{n}\right)^{nt}$$

This is a difficult limit to evaluate, but we can rewrite the limit as

$$\lim_{n\to\infty} P\left[\left(1 + \frac{r}{n}\right)^{n/r}\right]^{rt} = P\left[\lim_{n\to\infty}\left(1 + \frac{r}{n}\right)^{n/r}\right]^{rt}$$

Now let $m = n/r$. If $n \to \infty$, we must have $n/r = m \to \infty$ also (dividing n by a positive constant r does not prevent the result from growing arbitrarily large), so we can write

$$A(t) = P\left[\lim_{m\to\infty}\left(1 + \frac{1}{m}\right)^m\right]^{rt}$$

m	$\left(1 + \dfrac{1}{m}\right)^m$
10	2.593742
100	2.704814
1,000	2.716924
10,000	2.718146
100,000	2.718268
1,000,000	2.718280

Let's investigate the limit in this expression. In the table at the left we compute the value of $[1 + (1/m)]^m$, rounded to six decimal places, for increasing values of m. The values appear to be approaching a familiar number, namely $e \approx 2.71828$. (Again we see the mysterious number e arising in a natural way!)

In fact, if we let $u = 1/m$, then as m becomes larger and larger, u gets closer and closer to 0. Thus $u \to 0$ as $m \to \infty$, and we can write

$$\lim_{m \to \infty} \left(1 + \frac{1}{m}\right)^m = \lim_{u \to 0} (1 + u)^{1/u} = e$$

by Equation 6 in Section 3.5. Thus with continuous compounding of interest at interest rate r, the value of an investment account after t years is

(3)
$$A(t) = Pe^{rt}$$

Returning to the example of $1000 invested for three years at 6% interest, we see that with continuous compounding of interest the value of the investment will be

$$A(3) = \$1000e^{(0.06)3}$$

$$= \$1000e^{0.18} = \$1197.22$$

Notice how close this is to the amount we calculated for daily compounding, $1197.20. But the amount is easier to compute if we use continuous compounding, and we are assured that the amount is the maximum possible for the given interest rate.

▪ **EXAMPLE 2** **Compounded Interest on an Investment**

$25,000 is invested in a certificate of deposit account earning 3.2% interest.

(a) Compute the value of the account after five years if interest is compounded quarterly.

(b) Compute the account value after five years if interest is compounded continuously.

SOLUTION

(a) From Equation 2, the value of the account is

$$\$25,000\left(1 + \frac{0.032}{4}\right)^{4 \cdot 5} = \$25,000(1.008)^{20} = \$29,319.10$$

(b) From Equation 3, the value of the account is

$$\$25,000e^{(0.032)5} = \$25,000e^{0.16} = \$29,337.77 \qquad \blacksquare$$

▪ Exponential Growth

At the beginning of the section we looked at quantities that change at a constant percentage rate. In other words, the rate of change of a quantity is a percentage or multiple of the quantity itself:

$$A'(t) = k \cdot A(t)$$

where k is a constant. This is an example of a *differential equation*, that is, an equation that contains an unknown function and one or more of its derivatives. Its solution is a function or family of functions A. It is easy to see that the function $A(t) = Pe^{kt}$ is a solution, since $A'(t) = Pe^{kt} \cdot k = k \cdot A(t)$. In fact, in Section 6.4 we will show that all solutions to this differential equation are of the form

(4)
$$A(t) = Ce^{kt}$$

where C and k are constants. Thus whenever a quantity grows in proportion to its size (or, equivalently, at a constant percentage rate) it can be described by a function as given in (4). Notice that $A(0) = Ce^0 = C$, so C represents the initial quantity. The constant k plays the same role as r in Equation 3; it is called the **relative growth rate**.

The relative growth rate is not the same as an annual percentage increase. If a population of 50,000 is growing at an annual rate of 10%, then it is described by $50,000(1 + 0.10)^t$. If the population is growing with a relative growth rate of 0.10 (or 10%), then it is described by $50,000e^{0.10t}$.

Many quantities observable in nature grow (or decline) at rates proportional to their size, so Equation 4 is an appropriate model. For instance, if 1000 bacteria are reproducing at a rate of 300 bacteria per hour, it would be reasonable to say that when there are 2000 bacteria, their numbers will be increasing at a rate of 600 bacteria per hour. If the population is doubled, the rate of growth is doubled. Populations of other animals and plants often follow this behavior, as do quantities in some chemical reactions.

The *relative growth rate* measures the rate of change relative to the quantity present and is computed by dividing the growth rate by the present quantity:

$$\frac{f'(t)}{f(t)}$$

In the case of exponential growth or decay,

$$\frac{A'(t)}{A(t)} = \frac{Ce^{kt} \cdot k}{Ce^{kt}} = k$$

so the relative growth rate is constant.

▪ EXAMPLE 3 Exponential Growth of Bacteria

A lab experiment starts with a population of 4000 bacteria that is growing with a relative growth rate of 7.2% per hour. Find the population and its rate of change after three hours. How long will it be until the population reaches 18,000?

SOLUTION

The bacteria population is growing with a constant relative growth rate, so we can use Equation 4 to describe the population:

$$P(t) = 4000e^{0.072t}$$

where t is the number of hours elapsed. After three hours the population is

$$P(3) = 4000e^{0.072(3)} \approx 4964$$

The rate of change is

$$P'(t) = 4000e^{0.072t}(0.072) = 288e^{0.072t}$$

so $P'(3) = 288e^{0.072(3)} \approx 357.4$. Thus the population is increasing at a rate of about 357 bacteria per hour after three hours.

To find the time required for the population to reach 18,000 we solve $P(t) = 18{,}000$ for t:

$$4{,}000e^{0.072t} = 18{,}000$$

$$e^{0.072t} = \frac{18{,}000}{4{,}000} = \frac{9}{2}$$

$$\ln(e^{0.072t}) = \ln\tfrac{9}{2}$$

$$0.072t = \ln\tfrac{9}{2}$$

$$t = \frac{\ln\tfrac{9}{2}}{0.072} \approx 20.890$$

Thus it will take about 20.9 hours for the population to reach 18,000. ▪

▪ EXAMPLE 4 Modeling the World's Population

Assuming that the growth rate is proportional to population size, use the data in Table 1 to model the population of the world in the second half of the 20th century. What is the relative growth rate? How well does the model fit the data? Use the model to estimate the population in 1993 and to predict the population in the year 2017.

SOLUTION

Here we let $t = 0$ in the year 1950. Then in Equation 4 we have $C = 2560$ and

$$P(t) = 2560e^{kt}$$

We can use one value from the data to estimate the relative growth rate k. Keep in mind that each data value will give a different estimate for k, so it may be worthwhile to view different choices on a graph. If we use the population in the year 2000 to estimate k, we get

$$P(50) = 2560e^{50k} = 6080$$

$$e^{50k} = \tfrac{6080}{2560}$$

$$\ln(e^{50k}) = \ln\tfrac{6080}{2560}$$

$$50k = \ln\tfrac{6080}{2560}$$

$$k = \tfrac{1}{50}\ln\tfrac{6080}{2560} \approx 0.01730$$

According to this estimate, the relative growth rate is about 1.73% per year and the model is

(5) $$P(t) = 2560e^{0.01730t}$$

We estimate that the world population in 1993 was

$$P(43) = 2560e^{0.01730(43)} \approx 5387 \text{ million}$$

The model predicts that the population in 2017 will be

$$P(67) = 2560e^{0.01730(67)} \approx 8159 \text{ million}$$

TABLE 1

Year	Population (millions)
1950	2560
1960	3040
1970	3710
1980	4450
1990	5280
2000	6080

The graph in Figure 2 shows that the model is fairly accurate to date, so the estimate for 1993 is quite reliable. But the prediction for 2017 is riskier.

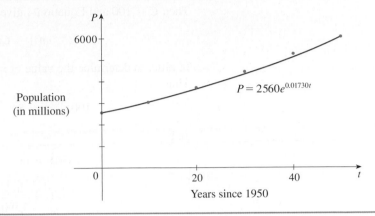

FIGURE 2

A model for world population growth in the second half of the 20th century

A graphing calculator (or computer) using regression instead of just two data points gives the model

$$P(t) = 2578.6(1.017765)^t$$

for the data in Example 4. This is also an exponential function, of course, but it is given in terms of an annual percentage increase rather than a continuous percentage increase. For this model we can see that the annual increase is 1.7765%, whereas in (5) we see that the relative growth rate is 0.01730 or 1.730%. In the project on page 204 we will convert between the two types of models in the context of interest-earning investment accounts.

■ Radioactive Decay

Radioactive substances decay by spontaneously emitting radiation. They have been found experimentally to erode at a rate proportional to the remaining mass. This means that the amount of material at a given time can be modeled by Equation 4: $A(t) = Ce^{kt}$. Again the constant C is the starting amount, and k is the relative growth rate. (Because the amount is decreasing, k will be negative.)

A radioactive substance continually loses a percentage of its mass but in theory it never completely disappears, so physicists express the rate of decay in terms of **half-life**, the time required for half of any given quantity to decay.

■ EXAMPLE 5

A Model for the Amount of a Radioactive Substance and Its Half-Life

The half-life of the radioactive material radium-226 ($^{226}_{88}\text{Ra}$) is 1590 years.

(a) A sample of radium-226 has a mass of 100 mg. Find a formula for the mass of $^{226}_{88}\text{Ra}$ that remains after t years.

(b) Find the mass after 1000 years correct to the nearest milligram.

(c) When will the mass be reduced to 30 mg?

(d) At what rate is the mass decreasing when 30 mg remains?

SOLUTION

(a) Let $m(t)$ be the mass of radium-226 (in milligrams) that remains after t years. Then $C = 100$ and Equation 4 gives

$$m(t) = Ce^{kt} = 100e^{kt}$$

In order to determine the value of k, we use the fact that $m(1590) = \frac{1}{2}(100)$. Thus

$$100e^{1590k} = 50 \qquad \text{so} \qquad e^{1590k} = \tfrac{1}{2}$$

and

$$1590k = \ln \tfrac{1}{2}$$

$$k = \frac{\ln \frac{1}{2}}{1590} \approx -0.00043594$$

Therefore $m(t) = 100e^{-0.00043594t}$

Note: Since the mass is halved every 1590 years, we could alternatively write an equivalent formula as

$$m(t) = 100 \times \left(\tfrac{1}{2}\right)^{t/1590}$$

(b) The mass after 1000 years is

$$m(1000) = 100e^{-0.00043594(1000)} \approx 64.7 \text{ mg}$$

(c) We want to find the value of t such that $m(t) = 30$, that is,

$$100e^{-0.00043594t} = 30 \qquad \text{or} \qquad e^{-0.00043594t} = 0.3$$

We solve this equation for t by taking the natural logarithm of both sides:

$$\ln(e^{-0.00043594t}) = \ln 0.3$$

$$-0.00043594t = \ln 0.3$$

Thus $t = \dfrac{\ln 0.3}{-0.00043594} \approx 2762 \text{ years}$

(d) We have $m(t) = 100e^{-0.00043594t}$, so

$$m'(t) = 100e^{-0.00043594t}(-0.00043594) = -0.043594e^{-0.00043594t}$$

and

$$m'(2762) = -0.043594e^{-0.00043594(2762)} \approx -0.013077 \text{ mg/year} \qquad ■$$

As a check on our work in Example 5, we use a graphing device to draw the graph of $m(t)$ in Figure 3, together with the horizontal line $m = 30$. These curves intersect when $t \approx 2800$, and this agrees with the answer to part (c).

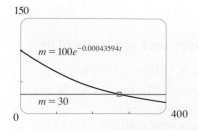

FIGURE 3

■ | Logistic Functions

Sometimes a population increases exponentially in its early stages but levels off eventually when it approaches a maximum population, called the **carrying capacity**, that environmental conditions can sustain in the long run. Thus, initially, the population increases at nearly a constant percentage rate and follows exponential growth. But the relative growth rate decreases as the population further increases (and turns negative if the population manages to surpass its carrying capacity). A more sophisticated function to model such behavior is called the **logistic function** with equation given by

In Section 6.4 we will derive this equation based on some initial assumptions.

(6)
$$P(t) = \frac{M}{1 + Ae^{-kt}}$$

where M is the carrying capacity, t is time, k is a constant, and A is a constant defined by

$$A = \frac{M - P_0}{P_0}$$

where P_0 is the initial population. The graph of a typical increasing logistic function is shown in Figure 4. You can see why the graph is sometimes referred to as "S-shaped." Note that the curve is concave upward at first but then changes to concave downward at an inflection point as the curve begins to level off.

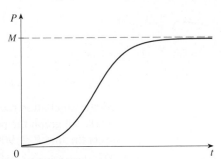

FIGURE 4

We have already seen examples of logistic functions in Exercises 49 and 50 in Section 1.5. The following example explores another.

■ EXAMPLE 6   A Logistic Model for a Population

A population is modeled by

$$P(t) = \frac{1000}{1 + 9e^{-0.08t}}$$

where t is the time elapsed, in years.

(a) What is the initial population? What is the population after 40 years? What is the carrying capacity?

(b) How long will it take for the population to reach 900?

(c) Compare the rates of change after 0, 40, and 80 years.

SOLUTION

(a) $P(0) = 1000/(1 + 9e^0) = 100$, so the initial population is 100. The population in 40 years is

$$P(40) = \frac{1000}{1 + 9e^{-3.2}} \approx 732$$

and comparing to Equation 6 we see that $M = 1000$, so the carrying capacity is 1000.

(b) The population reaches 900 when

$$\frac{1000}{1 + 9e^{-0.08t}} = 900$$

Solving this equation for t, we get

$$1 + 9e^{-0.08t} = \tfrac{10}{9}$$

$$9e^{-0.08t} = \tfrac{1}{9}$$

$$e^{-0.08t} = \tfrac{1}{81}$$

$$\ln(e^{-0.08t}) = \ln \tfrac{1}{81}$$

$$-0.08t = \ln \tfrac{1}{81}$$

$$t = -\frac{\ln \tfrac{1}{81}}{0.08} \approx 54.9$$

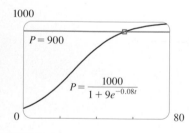

1000

$P = 900$

$$P = \frac{1000}{1 + 9e^{-0.08t}}$$

0   80

FIGURE 5

So the population reaches 900 in approximately 55 years. As a check on our work, we graph the population curve in Figure 5 and observe where it intersects the line $P = 900$. The cursor indicates that $t \approx 55$.

(c) The derivative is easier to calculate if we first rewrite P as

$$P(t) = 1000(1 + 9e^{-0.08t})^{-1}$$

Then, using the Chain Rule, we have

$$P'(t) = 1000(-1)(1 + 9e^{-0.08t})^{-2} \cdot \frac{d}{dt}(1 + 9e^{-0.08t})$$

Notice that the Chain Rule is again required in evaluating the derivative of $(1 + 9e^{-0.08t})$, giving

$$P'(t) = 1000(-1)(1 + 9e^{-0.08t})^{-2} \cdot 9e^{-0.08t}(-0.08)$$

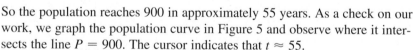

$$= 720(1 + 9e^{-0.08t})^{-2}e^{-0.08t} = \frac{720e^{-0.08t}}{(1 + 9e^{-0.08t})^2}$$

Then

$$P'(0) = \frac{720e^0}{(1 + 9e^0)^2} = \frac{720}{100} = 7.2$$

$$P'(40) = \frac{720e^{-3.2}}{(1 + 9e^{-3.2})^2} \approx 15.71$$

$$P'(80) = \frac{720e^{-6.4}}{(1 + 9e^{-6.4})^2} \approx 1.16$$

The initial growth rate is 7.2 members per year (when the population is 100 members). After 40 years the population is much larger, 732, and the growth rate is 15.71 members per year, more than double the initial growth rate. But after 80 years the population is $P(80) \approx 985$, very close to the carrying capacity, so the growth rate is much smaller: about 1.16 members per year. The derivative is graphed in Figure 6. You can see that the growth rate increases to a maximum around $t = 27$ and then decreases to almost 0.

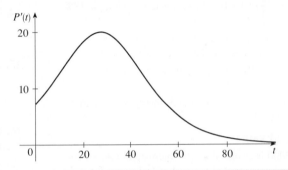

FIGURE 6

Prepare Yourself

1. (a) If $f(t) = 14e^{-0.2t}$, evaluate $f(4.5)$. Round to two decimal places.

  (b) Solve $f(t) = 4.5$. Round to two decimal places.

2. Solve for x (round to two decimal places).

  (a) $100(1.08)^x = 160$ **(b)** $5(0.8)^x = 3.1$

  (c) $45e^{0.3x} = 85$

3. (a) If $g(x) = \dfrac{240}{1 + 2e^{-0.4x}}$, evaluate $g(7.3)$. Round to two decimal places.

  (b) Solve $g(x) = 100$. Give both an exact answer and a decimal approximation rounded to two decimal places.

Exercises 3.6

1. Population   Write a formula for a population that is currently 46,500 and

  (a) increases by 2.4% each year.

  (b) decreases by 2.4% each year.

2. Gem value   Write a formula for the value of a gemstone that is currently worth $7825 if the value

  (a) increases 13.7% per year.

  (b) decreases 13.7% per year.

3. Investment value   A $5000 investment increases in value 6.2% per year.

  (a) Write an equation for a function that gives the value of the investment after t years.

  (b) What is the value of the investment after 7.5 years?

  (c) How long will it be until the investment is worth $8000?

4. Art value   A rare sculpture was purchased for $11.8 million and its value is expected to increase 14% per year.

  (a) Write an equation for the function that gives the value of the sculpture after t years.

(b) What is the value of the sculpture after 3.25 years?

(c) After how many years will the sculpture be worth $20 million?

5. Equipment depreciation   An automotive company purchases new robotic welding equipment for $28.6 million. The value of the equipment will be depreciated at a rate of 15% per year.

   (a) Write an equation for a function V that gives the value of the machinery after t years.

   (b) How long will it take for the equipment to be reduced to half its original value?

   (c) Compute and interpret $V'(5)$.

   (d) When will the value of the equipment be decreasing at a rate of $1 million per year?

6. Animal population   The eagle population in a state park is currently 1650 but is expected to decrease 18% per year.

   (a) Write an equation for the function that gives the number of eagles in the park t years from now.

   (b) Determine the time required for the population to be reduced to 1000.

   (c) What is the rate of change of the population after four years?

7. Investment value

   (a) If $3000 is invested at 5% interest, find the value of the investment at the end of five years if the interest is compounded **(i)** annually, **(ii)** quarterly, **(iii)** monthly, **(iv)** weekly, **(v)** daily, and **(vi)** continuously.

   (b) If the interest is compounded quarterly, how long will it take for the balance to double?

   (c) How long will it take for the value of the investment to double if interest is compounded continuously?

8. Debt value

   (a) If $500 is borrowed at 14% interest, find the amounts due at the end of two years if the interest is compounded **(i)** annually, **(ii)** quarterly, **(iii)** monthly, **(iv)** daily, **(v)** hourly, and **(vi)** continuously.

   (b) Suppose $500 is borrowed and the interest is compounded continuously. If $A(t)$ is the amount due after t years, where $0 \le t \le 2$, graph $A(t)$ for each of the interest rates 14%, 10%, and 6% on a common screen.

9. Investment value $16,000 is invested in an account that earns 4.3% interest.

   (a) Write an equation for a function V that gives the value of the account after t years if interest is compounded monthly.

   (b) Compute and interpret $V'(3.5)$.

10. Investment doubling time

   (a) How long will it take an investment to double in value if the interest rate is 6% compounded continuously?

   (b) What if the interest is compounded annually?

11. Investment value   Which option would you prefer for an investment, 5.1% interest compounded continuously or 5.25% interest compounded quarterly?

12. Investment value   Does an interest rate of 6.1% compounded continuously earn more than a rate of 6.2% compounded monthly?

13. Population   A population starts with 1300 members and increases with a relative growth rate of 17.2% per year. Write a formula that gives the population after t years. How large is the population after 7.5 years?

14. Population tripling time   How long does it take for a population that is growing with a constant relative growth rate of 10% per year to triple?

15. Animal population   A population of protozoa develops with a constant relative growth rate of 0.7944 per day. On day zero the population consists of two members. Find the population size after six days.

16. Bacteria population   A common inhabitant of human intestines is the bacterium *Escherichia coli* (*E. coli*). A cell of this bacterium in a nutrient-broth medium divides into two cells every 20 minutes. The initial population of a culture is 60 cells.

   (a) Find the relative growth rate.

   (b) Find an expression for the number of cells after t hours.

   (c) Find the number of cells after eight hours.

   (d) Find the rate of growth after eight hours.

   (e) When will the population reach 20,000 cells?

17. Bacteria population   A bacteria culture initially contains 100 cells and grows at a rate proportional to its size. After an hour the population has increased to 420.

   (a) Find an expression for the number of bacteria after t hours.

   (b) Find the number of bacteria after three hours.

   (c) Find the rate of growth after three hours.

   (d) When will the population reach 10,000?

18. Bacteria population   While testing for *Salmonella* in a sample of chicken at a meat-packing plant, an inspector finds a bacteria count of 15 colony-forming units per milliliter (CFU/mL). The sample is kept at a temperature of 80°F, and six hours later the count is 20,000 CFU/mL.

   (a) If the bacteria count is increasing at a constant relative growth rate, write a model for the bacteria count in the sample after t hours.

(b) What does the model predict the bacteria count will be after four hours? What is the growth rate then?

(c) According to the model, how long will it take for the bacteria count to reach ten million CFU/mL?

19. Virus population A population of viruses is increasing at a constant relative growth rate in the bloodstream of its host. After three days the population increases from 260 to 1720.

(a) Write a model for the virus population after t days.

(b) According to the model, how many viruses will be present after one week?

(c) At what rate is the virus population increasing after one week?

20. Bacteria population A bacteria culture grows with constant relative growth rate. After two hours there are 600 bacteria and after eight hours the count is 75,000.

(a) Find the initial population.

(b) Find an expression for the population after t hours.

(c) Find the number of cells after five hours.

(d) Find the rate of growth after five hours.

(e) When will the population reach 200,000?

21. World population The table gives estimates of the world population, in millions, from 1750 to 2000:

Year	Population	Year	Population
1750	790	1900	1650
1800	980	1950	2560
1850	1260	2000	6080

(a) Use the population figures for 1750 and 1800 to write an exponential growth model and predict the world population in 1900 and 1950. Compare with the actual figures.

(b) Use the population figures for 1850 and 1900 to write an exponential growth model and predict the world population in 1950. Compare with the actual population.

(c) Use the population figures for 1900 and 1950 to write an exponential growth model and predict the world population in 2000. Compare with the actual population and try to explain the discrepancy.

22. US population The table gives the population of the United States, in millions, for the years 1900–2000.

(a) Use the census figures for 1900 and 1910 to write an exponential growth model and predict the population in 2000. Compare with the actual figure and try to explain the discrepancy.

(b) Use the census figures for 1980 and 1990 to write an exponential growth model and predict the population

in 2000. Compare with the actual population. Then use this model to predict the population in the years 2020 and 2030.

(c) Graph both of the exponential functions in parts (a) and (b) together with a plot of the actual population. Are these models reasonable ones?

(d) Use a graphing calculator (or computer) with regression capabilities to find an exponential model for the data. What does the model predict for the population in 2020?

Year	Population	Year	Population
1900	76	1960	179
1910	92	1970	203
1920	106	1980	227
1930	123	1990	250
1940	131	2000	275
1950	150		

23. Radioactive decay The half-life of the radioactive material cesium-137 is 30 years. Suppose we have a 100-mg sample.

(a) Write a formula that gives the mass that remains after t years.

(b) How much of the sample remains after 100 years?

(c) After how long will only 1 mg remain?

(d) At what rate is the mass decreasing after 100 years?

24. Radioactive decay The radioactive substance Bismuth-210, ^{210}Bi, has a half-life of 5.0 days.

(a) A sample originally has a mass of 800 mg. Find a formula for the mass remaining after t days.

(b) Find the mass remaining after 30 days.

(c) When is the mass reduced to 10 mg?

(d) Sketch the graph of the mass function.

25. Carbon dating Scientists can determine the age of ancient objects by the method of *radiocarbon dating*. The bombardment of the upper atmosphere by cosmic rays converts nitrogen to a radioactive isotope of carbon, ^{14}C, with a half-life of about 5730 years. Vegetation absorbs carbon dioxide through the atmosphere and animal life assimilates ^{14}C through food chains. When a plant or animal dies, it stops replacing its carbon and the amount of ^{14}C begins to decrease through radioactive decay. Therefore the level of radioactivity must also decay exponentially. A parchment fragment was discovered that had about 74% as much ^{14}C radioactivity as does plant material on earth today. Estimate the age of the parchment.

26. Carbon dating In 1991 two mountain hikers discovered a human body frozen in ice. It was soon determined that the man, dubbed the "iceman," had been in the ice for quite some time. Tissue samples indicated that his body had 57.67% of the ^{14}C that is present in a living person. How long ago did the iceman die?

27. Radioactive decay   A 100-g sample of an isotope of sodium, ^{24}Na, decays to 30 g after 26 hours.

(a) Find the half-life of the isotope.

(b) How much of the sample remains after two hours?

(c) How long will it take for only 5 g of the sample to remain?

28. Radioactive decay   A sample of tritium-3 decayed to 94.5% of its original amount after a year.

(a) What is the half-life of tritium-3?

(b) How long would it take the sample to decay to 20% of its original amount?

29. Newton's Law of Cooling   When an object at one temperature is placed in an environment at a different temperature, the temperature of the object changes at a rate proportional to the *difference* between the temperatures. (This result is known as Newton's Law of Cooling.) Thus the temperature difference follows exponential growth or decay. Suppose a roast turkey is taken from an oven when its temperature has reached 185°F and is placed on a table in a room where the temperature is 75°F.

(a) If the temperature of the turkey is 150°F after half an hour, what is the temperature after 45 minutes?

(b) When will the turkey have cooled to 100°F?

(c) At what rate is the turkey cooling after half an hour?

30. Newton's Law of Cooling   A thermometer is taken from a room where the temperature is 20°C to the outdoors, where the temperature is 5°C. After one minute the thermometer reads 12°C.

(a) What will the reading on the thermometer be after one more minute?

(b) When will the thermometer read 6°C?

31. Animal population   The number of mountain lions in a wildlife preserve is modeled by

$$P(t) = \frac{1680}{1 + 4.2e^{-0.11t}}$$

where t is the number of years after January 1, 2010.

(a) What is the carrying capacity? How many mountain lions are there on January 1, 2010?

(b) According to the model, what is the population after 15 years?

(c) When does the model predict that the mountain lion population will reach 1500?

(d) Compute and interpret $P'(12)$.

32. Fish population   A lake is stocked with 1000 fish and the fish population is expected to follow the model

$$P(t) = \frac{17,000}{1 + 16e^{-0.7t}}$$

where t is the time elapsed, in years.

(a) What is the carrying capacity? What is the fish population after 2.5 years?

(b) How many years are required for the fish population to reach 12,000?

(c) What is the growth rate of the fish population after five years?

33. Animal population   In Exercise 49 of Section 1.5 we used the logistic function

$$P(t) = \frac{23.7}{1 + 4.8e^{-0.2t}}$$

to model an animal population, measured in thousands, t years after January 1, 2000. Compute and interpret $P'(8)$.

34. Market penetration   In Exercise 50 of Section 1.5 the percentage of households owning a certain product t years after 1980 is given by

$$g(t) = \frac{0.94}{1 + 2.5e^{-0.3t}}$$

Find $g'(12)$ and interpret your result.

35. Fish farming   The Pacific halibut fishery has been modeled by a logistic function B, where $B(t)$ is the biomass (the total mass of the members of the population) in kilograms at time t (measured in years), the carrying capacity is estimated to be 8×10^7 kg, and $k = 0.71$.

(a) If $B(0) = 2 \times 10^7$ kg, find the biomass a year later.

(b) How long will it take for the biomass to reach 4×10^7 kg?

36. Population   Suppose that a population grows according to a logistic model with carrying capacity 8000 and $k = 0.026$.

(a) If the initial population is 1400, write a formula for a model of the population after t years.

(b) Use your formula to predict the population in 40 years.

(c) Sketch a graph of the model. Be sure to use an appropriate domain to show the behavior of the curve.

37. World population   The population of the world was about 5.3 billion in 1990. Birth rates in the 1990s ranged from 35 to 40 million per year and death rates ranged from 15 to 20 million per year, so we estimate that the population was growing at a rate of about 20 million/year at that time. Let's assume that the carrying capacity for world population is 100 billion.

(a) Estimate the relative growth rate in 1990.

(b) Write a logistic model for the population using the answer from part (a) as the value for k. Use the model to estimate the world population in the year 2000 and compare with the actual population of 6.1 billion.

(c) Use the model to predict the world population in the years 2100 and 2500.

38. Fish population   Biologists stocked a lake with 400 fish and estimated the carrying capacity (the maximal population for the fish of that species in that lake) to be 10,000. The number of fish tripled in the first year.

(a) Assuming that the size of the fish population satisfies a logistic equation, find an expression for the size of the population after t years.

(b) How long will it take for the population to increase to 5000?

(c) How quickly is the fish population increasing after three years?

39. Spread of a rumor   Under certain circumstances a rumor spreads according to the logistic function

$$p(t) = \frac{1}{1 + ae^{-kt}}$$

where $p(t)$ is the proportion of the population that knows the rumor at time t and a and k are positive constants.

(a) Graph p for the case $a = 10$, $k = 0.5$ with t measured in hours. Use the graph to estimate how long it will take for 80% of the population to hear the rumor.

(b) For the values given in part (a), find the rate of spread of the rumor after six hours.

40. Retail sales   The estimated cumulative sales, in thousands, of a company's flagship product at various times are given in the table.

Year	Sales
2000	8.1
2002	15.8
2004	28.3
2006	43.5
2008	55.7
2010	64.2

(a) Use a graphing calculator (or computer) to find a logistic model for the sales $f(t)$ in thousands, where t is the year, with $t = 0$ corresponding to 2000.

(b) Use the model to compute $f'(7)$ and interpret your result.

41. Japan population   The table gives the midyear population of Japan, in thousands, from 1960 to 2005.

Year	Population	Year	Population
1960	94,092	1985	120,754
1965	98,883	1990	123,537
1970	104,345	1995	125,341
1975	111,573	2000	126,700
1980	116,807	2005	127,417

Use a graphing calculator to fit both an exponential function and a logistic function to these data. Graph the data points and both functions, and comment on the accuracy of the models. [*Hint*: Subtract 94,000 from each of the population figures. Then, after obtaining a model from your calculator, add 94,000 to get your final model. It might be helpful to choose $t = 0$ to correspond to 1960 or 1980.]

42. Yeast population   The table gives the number of yeast cells in a new laboratory culture.

Time (hours)	Yeast cells	Time (hours)	Yeast cells
0	18	10	509
2	39	12	597
4	80	14	640
6	171	16	664
8	336	18	672

(a) Plot the data and use the plot to estimate the carrying capacity for the yeast population.

(b) Use a graphing calculator (or computer) to find a logistic model for these data. What carrying capacity is this model using? Comment on how well your model fits the data.

(c) Use both the initial value and the value of k from the logistic model to write an exponential model for the data. Compare the graph to the logistic model; for how many hours is the exponential model accurate?

(d) Use your logistic model to estimate the number of yeast cells after seven hours.

Challenge Yourself

43. Insect population   If $B(t)$ is the number of bees in a colony t weeks from now and the bee population is decreasing at a constant percentage rate, which one of the following is true: $B'(4) = B'(8)$, $B'(4) < B'(8)$, or $B'(4) > B'(8)$? Explain.

44. Investment value   What is the lowest interest rate that, when compounded monthly, will earn as much interest as a 4% rate compounded continuously?

45. Spread of a rumor   One model for the spread of a rumor is that the rate of spread is proportional to the product of the fraction R of the population who have heard the rumor and the fraction who have not heard the rumor. Thus R satisfies the differential equation

$$R'(t) = k \cdot R(t) \cdot [1 - R(t)]$$

(a) Prove that the function

$$R(t) = \frac{R_0}{R_0 + (1 - R_0)e^{-kt}}$$

where R_0 is the initial fraction of the population who have already heard the rumor and t is the time in hours, satisfies the differential equation.

(b) A small town has 1000 inhabitants. At 8 AM, 80 people have heard a rumor. By noon half the town has heard it. At what time will 90% of the population have heard the rumor?

46. For a fixed value of M (say $M = 10$), the family of logistic functions given by Equation 6 depends on the initial population P_0 and the proportionality constant k. Graph several members of this family. How does the graph change when P_0 varies? How does it change when k varies?

47. Another model for a growth function for a limited population is given by the **Gompertz function**

$$P(t) = Me^{-\ln(M/P_0)e^{-ct}}$$

where M is the carrying capacity, P_0 is the initial population, and c is a constant.

(a) Graph the Gompertz growth function for $M = 1000$, $P_0 = 100$, and $c = 0.05$, and compare it with the graph of the function in Example 6. What are the similarities? What are the differences?

(b) If t represents the time elapsed in years, compute the population and its rate of change after 40 years.

PROJECT ▪ APR versus APY

Banking institutions often advertise two kinds of interest rates: an annual percentage rate (APR) along with an "APY" number that is a little higher. APY stands for Annual Percentage Yield and represents the effective annual return with compounding taken into consideration. If we invest $1000 at 6% interest compounded continuously, then after one year we will have

$$\$1000e^{0.06(1)} = \$1061.84$$

or an increase of 6.184%. Thus a 6% interest rate compounded continuously corresponds to an effective annual rate (APY) of about 6.184%.

In general, if we let r be the continuously compounded interest rate and r_y the effective annual rate, then, comparing Equations 1 and 3 in Section 3.6, we have

$$P(1 + r_y)^t = Pe^{rt} = P(e^r)^t$$

Then we must have

$$1 + r_y = e^r$$

so

$$r = \ln(1 + r_y)$$

1. What is the APY for a 3.75% interest rate compounded continuously?

2. What interest rate, to be compounded continuously, should be used to give an APY of 4.00%?

3. Write a formula that gives the APY for an interest rate r that is compounded n times per year.

   (a) Compute the APY for an interest rate of 4.5% that is compounded (i) quarterly or (ii) monthly.

   (b) What interest rate compounded monthly corresponds to an APY of 3.5%?

4. The concept of APY applies in other contexts as well. For example, an initial quantity C growing exponentially with a relative growth rate of 2.4% per year is described by $A = Ce^{0.024t}$, but after one year the quantity will have increased more than 2.4%.

   (a) Find the effective annual percentage increase for a population that grows with a constant relative growth rate of 2.4% per year.

   (b) If a radioactive substance loses 16% of its mass each week, what is the relative decay rate?

CHAPTER 3 REVIEW

■ CONCEPT CHECK

1. State each differentiation rule both in symbols and in words.
   (a) The Power Rule
   (b) The Constant Multiple Rule
   (c) The Sum Rule
   (d) The Difference Rule
   (e) The Product Rule
   (f) The Quotient Rule
   (g) The Chain Rule

2. State the derivative of each function.
   (a) $y = c$, where c is a constant
   (b) $y = x^n$
   (c) $y = e^x$
   (d) $y = a^x$
   (e) $y = \ln x$

3. (a) How is the number e defined?
   (b) Why is the natural exponential function $y = e^x$ used more often in calculus than the other exponential functions $y = a^x$?

   (c) Why is the natural logarithmic function $y = \ln x$ used more often in calculus than the other logarithmic functions $y = \log_a x$?

4. Give a formula to compute the balance on an interest-earning investment account after t years if the interest rate is r and the interest is compounded (a) annually, (b) n times per year, and (c) continuously.

5. What is the general formula for a quantity that grows at a rate proportional to itself?

6. (a) What does the graph of a logistic function look like?
   (b) Under what circumstances is this an appropriate model for population growth?

7. Explain how implicit differentiation works.

8. (a) What are fixed costs for a business? Variable costs?
   (b) How do we compute the average cost of producing q units of a good or service?
   (c) What is marginal cost? How do we compute it?
   (d) How do we determine the minimum average cost?

9. (a) What is marginal revenue? How is it computed?

(b) How is the profit function defined?

(c) How do we determine when profit is maximized?

10. What is a demand function? Why are demand functions typically decreasing functions?

11. What is a linear approximation? What is the connection with the tangent line?

Answers to the Concept Check can be found on the back endpapers.

▪ Exercises

1–34 ▪ Find the derivative of the function.

1. $f(x) = 5x^3 - 7x + 13$

2. $g(t) = \sqrt{t} + 3t^2 - 1$

3. $q = \sqrt[3]{r} + 6/r$

4. $A = \dfrac{2}{v^2} + v - 1$

5. $h(u) = 3e^u + 1/\sqrt{u}$

6. $y = 2^x + 5$

7. $E(x) = 2.3(1.06)^x$

8. $F(y) = 6y - \ln y$

9. $B(t) = 1 + 4\ln t$

10. $T = 2e^h(5h^2 + 3h)$

11. $C(a) = \sqrt{a}\,(e^a + 1)$

12. $y = \dfrac{e^x}{1 + x^2}$

13. $y = \dfrac{t}{1 - t^2}$

14. $g(v) = \dfrac{v^2 + 6v - 2}{v}$

15. $y = (x^4 - 3x^2 + 5)^3$

16. $y = \dfrac{3}{(2x^5 + x)^4}$

17. $A = 16e^{-2t}$

18. $g(t) = 3.8\ln(t^2 + t)$

19. $y = \ln(x^3 + 5)$

20. $y = e^{-t}(t^2 - 2t + 2)$

21. $y = 2x\sqrt{x^2 + 1}$

22. $y = \dfrac{e^{2x}}{1 + e^{-x}}$

23. $z = \sqrt{\dfrac{t}{t^2 + 4}}$

24. $y = \dfrac{3x - 2}{\sqrt{2x + 1}}$

25. $y = xe^{-1/x}$

26. $y = x^r e^{sx}$

27. $f(x) = 10^{x\sqrt{x-1}}$

28. $y = \ln(x^2 e^x)$

29. $A(r) = 6(\ln r)^4$

30. $P(t) = \dfrac{75}{1 + 4e^{-0.5t}}$

31. $y = 3^{x\ln x}$

32. $y = e^{\sqrt{1 + x^4}}$

33. $y = [\ln(x^2 + 1)]^3$

34. $y = \sqrt{t\ln(t^4)}$

35–36 ▪ Find f' and f''.

35. $f(t) = 500e^{0.65t}$

36. $f(x) = \dfrac{\ln x}{2x}$

37–38 ▪ Find the first and second derivatives of the function. Check to see that your answers are reasonable by comparing the graphs of f, f', and f''.

37. $f(x) = 2x - 5x^{3/4}$

38. $f(t) = \ln(t^2 + 1)$

39–40 ▪ Find dy/dx by implicit differentiation.

39. $xy^4 + x^2y = x + 3y$

40. $xe^y = y - 1$

---

41. If $f(t) = \sqrt{10 + 3t}$, find $f''(2)$.

42. Find the slope of the curve $y = 3(1.25^x)$ where $x = 3$. Round to three decimal places.

43–46 ▪ Find an equation of the tangent to the curve at the given point.

43. $y = (2 + x)e^{-x}$, $(0, 2)$

44. $y = \dfrac{x^2 - 1}{x^2 + 1}$, $(0, -1)$

45. $y = (3x - 2)^5$, $(1, 1)$

46. $x^2 + 4xy + y^2 = 13$, $(2, 1)$

---

47. (a) If $f(x) = x\sqrt{5 - x}$, find $f'(x)$.

(b) Find equations of the tangent lines to the curve $y = x\sqrt{5 - x}$ at the points $(1, 2)$ and $(4, 4)$.

(c) Illustrate part (b) by graphing the curve and tangent lines on the same screen.

(d) Check to see that your answer to part (a) is reasonable by comparing the graphs of f and f'.

48. Motion The distance, in feet, a moving object has traveled after t seconds is given by $s(t) = 4t^2/(t + 1)$. Find the speed and acceleration of the object after 3 seconds.

49. Cost function The cost, in dollars, of producing q units of a commodity is

$$C(q) = 920 + 2q - 0.02q^2 + 0.00007q^3$$

(a) Find the average cost function and compute the average per unit cost after producing 1500 units.

(b) Find the marginal cost function.

(c) Find $C'(100)$ and explain its meaning.

(d) Compare $C'(100)$ with the actual cost of producing the 101st item.

50. Food service   A caterer estimates that it costs

$$C(x) = 0.02x^2 + 8x + 375$$

dollars to provide dinner to x guests at a charity event.

(a) What are the fixed costs?

(b) How large an event minimizes the average cost per person?

51. Beverage production   A company estimates that the daily cost, in dollars, for producing q cans of its new energy drink is

$$C(q) = 380 + 0.32q + 0.0002q^2$$

The revenue, in dollars, the company earns from selling q cans is given by

$$R(q) = 1.36q - 0.0001q^2$$

(a) Write a function for the profit P the company earns after selling q cans of the drink each day.

(b) Write equations for the average revenue and marginal revenue functions.

(c) Find the number of cans the company should produce daily in order to maximize profit.

52. Software demand   The demand function for a particular mobile phone application is estimated to be $p = 28e^{-0.2q}$ where q is measured in thousands of copies and p is measured in dollars.

(a) What price should be chosen for the application in order to sell 8000 copies?

(b) If the application is priced at \$4.95 per copy, how many copies are expected to sell?

(c) Write an equation for a function that gives the revenue from selling q copies of the application.

53. Property value   A commercial property is purchased for \$2.6 million.

(a) Write a formula for the value of the property if the value increases 4.6% per year. What is the value after 3.5 years?

(b) Write a formula for the value of the property if the value decreases 4.6% per year. How long will it take for the value to decline to \$2 million?

54. Investment value

(a) If \$8000 is invested at 3.8% interest, find the value of the investment at the end of six years if the interest is compounded (**i**) annually, (**ii**) quarterly, (**iii**) monthly, (**iv**) weekly, (**v**) daily, and (**vi**) continuously.

(b) If the interest is compounded quarterly, how long will it take for the balance to reach \$10,000?

(c) How long will it take for the value of the investment to reach \$10,000 if interest is compounded continuously?

(d) If interest is compounded continuously, what is the instantaneous rate of change of the value at the end of six years?

55. Bacteria population   A bacteria culture contains 200 cells initially and grows at a rate proportional to its size. After half an hour the population has increased to 360 cells.

(a) Find the number of bacteria after t hours.

(b) Find the number of bacteria after four hours.

(c) Find the rate of growth after four hours.

(d) What is the relative growth rate?

(e) When will the population reach 10,000?

56. Insect population   An insect population is increasing at a constant relative growth rate in a community. It is estimated that after four weeks the population increased from 6000 to 13,400.

(a) Write a model for the insect population after t weeks.

(b) According to the model, what is the insect population after ten weeks?

(c) At what rate is the insect population increasing after ten weeks?

57. Drug concentration   Let $C(t)$ be the concentration of a drug in the bloodstream. As the body eliminates the drug, $C(t)$ decreases at a rate that is proportional to the amount of the drug that is present at the time. If the body eliminates half the drug in 30 hours, how long does it take to eliminate 90% of the drug?

58. Radioactive decay   Cobalt-60 has a half-life of 5.24 years.

(a) Find the mass that remains from a 100-mg sample after 20 years.

(b) How long would it take for the mass to decay to 1 mg?

59. Radioactive decay   A 2-ounce sample of a radioactive material decays to 0.8 ounce after 6.3 years.

(a) What is the half-life of the material?

(b) How long will it take for 90% of the material to decay?

60. World population

(a) The population of the world was 5.28 billion in 1990 and 6.07 billion in 2000. Find an exponential model for these data and use the model to predict the world population in the year 2020. Use $t = 0$ to represent 1990.

(b) According to the model in part (a), when will the world population exceed 10 billion?

(c) Use the data in part (a) to find a logistic model for the population. Assume a carrying capacity of 100 billion and use the same value of k from part (a). Then use the logistic model to predict the population in 2020. Compare with your prediction from the exponential model.

(d) According to the logistic model, when will the world population exceed 10 billion? Compare with your prediction in part (b).

61. Animal population   An animal population, in thousands, is modeled by

$$P(t) = \frac{285}{1 + 3.8e^{-0.08t}}$$

where t is the number of years after January 1, 2010.

(a) What is the carrying capacity? What is the population on January 1, 2010?

(b) According to the model, what is the population after 25 years?

(c) When does the model predict that the population will reach 200,000?

(d) Compute and interpret $P'(30)$.

62. Broadband internet   The table shows one study's year-end estimates of the percentage of Scottish households with Internet service that have a broadband connection.

Year	Percentage	Year	Percentage
2001	15	2006	69
2002	20	2007	77
2003	30	2008	82
2004	46	2009	85
2005	58	2010	87

(a) Use a graphing calculator (or computer) to find an exponential model and a logistic model for the data. Graph both models with the data. Which model appears more reasonable?

(b) Use each model to predict the broadband penetration of Scottish households with Internet service in 2015. Which is the more realistic prediction?

63. At what point on the curve $y = [\ln(x + 4)]^2$ is the tangent horizontal?

64. (a) Find an equation of the tangent to the curve $y = e^x$ that is parallel to the line $x - 4y = 1$.

(b) Find an equation of the tangent to the curve $y = e^x$ that passes through the origin.

65. Find the points on the ellipse $x^2 + 2y^2 = 1$ where the tangent line has slope 1.

66. If g is a differentiable function where $g(72) = 285.4$ and $g'(72) = -3.7$, use a linear approximation to estimate the value of $g(70)$ and $g(73.3)$.

67. Suppose that $h(x) = f(x)\,g(x)$ and $F(x) = f(g(x))$, where $f(2) = 3$, $g(2) = 5$, $g'(2) = 4$, $f'(2) = -2$, and $f'(5) = 11$. Find (a) $h'(2)$ and (b) $F'(2)$.

68. Drug concentration   The function

$$C(t) = K(e^{-at} - e^{-bt})$$

where a, b, and K are positive constants and $b > a$, can be used to model the concentration at time t of a drug injected into the bloodstream.

(a) Graph C for $K = 8$, $a = 0.25$, and $b = 1.5$.

(b) If the concentration is measured in micrograms per milliliter (μg/mL) and t is measured in hours, compute and interpret $C'(2)$.

(c) When is the rate at which the drug is cleared from circulation equal to 0?

4

The velocity of an ocean wave is a function of the length of the wave. (See Exercise 41 on page 278.) The calculus that you learn in this chapter will enable you to identify the optimal values of this and other functions. © EpicStockMedia/ Shutterstock

Applications of Differentiation

We have already seen many examples of applications of derivatives, but now that we know the differentiation rules we are in a better position to pursue the applications of differentiation in greater depth. We show how to analyze the behavior of families of functions, how to solve related rates problems (how to calculate rates that we can't measure from those that we can), and how to find the maximum or minimum value of a quantity.

4.1 Related Rates

If we are pumping air into a balloon, both the volume and the radius of the balloon are increasing and their rates of increase are related to each other. But it is much easier to measure directly the rate of increase of the volume than the rate of increase of the radius.

▪ Connecting Rates of Change

In a related rates problem the idea is to compute the rate of change of one quantity in terms of the rate of change of one or more other quantities (which may be more easily measured). We start by finding an equation that relates the different quantities. If we consider each of the quantities as a function of time, we can use the Chain Rule to differentiate both sides of the equation with respect to time. The result is an equation that connects the rates of change.

▪ EXAMPLE 1 Inflating a Balloon

Air is being pumped into a spherical balloon so that its volume increases at a rate of 100 cm³/s. How fast is the radius of the balloon increasing when the diameter is 50 cm?

SOLUTION

We start by identifying two things:

the *given information:*

the rate of increase of the volume of air is 100 cm³/s

and the *unknown:*

the rate of increase of the radius when the diameter is 50 cm

In order to express these quantities mathematically, we introduce some suggestive *notation:*

Let V be the volume of the balloon and let r be its radius.

The key thing to remember is that rates of change are derivatives. In this problem, the volume and the radius both change with time and so we can consider them both as functions of the time t. The rate of increase of the volume with respect to time is the derivative dV/dt, and the rate of increase of the radius is dr/dt. We can therefore restate the given and the unknown as follows:

$$\text{Given:} \qquad \frac{dV}{dt} = 100 \text{ cm}^3/\text{s}$$

Notice that, although dV/dt is constant, dr/dt is *not* constant.

$$\text{Unknown:} \qquad \frac{dr}{dt} \quad \text{when } r = 25 \text{ cm}$$

In order to connect dV/dt and dr/dt, we first relate V and r by the formula for the volume of a sphere:

$$V = \tfrac{4}{3}\pi r^3$$

If you have trouble using the Chain Rule correctly, try expressing the formula using function notation. Since V and r are both functions of time, we can write

$$V(t) = \tfrac{4}{3}\pi[r(t)]^3$$

Then, differentiating each side with respect to t and using the Chain Rule, we have

$$V'(t) = 4\pi[r(t)]^2 r'(t)$$

To extend the link between V and r to one that involves the rates of change, we differentiate each side of this equation implicitly with respect to the hidden variable t. Differentiating the left side of the equation we get dV/dt. To differentiate the right side, we need to use the Chain Rule:

$$\frac{dV}{dt} = \frac{dV}{dr}\frac{dr}{dt} = 4\pi r^2 \frac{dr}{dt}$$

We know that $r = 25$ (since the diameter is 50 cm) and $dV/dt = 100$, and substituting these values gives

$$100 = 4\pi(25)^2 \frac{dr}{dt}$$

Now we solve for the unknown quantity:

$$\frac{dr}{dt} = \frac{100}{4\pi(25)^2} = \frac{1}{25\pi}$$

Thus the radius of the balloon is increasing at the rate of $1/(25\pi) \approx 0.013$ cm/s at that moment.   ■

■ Strategy for Related Rates Problems

Before exploring additional examples of related rates, we summarize the strategy illustrated in Example 1.

STRATEGY

⊘ **Warning** A common error is to substitute the given numerical information (for quantities that vary with time) too early. This should be done only *after* the differentiation. (Step 7 follows Step 6.) For instance, in Example 1 we dealt with general values of r until we finally substituted $r = 25$ at the last stage. (If we had put $r = 25$ earlier, we would have gotten $dr/dt = 0$, which is clearly wrong.)

1. Read the problem carefully.

2. Draw a diagram if possible.

3. Introduce suggestive notation. Assign symbols to all quantities that are functions of time.

4. Express the given information and the required rate in terms of derivatives.

5. Write an equation that relates the various quantities (not the rates of change) of the problem.

6. Use the Chain Rule to differentiate both sides of the equation with respect to t.

7. Substitute the given information into the resulting equation and solve for the unknown rate.

Keep the steps of this strategy in mind for the following examples.

■ EXAMPLE 2 The Sliding Ladder Problem

A ladder 10 ft long rests against a vertical wall. If the bottom of the ladder slides away from the wall at a rate of 1 ft/s, how fast is the top of the ladder sliding down the wall when the bottom of the ladder is 6 ft from the wall?

SOLUTION

You may be tempted to guess that the top of the ladder must also be sliding at a rate of 1 ft/s, but in fact the top of the ladder slides down slowly at first and

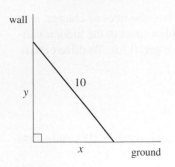

FIGURE 1

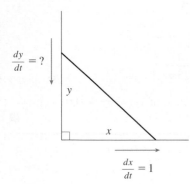

FIGURE 2

speeds up as the bottom moves farther from the wall. Following the outlined strategy, we first draw a diagram and label it as in Figure 1. Let x feet be the distance from the bottom of the ladder to the wall and y feet the distance from the top of the ladder to the ground. Note that x and y are both functions of t (time, measured in seconds).

We are given that $dx/dt = 1$ ft/s and we are asked to find dy/dt when $x = 6$ ft. (See Figure 2.) Next we need to establish a connection between x and y. From Figure 1 we see that a right triangle is formed, so the relationship between x and y is given by the Pythagorean Theorem:

$$x^2 + y^2 = 100$$

Differentiating each side with respect to t using the Chain Rule, we have

$$2x\frac{dx}{dt} + 2y\frac{dy}{dt} = 0$$

Note that when $x = 6$ the equation $x^2 + y^2 = 100$ gives $y = 8$, and we know $dx/dt = 1$, so substituting these values gives

$$2(6)(1) + 2(8)\frac{dy}{dt} = 0$$

Solving this equation for dy/dt, we obtain

$$16\frac{dy}{dt} = -12 \quad \Rightarrow \quad \frac{dy}{dt} = -\frac{12}{16} = -\frac{3}{4} \text{ ft/s}$$

The fact that dy/dt is negative means that the distance from the top of the ladder to the ground is *decreasing* at a rate of $\frac{3}{4}$ ft/s. In other words, the top of the ladder is sliding down the wall at a rate of $\frac{3}{4}$ ft/s at that moment.   ▪

▪ EXAMPLE 3   A Demand Equation

Demand functions were introduced in Section 3.2.

The demand equation for a company's memory chips is $(q^2 + 80)p = 10{,}000$, where p is the price of each chip and q is the number of chips, in thousands, that will be sold monthly. When the chips are priced at \$49.75 the company expects to sell 11,000 units monthly, but the price is decreasing at a rate of \$1.50 per week. Find the rate at which the demand for the memory chips is changing with respect to time.

SOLUTION

Both q and p are functions of time, so we differentiate each side of the demand equation with respect to t (we use the Product Rule on the left):

$$(q^2 + 80)\frac{dp}{dt} + p \cdot 2q\frac{dq}{dt} = 0$$

When $p = 49.75$ we have $q = 11$ and we are given that $dp/dt = -1.50$, so

$$(11^2 + 80)(-1.50) + (49.75)(2 \cdot 11)\frac{dq}{dt} = 0$$

Next we solve for dq/dt:

$$-301.5 + 1094.5\frac{dq}{dt} = 0$$

$$\frac{dq}{dt} = \frac{301.5}{1094.5} \approx 0.275$$

Thus the demand for the memory chips is increasing at a rate of about 275 chips per month. ■

It is not uncommon to see three (or more) related quantities in a related rates problem, as the next example shows.

■ EXAMPLE 4

Using the Pythagorean Theorem in a Related Rates Problem

Car A is traveling west at 50 mi/h and car B is traveling north at 60 mi/h. Both are headed for the intersection of the two roads. At what rate are the cars approaching each other when car A is 0.3 mi and car B is 0.4 mi from the intersection?

SOLUTION

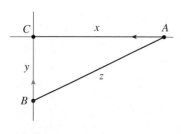

FIGURE 3

We draw Figure 3, where C is the intersection of the roads. At a given time t, let x be the distance from car A to C, let y be the distance from car B to C, and let z be the distance between the cars, where x, y, and z are measured in miles.

We are given that $dx/dt = -50$ mi/h and $dy/dt = -60$ mi/h. (The derivatives are negative because x and y are decreasing.) We are asked to find dz/dt, the rate of change (with respect to time) of the distance between the cars. We have a right triangle in Figure 3, so the equation that relates x, y, and z is given by the Pythagorean Theorem:

$$z^2 = x^2 + y^2$$

Differentiating each side with respect to t, we have

$$2z\frac{dz}{dt} = 2x\frac{dx}{dt} + 2y\frac{dy}{dt}$$

When $x = 0.3$ and $y = 0.4$, the equation $z^2 = x^2 + y^2$ gives $z = 0.5$. Substituting these values, along with $dx/dt = -50$ and $dy/dt = -60$, we have

$$2(0.5)\frac{dz}{dt} = 2(0.3)(-50) + 2(0.4)(-60)$$

$$\frac{dz}{dt} = -30 - 48 = -78 \text{ mi/h}$$

The cars are approaching each other at a rate of 78 mi/h. ■

▪ **EXAMPLE 5   Filling a Tank**

A water tank has the shape of an inverted circular cone with base radius 2 m and height 4 m. If water is being pumped into the tank at a rate of 2 m³/min, find the rate at which the water level is rising when the water is 3 m deep.

SOLUTION

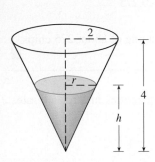

FIGURE 4

We first sketch the cone and label it as in Figure 4. Let V, r, and h be the volume of the water, the radius of the surface, and the height at time t, where t is measured in minutes.

We are given that $dV/dt = 2$ m³/min and we are asked to find dh/dt when h is 3 m. The quantities V and h are related by the equation

$$V = \tfrac{1}{3}\pi r^2 h$$

but it is very useful to express V as a function of h alone. In order to eliminate r, we use the similar triangles in Figure 4 to write

$$\frac{r}{h} = \frac{2}{4} \qquad r = \frac{h}{2}$$

and the expression for V becomes

$$V = \frac{1}{3}\pi \left(\frac{h}{2}\right)^2 h = \frac{\pi}{12} h^3$$

Now we can differentiate each side with respect to t:

$$\frac{dV}{dt} = \frac{\pi}{4} h^2 \frac{dh}{dt}$$

so

$$\frac{dh}{dt} = \frac{4}{\pi h^2} \frac{dV}{dt}$$

Substituting $h = 3$ m and $dV/dt = 2$ m³/min, we have

$$\frac{dh}{dt} = \frac{4}{\pi(3)^2} \cdot 2 = \frac{8}{9\pi}$$

The water level is rising at a rate of $8/(9\pi) \approx 0.28$ m/min.   ▪

▪ Prepare Yourself

1. A car is 40 miles east of a landmark and a truck is 60 miles south of the landmark. How far is the truck from the car?

2. Two planes leave an airport at the same time, one flying north and the other flying west. After two hours the northbound plane has traveled x miles while the westbound plane has traveled y miles. Write an expression for the distance between the planes at that time.

3. Find the derivative (with respect to x) of the function. Give your result in terms of $f(x)$ and $f'(x)$.
 (a) $y = [f(x)]^4$ (b) $y = x^2 + xf(x)$

4. If A and B are differentiable functions of t, find the derivative of the function with respect to t. Give your result in terms of $A(t)$, $A'(t)$, $B(t)$, and $B'(t)$.
 (a) $y = A(t) B(t)$ (b) $y = [A(t)]^2 + [B(t)]^2$
 (c) $y = [A(t)]^2 B(t)$

▪ Exercises 4.1

1. Cube volume   If V is the volume of a cube with edge length x and the cube expands as time passes, find dV/dt in terms of dx/dt.

2. Oil spill

(a) If A is the area of a circle with radius r and the circle expands as time passes, find dA/dt in terms of dr/dt.

(b) Suppose oil spills from a ruptured tanker and spreads in a circular pattern. If the radius of the oil spill increases at a constant rate of 1 m/s, how fast is the area of the spill increasing when the radius is 30 m?

3. Square area   Each side of a square is increasing at a rate of 6 cm/s. At what rate is the area of the square increasing when its area is 16 cm^2?

4. Rectangle area   The length of a rectangle is increasing at a rate of 8 cm/s and its width is increasing at a rate of 3 cm/s. When the length is 20 cm and the width is 10 cm, how fast is the area of the rectangle increasing?

5. Assume that x and y are functions of t. If $y = x^3 + 2x$ and $dx/dt = 5$, find dy/dt when $x = 2$.

6. Assume that x and y are positive functions of t. If $x^2 + y^2 = 25$ and $dy/dt = 6$, find dx/dt when $y = 4$.

7. Suppose x, y, and z are positive functions of t. If $z^2 = x^2 + y^2$, $dx/dt = 2$, and $dy/dt = 3$, find dz/dt when $x = 5$ and $y = 12$.

8. Motion   A particle moves along the curve $y = \sqrt{1 + x^3}$. As it reaches the point $(2, 3)$, the y-coordinate is increasing at a rate of 4 cm/s. How fast is the x-coordinate of the particle changing at that instant?

9. Tire production   The weekly cost C, in dollars, for a manufacturer to produce q automobile tires is given by

$$C = 2200 + 16q - 0.01q^2 \qquad 0 \leqslant q \leqslant 800$$

If 600 tires are currently being made per week but production levels are increasing at a rate of 40 tires/week, compute the rate of change of cost with respect to time.

10. Appliance production   A manufacturer of power adapters estimates that its total cost during one month to produce q units is $2500 + 4q + 0.005q^2$ dollars, and it will collect $16q - 0.002q^2$ dollars in revenue after selling the units.

(a) The manufacturer is currently producing 700 units per month but is ramping up its production at a rate of 30 units per month. Find the rate of change of the manufacturer's monthly cost (with respect to time).

(b) Compute the rate of change of the manufacturer's monthly profit.

11. Wine demand   The annual demand q for bottles of wine from a vineyard when the bottles are priced at p dollars each satisfies the equation $qe^{0.03p} = 5000$. The price is currently \$14 per bottle. Find the rate at which demand changes (with respect to time) if the price increases at a rate of \$1.20 per year.

12. Crop demand   A corn farmer is inclined to grow more corn if he can sell the crop at a higher price. The farmer estimates that the price p, in dollars, per ton of corn is related to the number of tons x he will produce annually by the equation $p = 210 + 3.7\sqrt{x}$. Currently the farmer is selling corn for \$265 per ton. If the price increases \$15/year, at what rate, with respect to time, will the farmer's corn production change?

13–16 ▪

(a) What quantities are given in the problem?

(b) What is the unknown?

(c) Draw a picture of the situation for any time t.

(d) Write an equation that relates the quantities.

(e) Finish solving the problem.

13. Melting snowball   If a snowball melts so that its surface area decreases at a rate of 1 cm^2/min, find the rate at which the diameter decreases when the diameter is 10 cm.

14. Navigation   At noon, ship A is 150 km west of ship B. Ship A is sailing east at 35 km/h and ship B is sailing north at 25 km/h. How fast is the distance between the ships changing at 4:00 PM?

15. Aviation   A plane flying horizontally at an altitude of 1 mi and a speed of 500 mi/h passes directly over a radar station. Find the rate at which the distance from the plane to the station is increasing when the plane is 2 mi away from the station.

16. Construction   A crane pulls up a steel beam from the ground at a rate of 4 ft/s. A foreman is on the ground 20 feet from the crane. When the beam is 30 ft above the ground, find the rate at which the distance from the beam to the foreman is changing.

---

17. Navigation   Two cars start moving from the same point. One travels south at 60 mi/h and the other travels west at 25 mi/h. At what rate is the distance between the cars increasing two hours later?

18. Boating   A boat is pulled into a dock by a rope attached to the bow of the boat and passing through a pulley on the dock that is 1 m higher than the bow of the boat. If the rope is pulled in at a rate of 1 m/s, how fast is the boat approaching the dock when it is 8 m from the dock?

19. **Navigation**   At noon, ship A is 100 km west of ship B. Ship A is sailing south at 35 km/h and ship B is sailing north at 25 km/h. How fast is the distance between the ships changing at 4:00 PM?

20. **Water storage**   Water is being pumped into an inverted conical tank at a constant rate. The tank has height 6 m and the diameter at the top is 4 m. If the water level is rising at a rate of 20 cm/min when the height of the water is 2 m, find the rate at which water is being pumped into the tank.

21. **Pressurized gas**   Boyle's Law states that when a sample of gas is compressed at a constant temperature, the pressure P and volume V satisfy the equation $PV = C$, where C is a constant. Suppose that at a certain instant the volume is 600 cm³, the pressure is 150 kPa, and the pressure is increasing at a rate of 20 kPa/min. At what rate is the volume decreasing at this instant?

22. **Pressurized air**   When air expands adiabatically (without gaining or losing heat), its pressure P and volume V are related by the equation $PV^{1.4} = C$, where C is a constant. Suppose that at a certain instant the volume is 400 cm³ and the pressure is 80 kPa and is decreasing at a rate of 10 kPa/min. At what rate is the volume increasing at this instant?

23. **Rocket launching**   A television camera is positioned 4000 ft from the base of a rocket launching pad. The mechanism for focusing the camera has to take into account the increasing distance from the camera to the rising rocket. Let's assume the rocket rises vertically and its speed is 600 ft/s when it has risen 3000 ft. How fast is the distance from the television camera to the rocket changing at that moment?

24. **Baseball**   A baseball diamond is a square with side 90 ft. A batter hits the ball and runs toward first base with a speed of 24 ft/s.

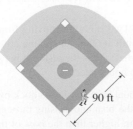

  (a) At what rate is his distance from second base decreasing when he is halfway to first base?

  (b) At what rate is his distance from third base increasing at the same moment?

25. **Construction**   Gravel is being dumped from a conveyor belt at a rate of 30 ft³/min, and its coarseness is such that it forms a pile in the shape of a cone whose base diameter and height are always equal. How fast is the height of the pile increasing when the pile is 10 ft high?

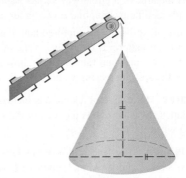

26. **Motion**   A particle is moving along the curve $y = \sqrt{x}$. As the particle passes through the point $(4, 2)$, its x-coordinate increases at a rate of 3 cm/s. How fast is the distance from the particle to the origin changing at this instant?

27. **Wind-chill index**   The wind-chill index is used to describe the apparent severity of cold weather combined with wind. The apparent temperature W has been modeled by the equation

$$W = 13.12 + 0.6215T - 11.37v^{0.16} + 0.3965Tv^{0.16}$$

where T is the actual temperature in °C and v is the wind speed in km/h. Suppose it is currently −15°C with wind blowing at 30 km/h. If the temperature is increasing at a rate of 2°C per hour and the wind speed is increasing at a rate of 4 km/h per hour, compute the rate at which the apparent temperature is changing.

28. **Cobb-Douglas production function**   In 1928 Charles Cobb and Paul Douglas published a model for the American economy during the early 20th century. The equation they used, known as the Cobb-Douglas production function, is $P = 1.01L^{0.75}K^{0.25}$ where P is the total production (the monetary value of all goods produced annually), L is the amount of labor (the total number of person-hours worked in a year), and K represents the amount of capital invested (the monetary worth of all machinery, equipment, and buildings). The values are measured as a percentage of the figures in 1899. In 1920, the total amount of labor was estimated at 194% (so $L = 194$) and the amount of capital invested was 407%. If the percentage of capital was increasing by 10 percentage points per year but labor was decreasing at a rate of 25 points per year, at what rate was production changing in 1920?

29. **Triangle area**   The altitude of a triangle is increasing at a rate of 1 cm/min while the area of the triangle is

increasing at a rate of 2 cm²/min. At what rate is the base of the triangle changing when the altitude is 10 cm and the area is 100 cm²?

30. Shadow motion   A street light is mounted at the top of a 15-ft-tall pole. A man 6 ft tall walks away from the pole with a speed of 5 ft/s along a straight path. How fast is the tip of his shadow moving when he is 40 ft from the pole?

31. Water level   A trough is 10 ft long and its ends have the shape of isosceles triangles that are 3 ft across at the top and have a height of 1 ft. If the trough is filled with water

at a rate of 12 ft³/min, how fast is the water level rising when the water is 6 inches deep?

32. Brain weight   Brain weight B as a function of body weight W in fish has been modeled by the power function $B = 0.007W^{2/3}$ where B and W are measured in grams. A model for body weight as a function of body length L (measured in centimeters) is $W = 0.12L^{2.53}$. If, over 10 million years, the average length of a certain species of fish evolved from 15 cm to 20 cm at a constant rate, how fast was this species' brain growing when the average length was 18 cm?

■ Challenge Yourself

33. Length of a sliding ladder   The top of a ladder slides down a vertical wall at a rate of 0.15 m/s. At the moment when the bottom of the ladder is 3 m from the wall, it slides away from the wall at a rate of 0.2 m/s. How long is the ladder?

34. Filling a swimming pool   A swimming pool is 20 ft wide, 40 ft long, 3 ft deep at the shallow end, and 9 ft deep at its deepest point. A cross-section is shown in the figure. If the pool is being filled at a rate of 0.8 ft³/min, how fast is the water level rising when the depth at the deepest point is 5 ft?

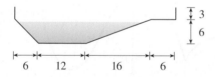

35. Pulley system   Two carts, A and B, are connected by a rope 39 ft long that passes over a pulley P (see the figure). The point Q is on the floor 12 ft directly beneath P and between the carts. Cart A is being pulled away from Q at a speed of 2 ft/s. How fast is cart B moving toward Q at the instant when cart A is 5 ft from Q?

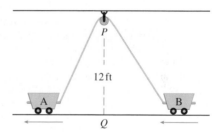

4.2 Maximum and Minimum Values

Some of the most important applications of differential calculus are *optimization problems*, in which we are required to find the optimal (best) way of doing something. Here are examples of such problems that we will solve in this chapter:

- What is the shape of a can that minimizes manufacturing costs?
- What is the maximum acceleration of a space shuttle? (This is an important question to the astronauts who have to withstand the effects of acceleration.)
- What is the radius of a contracted windpipe that expels air most rapidly during a cough?
- At what price should a business sell its products in order to maximize revenue?
- What is the maximum concentration of a drug in a patient's bloodstream?

These problems can be reduced to finding the maximum or minimum values of a function. Let's first explain exactly what we mean by maximum and minimum values; we will discuss two different kinds.

▪ Absolute and Local Maximum and Minimum Values

The *absolute maximum* of a function with a given domain is the largest output overall that the function produces, and the *absolute minimum* is the smallest.

> **(1) ▪ Definition**   Let c be a number in the domain of a function f. Then $f(c)$ is the
> - **absolute maximum** value of f on the domain if $f(c) \geq f(x)$ for all x in the domain.
> - **absolute minimum** value of f on the domain if $f(c) \leq f(x)$ for all x in the domain.

An absolute maximum or minimum is sometimes called a **global** maximum or minimum. The maximum and minimum values of f are called the **extreme values** of f.

Figure 1 shows the graph of a function f with absolute maximum at $x = 11$ and absolute minimum at $x = -5$. Note that $(11, f(11))$ is the highest point on the graph and $(-5, f(-5))$ is the lowest point.

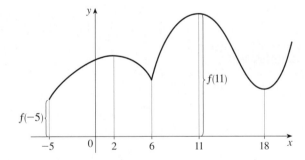

FIGURE 1

Minimum value $f(-5)$, maximum value $f(11)$

The second type of maximum or minimum value was defined in Section 2.4, namely *local maximum* and *local minimum* values. In Figure 1, $f(2)$ is a local maximum value of f because it is the largest function value for inputs near $x = 2$ [for instance, if we restrict our attention to the interval $(0, 4)$]. Likewise, $f(6)$ is a local minimum value of f because $f(6) \leq f(x)$ for x-values near $x = 6$ [in the interval $(5, 7)$, for example]. Looking at a continuous curve, we see that local maximum values occur at the tops of "hills" while local minimum values occur at the bottoms of "valleys." The function f in Figure 1 has another local maximum at $x = 11$ (this is also the absolute maximum) and a local minimum at $x = 18$. We now define these terms more formally.

> **(2) ▪ Definition**   The number $f(c)$ is a
> - **local maximum** value of f if $f(c) \geq f(x)$ when x is near c.
> - **local minimum** value of f if $f(c) \leq f(x)$ when x is near c.

Recall that if an interval is *open*, it means that the endpoints are not included. A *closed* interval includes both its endpoints.

In Definition 2 (and elsewhere), if we say that something is true **near** *c*, we mean that it is true on some open interval containing *c*. For example, in Figure 1, $f(18)$ is a local minimum because it's the smallest value of f on the interval (16, 19). [We could just as well use the interval (15, 20) or (17.9, 18.1).] It's not the absolute minimum because $f(x)$ takes on smaller values in other portions of the graph (near $x = -5$, for instance). Note that the value c in Definition 2 must be contained within an open interval, so local maximum or minimum values never occur at endpoints of a curve.

▪ EXAMPLE 1 An Absolute Maximum at an Endpoint

The graph of the function

$$f(x) = 3x^4 - 16x^3 + 18x^2 \qquad -1 \le x \le 4$$

is shown in Figure 2. You can see that $f(1) = 5$ is a local maximum, whereas the absolute maximum is $f(-1) = 37$. (This absolute maximum is not a local maximum because it occurs at an endpoint.) Also, $f(0) = 0$ is a local minimum and $f(3) = -27$ is both a local and an absolute minimum. Note that f has neither a local nor an absolute maximum at $x = 4$.

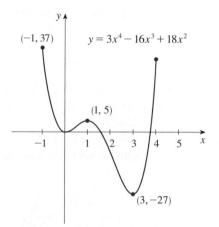

FIGURE 2

▪ EXAMPLE 2 An Infinite Number of Extreme Values

Figure 3 shows the graph of a model N for the number of daylight hours in Vancouver, Canada, t days after January 1 of any year. The function N takes on its absolute maximum value of approximately 16.2 infinitely many times (at about $t = 173, 538, 903, \ldots$). Each of these is also a local maximum value for the function. Likewise, N achieves its (local and absolute) minimum value of approximately 8.2 an infinite number of times.

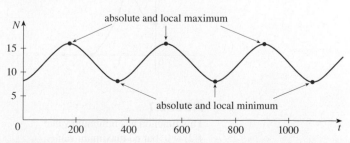

FIGURE 3
Repeating maximum and minimum values

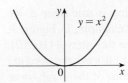

FIGURE 4
Minimum value 0, no maximum

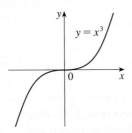

FIGURE 5
No minimum, no maximum

■ EXAMPLE 3 A Minimum but No Maximum

If $f(x) = x^2$, then $f(x) \geq f(0)$ because $f(0) = 0$ and $x^2 \geq 0$ for all x. Therefore $f(0) = 0$ is the absolute (and local) minimum value of f. This corresponds to the fact that the origin is the lowest point on the parabola $y = x^2$. (See Figure 4.) However, there is no highest point on the parabola and so this function has no maximum value. ■

■ EXAMPLE 4 No Extreme Values

From the graph of the function $f(x) = x^3$, shown in Figure 5, we see that this function has neither an absolute maximum value nor an absolute minimum value. In fact, it has no local extreme values either. ■

We have seen that some functions have extreme values, whereas others do not. The following theorem gives conditions under which a function is guaranteed to possess extreme values.

---

(3) ■ The Extreme Value Theorem If f is continuous on a closed interval, then it always attains an absolute maximum value and an absolute minimum value on that interval.

The Extreme Value Theorem is illustrated in Figure 6. Note that an extreme value can be taken on more than once. Although the Extreme Value Theorem is intuitively very plausible (just try to draw a continuous curve with two endpoints that doesn't achieve a maximum or minimum), it is difficult to prove and so we choose not to include the proof here.

FIGURE 6
Continuous curve on $[a, b]$,
maximum at c, minimum at d

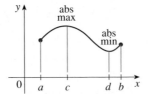

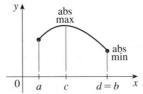

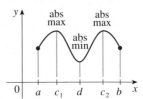

Figures 7 and 8 show that a function need not possess extreme values if either condition (continuity or closed interval) is omitted from the Extreme Value Theorem.

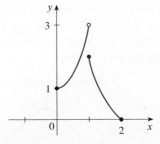

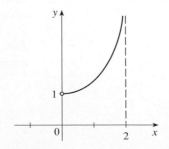

FIGURE 7
This function has minimum value
$f(2) = 0$, but no maximum value.

FIGURE 8
This continuous function g has
no maximum or minimum.

The function f whose graph is shown in Figure 7 is defined on the closed interval $[0, 2]$ but has no maximum value. [Notice that the range of f is $[0, 3)$. The function takes on values arbitrarily close to 3, but never actually attains the value 3. Thus there is no largest output.] This does not contradict the Extreme Value Theorem because f is not continuous. [Nonetheless, a discontinuous function *could* have maximum and minimum values. See Exercise 13(b).]

The function g shown in Figure 8 is continuous on the open interval $(0, 2)$ but has neither a maximum nor a minimum value. [The range of g is $(1, \infty)$. The function takes on arbitrarily large values.] This does not contradict the Extreme Value Theorem because the interval $(0, 2)$ is not closed.

▪ Finding Extreme Values

The Extreme Value Theorem says that a continuous function on a closed interval has a maximum value and a minimum value, but it does not tell us how to find these extreme values. We start by looking for local extreme values.

Figure 9 shows the graph of a function f with a local maximum at c and a local minimum at d. It appears that at the maximum and minimum points the tangent lines are horizontal and therefore each has slope 0. We know that the derivative is the slope of the tangent line, so it appears that $f'(c) = 0$ and $f'(d) = 0$. The following theorem says that this is always true for differentiable functions.

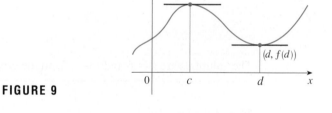

FIGURE 9

Fermat

Fermat's Theorem is named after Pierre Fermat (1601–1665), a French lawyer who took up mathematics as a hobby. Despite his amateur status, Fermat was one of the two inventors of analytic geometry (Descartes was the other). His methods for finding tangents to curves and maximum and minimum values (before the invention of limits and derivatives) made him a forerunner of Newton in the creation of differential calculus.

(4) ▪ Fermat's Theorem   If f has a local maximum or minimum at c, and if $f'(c)$ exists, then $f'(c) = 0$.

Fermat's Theorem can be proved using the definition of a derivative, although we don't include the proof here. Our intuition certainly suggests that the theorem is true. If we try to draw a curve with a local maximum, we will find that we always end up with a horizontal tangent, where the derivative is 0, or a cusp or corner, where the derivative doesn't exist.

Although Fermat's Theorem is very useful, we have to guard against reading too much into it. For instance, if $f(x) = x^3$, then $f'(x) = 3x^2$, so $f'(0) = 0$. But f has no maximum or minimum at 0, as you can see from its graph in Figure 10. The fact that $f'(0) = 0$ simply means that the curve $y = x^3$ has a horizontal tangent at $(0, 0)$. Instead of having a maximum or minimum at $(0, 0)$, the curve crosses its horizontal tangent there.

Thus, when $f'(c) = 0$, f doesn't necessarily have a maximum or minimum at c. The theorem simply states that *if* f has a local maximum or minimum, *then* the derivative (if it exists) must be 0 there.

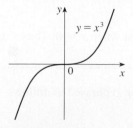

FIGURE 10
If $f(x) = x^3$, then $f'(0) = 0$ but f has no maximum or minimum.

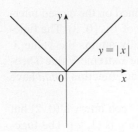

FIGURE 11
If $f(x) = |x|$, then $f(0) = 0$ is a minimum value, but $f'(0)$ does not exist.

We should bear in mind that there may be an extreme value where $f'(c)$ does not exist. For instance, the function $f(x) = |x|$ has its (local and absolute) minimum value at 0 (see Figure 11), but that value cannot be found by setting $f'(x) = 0$ because, as was shown in Example 5 in Section 2.4, $f'(0)$ does not exist.

Fermat's Theorem does suggest that we should at least *start* looking for extreme values of f at the numbers c where $f'(c) = 0$ or where $f'(c)$ does not exist. Such numbers are given a special name.

> **(5) ▪ Definition**   A **critical number** of a function f is a number c in the domain of f such that either $f'(c) = 0$ or $f'(c)$ does not exist.

Figure 12 shows a graph of the function A in Example 5. It supports our answer because there are horizontal tangents when $x = -2$ and when $x = 6$.

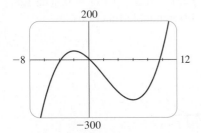

FIGURE 12

▪ EXAMPLE 5   Locating Critical Numbers

Find the critical numbers of $A(t) = t^3 - 6t^2 - 36t + 7$.

SOLUTION
First we compute the derivative:

$$A'(t) = 3t^2 - 6(2t) - 36 + 0 = 3t^2 - 12t - 36$$

There are no values of t at which $A'(t)$ is undefined, so we need only find the values where $A'(t) = 0$:

$$3t^2 - 12t - 36 = 0$$
$$3(t^2 - 4t - 12) = 0$$
$$3(t - 6)(t + 2) = 0$$
$$t - 6 = 0 \quad \text{or} \quad t + 2 = 0$$

The solutions are $t = 6$ and $t = -2$, so the critical numbers of A are 6 and -2. ▪

Figure 13 shows a graph of the function f in Example 6. There are no horizontal tangents but you can see a vertical tangent line at $x = 0$.

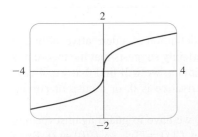

FIGURE 13

▪ EXAMPLE 6

A Critical Number Where the Derivative Doesn't Exist

Find the critical numbers of $f(x) = \sqrt[3]{x}$.

SOLUTION
The derivative of $f(x) = \sqrt[3]{x} = x^{1/3}$ is

$$f'(x) = \tfrac{1}{3}x^{-2/3} = \frac{1}{3(\sqrt[3]{x^2})}$$

Notice that $f'(x) = 0$ has no solution but $f'(x)$ does not exist when $x = 0$. Thus the only critical number is 0. ▪

In terms of critical numbers, Fermat's Theorem can be rephrased as follows (compare Definition 5 with Theorem 4):

> **(6) ▪** If f has a local maximum or minimum at c, then c is a critical number of f.

▪ The Closed Interval Method

To find an absolute maximum or minimum of a continuous function on a closed interval, we note that either it is also a local maximum or minimum [in which case it occurs at a critical number by (6)] or it occurs at an endpoint of the interval. Thus the following three-step procedure always works.

▪ The Closed Interval Method   To find the *absolute* maximum and minimum values of a continuous function f on a closed interval $[a, b]$:

1. Find the critical numbers of f in the interval (a, b) and compute the values of f at these numbers.

2. Find the values of f at the endpoints of the interval.

3. The largest of the output values from steps 1 and 2 is the absolute maximum value; the smallest of these values is the absolute minimum value.

▪ EXAMPLE 7   Using the Closed Interval Method

Find the absolute maximum and minimum values of the function

$$f(x) = x^3 - 3x^2 + 1 \qquad -\tfrac{1}{2} \le x \le 4$$

SOLUTION

Because f is continuous on the closed interval $\left[-\tfrac{1}{2}, 4\right]$, we can use the Closed Interval Method:

$$f(x) = x^3 - 3x^2 + 1$$

$$f'(x) = 3x^2 - 6x = 3x(x - 2)$$

Since $f'(x)$ exists for all x, the only critical numbers of f occur when $f'(x) = 0$, that is, $x = 0$ or $x = 2$. Notice that each of these critical numbers lies in the interval $\left(-\tfrac{1}{2}, 4\right)$. The values of f at these critical numbers are

$$f(0) = 1 \qquad f(2) = -3$$

The values of f at the endpoints of the interval are

$$f\left(-\tfrac{1}{2}\right) = \tfrac{1}{8} \qquad f(4) = 17$$

Comparing these four numbers, we see that the absolute maximum value is $f(4) = 17$ and the absolute minimum value is $f(2) = -3$.

Note that in this example the absolute maximum occurs at an endpoint, whereas the absolute minimum occurs at a critical number. The graph of f is sketched in Figure 14.   ▪

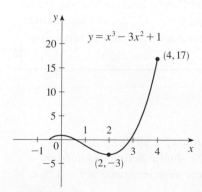

FIGURE 14

If you have a graphing calculator or a computer with graphing software, it is possible to estimate maximum and minimum values very easily. But, as the next example shows, calculus is needed to find the *exact* values.

■ EXAMPLE 8 Estimated and Exact Extreme Values

(a) Use a graphing device to estimate the absolute minimum and maximum values of the function $f(x) = xe^{-x^2}$, $-1 \leq x \leq 1$.

(b) Use calculus to find the exact minimum and maximum values.

SOLUTION

(a) Figure 15 shows a graph of f in the viewing rectangle $[-1, 1]$ by $[-0.6, 0.6]$. By moving the cursor close to the maximum point, we see that the y-coordinates don't change very much in the vicinity of the maximum. The absolute maximum value is about 0.43 and it occurs when $x \approx 0.7$. Similarly, by moving the cursor close to the minimum point, we see that the absolute minimum value is about -0.43 and it occurs when $x \approx -0.7$. It is possible to get more accurate estimates by zooming in toward the maximum or minimum point or by using a built-in maximum feature of a graphing calculator, but instead let's use calculus to get exact values.

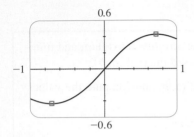

0.6

−1   1

−0.6

FIGURE 15

(b) The function $f(x) = xe^{-x^2}$ is continuous on $[-1, 1]$ so the Closed Interval Method applies. First we find the critical numbers of f. The derivative is

$$f'(x) = x \cdot e^{-x^2}(-2x) + e^{-x^2} \cdot 1 = e^{-x^2}(1 - 2x^2)$$

which is defined everywhere. Because e^{-x^2} is never zero, we have $f'(x) = 0$ when $1 - 2x^2 = 0$ and this occurs when $x^2 = \frac{1}{2}$ or $x = \pm 1/\sqrt{2}$. The values of f at these critical numbers are

$$f\left(1/\sqrt{2}\right) = \frac{1}{\sqrt{2}}\, e^{-(1/\sqrt{2})^2} = \frac{1}{\sqrt{2}}\, e^{-1/2} = \frac{1}{\sqrt{2e}} \approx 0.42888$$

and

$$f\left(-1/\sqrt{2}\right) = -\frac{1}{\sqrt{2}}\, e^{-(-1/\sqrt{2})^2} = -\frac{1}{\sqrt{2e}} \approx -0.42888$$

The values of f at the endpoints are

$$f(-1) = -e^{-1} \approx -0.36788 \qquad \text{and} \qquad f(1) = e^{-1} \approx 0.36788$$

Comparing these four numbers and using the Closed Interval Method, we see that the absolute minimum value is $f\left(-1/\sqrt{2}\right) = -1/\sqrt{2e}$ and the absolute maximum value is $f\left(1/\sqrt{2}\right) = 1/\sqrt{2e}$. The values from part (a) serve as a check on our work. ■

■ EXAMPLE 9 Rocket Acceleration

The Hubble Space Telescope was deployed on April 24, 1990, by the space shuttle *Discovery*. A model for the velocity of the shuttle during this mission, from liftoff at $t = 0$ until the solid rocket boosters were jettisoned at $t = 126$ s, is given by

$$v(t) = 0.001302t^3 - 0.09029t^2 + 23.61t - 3.083$$

in feet per second. Using this model, estimate the absolute maximum and minimum values of the *acceleration* of the shuttle between liftoff and the jettisoning of the boosters.

NASA

SOLUTION

We are asked for the extreme values not of the given velocity function, but rather of the acceleration function. So we first need to differentiate to find the acceleration:

$$a(t) = v'(t) = \frac{d}{dt}(0.001302t^3 - 0.09029t^2 + 23.61t - 3.083)$$

$$= 0.003906t^2 - 0.18058t + 23.61$$

We now apply the Closed Interval Method to the continuous function a on the interval $0 \leqslant t \leqslant 126$. Its derivative is

$$a'(t) = 0.007812t - 0.18058$$

The only critical number occurs when $a'(t) = 0$:

$$t_1 = \frac{0.18058}{0.007812} \approx 23.12$$

Evaluating $a(t)$ at the critical number and the endpoints, we have

$$a(0) = 23.61 \qquad a(t_1) \approx 21.52 \qquad a(126) \approx 62.87$$

So the maximum acceleration is about 62.87 ft/s^2 and the minimum acceleration is about 21.52 ft/s^2.   ■

■ Prepare Yourself

1. Solve the given equation.

(a) $4x^2 + x - 3 = 0$ **(b)** $\dfrac{a^2 - 2a}{(a+1)^2} = 0$

(c) $5t^2 - 5 = 0$ **(d)** $3 + \dfrac{1}{x} = 0$

2. Use the quadratic formula to solve the equation $2x^2 - x - 5 = 0$.

3. Factor $(1 + x^2)^{-1/2}$ from the expression $x^2(1 + x^2)^{-1/2} + 3(1 + x^2)^{1/2}$.

4. Factor completely: $t^{2/3}(t + 2) - 5t^{-1/3}(t + 2)^2$

5. Solve the given equation.

(a) $3xe^x + 2e^x = 0$ **(b)** $\ln x - 2 = 0$

(c) $2x(x^2 - 9)^3 = 0$

6. Find the derivative of the function.

(a) $f(x) = xe^{5x}$ **(b)** $f(t) = \dfrac{t + 3}{t^2 + 4}$

(c) $y = \sqrt{1 + \ln x}$ **(d)** $y = x\sqrt{1 + \ln x}$

■ Exercises 4.2

1. Explain the difference between an absolute minimum and a local minimum.

2. Suppose f is a continuous function defined on a closed interval $[a, b]$.

  (a) What theorem guarantees the existence of an absolute maximum value and an absolute minimum value for f?

  (b) What steps would you take to find those maximum and minimum values?

3–4 ■ For each of the numbers a, b, c, d, e, r, s, and t, state whether the function whose graph is shown has an absolute

maximum or minimum, a local maximum or minimum, or neither a maximum nor a minimum.

3.

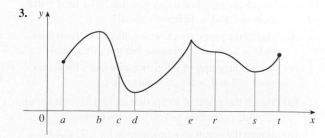

4.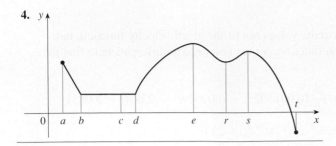

5–6 ▪ Use the graph to state the absolute and local maximum and minimum values of the function.

5.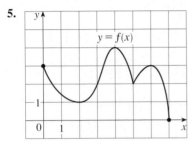

6.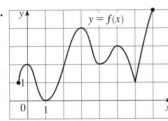

7–10 ▪ Sketch the graph of a function f that is continuous on $[1, 5]$ and has the given properties.

7. Absolute minimum at 2, absolute maximum at 3, local minimum at 4

8. Absolute minimum at 1, absolute maximum at 5, local maximum at 2, local minimum at 4

9. Absolute maximum at 5, absolute minimum at 2, local maximum at 3, local minima at 2 and 4

10. f has no local maximum or minimum, but 2 and 4 are critical numbers

11. (a) Sketch the graph of a function that has a local maximum at 2 and is differentiable at 2.
   (b) Sketch the graph of a function that has a local maximum at 2 and is continuous but not differentiable at 2.
   (c) Sketch the graph of a function that has a local maximum at 2 and is not continuous at 2.

12. (a) Sketch the graph of a function on $[-1, 2]$ that has an absolute maximum but no local maximum.
   (b) Sketch the graph of a function on $[-1, 2]$ that has a local maximum but no absolute maximum.

13. (a) Sketch the graph of a function on $[-1, 2]$ that has an absolute maximum but no absolute minimum.
   (b) Sketch the graph of a function on $[-1, 2]$ that is discontinuous but has both an absolute maximum and an absolute minimum.

14. (a) Sketch the graph of a function that has two local maximum values, one local minimum, and no absolute minimum.
   (b) Sketch the graph of a function that has three local minimum values, two local maximum values, and seven critical numbers.

15–22 ▪ Sketch the graph of f by hand and use your sketch to find the absolute and local maximum and minimum values of f. (Use the graphs and transformations of Chapter 1.)

15. $f(x) = 8 - 3x, \quad x \geq 1$ **16.** $f(x) = 3 - 2x, \quad x \leq 5$

17. $f(x) = x^2, \quad 0 < x < 2$

18. $f(x) = 1 + (x + 1)^2, \quad -2 \leq x < 5$

19. $f(x) = \ln x, \quad 0 < x \leq 2$ **20.** $f(x) = e^x$

21. $f(x) = 1 - \sqrt{x}$

22. $f(x) = \begin{cases} 4 - x^2 & \text{if } -2 \leq x < 0 \\ 2x - 1 & \text{if } 0 \leq x \leq 2 \end{cases}$

23–34 ▪ Find the critical numbers of the function.

23. $f(x) = 5x^2 + 4x$ **24.** $f(x) = x^3 + x^2 - x$

25. $f(x) = x^3 + 3x^2 - 24x$ **26.** $g(t) = |3t - 4|$

27. $s(t) = 3t^4 + 4t^3 - 6t^2$ **28.** $f(x) = x^3 + x^2 + x$

29. $f(x) = x \ln x$ **30.** $f(x) = xe^{2x}$

31. $g(y) = \dfrac{y - 1}{y^2 - y + 1}$ **32.** $h(p) = \dfrac{p - 1}{p^2 + 4}$

33. $F(x) = x^{4/5}(x - 4)^2$ **34.** $g(x) = \sqrt{1 - x^2}$

35–46 ▪ Find the absolute maximum and absolute minimum values of f on the given interval.

35. $f(x) = 3x^2 - 12x + 5, \quad [0, 3]$

36. $f(x) = x^3 - 3x + 1, \quad [0, 3]$

37. $f(x) = 2x^3 - 3x^2 - 12x + 1, \quad [-2, 3]$

38. $f(x) = x^3 - 6x^2 + 9x + 2, \quad [-1, 4]$

39. $f(x) = x^4 - 2x^2 + 3, \quad [-2, 3]$

40. $f(x) = (x^2 - 1)^3, \quad [-1, 2]$

41. $f(t) = t\sqrt{4 - t^2}, \quad [-1, 2]$

42. $f(x) = \dfrac{x}{x^2 + 4}, \quad [0, 3]$

43. $f(x) = xe^{-x^2/8}, \quad [-1, 4]$ **44.** $f(x) = x - \ln x, \quad \left[\frac{1}{2}, 2\right]$

45. $f(x) = \ln(x^2 + x + 1), \quad [-1, 1]$

46. $f(x) = e^{-x} - e^{-2x}, \quad [0, 1]$

---

47. Use a graph to estimate the critical numbers of $f(x) = |x^2 - 8| - 3x$ correct to two decimal places.

48. Use a graph to estimate the critical numbers of $f(x) = |x^3 - 3x^2 + 2|$ correct to one decimal place.

49–52 ▪

(a) Use a graph to estimate the absolute maximum and minimum values of the function to two decimal places.

(b) Use calculus to find the exact maximum and minimum values.

49. $f(x) = x^5 - x^3 + 2, \quad -1 \leqslant x \leqslant 1$

50. $f(x) = e^{x^3 - x}, \quad -1 \leqslant x \leqslant 0$

51. $g(t) = te^{-t}, \quad 0.5 \leqslant t \leqslant 4.5$

52. $F(v) = 2v^3 - 87v + 33, \quad -2.4 \leqslant v \leqslant 5.1$

53–56 ▪ Use a graphing calculator to estimate the absolute maximum and absolute minimum values of the function, accurate to three decimal places, for the given interval.

53. $f(x) = e^x/(1 + \sqrt{x}), \quad 0 \leqslant x \leqslant 1$

54. $g(x) = \ln \dfrac{\sqrt{x} + 1}{x^2 + 1}, \quad 0 \leqslant x \leqslant 2$

55. $p(t) = \sqrt[3]{t^2 - t - 1}, \quad -1 \leqslant t \leqslant 3$

56. $h(x) = x^{2/3} - 3x^{1/3} + 1, \quad -10 \leqslant x \leqslant 10$

---

57. Blood alcohol concentration After the consumption of an alcoholic beverage, the concentration of alcohol in the bloodstream (blood alcohol concentration, or BAC) surges as the alcohol is absorbed, followed by a gradual decline as the alcohol is metabolized. The function

$$C(t) = 1.2te^{-2.6t}$$

models the BAC, measured in mg/mL, in a test patient's bloodstream t hours after rapidly consuming 15 mL of alcohol. What is the maximum BAC during the first three hours? When does it occur?

58. Consumer demand A software developer released a subscription-based streaming music application for mobile phones that quickly attracted users with free trial subscriptions, but most users did not renew the service and the application lost popularity. The number of subscriptions, in thousands, t weeks after the release was given by

$$A(t) = 186t^3e^{-1.4t}$$

What was the highest number of subscribers to the service during the first six weeks?

59. Water volume Between 0°C and 30°C, the volume V, in cubic centimeters, of 1 kg of water at a temperature T is given approximately by the formula

$$V = 999.87 - 0.06426T + 0.0085043T^2 - 0.0000679T^3$$

Find the temperature at which water has its minimum volume.

60. Water level The water level, measured in feet above mean sea level, of Lake Lanier in the state of Georgia during 2008 can be modeled by the function

$$L(t) = 0.007653t^3 - 0.3327t^2 + 2.799t + 1050.3$$

where t is measured in months since January 1, 2008. Estimate when the water level was highest during 2008.

61. Consumer prices A model for the US average price of a pound of white sugar from 1993 to 2003 is given by the function

$$S(t) = -0.00003237t^5 + 0.0009037t^4 - 0.008956t^3$$
$$+ 0.03629t^2 - 0.04458t + 0.4074$$

where t is measured in years since August 1993. Estimate the times when sugar was cheapest and most expensive during the period 1993–2003.

62. Space shuttle launch On May 7, 1992, the space shuttle *Endeavour* was launched on mission STS-49, the purpose of which was to install a new perigee kick motor in an Intelsat communications satellite. The table gives the velocity data for the shuttle between liftoff and the jettisoning of the solid rocket boosters.

Event	Time (s)	Velocity (ft/s)
Launch	0	0
Begin roll maneuver	10	185
End roll maneuver	15	319
Throttle to 89%	20	447
Throttle to 67%	32	742
Throttle to 104%	59	1325
Maximum dynamic pressure	62	1445
Solid rocket booster separation	125	4151

(a) Use a graphing calculator or computer to find the cubic polynomial that best models the velocity of the shuttle for the time interval $0 \leqslant t \leqslant 125$. Then graph this polynomial.

(b) Find a model for the acceleration of the shuttle and use it to estimate the maximum and minimum values of the acceleration during the first 125 seconds.

▪ Challenge Yourself

63. If a and b are positive numbers, find the maximum value of $f(x) = x^a(1 - x)^b$, $0 \leqslant x \leqslant 1$.

64. A cubic function is a polynomial of degree 3; that is, it has the form $f(x) = ax^3 + bx^2 + cx + d$, where $a \neq 0$.

  (a) Show that a cubic function can have two, one, or no critical number(s). Give examples and sketches to illustrate the three possibilities.

  (b) How many local extreme values can a cubic function have?

65. Prove that the function

$$f(x) = x^{101} + x^{51} + x + 1$$

has neither a local maximum nor a local minimum.

66. Coughing force   When a foreign object lodged in the trachea (windpipe) forces a person to cough, the diaphragm thrusts upward causing an increase in pressure in the lungs. This is accompanied by a contraction of the trachea, making a narrower channel for the expelled air to flow through. For a given amount of air to escape in a fixed time, it must move faster through the narrower channel than the wider one. The greater the velocity of the airstream, the greater the force on the foreign object. X-rays show that the radius of the circular tracheal tube contracts to about two-thirds of its normal radius during a cough. According to a mathematical model of coughing, the velocity v of the airstream is related to the radius r of the trachea by the equation

$$v(r) = k(r_0 - r)r^2 \qquad \tfrac{1}{2}r_0 \leqslant r \leqslant r_0$$

where k is a constant and r_0 is the normal radius of the trachea. The restriction on r is due to the fact that the tracheal wall stiffens under pressure and a contraction greater than $\tfrac{1}{2}r_0$ is prevented (otherwise the person would suffocate).

  (a) Determine the value of r in the interval $\left[\tfrac{1}{2}r_0, r_0\right]$ at which v has an absolute maximum. How does this compare with experimental evidence?

  (b) What is the absolute maximum value of v on the interval?

  (c) Sketch the graph of v on the interval $[0, r_0]$.

4.3   Derivatives and the Shapes of Curves

In Section 2.4 we discussed how the signs of the first and second derivatives $f'(x)$ and $f''(x)$ influence the shape of the graph of f. Here we revisit those facts and use them, together with the differentiation formulas of Chapter 3, to explain the shapes of curves.

▪ Increasing and Decreasing Functions

We saw in Section 2.4 that a function increases where its derivative is positive and decreases where the derivative is negative.

Let's abbreviate the name of this test to the I/D Test.

> **▪ Increasing / Decreasing Test**
>
> **(a)** If $f'(x) > 0$ on an interval, then f is increasing on that interval.
>
> **(b)** If $f'(x) < 0$ on an interval, then f is decreasing on that interval.

We can use the I/D Test to identify intervals where a function is increasing or decreasing by first finding the critical numbers of the function, as the next example illustrates.

▪ EXAMPLE 1   Identifying Intervals of Increase or Decrease

Find where the function $f(x) = 3x^4 - 4x^3 - 12x^2 + 5$ is increasing and where it is decreasing.

SOLUTION

$$f'(x) = 12x^3 - 12x^2 - 24x = 12x(x - 2)(x + 1)$$

Notice that $f'(x) = 0$ when $x = 0$, $x = 2$, or $x = -1$, so the critical numbers are 0, 2, and -1. To use the I/D Test we have to know where $f'(x) > 0$ and where $f'(x) < 0$. This depends on the signs of the three factors of $f'(x)$, namely, $12x$, $x - 2$, and $x + 1$. We divide the domain into intervals whose endpoints are the critical numbers -1, 0, and 2, and arrange our work in a chart. A plus sign indicates that the given expression is positive, and a minus sign indicates that it is negative. The last column of the chart gives the conclusion based on the I/D Test. For instance, $f'(x) < 0$ for $0 < x < 2$, so f is decreasing on $(0, 2)$. (It would also be true to say that f is decreasing on the closed interval $[0, 2]$.)

Interval	$12x$	$x - 2$	$x + 1$	$f'(x)$	f
$x < -1$	$-$	$-$	$-$	$-$	decreasing on $(-\infty, -1)$
$-1 < x < 0$	$-$	$-$	$+$	$+$	increasing on $(-1, 0)$
$0 < x < 2$	$+$	$-$	$+$	$-$	decreasing on $(0, 2)$
$x > 2$	$+$	$+$	$+$	$+$	increasing on $(2, \infty)$

The graph of f shown in Figure 1 confirms the information in the chart.   ▪

Alternatively, we can choose a "test value" within each interval. For instance, 1 lies in the interval $(0, 2)$ and $f'(1) = -24$, so we know that f' is negative on the interval $(0, 2)$. We are justified in using this approach when f' is continuous, as it cannot change sign without passing through 0 (a critical number). Thus f' is either positive throughout the entire interval between critical numbers or negative on the entire interval.

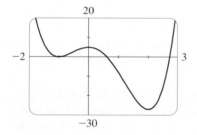

FIGURE 1

Recall from Section 4.2 that if f has a local maximum or minimum at c, then c must be a critical number of f (by Fermat's Theorem), but not every critical number gives rise to a maximum or a minimum. We therefore need a test that will tell us whether or not f has a local maximum or minimum at a critical number.

You can see from Figure 1 that $f(0) = 5$ is a local maximum value of f because f increases on $(-1, 0)$ and decreases on $(0, 2)$. Or, in terms of derivatives, $f'(x) > 0$ for $-1 < x < 0$ and $f'(x) < 0$ for $0 < x < 2$. In other words, the sign of $f'(x)$ changes from positive to negative at 0. This observation is the basis of the following test.

▪ **The First Derivative Test**   Suppose that c is a critical number of a continuous function f.

(a) If f' changes from positive to negative at c, then f has a local maximum at c.

(b) If f' changes from negative to positive at c, then f has a local minimum at c.

(c) If f' does not change sign at c (for example, if f' is positive on both sides of c or negative on both sides), then f has no local maximum or minimum at c.

The First Derivative Test is a consequence of the I/D Test. In part (a), for instance, since the sign of $f'(x)$ changes from positive to negative at c, f is increasing to the left of c and decreasing to the right of c. It follows that f has a local maximum at c.

It is easy to remember the First Derivative Test by visualizing diagrams such as those in Figure 2.

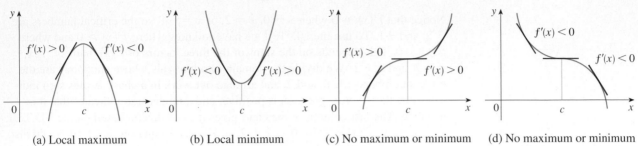

(a) Local maximum   (b) Local minimum   (c) No maximum or minimum   (d) No maximum or minimum

FIGURE 2

■ **EXAMPLE 2**   **Identifying Local Minimum and Maximum Values**

Find the local minimum and maximum values of the function f in Example 1.

SOLUTION

From the chart in the solution to Example 1 we see that $f'(x)$ changes from negative to positive at -1, so $f(-1) = 0$ is a local minimum value by the First Derivative Test. Similarly, f' changes from negative to positive at 2, so $f(2) = -27$ is also a local minimum value. As previously noted, $f(0) = 5$ is a local maximum value because $f'(x)$ changes from positive to negative at 0.   ■

■ Concavity

We have learned how to identify where a function is increasing and decreasing. But we can investigate *how* a function increases or decreases by looking at how the derivative values change.

We saw in Section 2.4 that if f' itself is an increasing function, then the slopes of the tangent lines of the curve $y = f(x)$ increase from left to right and the curve bends upward. We called such a curve *concave upward*. If f' is decreasing, the slopes decrease from left to right. The curve bends downward and we called it *concave downward*.

> ■ **Definition**   A function (or its graph) is called **concave upward** on an interval if f' is an increasing function on that interval. It is called **concave downward** on the interval if f' is decreasing there.

Concavity is independent of whether the curve is increasing or decreasing. Figure 3 shows the different combinations possible.

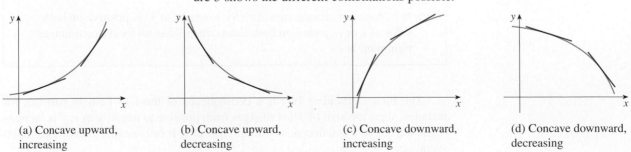

(a) Concave upward, increasing   (b) Concave upward, decreasing   (c) Concave downward, increasing   (d) Concave downward, decreasing

FIGURE 3

Parts (a) and (b) of Figure 3 show curves that are concave upward; you can see that the slopes of the tangent lines are increasing. [In (b) the slopes are negative and are getting closer to 0.] Similarly, in parts (c) and (d) the curves are concave downward and the slopes are decreasing. Notice that when the curve is concave upward it lies above its tangent lines, and when it is concave downward it lies below the tangent lines.

We also mentioned in Section 2.4 that a point where the concavity changes direction is called an *inflection point*.

▪ **Definition**   A point on a curve $y = f(x)$ is called an **inflection point** if f is continuous there and the curve changes from concave upward to concave downward or from concave downward to concave upward at that point.

■ **EXAMPLE 3**

Identifying Concavity and Inflection Points from a Graph

A function f is graphed in Figure 4. The curve begins with a negative slope and the curve bends upward. If we draw tangent lines along the curve, they would show increasing slopes as we look from left to right. The slopes turn positive and continue to increase until we reach $x = 6$. Thus f' is increasing and f is concave upward (we will abbreviate this as CU) on the interval $(1, 6)$. (We can also observe that the curve lies above its tangent lines on this interval.) For $6 < x < 10$ the curve bends downward. The slopes start out positive but decrease, turning negative at about $x = 8$ and continuing to decrease to $x = 10$, so f is concave downward (CD) on $(6, 10)$. The curve is concave upward on $(10, 13)$ because the slopes increase there. Again the curve bends upward. We have an inflection point at $x = 6$ because the concavity changes from upward to downward, and another at $x = 10$, where the concavity changes from downward to upward.

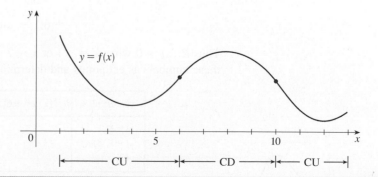

FIGURE 4

Let's see how the second derivative helps determine the intervals of concavity. Wherever the function f is concave upward, we know f' is increasing. This means that the derivative of f' is positive. Because $f'' = (f')'$, we know that f'' must be positive there. Similarly, if $f''(x)$ is negative, then f' is decreasing and f is concave downward. Thus we have the following test for concavity.

▪ **Concavity Test**

(a) If $f''(x) > 0$ for all x in an interval, then the graph of f is concave upward on that interval.

(b) If $f''(x) < 0$ for all x in an interval, then the graph of f is concave downward on that interval.

In view of the Concavity Test, there is an inflection point at any point where the second derivative changes sign (assuming that the function is continuous). Recall how we used the first derivative to find local extreme points and intervals of increase or decrease: We found where $f'(x) = 0$ or where $f'(x)$ was undefined, then we checked the intervals between these values for the sign of f'. Similar analysis with the second derivative allows us to find inflection points and intervals of concavity, as the next example illustrates.

▪ **EXAMPLE 4**

Identifying Concavity and Inflection Points from an Equation

Identify the intervals of concavity and inflection points for the function

$$R(x) = -0.5x^4 + x^3 + 6x^2 - 2x + 4$$

SOLUTION

We begin by calculating the first and second derivatives:

$$R'(x) = -2x^3 + 3x^2 + 12x - 2$$

$$R''(x) = -6x^2 + 6x + 12$$

We have possible inflection points where $R''(x) = 0$ or where $R''(x)$ is undefined. Here $R''(x)$ is defined everywhere, so we solve

$$-6x^2 + 6x + 12 = 0$$

$$-6(x^2 - x - 2) = 0$$

$$-6(x - 2)(x + 1) = 0$$

Thus $R''(x) = 0$ when $x = -1$ or $x = 2$. We divide the domain into intervals with these numbers as endpoints and determine the sign of R'' in each interval.

Interval	$f''(x) = -6(x - 2)(x + 1)$	Concavity
$(-\infty, -1)$	−	downward
$(-1, 2)$	+	upward
$(2, \infty)$	−	downward

Thus R is concave upward on $(-1, 2)$ and concave downward on the intervals $(-\infty, -1)$ and $(2, \infty)$. We have inflection points where $x = -1$ and where $x = 2$ because R'' changes sign at these values. The corresponding inflection points are $(-1, 10.5)$ and $(2, 24)$. The graph of R in Figure 5 serves as a visual check on our work. ▪

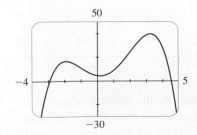

FIGURE 5

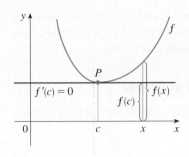

FIGURE 6
$f''(c) > 0$, f is concave upward

A consequence of the Concavity Test is the following test for maximum and minimum values, which can be used as an alternative to the First Derivative Test.

■ **The Second Derivative Test**   Suppose f'' is continuous near c.

(a) If $f'(c) = 0$ and $f''(c) > 0$, then f has a local minimum at c.

(b) If $f'(c) = 0$ and $f''(c) < 0$, then f has a local maximum at c.

For instance, part (a) is true because $f''(x)$ is positive near c and so f is concave upward near c. This means that the graph of f bends upward and lies *above* its horizontal tangent at c and so f has a local minimum at c. (See Figure 6.)

■ **EXAMPLE 5   Analyzing a Curve Using Derivatives**

Discuss the curve $y = x^4 - 4x^3$ with respect to concavity, points of inflection, and local maximum and minimum values. Use this information to sketch the curve.

SOLUTION

If $f(x) = x^4 - 4x^3$, then

$$f'(x) = 4x^3 - 12x^2 = 4x^2(x - 3)$$

$$f''(x) = 12x^2 - 24x = 12x(x - 2)$$

To find the critical numbers we set $f'(x) = 0$ [note that $f'(x)$ is defined everywhere] and obtain $x = 0$ and $x = 3$. To use the Second Derivative Test we evaluate f'' at these critical numbers:

$$f''(0) = 0 \qquad f''(3) = 36 > 0$$

Since $f'(3) = 0$ and $f''(3) > 0$, $f(3) = -27$ is a local minimum. Since $f''(0) = 0$, the Second Derivative Test gives no information about the critical number 0. But because $f'(x) < 0$ for $x < 0$ and also for $0 < x < 3$, the First Derivative Test tells us that f does not have a local maximum or minimum at 0.

Since $f''(x) = 0$ when $x = 0$ or 2, we divide the domain into intervals with these numbers as endpoints and complete the following chart.

Interval	$f''(x) = 12x(x - 2)$	Concavity
$(-\infty, 0)$	+	upward
$(0, 2)$	−	downward
$(2, \infty)$	+	upward

The point $(0, 0)$ is an inflection point since the curve changes from concave upward to concave downward there. Also $(2, -16)$ is an inflection point since the curve changes from concave downward to concave upward there.

Using the local minimum, the intervals of concavity, and the inflection points, we sketch the curve in Figure 7. ■

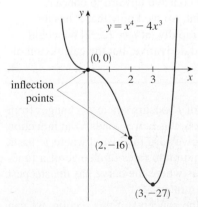

FIGURE 7

NOTE: The Second Derivative Test is inconclusive when $f''(c) = 0$. In other words, at such a point there might be a maximum, there might be a minimum, or there might be neither (as in Example 5). This test also fails when $f''(c)$ does not exist. In such cases the First Derivative Test must be used. In fact, even when both tests apply, the First Derivative Test is often the easier one to use.

▪ EXAMPLE 6   Finding the Greatest Rate of Change

A population of honeybees raised in an apiary started with 50 bees at time $t = 0$ and was modeled by the function

$$P(t) = \frac{75{,}200}{1 + 1503e^{-0.5932t}}$$

where t is the time in weeks, $0 \le t \le 25$. Use a graph to estimate the time at which the bee population was growing fastest. Then graph the derivative to obtain a more accurate estimate.

SOLUTION

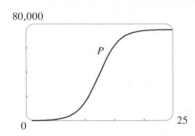

FIGURE 8

The population grows fastest when the population curve $y = P(t)$ has the steepest tangent line. From the graph of P in Figure 8, we estimate that the steepest tangent occurs when $t \approx 12$, so the bee population was growing most rapidly after about 12 weeks.

For a better estimate we calculate the derivative $P'(t)$, which is the rate of increase of the bee population. If we write P as

$$P(t) = 75{,}200(1 + 1503e^{-0.5932t})^{-1}$$

then

$$P'(t) = 75{,}200(-1)(1 + 1503e^{-0.5932t})^{-2} \cdot 1503e^{-0.5932t}(-0.5932)$$

$$= -\frac{67{,}046{,}785.92e^{-0.5932t}}{(1 + 1503e^{-0.5932t})^2}$$

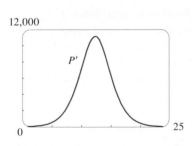

FIGURE 9

We graph P' in Figure 9 and observe that P' has its maximum value when $t \approx 12.3$. Thus the bee population was growing fastest after 12.3 weeks.

Note that P' has its maximum value when P' changes from increasing to decreasing. This happens when P changes from concave upward to concave downward, that is, when P has an inflection point. So we know that P has an inflection point where the values increase most rapidly, at $t \approx 12.3$. [We could have found this inflection point using the second derivative, but the calculation of $P''(t)$ is rather complicated.]   ▪

If f' is continuous, a local maximum value of f' occurs where f' changes from increasing to decreasing. As we saw in Example 6, this corresponds to an inflection point of f. Similarly, concavity changes from downward to upward where f' has a local minimum value. Thus the maximum or minimum rates of change of a function occur at inflection points. On a graph, this is where the curve has the steepest or most gradual incline or decline (on some open interval).

The next example shows that, armed with the concepts of this section, we can obtain valuable information about a function's behavior from just a graph of its derivative.

▪ **EXAMPLE 7**

Analyzing a Function's Behavior From a Graph of Its Derivative

If it is known that the graph of the derivative f' of a function is as shown in Figure 10, identify where f is increasing or decreasing, where it has local maximum or minimum values, where it is concave upward or downward, and where any inflection points are located.

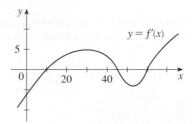

FIGURE 10

SOLUTION

From Figure 10 we see that $f'(x) = 0$ at $x = 10$, about $x = 45$, and $x = 60$, so the critical numbers of f are 10, 45, and 60. The values of $f'(x)$ are negative to the left of $x = 10$, and if we assume that this remains true for all $x < 10$, then f must be decreasing on the interval $(-\infty, 10)$. We see that $f'(x) > 0$ for $10 < x < 45$, so f is increasing on $(10, 45)$. Similarly, f is decreasing on $(45, 60)$ and increasing on $(60, \infty)$ [assuming that $f'(x)$ remains positive to the right of $x = 60$]. Because $f'(x)$ changes from negative to positive at $x = 10$, we know from the First Derivative Test that $f(10)$ is a local minimum. Note that f also has a local minimum at $x = 60$, and f must have a local maximum at $x = 45$ because the derivative changes from positive to negative there.

The graph of f is concave upward where the derivative is increasing. Looking at Figure 10, it appears that f' is increasing on $(-\infty, 30)$ and approximately $(53, \infty)$, so f is concave upward on those intervals. f' is decreasing on $(30, 53)$, so f is concave downward there. We have inflection points where f' changes from increasing to decreasing, at $x = 30$, or from decreasing to increasing, at $x = 53$. Notice that these are the locations of local maximum and local minimum values of the graph of $y = f'(x)$. ▪

▪ Prepare Yourself

1. Solve the given inequality.

  (a) $r^2 - 3r - 18 < 0$

  (b) $x^3 - 9x > 0$

  (c) $x^2 e^x - 4xe^x > 0$

  (d) $\dfrac{t - 4}{(t^2 + 2)^2} < 0$

  (e) $\dfrac{x \ln x - 6x}{x^2} > 0$

2. Find the critical numbers.

  (a) $f(x) = \dfrac{2 + \ln x}{x}$ **(b)** $g(t) = (t^2 + 2t)e^t$

3. Compute the second derivative of the function.

  (a) $B(t) = 3te^{-2t}$ **(b)** $y = \ln(x^3 + x)$

4. If $f(x) = \dfrac{x^2}{2x^2 + 1}$, is $f''(3)$ positive or negative?

▪ Exercises 4.3

1–2 ▪ Use the given graph of f to find the following.
(a) The intervals on which f is increasing.
(b) The intervals on which f is decreasing.
(c) The intervals on which f is concave upward.
(d) The intervals on which f is concave downward.
(e) The coordinates of the points of inflection.

1.

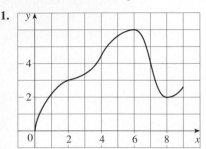

2.

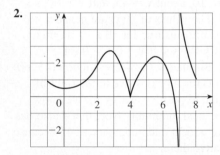

3. Suppose you are given a formula for a function f.
(a) How do you determine where f is increasing or decreasing?
(b) How do you determine where the graph of f is concave upward or concave downward?
(c) How do you locate inflection points?

4. (a) State the First Derivative Test.
(b) State the Second Derivative Test. Under what circumstances is it inconclusive? What do you do if it fails?

5–10 ▪ Use the First Derivative Test to find the local maximum and minimum values of the function.

5. $y = 3x^2 - 11x + 4$

6. $r(x) = 2.6x^2 - 7.28x + 4.9$

7. $M(t) = 4t^3 - 11t^2 - 20t + 7$

8. $g(u) = 0.2u^3 + 1.8u^2 + 141$

9. $f(x) = (\ln x)/\sqrt{x}$

10. $h(t) = 3t - e^t + 5$

11–14 ▪ Find the intervals of concavity and inflection points of the function.

11. $f(x) = x^4 - 4x^3 + 6x^2 - 1$

12. $V(x) = x^4 + 2x^3 - 36x^2 + 6$

13. $h(t) = -1.6t^3 + 0.9t^2 + 2.2t - 6.4$

14. $g(s) = s^2 - s \ln s$

15–24 ▪
(a) Find the intervals on which f is increasing or decreasing.
(b) Find the local maximum and minimum values of f.
(c) Find the intervals of concavity and the inflection points.

15. $f(x) = x^3 - 12x + 1$

16. $f(x) = 4x^3 - 3x^2 - 18x + 4$

17. $f(x) = x^4 - 2x^2 + 2$ **18.** $f(x) = x^4 - 4x - 1$

19. $f(x) = 5xe^{-0.2x}$ **20.** $f(x) = \dfrac{x^2}{x^2 + 3}$

21. $f(x) = xe^x$ **22.** $f(x) = x^2e^x$

23. $f(x) = (\ln x)/x$ **24.** $f(x) = x \ln x$

25–30 ▪
(a) Find the intervals of increase or decrease.
(b) Find the local maximum and minimum values.
(c) Find the intervals of concavity and the inflection points.
(d) Use the information from parts (a)–(c) to sketch the graph. Check your work with a graphing device if you have one.

25. $f(x) = 2x^3 - 3x^2 - 12x$ **26.** $g(x) = 200 + 8x^3 + x^4$

27. $h(x) = 3x^5 - 5x^3 + 3$ **28.** $h(x) = (x^2 - 1)^3$

29. $A(x) = x\sqrt{x + 3}$ **30.** $f(x) = \ln(x^4 + 27)$

31. In each part state the x-coordinates of the inflection points of f. Give reasons for your answers.
(a) The curve is the graph of f.
(b) The curve is the graph of f'.
(c) The curve is the graph of f''.

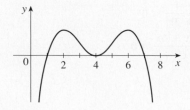

32. The graph of the first derivative f' of a function f is shown.

(a) On what intervals is f increasing? Explain.

(b) At what values of x does f have a local maximum or minimum? Explain.

(c) On what intervals is f concave upward or concave downward? Explain.

(d) What are the x-coordinates of the inflection points of f? Why?

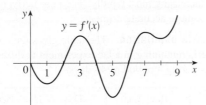

33–34 ■ The graph of the derivative f' of a continuous function f is shown.

(a) On what intervals is f increasing or decreasing?

(b) At what values of x does f have a local maximum or minimum?

(c) On what intervals is f concave upward or concave downward?

(d) State the x-coordinate(s) of the point(s) of inflection.

(e) Assuming that $f(0) = 0$, sketch a graph of f.

33.

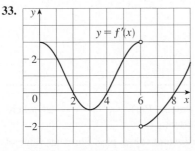

34.

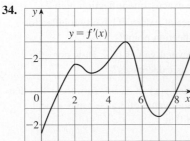

35–36 ■ Find the local maximum and minimum values of f using both the First and Second Derivative Tests. Which method do you prefer?

35. $f(x) = x^5 - 5x + 3$ **36.** $f(x) = \dfrac{x}{x^2 + 4}$

37. Suppose f'' is continuous on $(-\infty, \infty)$.

(a) If $f'(2) = 0$ and $f''(2) = -5$, what can you say about f?

(b) If $f'(6) = 0$ and $f''(6) = 0$, what can you say about f?

38. **(a)** Find the critical numbers of $f(x) = x^4(x - 1)^3$.

(b) What does the Second Derivative Test tell you about the behavior of f at these critical numbers?

(c) What does the First Derivative Test tell you?

39. Temperature Let $f(t)$ be the temperature at time t where you live and suppose that at time $t = 3$ you feel uncomfortably hot. How do you feel about the given data in each case?

(a) $f'(3) = 2, \qquad f''(3) = 4$

(b) $f'(3) = 2, \qquad f''(3) = -4$

(c) $f'(3) = -2, \qquad f''(3) = 4$

(d) $f'(3) = -2, \qquad f''(3) = -4$

40. Unemployment rate At one point during the economic downturn in 2009, an American reporter said that unemployment was "getting worse more slowly." If $U(t)$ gives the unemployment rate at time t, what can you say about the signs of U' and U'' when the reporter made the remark?

41. Company profit The chief financial officer of a company reported that during the previous quarter, profits continued to increase but at a slower rate. If $P(t)$ gives the company's profit at time t, what can you say about the signs of P' and P'' during that quarter?

42. Suppose $f(3) = 2$, $f'(3) = \frac{1}{2}$, and $f'(x) > 0$ and $f''(x) < 0$ for all x.

(a) Sketch a possible graph for f.

(b) How many solutions does the equation $f(x) = 0$ have? Why?

(c) Is it possible that $f'(2) = \frac{1}{3}$? Why?

43–44 ■

(a) Use a graph of f to estimate the maximum and minimum values. Then find the exact values.

(b) Estimate the value of x at which f increases most rapidly. Then find the exact value.

43. $f(x) = \dfrac{x + 1}{\sqrt{x^2 + 1}}$ **44.** $f(x) = x^2 e^{-x}$

45–46 ■

(a) Use a graph of f to give a rough estimate of the intervals of concavity and the coordinates of the points of inflection.

(b) Use a graph of f'' to give better estimates.

45. $f(x) = e^{-x} + x^3$ **46.** $f(x) = x^3(x - 2)^2$

47–48 ▪ Estimate the intervals of concavity, accurate to two decimal places, from a graph of f'.

47. $f(x) = \dfrac{8e^x}{1 + xe^x}$

48. $f(x) = \dfrac{(x^3 + 2)(x^2 + 5)}{(x^2 + 1)(x^2 + 4)}$

49. Drug concentration A *drug response curve* describes the level of medication in the bloodstream after a drug is administered. A surge function $S(t) = At^p e^{-kt}$ is often used to model the response curve, reflecting an initial surge in the drug level and then a more gradual decline. Graph the drug response curve for a particular drug where $A = 0.01$, $p = 4$, $k = 0.07$, and t is measured in minutes. Then use a graph of S' to estimate the times corresponding to the inflection points and explain their significance.

50. Use a graph of the derivative of the logistic function

$$P(t) = \dfrac{28}{1 + 4e^{-0.3t}}$$

to estimate (accurate to two decimal places) the t-value of the inflection point of P. What is the rate of change there? Verify that the y-coordinate of the inflection point is half the carrying capacity of P.

51. Market penetration For the period from 2000 to 2008, the percentage of households in Australia with at least one DVD player has been modeled by the function

$$f(t) = \dfrac{88.5}{1 + 17.7e^{-0.94t}}$$

where the time t is measured in years since midyear 2000, so $0 \le t \le 8$. Use a graph to estimate the time at which the number of households with DVD players was increasing most rapidly. Then use derivatives to give a more accurate estimate.

52. Bacteria population The table gives the number of bacteria remaining in a lab experiment at different times during a period of 45 days.

Days elapsed	Bacteria (thousands)	Days elapsed	Bacteria (thousands)
0	60.39	25	15.98
5	57.62	30	10.53
10	52.08	35	6.83
15	38.23	40	4.43
20	23.12	45	2.68

Use a graphing calculator to fit a logistic function to these data. Then graph the derivative of your model to determine when the number of bacteria was decreasing most rapidly.

▪ Challenge Yourself

53. Suppose the derivative of a function f is

$$f'(x) = (x + 1)^2(x - 3)^5(x - 6)^4$$

On what interval is f increasing?

54. For what values of the numbers a and b does the function

$$f(x) = axe^{bx^2}$$

have the maximum value $f(2) = 1$?

55. Show that a cubic function (a third-degree polynomial) always has exactly one point of inflection.

56. Consider the polynomial $P(x) = x^4 + cx^3 + x^2$, where c is a positive constant.
  (a) Graph P'' for several values of c. For what value of c does P have exactly one inflection point?
  (b) For what values of c does P have two inflection points? No inflection point?
  (c) Graph P for several values of c. How does the graph change as c decreases?

57. Find a cubic function $f(x) = ax^3 + bx^2 + cx + d$ that has a local maximum value of 3 at -2 and a local minimum value of 0 at 1.

4.4 Asymptotes

We have previously discussed asymptotes of curves informally, but in this section we study them in more detail and explore their connection to limits involving infinity.

■ | Infinite Limits and Vertical Asymptotes

x	$\dfrac{1}{x^2}$
± 1	1
± 0.5	4
± 0.2	25
± 0.1	100
± 0.05	400
± 0.01	10,000
± 0.001	1,000,000

In Example 6 in Section 2.2 we concluded that

$$\lim_{x \to 0} \frac{1}{x^2} \quad \text{does not exist}$$

by observing, from the table of values and the graph of $y = 1/x^2$ in Figure 1, that the values of $1/x^2$ can be made arbitrarily large by taking x close enough to 0. Thus the values of $f(x)$ do not approach a number, so $\lim_{x \to 0} (1/x^2)$ does not exist.

To indicate this kind of behavior we use the notation

$$\lim_{x \to 0} \frac{1}{x^2} = \infty$$

This does not mean that we are regarding ∞ as a number. Nor does it mean that the limit exists. It simply expresses the particular way in which the limit does not exist: $1/x^2$ grows larger and larger as $x \to 0$.

In general, we write symbolically

$$\lim_{x \to a} f(x) = \infty$$

to indicate that the values of $f(x)$ become larger and larger, beyond any finite value we might choose, as x approaches a.

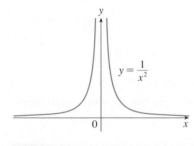

FIGURE 1

> **(1) ■ Definition** The notation
>
> $$\lim_{x \to a} f(x) = \infty$$
>
> means that the values of $f(x)$ become arbitrarily large as x approaches a.

The idea is that $f(x)$ can be made as large as we please by taking x sufficiently close to a (on either side of a, but not equal to a). Another notation for $\lim_{x \to a} f(x) = \infty$ is

$$f(x) \to \infty \qquad \text{as} \qquad x \to a$$

Again, the symbol ∞ is not a number, but the expression $\lim_{x \to a} f(x) = \infty$ is often read as

"the limit of $f(x)$, as x approaches a, is infinity"

or   "$f(x)$ becomes infinite as x approaches a"

or   "$f(x)$ increases without bound as x approaches a"

This definition is illustrated graphically in Figure 2.

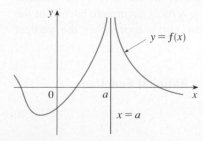

FIGURE 2
$$\lim_{x \to a} f(x) = \infty$$

When we say a number is "large negative," we mean that it is negative but its magnitude (absolute value) is large.

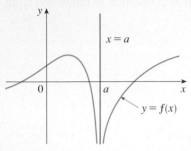

FIGURE 3
$\lim\limits_{x \to a} f(x) = -\infty$

Similarly, as shown in Figure 3,

$$\lim_{x \to a} f(x) = -\infty$$

means that the values of $f(x)$ become arbitrarily large negative as x approaches a.

The symbol $\lim_{x \to a} f(x) = -\infty$ can be read as "the limit of $f(x)$, as x approaches a, is negative infinity" or "$f(x)$ decreases without bound as x approaches a."

Similar definitions can be given for the one-sided infinite limits

$$\lim_{x \to a^-} f(x) = \infty \qquad \lim_{x \to a^+} f(x) = \infty$$

$$\lim_{x \to a^-} f(x) = -\infty \qquad \lim_{x \to a^+} f(x) = -\infty$$

remembering that "$x \to a^-$" means that we consider only values of x that are less than a, and similarly "$x \to a^+$" means that we consider only $x > a$. Illustrations of these four cases are given in Figure 4.

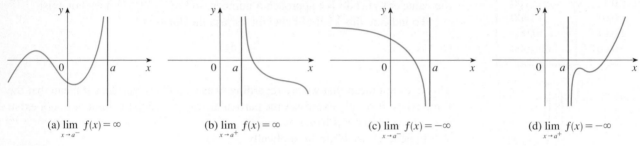

(a) $\lim\limits_{x \to a^-} f(x) = \infty$ (b) $\lim\limits_{x \to a^+} f(x) = \infty$ (c) $\lim\limits_{x \to a^-} f(x) = -\infty$ (d) $\lim\limits_{x \to a^+} f(x) = -\infty$

FIGURE 4

The following definition says that infinite limits are the necessary ingredients of vertical asymptotes.

---

(2) ▪ Definition   The line $x = a$ is called a **vertical asymptote** of the curve $y = f(x)$ if at least one of the following statements is true:

$$\lim_{x \to a} f(x) = \infty \qquad \lim_{x \to a^-} f(x) = \infty \qquad \lim_{x \to a^+} f(x) = \infty$$

$$\lim_{x \to a} f(x) = -\infty \qquad \lim_{x \to a^-} f(x) = -\infty \qquad \lim_{x \to a^+} f(x) = -\infty$$

For instance, the y-axis is a vertical asymptote of the curve $y = 1/x^2$ because $\lim_{x \to 0} (1/x^2) = \infty$. In Figure 4 the line $x = a$ is a vertical asymptote in each of the four cases shown. Notice that, in each case, the curve approaches the vertical asymptote without touching it.

▪ **EXAMPLE 1   Vertical Asymptote of a Rational Function**

Find $\lim\limits_{x \to 3^+} \dfrac{2x}{x - 3}$ and $\lim\limits_{x \to 3^-} \dfrac{2x}{x - 3}$.

SOLUTION

If x is close to 3 but larger than 3, then the denominator $x - 3$ is a small positive number and $2x$ is close to 6. So the quotient $2x/(x - 3)$ is a large *positive* number. Thus, intuitively, we see that

$$\lim_{x \to 3^+} \frac{2x}{x-3} = \infty$$

Likewise, if x is close to 3 but smaller than 3, then $x - 3$ is a small negative number but $2x$ is still a positive number (close to 6). So $2x/(x - 3)$ is a numerically large *negative* number. Thus

$$\lim_{x \to 3^-} \frac{2x}{x-3} = -\infty$$

The graph of the curve $y = 2x/(x - 3)$ is given in Figure 5. The line $x = 3$ is a vertical asymptote. ■

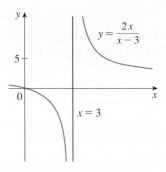

FIGURE 5

Two familiar functions whose graphs have vertical asymptotes are $y = \ln x$ and the reciprocal function $y = 1/x$. From Figure 6 we see that

(3)
$$\lim_{x \to 0^+} \ln x = -\infty$$

and so the line $x = 0$ (the y-axis) is a vertical asymptote. In fact, the same is true for $y = \log_a x$ provided that $a > 1$.

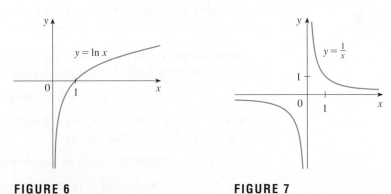

FIGURE 6   **FIGURE 7**

Figure 7 shows that

$$\lim_{x \to 0^-} \frac{1}{x} = -\infty \quad \text{and} \quad \lim_{x \to 0^+} \frac{1}{x} = \infty$$

and so the line $x = 0$ is a vertical asymptote for $y = 1/x$.

■ Limits at Infinity and Horizontal Asymptotes

In computing infinite limits, we let x approach a number and the result was that the values of y became arbitrarily large (positive or negative). Here we let x become arbitrarily large (positive or negative) and see what happens to y.

x	$f(x)$
0	−1
±1	0
±2	0.600000
±3	0.800000
±4	0.882353
±5	0.923077
±10	0.980198
±50	0.999200
±100	0.999800
±1000	0.999998

FIGURE 8

Let's begin by investigating the behavior of the function f defined by

$$f(x) = \frac{x^2 - 1}{x^2 + 1}$$

as x becomes large. The table at the left gives values of this function correct to six decimal places, and the graph of f has been drawn by a computer in Figure 8.

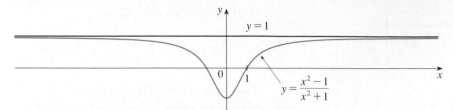

As x grows larger and larger you can see that the values of $f(x)$ get closer and closer to 1. This situation is expressed symbolically by writing

$$\lim_{x \to \infty} \frac{x^2 - 1}{x^2 + 1} = 1$$

In general, we have the following definition.

(4) ■ **Definition**   Let f be a function defined on some interval (a, ∞). Then

$$\lim_{x \to \infty} f(x) = L$$

means that the values of $f(x)$ approach L as x becomes large.

Here the idea is that we can make the values of $f(x)$ as close as we like to L by taking x sufficiently large. Another notation for $\lim_{x \to \infty} f(x) = L$ is

$$f(x) \to L \qquad \text{as } x \to \infty$$

Remember that the symbol ∞ does not represent a number. Nonetheless, the expression $\lim_{x \to \infty} f(x) = L$ is often read as

"the limit of $f(x)$, as x approaches infinity, is L"

or   "the limit of $f(x)$, as x becomes infinite, is L"

or   "the limit of $f(x)$, as x increases without bound, is L"

Geometric illustrations of Definition 4 are shown in Figure 9. Notice that there are many ways for the graph of f to approach the line $y = L$ as we look to the far right of each graph.

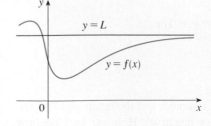 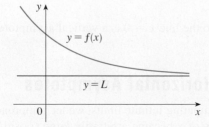 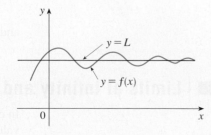

FIGURE 9   Examples illustrating $\lim_{x \to \infty} f(x) = L$

We can also look at limits as x becomes large negative. Referring back to Figure 8, we see that for numerically large negative values of x, the values of $f(x)$ are close to 1. By letting x decrease through negative values without bound, we can make $f(x)$ as close to 1 as we like. This is expressed by writing

$$\lim_{x \to -\infty} \frac{x^2 - 1}{x^2 + 1} = 1$$

In general, as shown in Figure 10, the notation

$$\lim_{x \to -\infty} f(x) = L$$

means that the values of $f(x)$ approach L as x becomes larger and larger negative, beyond any finite value we could choose.

Again, the symbol $-\infty$ does not represent a number, but the expression $\lim_{x \to -\infty} f(x) = L$ is often read as

"the limit of $f(x)$, as x approaches negative infinity, is L"

Just as infinite limits correspond to vertical asymptotes, limits at infinity correspond to *horizontal asymptotes*.

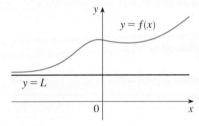

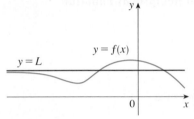

FIGURE 10

Examples illustrating $\lim_{x \to -\infty} f(x) = L$

> **(5)** ▪ **Definition**   The line $y = L$ is called a **horizontal asymptote** of the curve $y = f(x)$ if either
>
> $$\lim_{x \to \infty} f(x) = L \qquad \text{or} \qquad \lim_{x \to -\infty} f(x) = L$$

For instance, the curve illustrated in Figure 8 has the line $y = 1$ as a horizontal asymptote because

$$\lim_{x \to \infty} \frac{x^2 - 1}{x^2 + 1} = 1$$

The graph of the natural exponential function $y = e^x$ has the line $y = 0$ (the x-axis) as a horizontal asymptote. (The same is true of any exponential function with base $a > 1$.) In fact, from the graph in Figure 11 and the corresponding table of values, we see that

$$\lim_{x \to -\infty} e^x = 0$$

Notice from the table that the values of e^x approach 0 very rapidly.

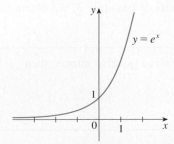

FIGURE 11

x	e^x
0	1.00000
-1	0.36788
-2	0.13534
-3	0.04979
-5	0.00674
-8	0.00034
-10	0.00005

▪ **EXAMPLE 2**   **Determining Limits at Infinity from a Graph**

Find the limits at infinity and asymptotes for the function f whose graph is shown in Figure 12.

SOLUTION

From the graph, it appears that both of the lines $x = -1$ and $x = 2$ are vertical asymptotes.

As x becomes large, it appears that $f(x)$ approaches 4. But as x decreases through negative values, $f(x)$ approaches 2. So

$$\lim_{x \to \infty} f(x) = 4 \qquad \text{and} \qquad \lim_{x \to -\infty} f(x) = 2$$

This means that both $y = 4$ and $y = 2$ are horizontal asymptotes. ▪

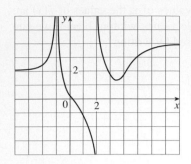

FIGURE 12

▪ **EXAMPLE 3**   **Limits at Infinity of the Reciprocal Function**

Find $\lim\limits_{x \to \infty} \dfrac{1}{x}$ and $\lim\limits_{x \to -\infty} \dfrac{1}{x}$.

SOLUTION

Observe that when x is large, $1/x$ is small. For instance,

$$\frac{1}{100} = 0.01 \qquad \frac{1}{10,000} = 0.0001 \qquad \frac{1}{1,000,000} = 0.000001$$

The larger x gets, the closer to 0 its reciprocal $1/x$ gets. In fact, by taking x large enough, we can make $1/x$ as close to 0 as we please. Therefore, according to Definition 4, we have

$$\lim_{x \to \infty} \frac{1}{x} = 0$$

Similar reasoning shows that when x is large negative, $1/x$ is small negative, so we also have

$$\lim_{x \to -\infty} \frac{1}{x} = 0$$

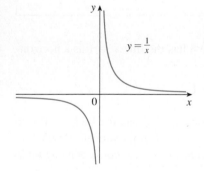

FIGURE 13

$\lim\limits_{x \to \infty} \dfrac{1}{x} = 0, \ \lim\limits_{x \to -\infty} \dfrac{1}{x} = 0$

It follows that the line $y = 0$ (the x-axis) is a horizontal asymptote of the curve $y = 1/x$. This is confirmed in the graph of $y = 1/x$ in Figure 13. ▪

It can be proved that the *Limit Laws listed in Section 2.2 are also valid if "$x \to a$" is replaced by "$x \to \infty$" or "$x \to -\infty$."* In particular, if we combine Law 6 with the results of Example 3, we obtain the following important rule for calculating limits.

(6) ▪   If n is a positive integer, then

$$\lim_{x \to \infty} \frac{1}{x^n} = 0 \qquad\qquad \lim_{x \to -\infty} \frac{1}{x^n} = 0$$

The following example demonstrates a technique that is helpful in evaluating limits at infinity of rational functions.

■ **EXAMPLE 4 Limit at Infinity of a Rational Function**

Evaluate

$$\lim_{x \to \infty} \frac{3x^2 - x - 2}{5x^2 + 4x + 1}$$

SOLUTION

As x becomes large, both numerator and denominator become large, so it isn't obvious what happens to their ratio. We need to do some preliminary algebra.

To evaluate the limit at infinity of any rational function, we first divide both the numerator and denominator by the highest power of x that occurs in the denominator. (We may assume that $x \neq 0$, since we are interested only in large values of x.) In this case the highest power of x in the denominator is x^2 and so, using the Limit Laws, we have

$$\lim_{x \to \infty} \frac{3x^2 - x - 2}{5x^2 + 4x + 1} = \lim_{x \to \infty} \frac{\dfrac{3x^2 - x - 2}{x^2}}{\dfrac{5x^2 + 4x + 1}{x^2}} = \lim_{x \to \infty} \frac{\dfrac{3x^2}{x^2} - \dfrac{x}{x^2} - \dfrac{2}{x^2}}{\dfrac{5x^2}{x^2} + \dfrac{4x}{x^2} + \dfrac{1}{x^2}}$$

$$= \lim_{x \to \infty} \frac{3 - \dfrac{1}{x} - \dfrac{2}{x^2}}{5 + \dfrac{4}{x} + \dfrac{1}{x^2}} = \frac{\lim_{x \to \infty}\left(3 - \dfrac{1}{x} - \dfrac{2}{x^2}\right)}{\lim_{x \to \infty}\left(5 + \dfrac{4}{x} + \dfrac{1}{x^2}\right)}$$

$$= \frac{\lim_{x \to \infty} 3 - \lim_{x \to \infty} \dfrac{1}{x} - 2 \lim_{x \to \infty} \dfrac{1}{x^2}}{\lim_{x \to \infty} 5 + 4 \lim_{x \to \infty} \dfrac{1}{x} + \lim_{x \to \infty} \dfrac{1}{x^2}}$$

$$= \frac{3 - 0 - 0}{5 + 0 + 0} \qquad \text{[by (6)]}$$

$$= \frac{3}{5}$$

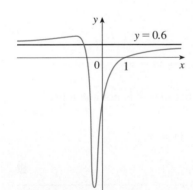

FIGURE 14

$$y = \frac{3x^2 - x - 2}{5x^2 + 4x + 1}$$

A similar calculation shows that the limit as $x \to -\infty$ is also $\frac{3}{5}$. Figure 14 illustrates the results of these calculations by showing how the graph of the given rational function approaches the horizontal asymptote $y = \frac{3}{5}$ or $y = 0.6$. ■

■ | **Infinite Limits at Infinity**

The notation
$$\lim_{x \to \infty} f(x) = \infty$$

is used to indicate that the values of $f(x)$ become large as x becomes large. Similar meanings are attached to the following symbols:

$$\lim_{x \to -\infty} f(x) = \infty \qquad \lim_{x \to \infty} f(x) = -\infty \qquad \lim_{x \to -\infty} f(x) = -\infty$$

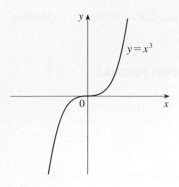

FIGURE 15

From Figures 11 and 15 we see that

$$\lim_{x \to \infty} e^x = \infty \qquad \lim_{x \to \infty} x^3 = \infty \qquad \lim_{x \to -\infty} x^3 = -\infty$$

Not only do limits like these tell us that the graph does not have a horizontal asymptote, but they also give us a sense of the "end behavior" of the curve.

▪ **EXAMPLE 5**   **Infinite Limit of a Rational Function**

Find $\displaystyle \lim_{x \to \infty} \frac{x^2 + x}{3 - x}$.

SOLUTION

We divide numerator and denominator by x (the highest power of x that occurs in the denominator):

$$\lim_{x \to \infty} \frac{x^2 + x}{3 - x} = \lim_{x \to \infty} \frac{\dfrac{x^2}{x} + \dfrac{x}{x}}{\dfrac{3}{x} - \dfrac{x}{x}} = \lim_{x \to \infty} \frac{x + 1}{\dfrac{3}{x} - 1} = -\infty$$

because $x + 1 \to \infty$ and $3/x - 1 \to 0 - 1 = -1$ as $x \to \infty$. ▪

▪ **EXAMPLE 6**   **Rewriting the Function Before Analyzing the Limit**

Find $\displaystyle \lim_{x \to \infty} (x^2 - x)$.

SOLUTION

It would be **wrong** to write

$$\lim_{x \to \infty} (x^2 - x) = \lim_{x \to \infty} x^2 - \lim_{x \to \infty} x = \infty - \infty$$

The Limit Laws can't be applied to infinite limits because ∞ is not a number ($\infty - \infty$ can't be defined). However, we can write

$$\lim_{x \to \infty} (x^2 - x) = \lim_{x \to \infty} x(x - 1) = \infty$$

because both x and $x - 1$ become arbitrarily large. ▪

▪ **EXAMPLE 7**   **Finding Horizontal Asymptotes**

Find the horizontal asymptotes, if any, of the graph of $y = e^{-x^2}$.

SOLUTION

First we write the function as $y = 1/e^{x^2}$. Now $x^2 \to \infty$ as $x \to \infty$, and since $\lim_{u \to \infty} e^u = \infty$, we have $e^{x^2} \to \infty$ as $x^2 \to \infty$. Finally, as in Example 3, if $e^{x^2} \to \infty$, then its reciprocal $1/e^{x^2} \to 0$. Thus

$$\lim_{x \to \infty} \frac{1}{e^{x^2}} = 0$$

Similarly, $x^2 \to \infty$ as $x \to -\infty$ so we have

$$\lim_{x \to -\infty} \frac{1}{e^{x^2}} = 0$$

This means that $y = e^{-x^2}$ has one horizontal asymptote, $y = 0$ (the x-axis).   ▪

▪ Prepare Yourself

1. (a) If $a = 1/10$, what is the value of $1/a^3$?

  (b) If $b = -1/100$, what is the value of $1/b^3$?

2. If x becomes larger and larger, describe what happens to the output values of the function.

  (a) $f(x) = -x^2$ **(b)** $g(x) = 1/x$

  (c) $h(x) = -1/x$ **(d)** $A(x) = e^x$

  (e) $B(x) = e^{-x}$

3. If c gets closer and closer to 5, what can you say about the values of $\dfrac{2c}{c-5}$? Does it make a difference whether $c < 5$ or $c > 5$?

4. If a is a positive number that gets closer and closer to 0, what happens to the values of $\ln a$?

5. Multiply and simplify: $\dfrac{4x^3 + x^2 - 2}{3x^3 + 3x^2} \cdot \dfrac{1/x^3}{1/x^3}$

▪ Exercises 4.4

1. Explain in your own words the meaning of each of the following.

  (a) $\lim\limits_{x \to 2} f(x) = \infty$ **(b)** $\lim\limits_{x \to 1^+} f(x) = -\infty$

  (c) $\lim\limits_{x \to \infty} f(x) = 5$ **(d)** $\lim\limits_{x \to -\infty} f(x) = 3$

2. (a) Can the graph of $y = f(x)$ intersect a vertical asymptote? Can it intersect a horizontal asymptote? Illustrate by sketching graphs.

  (b) How many horizontal asymptotes can the graph of $y = f(x)$ have? Sketch graphs to illustrate the possibilities.

3. For the function f whose graph is given, state the following.

  (a) $\lim\limits_{x \to 2} f(x)$ **(b)** $\lim\limits_{x \to -1^-} f(x)$

  (c) $\lim\limits_{x \to -1^+} f(x)$ **(d)** $\lim\limits_{x \to \infty} f(x)$

  (e) $\lim\limits_{x \to -\infty} f(x)$

  (f) The equations of the asymptotes

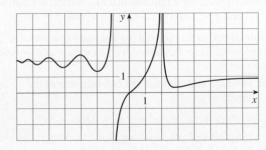

4. For the function g whose graph is given, state the following.

  (a) $\lim\limits_{x \to \infty} g(x)$ **(b)** $\lim\limits_{x \to -\infty} g(x)$

  (c) $\lim\limits_{x \to 3} g(x)$ **(d)** $\lim\limits_{x \to 0} g(x)$

  (e) $\lim\limits_{x \to -2^+} g(x)$

  (f) The equations of the asymptotes

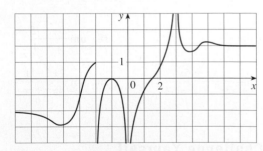

5–7 ▪ Sketch the graph of an example of a function f that satisfies all of the given conditions.

5. $\lim\limits_{x \to 0} f(x) = -\infty$, $\quad \lim\limits_{x \to -\infty} f(x) = 5$, $\quad \lim\limits_{x \to \infty} f(x) = -5$

6. $\lim\limits_{x \to 0^+} f(x) = \infty$, $\quad \lim\limits_{x \to 0^-} f(x) = -\infty$, $\quad \lim\limits_{x \to \infty} f(x) = 1$,

  $\lim\limits_{x \to -\infty} f(x) = 1$

7. $\lim\limits_{x \to 2} f(x) = -\infty$, $\quad \lim\limits_{x \to \infty} f(x) = \infty$, $\quad \lim\limits_{x \to -\infty} f(x) = 0$,

  $\lim\limits_{x \to 0^+} f(x) = \infty$, $\quad \lim\limits_{x \to 0^-} f(x) = -\infty$

8. Guess the value of the limit

$$\lim_{x \to \infty} \frac{x^2}{2^x}$$

by evaluating the function $f(x) = x^2/2^x$ for $x = 0, 1, 2, 3, 4, 5, 6, 7, 8, 9, 10, 20, 50,$ and 100. Then use a graph of f to support your guess.

9. Use a graph to estimate all the vertical and horizontal asymptotes of the curve

$$y = \frac{x^3}{x^3 - 2x + 1}$$

10. (a) Use a graph of

$$f(x) = \left(1 - \frac{2}{x}\right)^x$$

to estimate the value of $\lim_{x \to \infty} f(x)$ correct to two decimal places.

(b) Use a table of values of $f(x)$ to estimate the limit to four decimal places.

11–28 ▪ Find the limit.

11. $\lim_{x \to 4^+} \dfrac{3}{x - 4}$

12. $\lim_{x \to 4^-} \dfrac{3x}{x - 4}$

13. $\lim_{x \to -3^+} \dfrac{x + 2}{x + 3}$

14. $\lim_{x \to 5^-} \dfrac{e^x}{(x - 5)^3}$

15. $\lim_{x \to 1} \dfrac{2 - x}{(x - 1)^2}$

16. $\lim_{x \to 3^+} \ln(x^2 - 9)$

17. $\lim_{x \to \infty} \dfrac{2}{x^3}$

18. $\lim_{x \to \infty} \dfrac{3x + 5}{x - 4}$

19. $\lim_{x \to \infty} \dfrac{x^3 + 5x}{2x^3 - x^2 + 4}$

20. $\lim_{x \to \infty} \dfrac{2x^2 - 1}{4x^2 + x}$

21. $\lim_{p \to \infty} \dfrac{3p}{p^2 + 2p + 7}$

22. $\lim_{t \to -\infty} \dfrac{t^2 + 2}{t^3 + t^2 - 1}$

23. $\lim_{u \to \infty} \dfrac{4u^4 + 5}{(u^2 - 2)(2u^2 - 1)}$

24. $\lim_{b \to -\infty} \sqrt[3]{b}\,(b - 2)$

25. $\lim_{x \to -\infty} (x^4 + x^5)$

26. $\lim_{u \to \infty} (4u^3 - 2u^2)$

27. $\lim_{x \to \infty} \dfrac{x + x^3 + x^5}{1 - x^2 + x^4}$

28. $\lim_{t \to \infty} \dfrac{t^4 + 2.3t^2 - 1.7}{4t^2 + 9.9t + 4.2}$

29–32 ▪ Find the horizontal asymptotes, if any, of the curve.

29. $y = \dfrac{2x}{x^3 + 3}$

30. $y = \dfrac{4x^2}{3x^2 - 2}$

31. $y = \dfrac{1}{e^x + 1}$

32. $y = e^{2x} + 2e^x$

33–34 ▪ Find the horizontal and vertical asymptotes of each curve. Check your work by graphing the curve and estimating the asymptotes.

33. $y = \dfrac{2x^2 + x - 1}{x^2 + x - 2}$

34. $y = \dfrac{2e^x}{e^x - 5}$

35. Estimate the horizontal asymptote of the function

$$f(x) = \frac{3x^3 + 500x^2}{x^3 + 500x^2 + 100x + 2000}$$

by graphing f for $-10 \le x \le 10$. Then calculate the equation of the asymptote by evaluating the limit. How do you explain the discrepancy?

▪ Challenge Yourself

36. Evaluate

$$\lim_{x \to \infty} \frac{x + 3}{\sqrt{4x^2 + 1}}$$

Hint: Multiply numerator and denominator by $1/x$ but for the denominator write $1/x$ as $1/\sqrt{x^2}$.

37. Find the horizontal asymptotes of

$$y = \frac{x}{\sqrt{x^2 + 1}}$$

Determine the intervals of increase or decrease and concavity, and then sketch the graph.

38. Find a formula for a function that has vertical asymptotes $x = 1$ and $x = 3$ and horizontal asymptote $y = 1$.

39. As discussed in this section, by the *end behavior* of a function we mean the behavior of its values as $x \to \infty$ and as $x \to -\infty$.

(a) Describe and compare the end behavior of the functions

$$P(x) = 3x^5 - 5x^3 + 2x \qquad Q(x) = 3x^5$$

by graphing both functions in the viewing rectangles $[-2, 2]$ by $[-2, 2]$ and $[-10, 10]$ by $[-10,000, 10,000]$.

(b) Two functions are said to have the *same end behavior* if their ratio approaches 1 as $x \to \infty$. Show that P and Q have the same end behavior.

40. Show that

$$f(x) = \frac{x^3 + 2x^2 + 3}{x + 2} \quad \text{and} \quad g(x) = x^2$$

have the same end behavior. Then graph both functions on the same screen. Use a viewing window large enough to observe the end behavior.

41. Salinity level

(a) A tank contains 5000 L of pure water. Brine that contains 30 g of salt per liter of water is pumped into the tank at a rate of 25 L/min. Show that the concentration of salt t minutes later (in grams per liter) is

$$C(t) = \frac{30t}{200 + t}$$

(b) What happens to the concentration as $t \to \infty$?

4.5 Curve Sketching

So far we have learned to find maximum and minimum values of functions, intervals of increase and decrease, inflection points, concavity, vertical asymptotes, and the end behavior of graphs, including horizontal asymptotes. We now put all of this information together to sketch graphs that reveal the important features of functions.

■ Graphing by Hand

You might ask: Why don't we just use a graphing calculator or computer to graph a curve? Why do we need to use calculus?

It's true that modern technology is capable of producing very accurate graphs. But even the best graphing devices have to be used intelligently, and it can be easy to miss key details of a graph. How do we know if we have chosen an appropriate viewing window for the function, or if we should zoom in to a particular spot on the graph? The use of calculus enables us to discover the most interesting aspects of graphs and in many cases to calculate maximum and minimum points and inflection points *exactly* instead of approximately. We encourage you to use a graphing device as a check on your work, but be cautious in relying solely on these devices.

■ EXAMPLE 1   Sketching the Graph of a Polynomial Function

Sketch the graph of the function $f(x) = x^3 - 3x + 1$.

SOLUTION

$$f'(x) = 3x^2 - 3 = 3(x^2 - 1) = 3(x + 1)(x - 1)$$

and $f'(x) = 0$ when $x = \pm 1$, so the critical numbers are 1 and -1. We see that $f'(x) > 0$ when $x < -1$ or $x > 1$ and $f'(x) < 0$ for $-1 < x < 1$, so f is increasing on $(-\infty, -1)$ and $(1, \infty)$ and decreasing on $(-1, 1)$. By the First Derivative Test, $f(-1) = 3$ is a local maximum and $f(1) = -1$ is a local minimum. Since $f''(x) = 6x$, $f''(x) = 0$ when $x = 0$, $f''(x) < 0$ for $x < 0$, and $f''(x) > 0$ for $x > 0$. Thus f is concave upward on $(0, \infty)$, concave downward on $(-\infty, 0)$, and f has an inflection point at $x = 0$. We organize our findings in the following chart.

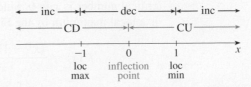

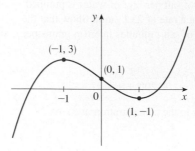

FIGURE 1

Note that

$$\lim_{x \to \infty} (x^3 - 3x + 1) = \lim_{x \to \infty} [x(x^2 - 3) + 1] = \infty$$

because both x and $(x^2 - 3)$ become arbitrarily large as $x \to \infty$. Similarly,

$$\lim_{x \to -\infty} (x^3 - 3x + 1) = \lim_{x \to -\infty} [x(x^2 - 3) + 1] = -\infty$$

So the graph does not have any horizontal asymptotes but we now know its end behavior. To draw the curve, we first plot the local maximum point $(-1, 3)$, the inflection point $(0, 1)$, and the local minimum point $(1, -1)$. We sketch the curve through these points, following the descriptions outlined in the chart and keeping the end behavior in mind, as shown in Figure 1. ▪

▪ EXAMPLE 2   Sketching the Graph of a Rational Function

Sketch the curve $y = \dfrac{2x^2}{x^2 - 1}$.

SOLUTION

First, the function is not defined when the denominator is 0, that is, when $x^2 - 1 = 0$, or $x = \pm 1$, so the domain is

$$\{x \mid x \neq \pm 1\}$$

Since the function is undefined when $x = \pm 1$, we compute the following limits:

$$\lim_{x \to 1^+} \frac{2x^2}{x^2 - 1} = \infty \qquad\qquad \lim_{x \to 1^-} \frac{2x^2}{x^2 - 1} = -\infty$$

$$\lim_{x \to -1^+} \frac{2x^2}{x^2 - 1} = -\infty \qquad\qquad \lim_{x \to -1^-} \frac{2x^2}{x^2 - 1} = \infty$$

Therefore the lines $x = 1$ and $x = -1$ are vertical asymptotes. Next we compute the limits at infinity:

$$\lim_{x \to \pm\infty} \frac{2x^2}{x^2 - 1} = \lim_{x \to \pm\infty} \frac{2x^2}{x^2 - 1} \cdot \frac{1/x^2}{1/x^2} = \lim_{x \to \pm\infty} \frac{2}{1 - 1/x^2} = 2$$

Therefore the line $y = 2$ is a horizontal asymptote (on both the left and right). This information about limits and asymptotes enables us to draw the preliminary sketch in Figure 2, showing the parts of the curve near the asymptotes.

Next we check the derivatives:

$$f'(x) = \frac{(x^2 - 1)(4x) - 2x^2 \cdot 2x}{(x^2 - 1)^2} = \frac{-4x}{(x^2 - 1)^2}$$

Since $f'(x) > 0$ when $x < 0$ $(x \neq -1)$ and $f'(x) < 0$ when $x > 0$ $(x \neq 1)$, f is increasing on $(-\infty, -1)$ and $(-1, 0)$ and decreasing on $(0, 1)$ and $(1, \infty)$. The only critical number is $x = 0$. Since f' changes from positive to negative at 0, $f(0) = 0$ is a local maximum by the First Derivative Test.

$$f''(x) = \frac{(x^2 - 1)^2(-4) + 4x \cdot 2(x^2 - 1)2x}{(x^2 - 1)^4} = \frac{12x^2 + 4}{(x^2 - 1)^3}$$

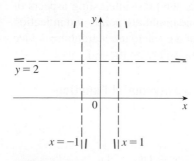

FIGURE 2

Preliminary sketch

We have shown the curve approaching its horizontal asymptote from above in Figure 2. This is confirmed by the intervals of increase and decrease.

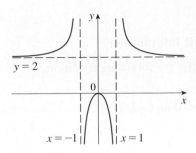

FIGURE 3
Finished sketch of $y = \dfrac{2x^2}{x^2 - 1}$

Since $12x^2 + 4 > 0$ for all x, we have

$$f''(x) > 0 \quad \text{when} \quad x^2 - 1 > 0 \quad \text{or} \quad x < -1, x > 1$$

and $f''(x) < 0$ when $-1 < x < 1$. Thus the curve is concave upward on the intervals $(-\infty, -1)$ and $(1, \infty)$ and concave downward on $(-1, 1)$. It has no point of inflection since 1 and -1 are not in the domain of f.

Combining all this information, we finish the sketch in Figure 3. ■

■ **EXAMPLE 3**

Sketching a Graph Involving an Exponential Function

Sketch a graph of the function $f(x) = e^{-x^2}$.

SOLUTION

We know from Example 7 in Section 4.4 that $\lim_{x \to \pm\infty} f(x) = 0$, so the x-axis is a horizontal asymptote. The derivative is $f'(x) = -2xe^{-x^2}$, and since e^{-x^2} is never 0, the only critical number is 0. We have $f'(x) > 0$ for $x < 0$ and $f'(x) < 0$ for $x > 0$, so f is increasing on $(-\infty, 0)$ and decreasing on $(0, \infty)$. Because f' changes from positive to negative at $x = 0$, $f(0) = 1$ is a local maximum by the First Derivative Test. (Here it must also be the absolute maximum.)

$$f''(x) = -2x\left(-2xe^{-x^2}\right) + e^{-x^2}(-2) = 2(2x^2 - 1)e^{-x^2}$$

and since $e^{-x^2} > 0$, $f''(x) > 0$ when

$$2x^2 - 1 > 0$$

$$x^2 > \tfrac{1}{2}$$

$$x < -\frac{1}{\sqrt{2}} \quad \text{or} \quad x > \frac{1}{\sqrt{2}}$$

Similarly, $f''(x) < 0$ when $-1/\sqrt{2} < x < 1/\sqrt{2}$. Thus f is concave up on $\left(-\infty, -1/\sqrt{2}\right)$, $\left(1/\sqrt{2}, \infty\right)$ and concave down on $\left(-1/\sqrt{2}, 1/\sqrt{2}\right)$; f has inflection points $\left(-1/\sqrt{2}, e^{-1/2}\right)$, $\left(1/\sqrt{2}, e^{-1/2}\right)$. We summarize these results in the following chart and use it to help sketch the graph in Figure 4.

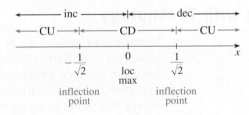

Notice that the functions in Examples 2 and 3 are even $[f(-x) = f(x)]$. We could have taken advantage of this fact as a shortcut in drawing the graph by determining the shape of the graph for $x \geqslant 0$ and then reflecting the graph about the y-axis.

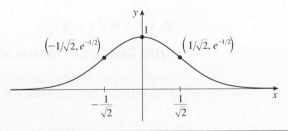

FIGURE 4 ■

▪ EXAMPLE 4

Sketching a Graph from Derivative and Limit Information

Sketch a possible graph of a function f that satisfies the following conditions:

(i) $f'(x) > 0$ on $(-\infty, 1)$, $f'(x) < 0$ on $(1, \infty)$

(ii) $f''(x) > 0$ on $(-\infty, -2)$ and $(2, \infty)$, $f''(x) < 0$ on $(-2, 2)$

(iii) $\lim\limits_{x \to -\infty} f(x) = -2$, $\lim\limits_{x \to \infty} f(x) = 0$

SOLUTION

Condition (i) tells us that f is increasing on $(-\infty, 1)$ and decreasing on $(1, \infty)$. This means we must have a local maximum value at $x = 1$. Condition (ii) says that f is concave upward on $(-\infty, -2)$ and $(2, \infty)$, and concave downward on $(-2, 2)$. Thus we have inflection points at $x = -2$ and $x = 2$. From condition (iii) we know that the graph of f has two horizontal asymptotes: $y = -2$ (on the left) and $y = 0$ (on the right).

We first draw the horizontal asymptote $y = -2$ as a dashed line (see Figure 5). We then draw the graph of f approaching this asymptote at the far left, increasing to its maximum point at $x = 1$ and decreasing toward the x-axis as $x \to \infty$. We also make sure that the graph has inflection points when $x = -2$ and 2. Notice that the curve bends upward for $x < -2$ and $x > 2$, and bends downward when x is between -2 and 2.

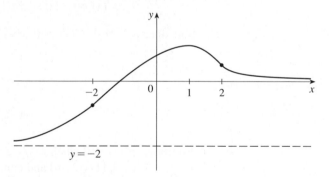

FIGURE 5

▪ | Graphing with Technology

When we use technology to graph a curve, our strategy is different from that in Examples 1–4. Here we *start* with a graph produced by a graphing calculator or computer and then we refine it. We use calculus to make sure that we reveal all the important features of the curve. And with the use of graphing devices we can tackle curves that would be far too complicated to consider without technology.

▪ EXAMPLE 5

Using Calculus in Tandem with Technology to Graph a Function

Graph the polynomial $f(x) = 2x^6 + 3x^5 + 3x^3 - 2x^2$. Use the graphs of f' and f'' to estimate all maximum and minimum points and intervals of concavity.

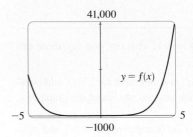

FIGURE 6

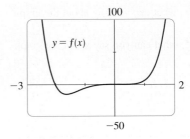

FIGURE 7

SOLUTION

If we specify a domain but not a range, many graphing devices will deduce a suitable range from the values computed. Figure 6 shows the plot from one such device if we specify that $-5 \le x \le 5$. Although this viewing rectangle is useful for showing the end behavior of the graph, it is obviously hiding some finer detail. So we change to the viewing rectangle $[-3, 2]$ by $[-50, 100]$ shown in Figure 7.

From Figure 7 it appears that there is an absolute minimum value of about -15.33 when $x \approx -1.62$ (by using the cursor) and f is decreasing on $(-\infty, -1.62)$ and increasing on $(-1.62, \infty)$. Also there appears to be a horizontal tangent at the origin and inflection points when $x = 0$ and when x is somewhere between -2 and -1.

Now let's try to confirm these impressions using calculus. We differentiate and get

$$f'(x) = 12x^5 + 15x^4 + 9x^2 - 4x$$

$$f''(x) = 60x^4 + 60x^3 + 18x - 4$$

When we graph f' in Figure 8 we see that $f'(x)$ changes from negative to positive when $x \approx -1.62$; this confirms (by the First Derivative Test) the minimum value that we found earlier. But, perhaps to our surprise, we also notice that $f'(x)$ changes from positive to negative when $x = 0$ and from negative to positive when $x \approx 0.35$. This means that f has a local maximum at 0 and a local minimum when $x \approx 0.35$, but these were hidden in Figure 7. Indeed, if we now zoom in toward the origin in Figure 9, we see what we missed before: a local maximum value of 0 when $x = 0$ and a local minimum value of about -0.1 when $x \approx 0.35$.

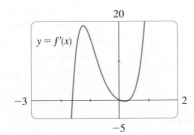

FIGURE 8

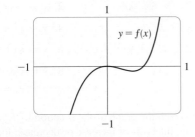

FIGURE 9

What about concavity and inflection points? From Figures 7 and 9 there appear to be inflection points when x is a little to the left of -1 and when x is a little to the right of 0. But it's difficult to determine inflection points from the graph of f, so we graph the second derivative f'' in Figure 10. We see that f'' changes from positive to negative when $x \approx -1.23$ and from negative to positive when $x \approx 0.19$. So, correct to two decimal places, f is concave upward on $(-\infty, -1.23)$ and $(0.19, \infty)$ and concave downward on $(-1.23, 0.19)$. The inflection points are $(-1.23, -10.18)$ and $(0.19, -0.05)$.

We have discovered that no single graph reveals all the important features of this polynomial. But Figures 7 and 9, when taken together, do provide an accurate picture. ■

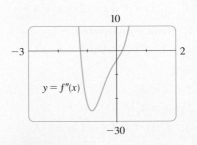

FIGURE 10

▪ Prepare Yourself

1. What is the domain of $y = \dfrac{x}{x^2 - 4}$?

2. Find the critical numbers of $L(t) = t^3 - 3t^2 - 9t$.

3. Find the local maximum and minimum values of
$g(x) = \dfrac{x^2}{x^2 + 1}$ as well as the intervals of increase or decrease.

4. Find the inflection points and intervals of concavity of $f(w) = 3w^4 - 2w^3 + 1$.

5. Find the vertical and horizontal asymptotes of the function
$R(t) = \dfrac{2t^2 + 1}{t^2 + 7t}$.

6. If $f'(2) = 0$ and $f''(2) = -1$, what can you say about the graph of f at $x = 2$?

7. (a) If $g'(1) = 0$, and $g'(x)$ is negative for $x < 1$ and positive for $x > 1$, what can you say about the graph of g at $x = 1$?

   (b) If $g''(1) = 0$, and $g''(x)$ is negative for $x < 1$ and positive for $x > 1$, what can you say about the graph of g at $x = 1$?

8. (a) If $\lim_{x \to 5} f(x) = \infty$, what can you say about the graph of f?

   (b) If $\lim_{x \to \infty} f(x) = 5$, what can you say about the graph of f?

▪ Exercises 4.5

1–8 ▪

(a) Find the intervals of increase or decrease.

(b) Find the local maximum and minimum values.

(c) Find the intervals of concavity and the inflection points.

(d) Determine the end behavior of the graph.

(e) Use the information from parts (a)–(d) to sketch the graph. Check your work with a graphing device if you have one.

1. $y = 2x^2 - 8x + 3$ **2.** $y = -\frac{1}{2}x^2 + 3x - 1$

3. $y = x^3 + x$ **4.** $y = x^3 + 6x^2 + 9x$

5. $y = 2 - 15x + 9x^2 - x^3$ **6.** $y = 8x^2 - x^4$

7. $y = x^4 + 4x^3$ **8.** $y = x(x + 2)^3$

9–17 ▪ Sketch the graph of the function by first considering the domain, intervals of increase or decrease, local extreme points, concavity and inflection points, and end behavior, including any vertical or horizontal asymptotes.

9. $y = \dfrac{x}{x - 1}$ **10.** $y = \dfrac{x^2 - 4}{x^2 - 2x}$

11. $y = \dfrac{1}{x^2 - 9}$ **12.** $y = \dfrac{x^2}{x^2 + 9}$

13. $y = \dfrac{x - 1}{x^2}$ **14.** $y = 2\sqrt{x} - x$

15. $h(a) = \sqrt{a}\,(a^2 - 4a)$ **16.** $y = e^{2x} - e^x$

17. $y = 1/(1 + e^{-x})$

18–19 ▪ Use a table of values to estimate $\lim_{x \to \infty} f(x)$ and $\lim_{x \to -\infty} f(x)$. Then sketch the graph of the function by considering the domain, intervals of increase or decrease, local extreme points, concavity and inflection points, and vertical and horizontal asymptotes.

18. $f(x) = \dfrac{\ln x^2}{x^2}$ **19.** $f(x) = xe^{-x^2}$

20. (a) What is the domain of $g(t) = 1 + \dfrac{1}{t} + \dfrac{1}{t^2}$? Find the intervals of increase or decrease, local extreme points, concavity, and inflection points.

   (b) Express g as a single fraction and identify all the asymptotes.

   (c) Sketch the graph of g by hand. Then compare to a graph produced by a graphing calculator or computer using an appropriate viewing window.

21–28 ▪ Sketch the graph of a function that satisfies all of the given conditions.

21. $f'(0) = f'(4) = 0$, $f'(x) > 0$ if $x < 0$,
$f'(x) < 0$ if $0 < x < 4$ or if $x > 4$,
$f''(x) > 0$ if $2 < x < 4$, $f''(x) < 0$ if $x < 2$ or $x > 4$

22. $f'(x) > 0$ for all $x \neq 1$,   vertical asymptote $x = 1$,
$f''(x) > 0$ if $x < 1$ or $x > 3$, $f''(x) < 0$ if $1 < x < 3$

23. $f'(0) = f'(2) = f'(4) = 0$,
$f'(x) > 0$ if $x < 0$ or $2 < x < 4$,
$f'(x) < 0$ if $0 < x < 2$ or $x > 4$,
$f''(x) > 0$ if $1 < x < 3$, $f''(x) < 0$ if $x < 1$ or $x > 3$

24. $f'(x) > 0$ if $-2 < x < 2$, $f'(x) < 0$ if $x > 2$ or $x < -2$,
$f'(2) = 0$, $\lim_{x \to \infty} f(x) = 1$, $\lim_{x \to -\infty} f(x) = 1$,
$f''(x) < 0$ if $0 < x < 3$, $f''(x) > 0$ if $x > 3$

25. $f'(x) > 0$ if $x \neq 2$, f is concave upward for $x < 2$ and concave downward for $x > 2$, f has inflection point $(2, 5)$, $\lim\limits_{x \to \infty} f(x) = 8$, $\lim\limits_{x \to -\infty} f(x) = 0$

26. $f'(x) > 0$ if $x \neq 2$, f is concave upward for $x < 2$ and concave downward for $x > 2$,
$\lim\limits_{x \to 2^-} f(x) = \infty$, $\lim\limits_{x \to 2^+} f(x) = -\infty$,
$\lim\limits_{x \to \infty} f(x) = \infty$, $\lim\limits_{x \to -\infty} f(x) = 0$

27. $f'(5) = 0$, $f'(x) < 0$ when $x < 5$,
$f'(x) > 0$ when $x > 5$, $f''(2) = 0$, $f''(8) = 0$,
$f''(x) < 0$ when $x < 2$ or $x > 8$,
$f''(x) > 0$ for $2 < x < 8$,
$\lim\limits_{x \to \infty} f(x) = 3$, $\lim\limits_{x \to -\infty} f(x) = 3$

28. $f(0) = 4$, $f'(x) < 0$ for all x, $f''(x) > 0$ for all x,
$\lim\limits_{x \to \infty} f(x) = 2$, $\lim\limits_{x \to -\infty} f(x) = \infty$

29–32 ▪ Use a graphing device to produce graphs of f that reveal all the important aspects of the curve. In particular, you should use graphs of f' and f'' to estimate the intervals of increase and decrease, extreme values, intervals of concavity, and inflection points.

29. $f(x) = 4x^4 - 32x^3 + 89x^2 - 95x + 29$

30. $f(x) = x^6 - 15x^5 + 75x^4 - 125x^3 - x$

31. $f(x) = x^6 - 10x^5 - 400x^4 + 2500x^3$

32. $f(x) = 3x^6 - 8x^5 + 0.2e^x$

33–37 ▪ Describe how the graph of f varies as c varies. Graph several members of the family to illustrate the trends that you discover. In particular, you should investigate how maximum and minimum points and inflection points move when c changes. You should also identify any transitional values of c at which the basic shape of the curve changes.

33. $f(x) = x^4 + cx^2$

34. $f(x) = x^3 + cx$

35. $f(x) = \dfrac{cx}{1 + c^2 x^2}$

36. $f(x) = \ln(x^2 + c)$

37. $f(x) = xe^{-cx}$

Challenge Yourself

38. Investigate the family of curves given by the equation $f(x) = x^4 + cx^2 + x$. Start by determining the transitional value of c at which the number of inflection points changes. Then graph several members of the family to see what shapes are possible. There is another transitional value of c at which the number of critical numbers changes. Try to discover it graphically. Then prove what you have discovered.

39. Use the methods of this section to sketch the curve $y = x^3 - 3a^2 x + 2a^3$, where a is a positive constant. What do the members of this family of curves have in common? How do they differ from each other?

40. Raindrop velocity   It can be shown, under certain assumptions, that the velocity $v(t)$ of a falling raindrop at time t is
$$v(t) = v^*(1 - e^{-gt/v^*})$$
where g is the acceleration due to gravity and v^* is the *terminal velocity* of the raindrop.

(a) Find $\lim\limits_{t \to \infty} v(t)$.

(b) Graph $v(t)$ if $v^* = 1$ m/s and $g = 9.8$ m/s². How long does it take for the velocity of the raindrop to reach 99% of its terminal velocity?

41. Spread of a rumor   A model for the spread of a rumor is given by the equation
$$p(t) = \frac{1}{1 + ae^{-kt}}$$
where $p(t)$ is the proportion of the population that knows the rumor at time t and a and k are positive constants.

(a) When will half the population have heard the rumor?

(b) When is the rate of spread of the rumor greatest?

(c) Sketch the graph of p.

42. Bell curves   The family of bell-shaped curves
$$y = \frac{1}{\sigma\sqrt{2\pi}} e^{-(x-\mu)^2/(2\sigma^2)}$$
occurs in probability and statistics, where it is called the *normal density function*. The constant μ is called the *mean* and the positive constant σ is called the *standard deviation*. For simplicity, let's scale the function so as to remove the factor $1/(\sigma\sqrt{2\pi})$ and let's analyze the special case where $\mu = 0$. So we study the function
$$f(x) = e^{-x^2/(2\sigma^2)}$$

(a) Find the horizontal asymptote, maximum value, and inflection points of f.

(b) What role does σ play in the shape of the curve?

(c) Illustrate by graphing four members of this family on the same screen.

4.6 Optimization

The methods we have learned in this chapter for finding extreme values have practical applications in many areas of life. A businessperson wants to minimize costs and maximize profits. A traveler wants to minimize transportation time. Fermat's Principle in optics states that light follows the path that takes the least time. In this section and the next we solve such problems as maximizing areas, volumes, and profits and minimizing distances, times, and costs.

In solving such practical problems the greatest challenge is often to convert the word problem into a mathematical optimization problem by setting up the function that is to be maximized or minimized. The following problem-solving principles will be helpful in achieving this goal:

STEPS IN SOLVING OPTIMIZATION PROBLEMS

1. **Understand the Problem**   The first step is to read the problem carefully until it is clearly understood. Ask yourself: What is the unknown? What are the given quantities? What are the given conditions?

2. **Draw a Diagram**   In most problems it is useful to draw a diagram and identify the given and required quantities on the diagram.

3. **Introduce Notation**   Assign a symbol to the quantity that is to be maximized or minimized (let's call it Q for now). Also select symbols $(a, b, c, \ldots, x, y)$ for other unknown quantities and label the diagram with these symbols. It may help to use initials as suggestive symbols—for example, A for area, h for height, t for time.

4. Express Q in terms of some of the other symbols from Step 3.

5. If Q has been expressed as a function of more than one variable in Step 4, use the given information to find relationships (in the form of equations) among these variables. Then use these equations to eliminate all but one of the variables in the expression for Q. Thus Q will be expressed as a function of *one* variable x, so we can write $Q = f(x)$. State the domain of this function.

6. Use the methods of Sections 4.2 and 4.3 to find the *absolute* maximum or minimum value of f. In particular, if the domain of f is a closed interval, then the Closed Interval Method in Section 4.2 can be used.

▪ EXAMPLE 1 Maximizing Area

A farmer has 2400 ft of fencing and wants to fence off a rectangular field that borders a straight river. He needs no fence along the river. What are the dimensions of the field that has the largest area?

SOLUTION 1

▪ Understand the problem

In order to get a feeling for what is happening in this problem, let's experiment with some special cases. Figure 1 (not to scale) shows three possible ways of laying out the 2400 ft of fencing. We see that when we try shallow, wide fields or deep, narrow fields, we get relatively small areas. It seems plausible that there is some intermediate configuration that produces the largest area.

▪ Draw diagrams

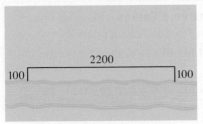

Area = 100 · 2200 = 220,000 ft²

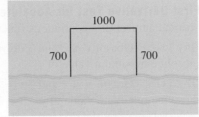

Area = 700 · 1000 = 700,000 ft²

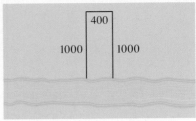

Area = 1000 · 400 = 400,000 ft²

FIGURE 1

■ Introduce notation

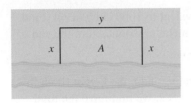

FIGURE 2

Figure 2 illustrates the general case. We wish to maximize the area A of the rectangle. Let x and y be the depth and width of the rectangle (in feet). Then we express A in terms of x and y:

$$A = xy$$

We want to express A as a function of just one variable, so we eliminate y by expressing it in terms of x. To do this we use the given information that the total length of the fencing is 2400 ft. Thus

$$2x + y = 2400$$

From this equation we have $y = 2400 - 2x$, which gives

$$A = xy = x(2400 - 2x) = 2400x - 2x^2$$

Note that $x \geq 0$ (x can't be negative) and $x \leq 1200$ (otherwise we won't have enough fencing to surround the field). So the function that we wish to maximize is

$$A(x) = 2400x - 2x^2 \qquad 0 \leq x \leq 1200$$

The derivative is $A'(x) = 2400 - 4x$, so to find the critical numbers we solve the equation

$$2400 - 4x = 0$$

which gives $x = 600$. As we saw in Section 4.2, the maximum value of A must occur either at this critical number or at an endpoint of the interval. Since $A(0) = 0$, $A(600) = 720,000$, and $A(1200) = 0$, the Closed Interval Method gives the maximum value as $A(600) = 720,000$.

[Alternatively, we could have observed that $A''(x) = -4 < 0$ for all x, so A is always concave downward and the local maximum at $x = 600$ must be an absolute maximum.]

Thus the rectangular field should be 600 ft deep and 1200 ft wide. ■

Because the domain of the function in Example 1 consisted of a closed interval, we were able to use the Closed Interval Method from Section 4.2 to find the absolute maximum. If, instead, the domain of the function is not a closed interval, we can use the following variant of the First Derivative Test (which applies only to *local* maximum or minimum values) to find the absolute maximum and minimum values.

▪ **First Derivative Test for Absolute Extreme Values**   Suppose that c is a critical number of a continuous function f defined on an interval.

(a) If $f'(x) > 0$ for all $x < c$ and $f'(x) < 0$ for all $x > c$, then $f(c)$ is the absolute maximum value of f.

(b) If $f'(x) < 0$ for all $x < c$ and $f'(x) > 0$ for all $x > c$, then $f(c)$ is the absolute minimum value of f.

We use this derivative test in the next example.

▪ **EXAMPLE 2   Minimizing Cost**

A cylindrical can is to be made to hold 1 L (equivalent to 1000 cm³) of oil. Find the dimensions that will minimize the cost of the metal to manufacture the can.

SOLUTION

Draw the diagram as in Figure 3, where r is the radius and h the height (both in centimeters). In order to minimize the cost of the metal, we minimize the total surface area of the cylinder (top, bottom, and sides). From Figure 4 we see that the sides are made from a rectangular sheet with dimensions $2\pi r$ (the circumference of a circle) and h. The top and bottom are each circular with area πr^2. So the surface area is

$$A = 2\pi r^2 + 2\pi r h$$

Remember that we need to express A in terms of just one variable. The volume is given as 1 L, which we write as 1000 cm³ to be consistent with the units of measure. The volume of a cylinder is given by $\pi r^2 h$, so we must have

$$\pi r^2 h = 1000$$

which we can write as $h = 1000/(\pi r^2)$. We substitute this expression for h in the formula for A to express A in terms of r only:

$$A = 2\pi r^2 + 2\pi r\left(\frac{1000}{\pi r^2}\right) = 2\pi r^2 + \frac{2000}{r}$$

We must have $r > 0$ for the can to have volume, but there is no theoretical maximum value for r. Therefore the function that we want to minimize is

$$A(r) = 2\pi r^2 + \frac{2000}{r} \qquad r > 0$$

To find the critical numbers, we differentiate:

$$A'(r) = 4\pi r + 2000(-r^{-2}) = 4\pi r - \frac{2000}{r^2}$$

$$= \frac{4\pi r^3}{r^2} - \frac{2000}{r^2} = \frac{4(\pi r^3 - 500)}{r^2}$$

The derivative is defined for all values of r in the domain and $A'(r) = 0$ when $\pi r^3 = 500$, so the only critical number is $r = \sqrt[3]{500/\pi}$.

FIGURE 3

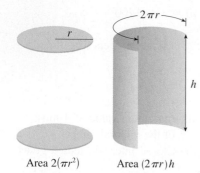

Area $2(\pi r^2)$ Area $(2\pi r)h$

FIGURE 4

TEC Module 4.6 takes you through five additional optimization problems, including animations of the physical situations.

Since the domain of A is $(0, \infty)$, we can't use the Closed Interval Method, as we did in Example 1. But we can observe that $A'(r) < 0$ for $r < \sqrt[3]{500/\pi}$ and $A'(r) > 0$ for $r > \sqrt[3]{500/\pi}$, so A is decreasing for *all* r to the left of the critical number and increasing for *all* r to the right. Thus, as stated in the First Derivative Test for Absolute Extreme Values, $r = \sqrt[3]{500/\pi}$ must give rise to an *absolute* minimum.

[Alternatively, we could argue that $A(r) \to \infty$ as $r \to 0^+$ and $A(r) \to \infty$ as $r \to \infty$, so there must be a minimum value of $A(r)$, which must occur at the critical number. See Figure 5.]

The value of h corresponding to $r = \sqrt[3]{500/\pi}$ is

$$h = \frac{1000}{\pi r^2} = \frac{1000}{\pi(500/\pi)^{2/3}} = \frac{2 \cdot 500}{\pi^{1/3}(500)^{2/3}}$$

$$= \frac{2(500)^{1/3}}{\pi^{1/3}} = 2\sqrt[3]{\frac{500}{\pi}}$$

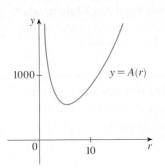

$y = A(r)$

FIGURE 5

Notice that this is twice the value of r. Thus, to minimize the cost of the can, the radius should be $\sqrt[3]{500/\pi} \approx 5.42$ cm and the height should be equal to twice the radius, namely the diameter. ∎

▪ **EXAMPLE 3**   **Optimal Swimming Speed of Migrating Fish**

For a fish swimming at a speed v relative to the water, the energy expenditure per unit time is proportional to v^3. It is believed that migrating fish try to minimize the total energy required to swim a fixed distance. If the fish are swimming against a current u ($u < v$), then the time required to swim a distance L is $L/(v - u)$ and the total energy E required to swim the distance is given by

$$E(v) = av^3 \cdot \frac{L}{v - u}$$

where a is a positive constant. Determine the value of v that minimizes E.

SOLUTION
We have

$$E(v) = aL \cdot \frac{v^3}{v - u}$$

and so

$$E'(v) = aL \cdot \frac{(v - u) \cdot 3v^2 - v^3 \cdot 1}{(v - u)^2} = aL \cdot \frac{2v^3 - 3uv^2}{(v - u)^2} = aL \cdot \frac{v^2(2v - 3u)}{(v - u)^2}$$

Then $E'(v) = 0$ when $v^2(2v - 3u) = 0 \iff v = \frac{3}{2}u$ (since $v \neq 0$). The domain is (u, ∞); $E'(v) < 0$ for $u < v < \frac{3}{2}u$ and $E'(v) > 0$ for $v > \frac{3}{2}u$, so the absolute minimum of E occurs when $v = \frac{3}{2}u$. A graph of E is shown in Figure 6.

Note: This result has been verified experimentally; migrating fish swim against a current at a speed 50% greater than the speed of the current. ∎

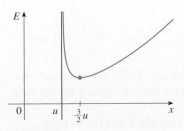

FIGURE 6

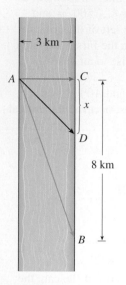

3 km

A   *C*

x

D

8 km

B

FIGURE 7

■ **EXAMPLE 4** **Minimizing Time**

A man launches his boat from point *A* on a bank of a straight river, 3 km wide, and wants to reach point *B*, 8 km downstream on the opposite bank, as quickly as possible (see Figure 7). He could row his boat directly across the river to point *C* and then run to *B*, or he could row directly to *B*, or he could row to some point *D* between *C* and *B* and then run to *B*. If he can row 6 km/h and run 8 km/h, where should he land to reach *B* as soon as possible? (We assume that the speed of the water is negligible compared with the speed at which the man rows.)

SOLUTION

If we let *x* be the distance from *C* to *D*, then the running distance is $8 - x$, the distance from *D* to *B*. We use the Pythagorean Theorem to determine that the rowing distance (from *A* to *D*) is $\sqrt{x^2 + 9}$. We use the equation

$$\text{time} = \frac{\text{distance}}{\text{rate}}$$

Then the rowing time is $\sqrt{x^2 + 9}/6$ and the running time is $(8 - x)/8$, so the total time *T* as a function of *x* is

$$T(x) = \frac{\sqrt{x^2 + 9}}{6} + \frac{8 - x}{8}$$

The domain of this function *T* is [0, 8]. Notice that if $x = 0$ he rows to *C* and if $x = 8$ he rows directly to *B*. We can write

$$T(x) = \tfrac{1}{6}(x^2 + 9)^{1/2} + \tfrac{1}{8}(8 - x)$$

so the derivative of *T* is

$$T'(x) = \tfrac{1}{6} \cdot \tfrac{1}{2}(x^2 + 9)^{-1/2} \cdot 2x + \tfrac{1}{8}(-1) = \frac{x}{6\sqrt{x^2 + 9}} - \frac{1}{8}$$

Thus, using the fact that $x \geqslant 0$, we have

$$T'(x) = 0 \iff \frac{x}{6\sqrt{x^2 + 9}} = \frac{1}{8} \iff 4x = 3\sqrt{x^2 + 9}$$

$$\iff 16x^2 = 9(x^2 + 9) \iff 7x^2 = 81$$

$$\iff x = \frac{9}{\sqrt{7}}$$

The only critical number is $x = 9/\sqrt{7}$. Following the Closed Interval Method, we check whether the minimum occurs at this critical number or at an endpoint of the domain [0, 8]:

$$T(0) = 1.5 \qquad T\left(\frac{9}{\sqrt{7}}\right) = 1 + \frac{\sqrt{7}}{8} \approx 1.33 \qquad T(8) = \frac{\sqrt{73}}{6} \approx 1.42$$

Since the smallest of these values of *T* occurs when $x = 9/\sqrt{7}$, the absolute minimum value of *T* must occur there. Figure 8 illustrates this calculation by showing the graph of *T*.

Thus the man should land the boat at a point $x = 9/\sqrt{7}$ km (≈ 3.4 km) downstream from his starting point. ■

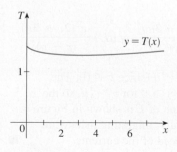

T

$y = T(x)$

1

0   2   4   6   *x*

FIGURE 8

▪ Prepare Yourself

1. If the legs of a right triangle measure x and $20 - x$, write an expression for the length of the hypotenuse.

2. If the hypotenuse of a right triangle has length $2x + 5$ and one of the legs measures $x - 2$, what is the length of the other leg?

3. If the volume of a rectangular box is 840 in³ and the base has width x inches and length $x + 4$ inches, write an expression for the height of the box.

4. What is the slope of a line through the origin and the point (a, b)?

5. Solve for x: $3x - \dfrac{4}{x^2} = 0$

6. If $A = \dfrac{p + m}{cm^2 - bm}$, find $\dfrac{dA}{dm}$.

7. A runner jogs at 3 mi/h for x miles and then walks at 1.5 mi/h for $x - 2$ miles. Write an expression for the total time of the trip.

▪ Exercises 4.6

1. Consider the following problem: Find two numbers whose sum is 23 and whose product is a maximum.

    (a) Make a table of values, like the following one, so that the sum of the numbers in the first two columns is always 23. On the basis of the evidence in your table, estimate the answer to the problem.

First number	Second number	Product
1	22	22
2	21	42
3	20	60
⋮	⋮	⋮

    (b) Use calculus to solve the problem and compare with your answer to part (a).

2. Find two numbers whose difference is 100 and whose product is a minimum.

3. Find two positive numbers whose product is 100 and whose sum is a minimum.

4. Find a positive number such that the sum of the number and its reciprocal is as small as possible.

5. **Rectangle area**   Find the dimensions of a rectangle with perimeter 100 m whose area is as large as possible.

6. **Rectangle perimeter**   Find the dimensions of a rectangle with area 1000 m² whose perimeter is as small as possible.

7. **Fenced area**   Consider the following problem: A farmer with 750 ft of fencing wants to enclose a rectangular area and then divide it into four pens with fencing parallel to one side of the rectangle. What is the largest possible total area of the four pens?

    (a) Draw several diagrams illustrating the situation, some with shallow, wide pens and some with deep, narrow pens. Find the total areas of these configurations. Does it appear that there is a maximum area? If so, estimate it.

    (b) Draw a diagram illustrating the general situation. Introduce notation and label the diagram with your symbols.

    (c) Write an expression for the total area.

    (d) Use the given information to write an equation that relates the variables.

    (e) Use part (d) to write the total area as a function of one variable.

    (f) Finish solving the problem and compare the answer with your estimate in part (a).

8. **Box design**   Consider the following problem: A box with an open top is to be constructed from a square piece of cardboard, 3 ft wide, by cutting out a square from each of the four corners and bending up the sides. Find the largest volume that such a box can have.

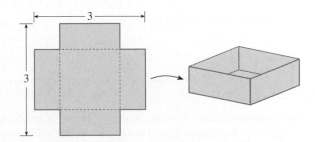

    (a) Draw several diagrams to illustrate the situation, some short boxes with large bases and some tall boxes with small bases. Find the volumes of several such boxes. Does it appear that there is a maximum volume? If so, estimate it.

    (b) Draw a diagram illustrating the general situation. Introduce notation and label the diagram with your symbols.

(c) Write an expression for the volume.

(d) Use the given information to write an equation that relates the variables.

(e) Use part (d) to write the volume as a function of one variable.

(f) Finish solving the problem and compare the answer with your estimate in part (a).

9. Box design If 1200 cm² of material is available to make a box with a square base and an open top, find the largest possible volume of the box.

10. Box design A box with a square base and open top must have a volume of 32,000 cm³. Find the dimensions of the box that minimize the amount of material used.

11. Rectangle area

(a) Show that of all the rectangles with a given area, the one with smallest perimeter is a square.

(b) Show that of all the rectangles with a given perimeter, the one with greatest area is a square.

12. Container design A rectangular storage container with an open top is to have a volume of 10 m³. The length of its base is twice the width. Material for the base costs \$10 per square meter. Material for the sides costs \$6 per square meter. Find the cost of materials for the cheapest such container.

13. Can design A cylindrical can without a top is made to contain 1 gallon (231 in³) of liquid. Find the dimensions that will minimize the cost of the metal to make the can.

14. Rectangle area Find the dimensions of the rectangle of largest area that has its base on the x-axis and its other two vertices above the x-axis and lying on the parabola $y = 8 - x^2$.

15. Crop yield A model used for the yield Y of an agricultural crop as a function of the nitrogen level N in the soil (measured in appropriate units) is

$$Y = \frac{kN}{1 + N^2}$$

where k is a positive constant. What nitrogen level gives the best yield?

16. Photosynthesis The rate, in mg carbon/m³/h, at which photosynthesis takes place for a species of phytoplankton is modeled by the function

$$P = \frac{100I}{I^2 + I + 4}$$

where I is the light intensity (measured in thousands of foot-candles). For what light intensity is P a maximum?

17. Electrical power If a resistor of R ohms is connected across a battery of E volts with internal resistance r ohms,

then the power, in watts, in the external resistor is

$$P = \frac{E^2 R}{(R + r)^2}$$

If E and r are fixed but R varies, what is the maximum value of the power?

18. Drug concentration The function

$$f(t) = C(e^{-at} - e^{-bt})$$

where a, b, and C are positive numbers and $b > a$, has been used to model the concentration of a drug injected into the bloodstream at time $t = 0$.

(a) Show that the critical number of f is $\dfrac{\ln(a/b)}{a - b}$.

(b) A drug is injected into a patient; the concentration, in mg/L, after t minutes is given by $f(t)$, where $a = 0.04$, $b = 0.5$, and $C = 11.2$. Graph f and find the maximum concentration of the drug in the patient's bloodstream.

19. Find, correct to three decimal places, the point on the curve $y = \ln x + e^x$ at which the tangent line has the smallest slope.

20. At which points on the curve $y = 1 + 40x^3 - 3x^5$ does the tangent line have the largest slope?

21. Cup design A cone-shaped drinking cup is made from a circular piece of paper of radius 4 inches by cutting out a sector and joining the two straight edges. Find the maximum capacity of such a cup.

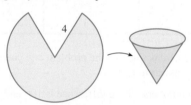

22. Cup design A cone-shaped paper drinking cup is to be made to hold 27 cm³ of water. Find the height and radius of the cup that will use the smallest amount of paper.

23. Combined area A piece of wire 10 m long is cut into two pieces. One piece is bent into a square and the other is bent into an equilateral triangle. How should the wire be cut so that the total area enclosed is **(a)** a maximum? **(b)** A minimum?

24. Navigation A boat leaves a dock at 2:00 PM and travels due south at a speed of 20 km/h. Another boat has been heading due east at 15 km/h and reaches the same dock at 3:00 PM. At what time were the two boats closest together?

25. Construction costs An oil refinery is located on the north bank of a straight river that is 2 km wide. A pipeline is to be constructed from the refinery to storage tanks located on the south bank of the river 6 km east of the refinery. The cost of laying pipe is \$400,000/km over land to a point P on the north bank and \$800,000/km under the river to the tanks. To minimize the cost of the pipeline, where should P be located?

26. Construction costs Suppose the refinery in Exercise 25 is located 1 km north of the river. Where should P be located?

27. Window design A Norman window has the shape of a rectangle surmounted by a semicircle. (Thus the diameter of the semicircle is equal to the width of the rectangle.) If the perimeter of the window is 30 ft, find the dimensions of the window so that the greatest possible amount of light is admitted.

© Brooks Cole / Cengage Learning

28. Fuel consumption The graph shows the fuel consumption c of a car (measured in gallons per hour) as a function of the speed v of the car. At very low speeds the engine runs inefficiently, so initially c decreases as the speed increases. But at high speeds the fuel consumption increases. You can see that $c(v)$ is minimized for this car when $v \approx 30$ mi/h. However, for fuel efficiency, what must be minimized is not the consumption in gallons per hour but rather the fuel consumption in gallons *per mile*. Let's call this consumption G. Using the graph, estimate the speed at which G has its minimum value.

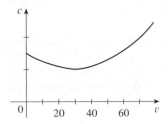

Hint: Looking at units,

$$(\text{gallons/hour})/(\text{miles/hour}) = (\text{gallons/mile})$$

so $G = c/v$. Geometrically we can interpret c/v as the slope of the line from the origin to the point (v, c).

■ Challenge Yourself

29. Find an equation of the line through the point $(3, 5)$ that cuts off the least area from the first quadrant.

30. Illumination The illumination of an object by a light source is directly proportional to the strength of the source and inversely proportional to the square of the distance from the source. If two light sources, one three times as strong as the other, are placed 10 ft apart, where should an object be placed on the line between the sources so as to receive the least illumination?

31. Flight path of birds Ornithologists have determined that some species of birds tend to avoid flights over large bodies of water during daylight hours. It is believed that more energy is required to fly over water than land because air generally rises over land and falls over water during the day. A bird with these tendencies is released from an island that is 5 km from the nearest point B on a straight shoreline, flies to a point C on the shoreline, and then flies along the shoreline to its nesting area D. Assume that the bird instinctively chooses a path that will minimize its energy expenditure. Points B and D are 13 km apart.

(a) In general, if it takes 1.4 times as much energy to fly over water as land, to what point C should the bird fly in order to minimize the total energy expended in returning to its nesting area?

(b) Let W and L denote the energy (in joules) per kilometer flown over water and land, respectively. What would a large value of the ratio W/L mean in terms of the bird's flight? What would a small value mean? Determine the ratio W/L corresponding to the minimum expenditure of energy.

(c) What should the value of W/L be in order for the bird to fly directly to its nesting area D? What should the value of W/L be for the bird to fly to B and then along the shore to D?

(d) If the ornithologists observe that birds of a certain species reach the shore at a point 4 km from B, how many times more energy does it take a bird to fly over water than land?

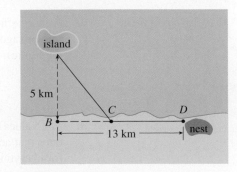

32. Lumber A rectangular beam will be cut from a cylindrical log of radius 10 inches.

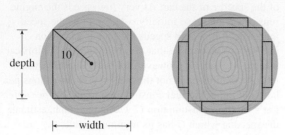

(a) Show that the beam of maximal cross-sectional area is a square.

(b) Four rectangular planks will be cut from the four sections of the log that remain after cutting the square beam. Determine the dimensions of the planks that will have maximal cross-sectional area.

(c) Suppose that the strength of a rectangular beam is proportional to the product of its width and the square of its depth. Find the dimensions of the strongest beam that can be cut from the cylindrical log.

33. Minimizing length The upper right-hand corner of a piece of paper, 12 in. by 8 in., as in the figure, is folded over to the bottom edge. How would you fold it so as to minimize the length of the fold? In other words, how would you choose x to minimize y?

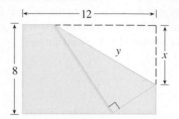

P R O J E C T ▪ The Shape of a Can

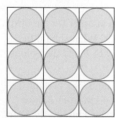

Discs cut from squares

Discs cut from hexagons

In this project we investigate the most economical shape for a can. We first interpret this to mean that the volume V of a cylindrical can is given and we need to find the height h and radius r that minimize the cost of the metal to make the can (see the figure). If we disregard any waste metal in the manufacturing process, then the problem is to minimize the surface area of the cylinder. We solved this problem in Example 2 in Section 4.6 and we found that $h = 2r$; that is, the height should be the same as the diameter. But if you go to your cupboard or your supermarket with a ruler, you will discover that the height is usually greater than the diameter and the ratio h/r varies from 2 up to about 3.8. Let's see if we can explain this phenomenon.

1. The material for the cans is cut from sheets of metal. The cylindrical sides are formed by bending rectangles; these rectangles are cut from the sheet with little or no waste. But if the top and bottom discs are cut from squares of side $2r$ (as in the figure), this leaves considerable waste metal, which may be recycled but has little or no value to the can makers. If this is the case, show that the amount of metal used is minimized when

$$\frac{h}{r} = \frac{8}{\pi} \approx 2.55$$

2. A more efficient packing of the discs is obtained by dividing the metal sheet into hexagons and cutting the circular lids and bases from the hexagons (see the figure). Show that if this strategy is adopted, then

$$\frac{h}{r} = \frac{4\sqrt{3}}{\pi} \approx 2.21$$

3. The values of h/r that we found in Problems 1 and 2 are a little closer to the ones that actually occur on supermarket shelves, but they still don't account for everything. If we look more closely at some real cans, we see that the lid

and the base are formed from discs with radius larger than r that are bent over the ends of the can. If we allow for this we would increase h/r. More significantly, in addition to the cost of the metal we need to incorporate the manufacturing of the can into the cost. Let's assume that most of the expense is incurred in joining the sides to the rims of the cans. If we cut the discs from hexagons as in Problem 2, then the total cost is proportional to

$$4\sqrt{3}\, r^2 + 2\pi rh + k(4\pi r + h)$$

where k is the reciprocal of the length that can be joined for the cost of one unit area of metal. Show that this expression is minimized when

$$\frac{\sqrt[3]{V}}{k} = \sqrt[3]{\frac{\pi h}{r}} \cdot \frac{2\pi - h/r}{\pi h/r - 4\sqrt{3}}$$

4. Plot $\sqrt[3]{V}/k$ as a function of $x = h/r$ and use your graph to argue that when a can is large or joining is cheap, we should make h/r approximately 2.21 (as in Problem 2). But when the can is small or joining is costly, h/r should be substantially larger.

5. Our analysis shows that large cans should be almost square but small cans should be tall and thin. Take a look at the relative shapes of the cans in a supermarket. Is our conclusion usually true in practice? Are there exceptions? Can you suggest reasons why small cans are not always tall and thin?

4.7 Optimization in Business and Economics

We begin this section by revisiting cost, revenue, profit, and demand functions now that we have more experience with derivatives and what they tell us about a function. Then we look at a way to measure how much price influences demand (called *elasticity*) and how a business can handle its inventory in an optimal way.

▪ Marginal and Average Cost

In Section 3.2 we introduced the idea of marginal cost. Recall that if $C(q)$, the **cost function**, is the cost of producing q units of a certain product, then the **marginal cost** is $C'(q)$, the rate of change of C with respect to q.

The graph of a typical cost function is shown in Figure 1. The marginal cost $C'(q)$ is the slope of the tangent to the cost curve at $(q, C(q))$. Notice that the cost curve is initially concave downward (the marginal cost is decreasing), as we previously described, but eventually there is an inflection point and the cost curve becomes concave upward.

The **average cost function** c is defined as

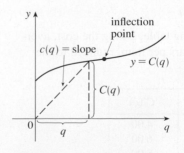

FIGURE 1 Cost function

(1)
$$c(q) = \frac{C(q)}{q}$$

and represents the cost per unit when q units are produced. We sketch a typical

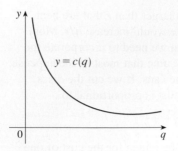

FIGURE 2
Average cost function

average cost function in Figure 2 by noting that $C(q)/q$ is the slope of the line that joins the origin to the point $(q, C(q))$ in Figure 1. Here this slope decreases at first as q increases but then it begins increasing. Thus it appears that there will be an absolute minimum in the graph of c. In Section 3.2, we reasoned that this minimum occurs when marginal cost is equal to average cost. Let's look at this from the perspective of derivatives.

The absolute minimum of c must occur at a critical number, which we locate by using the Quotient Rule to differentiate Equation 1:

$$c'(q) = \frac{q \cdot C'(q) - C(q)}{q^2}$$

Now $c'(q) = 0$ when $qC'(q) - C(q) = 0$ and this gives

$$C'(q) = \frac{C(q)}{q} = c(q)$$

Thus we have established the following.

> ▪ When average cost is a minimum, the marginal cost and average cost are equal.

▪ EXAMPLE 1   Marginal Cost and Minimizing Average Cost

See Section 3.2 for an explanation of why it is reasonable to model a cost function by a polynomial.

A company estimates that the cost, in dollars, of producing q items is

$$C(q) = 2600 + 2q + 0.001q^2$$

(a) Find the cost, average cost, and marginal cost of producing 1000 items, 2000 items, and 3000 items.

(b) At what production level will the average cost be lowest, and what is this minimum average cost?

SOLUTION

(a) The average cost function is

$$c(q) = \frac{C(q)}{q} = \frac{2600}{q} + 2 + 0.001q$$

The marginal cost function is

$$C'(q) = 2 + 0.002q$$

We use these expressions to fill in the following table, giving the cost, average cost, and marginal cost (in dollars, or dollars per item, rounded to the nearest cent).

q	$C(q)$	$c(q)$	$C'(q)$
1000	5,600.00	5.60	4.00
2000	10,600.00	5.30	6.00
3000	17,600.00	5.87	8.00

(b) One method is to use the techniques of this chapter to find the absolute minimum of $c(q)$ on the domain $(0, \infty)$. The derivative of the average cost function is

$$c'(q) = -\frac{2600}{q^2} + 0.001$$

We have a critical number when $c'(q) = 0$, so we solve

$$-\frac{2600}{q^2} + 0.001 = 0$$

$$0.001 = \frac{2600}{q^2}$$

$$0.001q^2 = 2600$$

$$q^2 = 2,600,000$$

$$q = \sqrt{2,600,000} \approx 1612$$

Now $c''(q) = 5200/q^3$ and $c''(q) > 0$ for $q > 0$, so c is concave upward on its entire domain. Thus $q \approx 1612$ gives both a local and absolute minimum, and the minimum average cost is

$$c(1612) = \frac{2600}{1612} + 2 + 0.001(1612) = \$5.22/\text{item}$$

Alternatively, we can determine where marginal cost and average cost are equal. (See Figure 3.) ▪

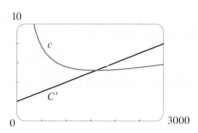

FIGURE 3

Figure 3 shows the graphs of the marginal cost function C' and average cost function c in Example 1. Notice that c has its minimum value when the two graphs intersect.

▪ Revenue, Profit, and Demand Functions

In Section 3.2 we studied **revenue functions** R, which give the total revenue collected after q units are sold. Recall that the **marginal revenue function** is defined as R', the rate of change of revenue with respect to the number of units sold. We also defined the **profit function** P as revenue minus cost:

$$P(q) = R(q) - C(q)$$

The **marginal profit function** is P', the rate of change of profit with respect to the number of units sold.

A natural goal for any business is to maximize profit. A maximum value of P must occur at a critical point of P, that is, when the marginal profit is 0. But if

$$P'(q) = R'(q) - C'(q) = 0$$

then $R'(q) = C'(q)$

Therefore, when the profit is maximized, marginal revenue and marginal cost are equal. This is precisely the observation we made in Section 3.2. To ensure that this condition gives a maximum, we could use the Second Derivative Test. Notice that

$$P''(q) = R''(q) - C''(q) < 0$$

when $R''(q) < C''(q)$

This condition says that the rate of increase of marginal revenue is less than the rate of increase of marginal cost. Thus we have established the following.

> ■ Profit is maximized when $R'(q) = C'(q)$ and $R''(q) < C''(q)$.

■ EXAMPLE 2   Maximizing Profit

In Examples 3 and 4 of Section 3.2 we had the cost function
$C(q) = 10{,}000 + 5q + 0.01q^2$ and the revenue function $R(q) = 48q - 0.012q^2$
for producing q chairs. The profit function is

$$P(q) = R(q) - C(q)$$
$$= (48q - 0.012q^2) - (10{,}000 + 5q + 0.01q^2)$$
$$= 43q - 0.022q^2 - 10{,}000$$

If we want to find the production level at which profit is maximized, one way is to look at the derivative: $P'(q) = 43 - 0.044q = 0$ when

$$q = \frac{43}{0.044} \approx 977$$

Thus $q \approx 977$ is a critical number of P. Because $P''(q) = -0.044 < 0$, P is concave downward everywhere and $P(977) \approx 11{,}011$ is the absolute maximum of the profit function. (Alternatively, we could use the First Derivative Test.)

In Section 3.2 we found the same result by finding the value of q where $R'(q) = C'(q)$. Also $C'(q) = 5 + 0.02q \Rightarrow C''(q) = 0.02$ and $R''(q) = -0.024$, so $R''(977) < C''(977)$. This verifies that the manufacturer should produce 977 chairs to earn the maximum profit.   ■

Recall from Section 3.2 that a **demand function** relates the price per unit of a good and the number of units that can be sold. Typically we use q, the number of units, as the input and then $p = D(q)$ is the price per unit. The corresponding revenue is

$$R(q) = q \cdot p = q \cdot D(q)$$

Normally, lowering the price of a product will allow a business to sell more units of that product, but is it more advantageous to sell fewer items at higher prices, or more items at lower prices? If we have information regarding consumers' interest and willingness to pay for an item, we can formulate a demand function and use it to determine the price for an item that will maximize revenue.

■ EXAMPLE 3

Writing a Demand Function and Maximizing Revenue

A new cell phone has been selling 200 units per week at an electronics store at a price of $220 each. A market survey indicates that for each $10 rebate offered to buyers, the number of units sold will increase by 20 per week. Find the demand function and the revenue function for this phone. How large a rebate should the store offer to maximize its revenue?

SOLUTION

If q is the number of cell phones sold per week, then the weekly increase in sales (beyond the usual 200) is $q - 200$. For each increase of 20 phones sold, the price is decreased by $10. Thus for each additional unit sold, the decrease in price will be $\frac{1}{20} \times 10$ and the demand function is

$$D(q) = 220 - \tfrac{10}{20}(q - 200) = 320 - \tfrac{1}{2}q$$

Revenue is given by

$$R(q) = q \cdot D(q) = 320q - \tfrac{1}{2}q^2$$

Since $R'(q) = 320 - q$, we see that $R'(q) = 0$ when $q = 320$. This value of q gives an absolute maximum by the First Derivative Test (or simply by observing that the graph of R is a parabola that opens downward). The corresponding price is

$$D(320) = 320 - \tfrac{1}{2}(320) = 160$$

and the rebate is $220 - 160 = 60$. Therefore, to maximize revenue, the store should offer a rebate of $60.   ∎

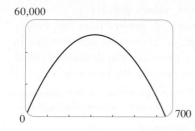

FIGURE 4

Figure 4 shows the revenue function from Example 3. The curve shows a maximum at about $q = 320$, which agrees with our computations.

▪ Elasticity

We have seen that a demand function provides the correlation between the price of a product and the number of consumers willing to buy that product. It is helpful to have a way to measure how strongly a change in price influences a change in demand for different demand functions. In general, we would expect the raising of price to result in a fall of demand; for some goods the change could be drastic, while for others the demand hardly changes. If hotel rates climb 20%, say, many consumers might choose to forgo travel and there could be a significant decrease in room reservations. Compare this to gasoline, where recent history has shown that a rise of 20% in the price does not affect the demand very much (in the short term at least).

To measure how much demand changes in response to a change in price, it is much more useful to compare relative (or equivalently, percentage) changes in these quantities rather than the actual changes. Recall that if x changes by an amount Δx, the relative change is $\Delta x / x$. Economists measure the effect of price changes on demand by computing the ratio

(2) $$\frac{\text{relative change in demand}}{\text{relative change in price}} = \frac{\Delta q / q}{\Delta p / p} = \frac{p/q}{\Delta p / \Delta q}$$

If we allow $\Delta q \to 0$ (and consequently $\Delta p \to 0$), then $\Delta p / \Delta q$ becomes dp/dq; we can now compute this ratio for a specific point on the demand curve. Economists refer to such ratios as *elasticity* and use them in a variety of contexts to measure the degree of change one quantity causes in another.

Elasticity of demand can be expressed as a function of price rather than demand. If we reformulate the demand function in the form $q = f(p)$, then the ratio in (2) can be written as

$$\frac{\Delta q / \Delta p}{q/p}$$

After taking the limit as $\Delta p \to 0$ (and including a negative sign), we get

$$E(p) = -\frac{dq/dp}{q/p} = -\frac{pf'(p)}{f(p)}$$

(3) ▪ **Definition**   The **elasticity of demand** E for a product whose demand q corresponds to the price $p = D(q)$ is given by

$$E(q) = -\frac{p/q}{dp/dq} = -\frac{D(q)}{qD'(q)}$$

NOTE: Typically, positive numbers are used to measure elasticity. Because demand functions are normally decreasing, dp/dq is negative and hence $\dfrac{p/q}{dp/dq}$ is negative. Thus we actually measure the absolute value of the ratio; the negative signs in Definition 3 serve to make the negative ratio values positive.

If $E(q) > 1$, then the relative change in demand is greater than the relative change in price and the demand is called **elastic**. If $E(q) < 1$, then changes in demand are proportionally smaller than changes in price and the demand is called **inelastic**. An elasticity of 1 is called **unit elastic**. In general, the larger the value of E, the more a change in price affects demand.

▪ EXAMPLE 4   Computing Elasticity of Demand

A tool company estimates that the monthly demand q for their power drill is related to the price p for each drill by $p = 185 - 0.06q$. Compute the elasticity of demand for drill prices of $50 and $95.

SOLUTION

If the price is $50 we solve $p = 50$ for q to find the corresponding demand:

$$185 - 0.06q = 50$$
$$-0.06q = -135$$
$$q = \frac{-135}{-0.06} = 2250$$

We see that $dp/dq = -0.06$ so by Equation 3 we have

$$E(2250) = -\left.\frac{p/q}{dp/dq}\right|_{q=2250} = -\frac{50/2250}{-0.06} \approx 0.37037$$

Thus the elasticity for a price of $50 is about 0.37. This means that changes in demand are *proportionally* about 37% of changes in price and the demand is inelastic.

Similarly, for a price of $95 we solve

$$185 - 0.06q = 95$$
$$-0.06q = -90$$
$$q = \frac{-90}{-0.06} = 1500$$

and

$$E(1500) = -\left.\frac{p/q}{dp/dq}\right|_{q=1500} = -\frac{95/1500}{-0.06} \approx 1.0556$$

The elasticity for a price of $95 is approximately 1.06, and the demand at this level is elastic.   ▪

For an elastic demand ($E > 1$), an increase in price produces a proportionally larger decrease in demand. Because revenue is the product of demand and price, this results in a decline in revenue. (The higher price collected per item does not make up for lower number of items sold.) If, instead, demand is inelastic ($E < 1$), then a price increase causes a proportionally smaller decrease in demand and revenue increases. (The higher price more than compensates for the decrease in number of units sold.) If $E = 1$, then changes in price and demand are proportional and revenue remains constant.

In Example 4, $E < 1$ when the drills are priced at $50, so the tool company would increase its revenue by raising prices. When the drills are priced at $95 we have $E > 1$, so it would be advantageous for the company to lower the drill price.

Now that we see a connection between elasticity and revenue, we ask the question: When is revenue maximized? When $E > 1$, revenue is increased by lowering prices, and when $E < 1$, revenue is increased by raising prices. It seems reasonable then to conclude that to maximize revenue, prices should be adjusted until the elasticity is 1. In Exercise 41 we ask you to prove this result mathematically.

Demand q	Elasticity $E(q)$	Revenue $R(q)$
elastic	$E(q) > 1$	increases if prices are lowered, increasing demand
inelastic	$E(q) < 1$	increases if prices are raised, decreasing demand
unit elastic	$E(q) = 1$	at maximum

■ EXAMPLE 5   Elasticity and Maximizing Revenue

The demand function for a manufacturer's product is $D(q) = 75e^{-0.05q}$. Write a formula for the elasticity of demand E and determine the price per unit that maximizes revenue.

SOLUTION

The derivative of demand is $D'(q) = 75e^{-0.05q}(-0.05) = -3.75e^{-0.05q}$ and, by Definition 3,

$$E(q) = -\frac{D(q)}{qD'(q)} = -\frac{75e^{-0.05q}}{q(-3.75e^{-0.05q})} = \frac{20}{q}$$

Revenue is maximized when $E(q) = 1$, that is, when $q = 20$. The corresponding price is

$$D(20) = 75e^{-0.05(20)} = 75e^{-1} = \frac{75}{e} \approx 27.59$$

To maximize revenue, the price per unit should be $27.59.

Note that we could also determine the maximum revenue by computing $R(q) = q \cdot D(q)$ and using the technique of Example 3.   ■

Figure 5 shows the demand curve from Example 5. Notice that the area of the shaded rectangle is the revenue when $q = 20$ and $p = \$27.59$.

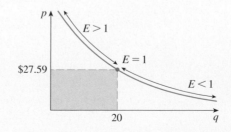

FIGURE 5

■ | Managing Inventory

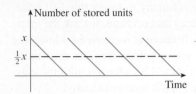

▲ Number of stored units

x

$\frac{1}{2}x$

Time

FIGURE 6

Figure 6 shows the inventory level when perfectly timed shipments of x units sell at a steady rate. We see that the average number of units on hand is $\frac{1}{2}x$.

Any business that purchases products for resale must manage the ordering and storage of the products. This can be a complex problem, with many factors coming into play. Here we look at a simplified situation.

Suppose a retailer sells a product at a steady rate all year. The managers want to be sure never to run out of stock, but they don't want to order too many products at a time because storing a large number of items becomes costly. They also don't want to place a large number of orders throughout the year because each order adds associated costs (processing paperwork, shipping, handling). The goal is to find the optimal order size and frequency.

If the retailer sells A items per year at a constant rate and places equally sized orders of x items during the year, then A/x orders will be placed annually. Let's assume perfect timing of the orders, in the sense that a new order arrives just as the previous stock of items is depleted. Because the items are sold at a steady rate, the retailer is storing *on average* half of the number of products in each order: $\frac{1}{2}x$. (Half of the time they have more than this, half of the time they have less; see Figure 6.) If we know the storage and ordering costs, we can use calculus to minimize the total cost, as we show in the next example.

■ EXAMPLE 6 Minimizing Inventory Costs

Suppose an online shop sells 5000 watches per year at a steady rate. The owner estimates that it costs $2.40 per year to store each watch and each time a shipment is ordered the shop incurs costs of $82. How often should the owner order a new shipment, and how many watches should be in each one?

SOLUTION

Let x be the number of watches in each shipment, so that $5000/x$ orders will be placed over one year. The average number of watches that need to be stored during the year is $\frac{1}{2}x$, and the total associated costs are

$$C(x) = \$82(5000/x) + \$2.40(x/2) = \frac{410{,}000}{x} + 1.2x$$

Thus we wish to find the minimum value of $C(x)$ for $x > 0$. We have

$$C'(x) = 410{,}000(-x^{-2}) + 1.2 = -\frac{410{,}000}{x^2} + 1.2$$

and $C'(x) = 0$ for

$$-\frac{410{,}000}{x^2} + 1.2 = 0$$

$$1.2 = \frac{410{,}000}{x^2}$$

$$x^2 = \frac{410{,}000}{1.2}$$

$$x = \pm\sqrt{\frac{410{,}000}{1.2}} \approx \pm 584.52$$

Thus the only critical number in the domain is approximately 585. Since $C'(x) < 0$ on $(0, 585)$ and $C'(x) > 0$ on $(585, \infty)$, the local and absolute minimum value of $C(x)$ occurs for $x \approx 585$, so the shop should place orders of 585 watches. There will be $5000/585 \approx 8.55$ orders placed per year, or about one every 1.4 months (1 month 12 days). ▪

▪ Prepare Yourself

1–4 ▪ Find the first and second derivative of the given function.

1. $f(x) = 3250 + 4.2x - 0.3x^2 + 0.002x^3$

2. $P(t) = 12e^{-0.4t}$

3. $g(x) = 60 \ln(4x)$ **4.** $h(a) = 5 + \sqrt{3a}$

5. Write an equation for a linear function g where $g(6000) = 16$ and $g(7200) = 12$.

6. Solve the given equation for x. Round your answer to two decimal places.

(a) $x^{0.7} = 18$ (b) $15e^{0.48x} = 113$

▪ Exercises 4.7

1. Cost function   A manufacturer keeps precise records of the cost $C(q)$ of making q items and produces the graph of the cost function shown in the figure.

(a) Explain why $C(0) > 0$.

(b) What is the significance of the inflection point?

(c) Use the graph of C to sketch the graph of the marginal cost function.

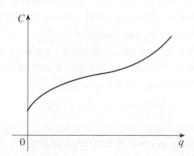

2. Cost function   The graph of a cost function C is given.

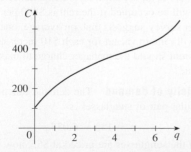

(a) Draw a careful sketch of the marginal cost function.

(b) Use the geometric interpretation of the average cost $c(q)$ as a slope (see Figure 1) to draw a careful sketch of the average cost function.

(c) Estimate the value of q for which $c(q)$ is a minimum. How are the average cost and the marginal cost related at that value of q?

3. Average and marginal cost   The average cost of producing q units of a commodity is $c(q) = 21.4 - 0.002q$. Find the marginal cost at a production level of 1000 units. In practical terms, what is the meaning of your answer?

4. Profit   The figure shows graphs of the cost and revenue functions reported by a manufacturer.

(a) Identify on the graph the value of q for which the profit is maximized.

(b) Sketch a graph of the profit function.

(c) Sketch a graph of the marginal profit function.

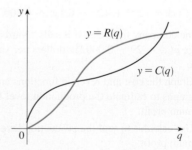

5–6 ▪ **Minimizing average cost**   For each cost function (given in dollars), find (a) the cost, average cost, and marginal cost at a production level of 1000 units; (b) the production level that will minimize the average cost; and (c) the minimum average cost.

5. $C(q) = 16{,}000 + 200q + 4q^{3/2}$

6. $C(q) = 10{,}000 + 340q - 0.3q^2 + 0.0001q^3$

7–8 ▪ Minimizing average cost   A cost function is given.

(a) Find the average cost and marginal cost functions.

(b) Use graphs of the functions in part (a) to estimate the production level that minimizes the average cost.

(c) Use calculus to find the minimum average cost.

(d) Find the minimum value of the marginal cost.

7. $C(q) = 3700 + 5q - 0.04q^2 + 0.0003q^3$

8. $C(q) = 339 + 25q - 0.09q^2 + 0.0004q^3$

9. Maximizing profit   If a company's cost function for a product is

$$C(q) = 2250 + 3.5q + 0.004q^2$$

and the revenue function is $R(q) = 12.2q - 0.002q^2$, find the production level that maximizes profit.

10. Toy production   A toy manufacturer's cost for producing q units of a game is given by $C(q) = 1480 + 3.8q + 0.0006q^2$. If the demand for the game is given by

$$p = 8.6 - \tfrac{1}{440}q$$

how many games should be produced to maximize profit?

11–12 ▪ Marginal cost   Find the production level at which the marginal cost function starts to increase.

11. $C(q) = 0.0008q^3 - 0.72q^2 + 325.3q + 78,000$

12. $C(q) = 0.001q^3 - 0.84q^2 + 416q + 55,000$

13. Fabric production   The cost, in dollars, of producing x yards of a certain fabric is

$$C(x) = 1200 + 12x - 0.1x^2 + 0.0005x^3$$

and the company finds that if it sells x yards, it can charge $p(x) = 29 - 0.00021x$ dollars per yard for the fabric.

(a) Graph the cost and revenue functions and use the graphs to estimate the production level for maximum profit.

(b) Use calculus to find the production level for maximum profit.

14. Aircraft manufacturing   An aircraft manufacturer wants to determine the best selling price for a new airplane. The company estimates that the initial cost of designing the airplane and setting up the factories in which to build it will be 500 million dollars. The additional cost of manufacturing the planes can be modeled by the function $m(x) = 20x - 5x^{3/4} + 0.01x^2$ where x is the number of aircraft produced and m is the manufacturing cost, in millions of dollars. The company

estimates that if it charges a price p, in millions of dollars, for each plane, the number of planes x it will be able to sell satisfies the equation $10x + 77p = 3200$.

(a) Find the cost, demand, and revenue functions.

(b) Find the production level and the associated selling price of the aircraft that maximizes profit.

15. Game attendance   A baseball team plays in a stadium that holds 55,000 spectators. With ticket prices at $10, the average attendance had been 27,000. When ticket prices were lowered to $8, the average attendance rose to 33,000.

(a) Find the demand function, assuming that it is linear.

(b) How should ticket prices be set to maximize revenue?

16. Retail sales   During the summer months Terry makes and sells necklaces on the beach. Last summer he sold the necklaces for $10 each and his sales averaged 20 per day. When he increased the price by $1, he found that he lost two sales per day.

(a) Find the demand function, assuming that it is linear.

(b) If the material for each necklace costs Terry $6, what should the selling price be to maximize his profit?

17. Retail sales   A manufacturer has been selling 1000 televisions a week at $450 each. A market survey indicates that for each $10 rebate offered to the buyer, the number of sets sold will increase by 100 per week.

(a) Find the demand function.

(b) How large a rebate should the company offer the buyer in order to maximize its revenue?

(c) If its weekly cost function is

$$C(q) = 68,000 + 150q$$

how should the manufacturer set the size of the rebate in order to maximize profit?

18. Property management   The manager of a 100-unit apartment complex knows from experience that all units will be occupied if the rent is $800 per month. A market survey suggests that, on average, one additional unit will remain vacant for each $10 increase in rent. What rent should the manager charge to maximize revenue?

19. Elasticity of demand   The demand function for a particular pair of sunglasses is

$$p = 155 - 0.035q$$

(a) If the sunglasses are priced at $65, how many pairs can be sold?

(b) Compute the elasticity of demand when the sunglasses are priced at $65 and interpret your result. At this price, is the demand elastic or inelastic?

20. Elasticity of demand The price p, in thousands of dollars, that an automobile manufacturer charges for its sports car is related to the quantity q that can be sold by the equation $pq^{0.85} = 81{,}000$.

(a) How many cars will sell when they are priced at $56,000?

(b) Compute the elasticity of demand when the price is $56,000 and interpret your result.

21. Elasticity of demand Do you think that the elasticity of demand for tablet computers is currently greater than one? Explain why or why not.

22. Elasticity of demand One published study estimated that the elasticity of demand for beef (at current prices) is 0.67. What can we say about the demand for beef?

23–26 ■ **Elasticity of demand** An equation relating demand q and price p is given. Compute the elasticity of demand for the given price. Is the demand elastic or inelastic?

23. $1.25p + q = 860, \quad p = 225$

24. $pe^{0.4q} = 61, \quad p = 14$

25. $q + 80\ln(2p) = 600, \quad p = 400$
[*Hint:* Use the alternate formula for E as a function of p given in the margin note on page 269.]

26. $q = 200 - \sqrt{125p}, \quad p = 175$

---

27. Elasticity of demand For the demand function $D(q) = 45e^{-0.4q}$ (measured in dollars), at what price is the elasticity of demand equal to one? Give an example of a price where the demand is elastic and another where it is inelastic.

28. Elasticity of demand Suppose demand q and price p are related by the equation

$$p = 540 - 18\sqrt{q}$$

For what value(s) of p is the demand unit elastic? Elastic? Inelastic?

29. Unit elasticity Show that for any demand curve of the form $p = M/q$, where M is a positive constant, we have $E(q) = 1$ and therefore revenue is constant for every price point on the curve.

30. Unit elasticity If demand is given by the linear function $p = mq + b$, where $m < 0$ and $b > 0$, show that the elasticity of demand equals 1 at the midpoint between the q- and p-intercepts.

31. Elasticity and revenue A manufacturer of utility trucks estimates that the demand function for their smallest model truck is $p = 58{,}000 - 32q$. They currently charge $28,000 for each truck.

(a) What is the elasticity of demand at this price point?

(b) To collect more revenue, should the manufacturer raise prices or lower prices for this truck model?

(c) What price gives the maximum revenue?

32. Elasticity and revenue An electronics company currently charges $53 for their mobile phone Bluetooth headset. They estimate that the price p (measured in dollars) and demand q (measured in thousands) are related by the equation $pe^{0.1q} = 116$. What is the elasticity of demand at the current price? To collect more revenue, should the owner raise the price or lower the price? What price gives the maximum revenue?

33. Elasticity and revenue If the demand function for a particular purse is $p = 150 - 4\sqrt{q}$, use elasticity to find the price and corresponding quantity that maximize revenue.

34. Elasticity and revenue An art gallery has determined that the demand function for a limited edition print they are preparing is $p = 475e^{-0.02q}$.

(a) Use elasticity to find the price for the print that maximizes revenue.

(b) Find the price that maximizes revenue using the derivative of the revenue function. Which method do you prefer?

35. Elasticity and revenue Before a company started selling a new digital camera, its marketing team surveyed consumers regarding whether or not they would consider buying the camera at different price levels. The table lists their estimates for the number of cameras that will sell at a particular price.

Price	Units sold (thousands)
$125	51.6
$150	43.4
$175	37.6
$200	31.6
$225	26.3

(a) Use a graphing calculator to find an exponential model of the demand function $p = D(q)$ for the camera.

(b) Use the model to estimate how many cameras will sell if the camera is priced at $189.

(c) According to your model, what is the elasticity of demand at the $189 price point? Does your answer suggest that the camera should be priced higher or lower?

(d) Compute the total revenue using the values in the table. Then estimate the camera price that will bring in the most revenue.

(e) Use elasticity to find the price for the camera that will bring the greatest revenue. How does it compare to your estimate in part (d)?

36. Inventory costs   A restaurant sells 825 bottles of wine from a particular winery annually. They order directly from the winery and sell the bottles at a constant rate throughout the year. It costs the restaurant $3.50 per year to store each bottle and $115 in fixed costs per order. To minimize costs, how often should the restaurant order wine and how many bottles should each order contain?

37. Inventory costs   A supermarket manager estimates that a total of 800 cases of soup will be sold at a steady rate during the coming year and it costs $4 to store a case for a year. He also estimates that the handling cost for each delivery is $100. What is the optimal reorder quantity that minimizes costs?

38. Inventory costs   A lumber yard sells pine 2 by 4 boards at a constant rate. Assume that it costs the yard $0.65 per board annually for storage, and there is a total cost of $450 each time the yard has a supply delivered. If the yard sells 15,000 boards per year, how often should the yard replenish its supply in order to minimize costs?

39. Inventory costs   An insurance company has custom envelopes printed for correspondence with its clients. It uses 740 boxes of envelopes annually at a steady rate. It costs $225 for delivery and processing (in addition to the printing costs) each time the company places an order, and it costs $4 per box for annual storage. To minimize costs, how many boxes of envelopes should the company order, and how often should they place an order?

■ Challenge Yourself

40. Financial transaction costs   Suppose a person has an amount A of spending money deposited to her savings account each month, where it earns interest at a monthly rate R. Assume she spends the entire amount throughout the month at a steady rate. When she makes cash withdrawals from the account, she incurs a transaction cost T (a combination of bank fees and the cost of her time). She would save money by making fewer withdrawals, but the more money she leaves in the account the more interest she earns. Suppose she makes n cash withdrawals for the same amount during the month. Then her average cash balance at any given time is $A/(2n)$. (Why?) Find the value of n that minimizes total costs (transaction costs and lost interest), and then show that the optimal average cash balance is $\sqrt{AT/(2R)}$.

41. In this exercise we prove that revenue is maximized when elasticity is 1.
   (a) Show that $R'(q) = q \cdot D'(q)[1 - E(q)]$.
   (b) Show that $R'(q) < 0$ when $0 < E(q) < 1$ and $R'(q) > 0$ when $E(q) > 1$.
   (c) Explain why we can conclude that $R(q)$ is maximized for $E(q) = 1$.

42. Consider the tangent line to a demand curve $p = D(q)$ at a point P. Let A be the point where the tangent line intersects the q-axis and B the point where it intersects the p-axis. Show that the elasticity at P is the ratio of the distance from P to A to the distance from P to B.

CHAPTER 4 REVIEW

■ CONCEPT CHECK

1. Explain the difference between an absolute maximum and a local maximum. Illustrate with a sketch.

2. (a) What does the Extreme Value Theorem say?
   (b) Explain how the Closed Interval Method works.

3. (a) Define a critical number of f.
   (b) What is the connection between critical numbers and local extreme values?

4. (a) State the Increasing/Decreasing Test.
   (b) What does it mean to say that f is concave upward on an interval?

(c) State the Concavity Test.
(d) What are inflection points? How do you find them?

5. (a) State the First Derivative Test.
   (b) State the Second Derivative Test.
   (c) What are the relative advantages and disadvantages of these tests?

6. Explain what each of the following means and illustrate with a sketch. What does the limit tell us about asymptotes of the graph?
   (a) $\lim\limits_{x \to 2^-} f(x) = \infty$ **(b)** $\lim\limits_{x \to \infty} f(x) = 5$

7. (a) Give an example of a familiar function whose graph has a vertical asymptote.

(b) Give an example of a familiar function whose graph has a horizontal asymptote. Can you think of one that has two different horizontal asymptotes?

8. If you have a graphing calculator or computer, why do you need calculus to graph a function?

9. (a) What are related rates problems? Describe a strategy for solving them.

(b) What are optimization problems? Describe a strategy for solving them.

10. What is the difference between average cost and marginal cost? How are they related when average cost is minimized?

11. What is a demand function? How do we use it to compute revenue?

12. (a) What is elasticity of demand? How do we compute it?

(b) What does it mean when elasticity of demand is equal to one? Greater than one? Less than one?

(c) How do we use elasticity of demand to determine maximum revenue?

Answers to the Concept Check can be found on the back endpapers.

■ Exercises

1–6 ■ Find the local and absolute extreme values of the function on the given interval.

1. $f(x) = x^3 - 6x^2 + 9x + 1$, $[2, 4]$

2. $f(x) = (x^2 + 2x)^3$, $[-2, 1]$

3. $g(t) = 4t^3 - 3t + 2$, $[-1, 1]$

4. $f(x) = x\sqrt{1 - x}$, $[-1, 1]$

5. $f(x) = \dfrac{3x - 4}{x^2 + 1}$, $[-2, 2]$

6. $f(x) = (\ln x)/x^2$, $[1, 3]$

---

7. Use a graphing calculator to estimate the absolute maximum and absolute minimum values, accurate to two decimal places, of $f(x) = \dfrac{x^2 + 1}{e^{\sqrt{x^2+1}}}$ for $0 \le x \le 4$.

8. Use the given graph of f to estimate the intervals of increase or decrease, local extreme points, intervals of concavity, inflection points, and $\lim_{x\to\infty} f(x)$.

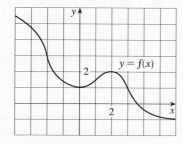

9–14 ■

(a) Find the intervals of increase or decrease.

(b) Find the local maximum and minimum values.

(c) Find the intervals of concavity and the inflection points.

9. $N(t) = t^3 - 3t^2 - 24t + 5$

10. $L = 4.3s^3 - 1.9s^2 - 8.8s + 13.1$

11. $f(x) = x + \sqrt{1 - x}$

12. $B(r) = \dfrac{r^2}{r^2 + 2}$

13. $y = xe^{4x}$

14. $y = t(1 - \ln t)$

---

15. The graph of the *derivative* f' of a function f is given.

(a) On what intervals is f increasing or decreasing?

(b) At what values of x does f have a local maximum or minimum?

(c) Where is f concave upward or downward?

(d) At what values of x does f have an inflection point?

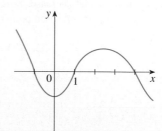

16. If the curve shown in Exercise 15 is a graph of the *second derivative* f'' of a function f, where is f concave upward? Downward? At what values of x does f have an inflection point?

17–22 ■ Find the limit.

17. $\lim_{x\to2^-} \dfrac{2x + 6}{x - 2}$

18. $\lim_{a\to-1} \dfrac{e^{2a}}{(a + 1)^2}$

19. $\lim_{x\to\infty} \dfrac{8x^2 - 5x}{2x^2 + x + 1}$

20. $\lim_{x\to-\infty} \dfrac{1 - 2x^2 - x^4}{5 + x - 3x^4}$

21. $\lim_{u\to\infty} \ln(2^u + u)$

22. $\lim_{x\to\infty} e^{x - x^2}$

23–28 ▪
(a) Find the vertical and horizontal asymptotes, if any.
(b) Find the intervals of increase or decrease.
(c) Find the local maximum and minimum values.
(d) Find the intervals of concavity and the inflection points.
(e) Use the information from parts (a)–(d) to sketch the graph of f. Check your work with a graphing device.

23. $f(x) = 2 - 2x - x^3$ **24.** $f(x) = x^4 + 4x^3$

25. $H(w) = \dfrac{3w}{2w - 4}$ **26.** $f(x) = \dfrac{1}{1 - x^2}$

27. $y = \ln(x^2 + 4)$ **28.** $y = e^{2x - x^2}$

29–30 ▪ Sketch the graph of a function that satisfies all of the given conditions.

29. $f'(3) = 0$, $f'(x) > 0$ when $x < 3$,
$f'(x) < 0$ when $x > 3$, $f''(-1) = 0$,
$f''(x) > 0$ when $x < -1$, $f''(x) < 0$ when $x > -1$,
$\lim_{x \to -\infty} f(x) = 2$, $\lim_{x \to \infty} f(x) = -\infty$

30. $f'(4) = 0$, $f'(x) > 0$ when $x < 1$ or $x > 4$,
$f'(x) < 0$ when $1 < x < 4$, $f''(x) > 0$ when $x \neq 1$,
$\lim_{x \to 1} f(x) = \infty$

31–34 ▪ Produce graphs of f that reveal all the important aspects of the curve. Use graphs of f' and f'' to estimate the intervals of increase and decrease, extreme values, intervals of concavity, and inflection points.

31. $f(x) = \dfrac{x^2 - 1}{x^3}$ **32.** $g(v) = 2^v - v^3$

33. $f(x) = 3x^6 - 5x^5 + x^4 - 5x^3 - 2x^2 + 2$

34. $f(x) = \dfrac{x^3 - x}{x^2 + x + 3}$

35. Graph $f(x) = e^{-1/x^2}$ in a viewing rectangle that shows all the main aspects of this function. Estimate the inflection points. Then use calculus to find them exactly.

36. (a) Graph the function $f(x) = 1/(1 + e^{1/x})$.
(b) Explain the shape of the graph by computing the limits of $f(x)$ as x approaches ∞, $-\infty$, 0^+, and 0^-.
(c) Use a graph of f' to estimate the coordinates of the inflection points.

37. Investigate the family of curves given by
$$f(x) = x^4 + x^3 + cx^2$$
In particular you should determine the transitional value of c at which the number of critical numbers changes and the transitional value at which the number of inflection points changes. Illustrate the various possible shapes with graphs.

38. Investigate the family of functions $f(x) = cxe^{-cx^2}$. What happens to the maximum and minimum points and the inflection points as c changes? Illustrate your conclusions by graphing several members of the family.

39. Prison population   The number of female prisoners in the state of Iowa for the years 1991–2008 can be modeled by
$$P(t) = \dfrac{825}{1 + 2.97e^{-0.224t}}$$
where t is the number of years after December 31, 1990. Use a graph of P' to estimate when, according to this model, the female prisoner population in Iowa was increasing most rapidly.

40. Drug concentration   Suppose the concentration (measured in mg/mL) of a drug evident in a patient's bloodstream t minutes after the drug is administered is given by
$$f(t) = 2.3te^{-0.14t}$$
When is the concentration highest?

41. Wave velocity   The velocity of a wave of length L in deep water is
$$v = K\sqrt{\dfrac{L}{C} + \dfrac{C}{L}}$$
where K and C are known positive constants. What is the length of the wave that gives the minimum velocity?

42. Assume that r and s are positive functions of t. If $r^2 + s^4 = 25$ and $dr/dt = -3$, find ds/dt when $s = 2$.

43. Consumer demand   The monthly demand q, measured in thousands, for a manufacturer's recordable DVD discs is related to the price per disc, in dollars, by the equation $p = 1.3e^{-0.04q}$. Currently the manufacturer is producing 30,000 discs per month. Find the rate at which demand will change with respect to time if the price increases at a rate of $0.05 per month.

44. Water level   A paper cup has the shape of a cone with height 10 cm and radius 3 cm (at the top). If water is poured into the cup at a rate of 2 cm^3/s, how fast is the water level rising when the water is 5 cm deep?

45. Rising balloon   A balloon is rising at a constant speed of 5 ft/s. A boy is cycling along a straight road at a speed of 15 ft/s. When he passes under the balloon, it is 45 ft above him. How fast is the distance between the boy and the balloon increasing 3 seconds later?

46. Ski ramp   A waterskier skis over the ramp shown in the figure at a speed of 30 ft/s. How fast is she rising as she leaves the ramp?

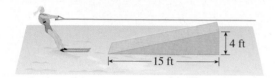

47. Find two positive integers such that the sum of the first number and four times the second number is 1000 and the product of the numbers is as large as possible.

48. Find the point on the curve $y = 8/x$ that is closest to the point $(3, 0)$.

49. Fenced area   The owner of a ranch wants to build a fence to enclose a rectangular plot of land. A large building will border one side, so only three sides need fencing, but the owner wants to construct an extra fence down the middle of the area, perpendicular to the building, to split the plot into two pens. If the owner has enough materials to build 2100 feet of fence, what dimensions should he use to enclose the largest area?

50. Container design   A confectioner wants to sell its new candies in a clear plastic circular cylinder with metal top and bottom caps. The plastic material costs $0.40 per square foot and the metal costs $1.20 per square foot. If the packaging must have a volume of 0.05 ft^3, what dimensions should be used to minimize the cost of materials?

51. Bicycle manufacturing   A manufacturer of bicycle helmets estimates that the monthly cost to produce q helmets is

$$C(q) = 6200 + 7.3q + 0.002q^2$$

dollars and the expected revenue, in dollars, is

$$R(q) = 31q - 0.003q^2$$

(a) What are the cost, marginal cost, average cost, and profit if 2000 helmets are produced monthly?

(b) How many helmets should be produced each month in order to minimize the average cost of each helmet?

(c) Find the number of helmets that should be produced each month in order to maximize profit.

52. Cost, demand, and profit   A manufacturer determines that the cost of making q units of a commodity is

$$C(q) = 1800 + 25q - 0.2q^2 + 0.001q^3$$

and the demand function is

$$p = D(q) = 48.2 - 0.03q$$

(a) Graph the cost and revenue functions and use the graphs to estimate the production level for maximum profit.

(b) Use calculus to find the production level for maximum profit.

(c) Estimate the production level that minimizes the average cost.

53. Game attendance   A hockey team plays in an arena with a seating capacity of 15,000 spectators. With the ticket price set at $12, average attendance at a game has been 11,000. A market survey indicates that for each dollar the ticket price is lowered, average attendance will increase by 1000. How should the owners of the team set the ticket price to maximize their revenue from ticket sales?

54. Elasticity of demand   A satellite television company is airing a pay-per-view boxing match and is trying to decide how much to charge for the event. The marketing department estimates that the demand function for the program is $p = 62e^{-0.04q}$ where q is the number of viewers, in thousands, who will pay p dollars for the event.

(a) If the company charges $29.95 for the boxing match, how many viewers will purchase it?

(b) Compute the elasticity of demand for the price $29.95. Is demand elastic or inelastic at this price? Would lowering the price increase revenue?

(c) Compute the elasticity of demand for the price $19.95. Is demand elastic or inelastic at this price? Would lowering the price increase revenue?

(d) Determine the price for the event that will bring in the most revenue.

55. Elasticity and revenue   The demand function for a product is $D(q) = 308 - \frac{1}{24}q$.

(a) Write a formula for the elasticity of demand E of the product as a function of q.

(b) For what prices is $E(q) < 1$? When is $E(q) > 1$?

(c) Use your formula from part (a) to determine the price that will produce maximum revenue.

(d) Write a revenue function R as a function of q and use R' to find the maximum revenue. Does your result agree with your answer from part (c)?

56. Inventory costs   A news organization has its daily newspaper printed in broadsheet format. The printer purchases large rolls of paper (1800 lb) and uses the paper at a steady rate. It costs $34 annually to store each roll of paper and $650 each time the printer receives a shipment. If 235 rolls of paper are used each year, how often should the printer order paper and how many rolls should each order contain in order to minimize costs?

© Lester Lefkowitz / CORBIS

5

In Example 3 in Section 5.3 you will see how to use power consumption data and an integral to compute the amount of energy used in one day in San Francisco.   © Nathan Jaskowiak / Shutterstock

Integrals

In earlier chapters we used the concepts of slope and tangent lines to develop the derivative, the central idea of differential calculus. In particular, we used derivatives to get a marginal cost function from a total cost function. In this chapter, we start with marginal cost and try to recover information about the total cost. We find that this information is linked to the area enclosed by a graph, which leads us to the idea of a definite integral, the basic concept of integral calculus. We will see that integrals can be used to solve a wide variety of problems such as determining the distance traveled from a velocity function or finding the amount of a drug administered by looking at the rate of dosage.

There is a connection between integral calculus and differential calculus. The Fundamental Theorem of Calculus relates the integral to the derivative, and we will see in this chapter that it greatly simplifies the solution of many problems.

5.1 Cost, Area, and the Definite Integral

Consider a good produced by a company. In Chapter 3 we were able to use information about the total production cost to determine the marginal cost, that is, the rate at which cost changes with respect to the number of units produced. Now we look at the reverse question: If the marginal cost is known at any production level, can we find the total cost to produce a certain number of goods?

▪ Computing Total Cost

If the marginal cost of a good remains constant, then computing the total production cost is easy: Making 500 units of a good that costs \$10 per unit to make will cost a total of $10 \times 500 = \$5000$. But if the marginal cost varies, the task of finding the total cost is not as straightforward. We investigate the problem in the following example.

▪ EXAMPLE 1 Estimating a Total Cost

Suppose a soft drink manufacturer estimates that the marginal cost for its cola at different production levels is as given in the following table. Estimate the cost, beyond the fixed costs, to produce the first 600 cases.

Production level (cases)	0	100	200	300	400	500	600
Marginal cost (\$/case)	8.00	6.23	4.92	4.07	3.68	3.75	4.28

SOLUTION

The marginal cost varies, so we estimate the total cost in batches. We can estimate the cost of producing the first 100 cases of cola by taking the initial marginal cost of \$8.00 as the marginal cost for the entire batch:

$$(8.00 \text{ dollars/case}) \times 100 \text{ cases} = \$800$$

After producing 100 units the marginal cost is 6.23, so we estimate the total cost of producing the next batch of 100 units as

(1) $(6.23 \text{ dollars/case}) \times 100 \text{ cases} = \$623$

We can estimate the total cost for producing 600 cases by adding similar estimates for the remaining 100-unit batches:

$$(8.00 \times 100) + (6.23 \times 100) + (4.92 \times 100)$$
$$+ (4.07 \times 100) + (3.68 \times 100) + (3.75 \times 100) = \$3065$$

We could just as well have used the ending marginal cost of each batch to make our cost estimates rather than the initial marginal cost. Then our estimate becomes

$$(6.23 \times 100) + (4.92 \times 100) + (4.07 \times 100)$$
$$+ (3.68 \times 100) + (3.75 \times 100) + (4.28 \times 100) = \$2693$$

It is likely that the actual cost is somewhere between these two estimates. ▪

In Example 1, if we had wanted a more accurate estimate, we could have estimated the marginal cost at every 20 cases, or even every case, and used smaller batches. Because a smaller number of units results in a smaller variance in marginal cost for each batch, it seems plausible that the smaller the batch size we use, the closer our estimate is to the actual cost.

In general, suppose the marginal cost $M(q)$ of a good is known at production levels $q = a$ through $q = b$ ($a < b$). We can split the total number of goods ($b - a$) into n batches of equal size Δq and determine the marginal cost at some production level within each batch. For convenience let's call the starting production level q_0 (so $q_0 = a$); we'll say the first batch ends at q_1, and so on. (The last batch ends at q_n, where $q_n = b$.)

If we use a large number of batches, then each batch consists of a small number of units, so the marginal cost should be relatively constant throughout each batch. For the first batch we could say the marginal cost is approximately $M(q_1)$, the ending marginal cost for that batch, and so the cost is $M(q_1) \cdot \Delta q$ (cost per unit times number of units). Similarly, the cost for the second batch is about $M(q_2) \cdot \Delta q$, and the total cost to increase production from $q = a$ to $q = b$ is

We could have used the initial marginal cost for each batch in (2) or the marginal cost at some intermediate production level. Although these choices typically give slightly different estimates for the cost, all such computations get closer to the actual cost as n becomes large, and thus the cost computation in Equation 3 gives the same result regardless of where we choose to measure the marginal cost in each batch.

(2) $$M(q_1)\,\Delta q + M(q_2)\,\Delta q + \cdots + M(q_n)\,\Delta q$$

The more batches we use (each consisting of fewer units), the more accurate our estimates become, so it seems plausible that the *exact* cost is the *limit* of such expressions:

(3) $$\text{cost} = \lim_{n\to\infty}\left[M(q_1)\,\Delta q + M(q_2)\,\Delta q + \cdots + M(q_n)\,\Delta q\right]$$

We will see in Section 5.2 that this is indeed true.

■ Interpreting Cost Graphically

In Figure 1, we have drawn a possible graph of the marginal cost function M from Example 1. We can visualize the cost calculation we made by drawing rectangles whose widths are the batch size and whose heights are the initial marginal cost of the batch. For instance, the second rectangle in Figure 1 is drawn from $q = 100$ to $q = 200$ with a height of $M(100) = 6.23$. The area of the rectangle is $6.23 \times 100 = 623$, our estimate in (1) for the cost of raising production from 100 to 200 cases. In fact, the area of each rectangle can be interpreted as a cost because the height represents the cost per unit and the width represents a number of units.

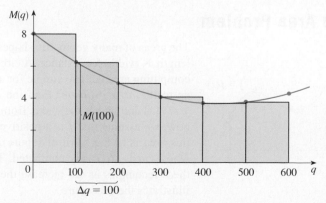

FIGURE 1

The area of the second rectangle is $M(100) \times 100$, the approximate cost for $100 \le q \le 200$.

The combined area of these six rectangles is 3065, which is our initial estimate for the total cost in Example 1.

Because we used the initial marginal cost of each batch to estimate the total cost, the height of each rectangle in Figure 1 is determined by the marginal cost curve at the left edge of the rectangle. Figure 2 shows the situation if we use the marginal cost at the end of each batch for our calculations. Notice that the height of each rectangle is determined by the marginal cost curve at the right edge. The combined areas of these rectangles is 2693, as we calculated when making the second cost estimate in Example 1.

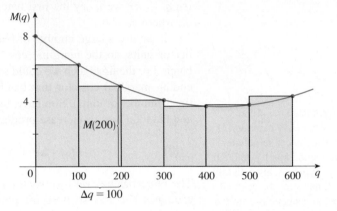

FIGURE 2

According to Equation 3, we get the exact cost in Example 1 by increasing the number of batches n and determining the limit as $n \to \infty$. But as n increases, the number of corresponding rectangles in Figure 2 increases while the widths decrease, resulting in a combined area that more closely matches the area of the region beneath the marginal cost curve. This suggests that there is a link between the exact cost when production levels are raised and what we intuitively think of as the area underneath the graph of marginal cost. We will see that this is true.

It turns out that the area under a curve represents the answers to a wide variety of questions. For example, if we draw rectangles similar to those in Figures 1 or 2 under a velocity curve, then the height of each rectangle represents a velocity and the width is a period of time. Thus the area of each rectangle can be interpreted as velocity times time, which gives a distance. This suggests that the area under the graph of a velocity function is the total distance traveled.

But how do we compute the precise area under a curve?

■ The Area Problem

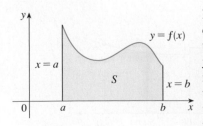

FIGURE 3

The areas of many geometric shapes are easily computed. The area of a rectangle is length × width, for instance. A circle's area is π times the square of the radius. But computing area isn't so simple for a region like that shown in Figure 3. This brings us to the *area problem*: Find the area of the region S that lies under the curve $y = f(x)$ and above the x-axis from $x = a$ to $x = b$, as illustrated in Figure 3. (For now, we assume that f is a positive continuous function.) The strategy to compute this area is in the spirit of Figure 2: We approximate the region S by rectangles, whose areas are easily computed. Then we take the limit of the combined area of these rectangles as we increase the number of rectangles. The following example illustrates the procedure.

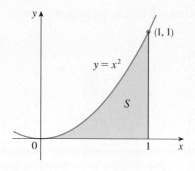

FIGURE 4

■ EXAMPLE 2 Estimating an Area

Use rectangles to estimate the area under the parabola $y = x^2$ from 0 to 1 (the region S illustrated in Figure 4).

SOLUTION

We first notice that the area of S must be somewhere between 0 and 1 because S is contained in a square with side length 1, but we can certainly do better than that. Suppose we divide S into four strips S_1, S_2, S_3, and S_4 by drawing the vertical lines $x = \frac{1}{4}$, $x = \frac{1}{2}$, and $x = \frac{3}{4}$ as in Figure 5(a).

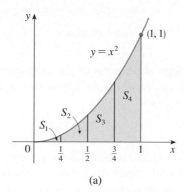

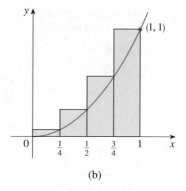

FIGURE 5   (a)   (b)

We can approximate each strip by a rectangle whose base is the same as the strip and whose height is the same as the right edge of the strip [see Figure 5(b)]. In other words, the heights of these rectangles are the values of the function $f(x) = x^2$ at the right endpoints of the subintervals $\left[0, \frac{1}{4}\right]$, $\left[\frac{1}{4}, \frac{1}{2}\right]$, $\left[\frac{1}{2}, \frac{3}{4}\right]$, and $\left[\frac{3}{4}, 1\right]$.

The notation R_4 indicates that we are using the right endpoint of each subinterval to compute the height of the rectangle and that we are using 4 subintervals. Similarly, L_4 indicates that we are using the left endpoint of each subinterval.

Each rectangle has width $\frac{1}{4}$ and the heights are $\left(\frac{1}{4}\right)^2$, $\left(\frac{1}{2}\right)^2$, $\left(\frac{3}{4}\right)^2$, and 1^2. If we let R_4 be the sum of the areas of these approximating rectangles, we get

$$R_4 = \left(\tfrac{1}{4}\right)^2 \cdot \tfrac{1}{4} + \left(\tfrac{1}{2}\right)^2 \cdot \tfrac{1}{4} + \left(\tfrac{3}{4}\right)^2 \cdot \tfrac{1}{4} + 1^2 \cdot \tfrac{1}{4} = \tfrac{15}{32} = 0.46875$$

Instead of using the rectangles in Figure 5(b) we could use the smaller rectangles in Figure 6 whose heights are the values of f at the *left* endpoints of the subintervals. (The leftmost rectangle has collapsed because its height is 0.) The sum of the areas of these approximating rectangles is

$$L_4 = 0^2 \cdot \tfrac{1}{4} + \left(\tfrac{1}{4}\right)^2 \cdot \tfrac{1}{4} + \left(\tfrac{1}{2}\right)^2 \cdot \tfrac{1}{4} + \left(\tfrac{3}{4}\right)^2 \cdot \tfrac{1}{4} = \tfrac{7}{32} = 0.21875$$

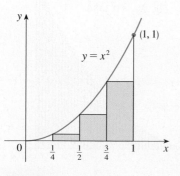

FIGURE 6

We see from Figures 5(b) and 6 that the area of S is smaller than R_4 but larger than L_4, so we can safely say that the area A lies between these values:

$$0.21875 < A < 0.46875$$

To get more accurate estimates, we can repeat this procedure with a larger number of strips. Figure 7 shows what happens when we divide the region S into eight strips of equal width.

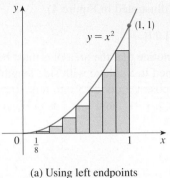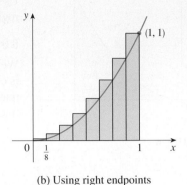

FIGURE 7

Approximating S with eight rectangles

(a) Using left endpoints   (b) Using right endpoints

The sum of the areas of the rectangles in Figure 7(a) is $L_8 = 0.2734375$ while the sum from Figure 7(b) is $R_8 = 0.3984375$, so we can refine our estimate for A:

$$0.2734375 < A < 0.3984375$$

n	L_n	R_n
10	0.2850000	0.3850000
20	0.3087500	0.3587500
30	0.3168519	0.3501852
50	0.3234000	0.3434000
100	0.3283500	0.3383500
1000	0.3328335	0.3338335

We could obtain better estimates by increasing the number of strips. The table at the left shows the results of similar calculations (with a computer) using n rectangles whose heights are found with left endpoints (L_n) or right endpoints (R_n). In particular, we see by using 50 strips that the area lies between 0.3234 and 0.3434. With 1000 strips we narrow it down even more: the area lies between 0.3328335 and 0.3338335. A good estimate is obtained by averaging these numbers:

$$\text{area} \approx 0.3333335$$

Figure 8 shows how the rectangles become a better and better approximation to the region S as n increases and so R_n (and L_n) gets closer and closer to the true area A.

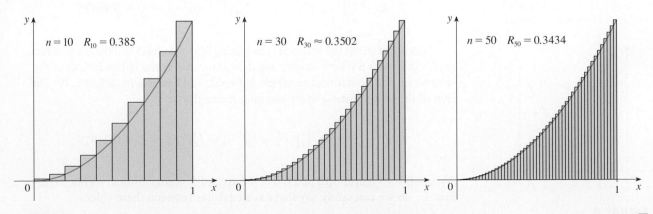

FIGURE 8

TEC In Visual 5.1A you can create pictures like those in Figure 8 for other values of n.

Let's apply the idea of Example 2 to the more general region S of Figure 3. We start by subdividing S into n strips $S_1, S_2, \ldots, S_n$ of equal width as in Figure 9.

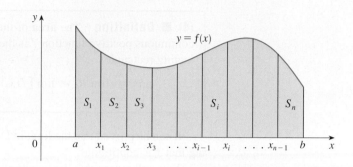

FIGURE 9

The width of the interval $[a, b]$ is $b - a$, so the width of each of the n strips is

$$\Delta x = \frac{b - a}{n}$$

These strips divide the interval $[a, b]$ into n subintervals

$$[x_0, x_1], \qquad [x_1, x_2], \qquad [x_2, x_3], \qquad \ldots, \qquad [x_{n-1}, x_n]$$

where $x_0 = a$ and $x_n = b$.

Let's approximate the ith strip S_i by a rectangle with width Δx and height $f(x_i)$, which is the value of f at the right endpoint (see Figure 10). Then the area of the ith rectangle is $f(x_i) \Delta x$. What we think of intuitively as the area of S is approximated by the sum of the areas of these rectangles, which is

$$R_n = f(x_1) \Delta x + f(x_2) \Delta x + \cdots + f(x_n) \Delta x$$

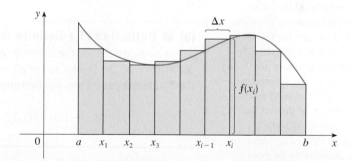

FIGURE 10

Figure 11 shows this approximation for $n = 2$, 4, 8, and 12. Notice that this approximation appears to become better and better as the number of strips increases, that is, as $n \to \infty$. Therefore we define the area of the region S in the following way.

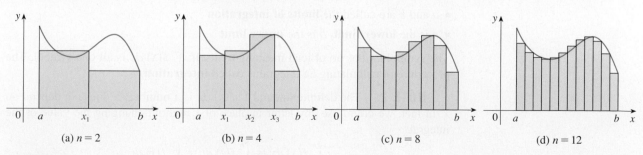

(a) $n = 2$ (b) $n = 4$ (c) $n = 8$ (d) $n = 12$

FIGURE 11

> **(4) ▪ Definition**   The **area** of the region S that lies under the graph of the continuous positive function f is the limit of the sum of the areas of approximating rectangles:
>
> $$\text{area} = \lim_{n \to \infty} R_n = \lim_{n \to \infty} \left[f(x_1)\,\Delta x + f(x_2)\,\Delta x + \cdots + f(x_n)\,\Delta x \right]$$

It can be proved that the limit in Definition 4 always exists, since we are assuming that f is continuous.

You will notice that the height of each rectangle in Figure 11 is determined by the function value at the right endpoint of each subinterval. We could have just as easily used the left endpoint of each subinterval, or a point within each subinterval. Although changing the points we use in each interval produces slightly different area estimates for a finite number of rectangles, the value of the limit as $n \to \infty$ is always the same, regardless of which points we use.

▪ The Definite Integral

Comparing Equation 3 with Definition 4, we see that limits of the form

$$\lim_{n \to \infty} \left[f(x_1)\,\Delta x + f(x_2)\,\Delta x + \cdots + f(x_n)\,\Delta x \right]$$

arise when we compute an area or a production cost. It turns out that this same type of limit occurs in a wide variety of situations even when f is not necessarily a positive function. We therefore give this type of limit a special name and notation.

Although we have defined the definite integral by using the right endpoint of each subinterval, we could also define it using the left endpoint, the midpoint, or any other point within the subinterval. Because we have assumed that f is continuous, it can be proved that the limit in Definition 5 always exists and gives the same value no matter how we choose the points from each subinterval.

> **(5) ▪ Definition of a Definite Integral**   If f is a continuous function for $a \le x \le b$, we divide the interval $[a, b]$ into n subintervals of equal width $\Delta x = (b - a)/n$. We let $x_0\ (= a)$, $x_1, x_2, \ldots, x_n\ (= b)$ be the endpoints of these subintervals. Then the **definite integral of f from a to b** is
>
> $$\int_a^b f(x)\,dx = \lim_{n \to \infty} \left[f(x_1)\,\Delta x + f(x_2)\,\Delta x + \cdots + f(x_n)\,\Delta x \right]$$

NOTE 1:   The symbol $\int$ was introduced by Leibniz and is called an **integral sign**. It is an elongated S and was chosen because an integral is a limit of sums.

In the notation $\int_a^b f(x)\,dx$,

- $f(x)$ is called the **integrand**
- a and b are called the **limits of integration**
- a is the **lower limit**, b is the **upper limit**

The symbol dx has no official meaning by itself; $\int_a^b f(x)\,dx$ is all one symbol. The procedure of calculating an integral is called **integration**.

NOTE 2:   The definite integral $\int_a^b f(x)\,dx$ is a number; it does not depend on x. In fact, we could use any letter in place of x without changing the value of the integral:

$$\int_a^b f(x)\,dx = \int_a^b f(t)\,dt = \int_a^b f(r)\,dr$$

NOTE 3: The sum

$$f(x_1)\,\Delta x + f(x_2)\,\Delta x + \cdots + f(x_n)\,\Delta x$$

that occurs in Definition 5 is an example of a **Riemann sum**, named after the German mathematician Bernhard Riemann (1826–1866). Note that the function f can be evaluated at any point within each subinterval in a Riemann sum, although we often choose the right endpoint for convenience.

Interpreting Sums and Integrals

We know that if f happens to be positive, then the Riemann sum can be interpreted as a sum of areas of approximating rectangles (see Figure 10). By comparing Definitions 4 and 5, we see that the definite integral $\int_a^b f(x)\,dx$ can be interpreted as the area under the curve $y = f(x)$ from a to b, as in Figure 3.

If f takes on both positive and negative values, as in Figure 12, then the Riemann sum is the sum of the areas of the rectangles that lie above the x-axis and the *negatives* of the areas of the rectangles that lie below the x-axis (the areas of the blue rectangles *minus* the areas of the gold rectangles). When we take the limit of such Riemann sums, we get the situation illustrated in Figure 13. A definite integral can be interpreted as a **net area**, that is, a difference of areas.

Riemann

Bernhard Riemann received his Ph.D. under the direction of the legendary Gauss at the University of Göttingen and remained there to teach. Gauss, who was not in the habit of praising other mathematicians, spoke of Riemann's "creative, active, truly mathematical mind and gloriously fertile originality." The definition (5) of an integral that we use is due to Riemann. He also made major contributions to the theory of functions of a complex variable, mathematical physics, number theory, and the foundations of geometry. Riemann's broad concept of space and geometry turned out to be the right setting, 50 years later, for Einstein's general relativity theory. Riemann's health was poor throughout his life, and he died of tuberculosis at the age of 39.

▪ Definite Integral as a Net Area

$$\int_a^b f(x)\,dx = A_1 - A_2$$

where A_1 is the area of the region above the x-axis and below the graph of f, and A_2 is the area of the region below the x-axis and above the graph of f.

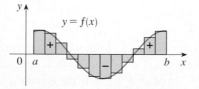

FIGURE 12
$f(x_1)\,\Delta x + f(x_2)\,\Delta x + \cdots + f(x_n)\,\Delta x$
is an approximation to the net area.

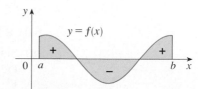

FIGURE 13
$\int_a^b f(x)\,dx$ is the net area.

▪ EXAMPLE 3 Examining Net Area

A function f is graphed in Figure 14. Is the value of $\int_1^6 f(x)\,dx$ positive, negative, or zero? What about $\int_0^4 f(x)\,dx$?

SOLUTION

The integral $\int_1^6 f(x)\,dx$ represents the net area between the graph of $y = f(x)$ and the x-axis for $1 \le x \le 6$. Because the curve encloses more area below the x-axis

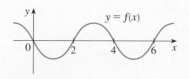

FIGURE 14

than above the x-axis for $1 \le x \le 6$ (see Figure 15), the net area is negative and $\int_1^6 f(x)\, dx < 0$. For $0 \le x \le 4$ however, notice from Figure 16 that the region below the x-axis and above the curve appears to have the same area as the region above the x-axis and below the curve. So we estimate that $\int_0^4 f(x)\, dx = 0$.

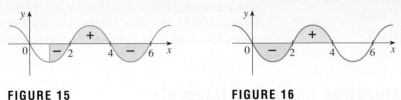

FIGURE 15   **FIGURE 16**   ▪

NOTE 4:   Although we have defined $\int_a^b f(x)\, dx$ by dividing $[a, b]$ into subintervals of equal width, there are situations in which it is advantageous to work with subintervals of unequal width. For instance, in Exercise 6 we give velocity data provided by NASA that are not equally spaced, and we ask you to estimate the distance traveled. And there are methods for estimating definite integrals that take advantage of unequal subintervals.

▪ Evaluating Integrals

Evaluating a definite integral using Definition 5 is usually no easy task. (See Exercises 37–38 if you are feeling up to the challenge.) We will learn a much easier method for the evaluation of integrals in the next section. In the meantime, the connection between the definite integral of a function and the net area enclosed by its graph allows us to use geometry to evaluate certain integrals.

▪ E X A M P L E 4   Using Geometry to Evaluate Integrals

Evaluate the following integrals by interpreting each in terms of areas.

(a) $\displaystyle\int_0^1 \sqrt{1 - x^2}\, dx$ **(b)** $\displaystyle\int_0^3 (x - 1)\, dx$

SOLUTION

(a) Since $f(x) = \sqrt{1 - x^2} \ge 0$, we can interpret this integral as the area under the curve $y = \sqrt{1 - x^2}$ from 0 to 1. But, squaring both sides, we get $y^2 = 1 - x^2$ or $x^2 + y^2 = 1$, the equation of a circle of radius 1 with center the origin. The graph of f is the quarter circle in the first quadrant as shown in Figure 17. The well-known formula for the area of a circle is πr^2, so

$$\int_0^1 \sqrt{1 - x^2}\, dx = \tfrac{1}{4}\pi(1)^2 = \frac{\pi}{4} \approx 0.7854$$

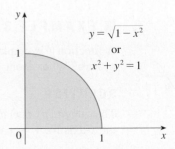

$$y = \sqrt{1 - x^2}$$
or
$$x^2 + y^2 = 1$$

FIGURE 17

(b) The graph of $y = x - 1$ is the line with slope 1 shown in Figure 18. We compute the integral as the difference of the areas of the two triangles (recall that the area of a triangle is $\frac{1}{2}$ base $\times$ height):

$$\int_0^3 (x - 1)\, dx = A_1 - A_2 = \tfrac{1}{2}(2 \cdot 2) - \tfrac{1}{2}(1 \cdot 1) = 1.5$$

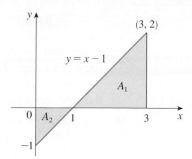

FIGURE 18

▪ The Midpoint Rule

If a definite integral can't be evaluated directly or geometrically, we can still *estimate* its value. Because a definite integral is defined in (5) as the limit of Riemann sums, a Riemann sum can be used as an approximation to a definite integral. We used the right endpoint of each subinterval in Definition 5 for convenience, but we could have used any point from each subinterval. In practice, we usually get better estimates by using the midpoint of each subinterval in the computation. (We can improve a Riemann sum approximation by increasing n, but midpoints usually give more accurate results than left or right endpoints for the same number of subintervals.) Using midpoints gives the following approximation.

TEC Module 5.1 shows how the Midpoint Rule estimates improve as n increases.

▪ **Midpoint Rule**

$$\int_a^b f(x)\, dx \approx \Delta x \left[f(\bar{x}_1) + \cdots + f(\bar{x}_n) \right]$$

where

$$\Delta x = \frac{b - a}{n}$$

and

$$\bar{x}_i = \text{midpoint of } [x_{i-1}, x_i] = \tfrac{1}{2}(x_{i-1} + x_i)$$

▪ **EXAMPLE 5   Estimating an Integral with Riemann Sums**

Approximate $\int_0^3 (x^3 - 6x)\, dx$ by a Riemann sum with 6 subintervals using **(a)** right endpoints and **(b)** midpoints.

SOLUTION

With $n = 6$ subintervals the interval width is

$$\Delta x = \frac{b - a}{n} = \frac{3 - 0}{6} = \frac{1}{2}$$

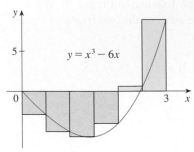

FIGURE 19

(a) The right endpoints are $x_1 = 0.5$, $x_2 = 1.0$, $x_3 = 1.5$, $x_4 = 2.0$, $x_5 = 2.5$, and $x_6 = 3.0$. So the Riemann sum is

$$R_6 = f(0.5)\Delta x + f(1.0)\Delta x + f(1.5)\Delta x + f(2.0)\Delta x + f(2.5)\Delta x + f(3.0)\Delta x$$

$$= \tfrac{1}{2}(-2.875 - 5 - 5.625 - 4 + 0.625 + 9)$$

$$= -3.9375$$

Notice that f is not a positive function and so the Riemann sum does not represent a sum of areas of rectangles. But it does represent the sum of the areas of the blue rectangles (above the x-axis) minus the sum of the areas of the gold rectangles (below the x-axis) in Figure 19.

(b) The midpoints of the subintervals are $\bar{x}_1 = 0.25$, $\bar{x}_2 = 0.75$, $\bar{x}_3 = 1.25$, $\bar{x}_4 = 1.75$, $\bar{x}_5 = 2.25$, and $\bar{x}_6 = 2.75$. (See Figure 20.)

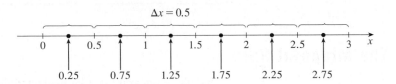

FIGURE 20

The Midpoint Rule gives

$$M_6 = f(0.25)\Delta x + f(0.75)\Delta x + f(1.25)\Delta x + f(1.75)\Delta x + f(2.25)\Delta x + f(2.75)\Delta x$$

$$= \tfrac{1}{2}(-1.484375 - 4.078125 - 5.546875 - 5.140625 - 2.109375 + 4.296875)$$

$$= -7.03125$$

Figure 21 shows the rectangles corresponding to the Midpoint Rule estimate. ▪

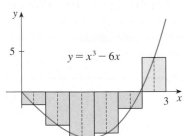

FIGURE 21

It turns out that the true value of the integral in Example 5 is -6.75, as we will see in Section 5.2. So here the approximation using midpoints is much more accurate than the one using right endpoints. For increased accuracy we could use a larger number of subintervals. Here if we apply the Midpoint Rule with 40 subintervals, we get the picture in Figure 22. The approximation $M_{40} \approx -6.7563$ is very close to the true value -6.75.

TEC In Visual 5.1B you can compare left, right, and midpoint approximations to the integral in Example 5 for different values of n.

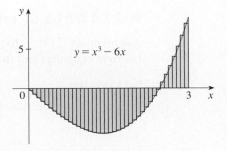

FIGURE 22
$M_{40} \approx -6.7563$

▪ Exercises 5.1

1. Total cost   The table shows the marginal cost for a company's product at various production levels.

Units	Marginal cost ($/unit)	Units	Marginal cost ($/unit)
0	3.22	7000	2.14
1000	2.89	8000	2.13
2000	2.76	9000	2.15
3000	2.71	10,000	2.19
4000	2.55	11,000	2.33
5000	2.40	12,000	2.61
6000	2.26	13,000	2.83

(a) Estimate the cost (not including fixed costs) to produce the first 12,000 units by considering batches of 2000 units. Use the initial marginal cost for each batch in your calculations.

(b) Use batches of 1000 units and the ending marginal cost for each batch to estimate the cost of raising the production level from 4000 to 11,000 units.

2. Total cost   A manufacturer estimates that the marginal cost for its latest product is

$$M(q) = 11.2 - 0.03q + 0.000032q^2$$

(a) Estimate the cost (beyond fixed costs) the manufacturer incurs if the production level is raised from 300 to 800 units using 100-unit batches and the initial marginal cost of each batch.

(b) Improve your estimate for the cost in part (a) by using 50-unit batches.

3. Running distance   The speed of a runner increased steadily during the first three seconds of a race. Her speed at half-second intervals is given in the table. Estimate the distance that she traveled during these three seconds by using the velocity at the end of each half-second time interval as the velocity during that interval.

t (s)	0	0.5	1.0	1.5	2.0	2.5	3.0
v (ft/s)	0	6.2	10.8	14.9	18.1	19.4	20.2

4. Cycling distance   Speedometer readings for a motorcycle at 12-second intervals are given in the table.

t (s)	0	12	24	36	48	60
v (ft/s)	30	28	25	22	24	27

(a) Estimate the distance traveled by the motorcycle during this time period using the velocities at the beginning of the time intervals.

(b) Give another estimate using the velocities at the end of the time periods.

(c) Are you able to judge whether your estimates in parts (a) and (b) are too high or too low? Explain.

5. Car racing distance   The table shows speedometer readings at 10-second intervals during a 1-minute period for a car racing at the Daytona International Speedway in Florida.

Time (s)	Velocity (mi/h)
0	182.9
10	168.0
20	106.6
30	99.8
40	124.5
50	176.1
60	175.6

(a) Estimate the distance the race car traveled during the 1-minute period using the velocities at the end of the time intervals. Illustrate your estimate by drawing a possible velocity curve and rectangles corresponding to your calculations.
[*Hint*: 1 mi/h = (5280/3600) ft/s].

(b) Repeat part (a) using 3 subintervals and the velocity at the *midpoint* of each subinterval.

6. Space shuttle altitude   When we estimate distances from velocity data, it is sometimes necessary to use times $t_0, t_1, t_2, t_3, \ldots,$ that are not equally spaced. We can still estimate distances using the (unequal) time periods $\Delta t_1 = t_1 - t_0, \Delta t_2 = t_2 - t_1, \ldots.$ For example, on May 7, 1992, the space shuttle *Endeavour* was launched on mission STS-49, the purpose of which was to install a new perigee kick motor in an Intelsat communications satellite. The table, provided by NASA, gives the velocity data for the shuttle between liftoff and the jettisoning of the solid rocket boosters.

Event	Time (s)	Velocity (ft/s)
Launch	0	0
Begin roll maneuver	10	185
End roll maneuver	15	319
Throttle to 89%	20	447
Throttle to 67%	32	742
Throttle to 104%	59	1325
Maximum dynamic pressure	62	1445
Solid rocket booster separation	125	4151

Use these data to estimate the height above the earth's surface of the space shuttle *Endeavour*, 62 seconds after liftoff.

7. Oil spill   Oil leaked from a tank at a rate of $r(t)$ liters per hour. The rate decreased as time passed and values of the rate at two-hour time intervals are shown in the table. Estimate the total amount of oil that leaked out.

t (h)	0	2	4	6	8	10
$r(t)$ (L/h)	8.7	7.6	6.8	6.2	5.7	5.3

8. Driving distance   The velocity graph of a car accelerating from rest to a speed of 120 km/h over a period of 30 seconds is shown. Estimate the distance traveled during this period.

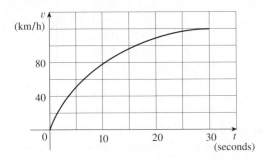

9. (a) By reading values from the given graph of f, use four rectangles with heights corresponding to the right endpoint of each subinterval to estimate the area under the graph of f from $x = 0$ to $x = 8$. Sketch the rectangles that you use. Is your estimate higher or lower than the actual area?

(b) Repeat part (a) using left endpoints.

(c) Find a new estimate (using right endpoints) with eight rectangles.

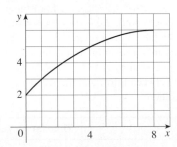

10. (a) Use six rectangles to find estimates of each type for the area under the given graph of f from $x = 0$ to $x = 12$.

  (i) L_6 (use left endpoints)

  (ii) R_6 (use right endpoints)

  (iii) M_6 (use midpoints)

(b) Is L_6 an underestimate or overestimate of the true area?

(c) Is R_6 an underestimate or overestimate of the true area?

(d) Which of the numbers L_6, R_6, or M_6 gives the best estimate? Explain.

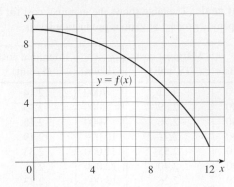

11. (a) Estimate the area under the graph of $f(x) = 1/x$ from $x = 1$ to $x = 5$ using four approximating rectangles and right endpoints. Sketch the graph and the rectangles. Is your estimate an underestimate or an overestimate?

(b) Repeat part (a) using left endpoints.

12. (a) Estimate the area under the graph of $f(x) = 25 - x^2$ from $x = 0$ to $x = 5$ using five approximating rectangles and right endpoints. Sketch the graph and the rectangles. Is your estimate an underestimate or an overestimate?

(b) Repeat part (a) using left endpoints.

13. (a) Estimate the area under the graph of $f(x) = 1 + x^2$ from $x = -1$ to $x = 2$ using three rectangles and right endpoints. Then improve your estimate by using six rectangles. Sketch the curve and the approximating rectangles.

(b) Repeat part (a) using left endpoints.

(c) Repeat part (a) using midpoints.

(d) From your sketches in parts (a)–(c), which appears to be the best estimate?

14. (a) Graph the function $f(x) = e^{-x^2}$, $-2 \le x \le 2$.

(b) Estimate the area under the graph of f using four approximating rectangles and using

  (i) right endpoints.

  (ii) midpoints.

  In each case sketch the curve and the rectangles.

(c) Improve your estimates in part (b) by using eight rectangles.

15. For the function g whose graph is given, determine whether each of the following integrals is positive, negative, or 0. Explain your reasoning.

(a) $\displaystyle\int_0^4 g(t)\, dt$ **(b)** $\displaystyle\int_6^8 g(t)\, dt$

(c) $\displaystyle\int_4^{10} g(t)\, dt$ **(d)** $\displaystyle\int_0^8 g(t)\, dt$

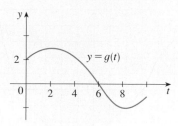

$y = g(t)$

16. If g is the function whose graph is shown in Exercise 15, determine if each of the following is true or false. Explain your reasoning.

(a) $\int_0^4 g(t)\, dt > 8$ (b) $\int_4^8 g(t)\, dt > 4$

(c) $\int_6^{10} g(t)\, dt < -2$

17. Each of the regions A, B, and C enclosed by the graph of f and the x-axis has area 3. Find the value of each of the following definite integrals.

(a) $\int_{-4}^0 f(x)\, dx$ (b) $\int_{-4}^2 f(x)\, dx$

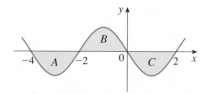

18. The graph of g consists of two straight lines and a semi-circle. Use it to evaluate each integral.

(a) $\int_0^2 g(x)\, dx$ (b) $\int_2^6 g(x)\, dx$ (c) $\int_0^7 g(x)\, dx$

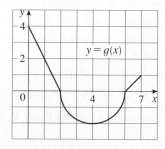

$y = g(x)$

19. The graph of f is shown. Evaluate each integral by interpreting it in terms of areas.

(a) $\int_0^2 f(x)\, dx$

(b) $\int_0^5 f(x)\, dx$

(c) $\int_5^7 f(x)\, dx$

(d) $\int_0^9 f(x)\, dx$

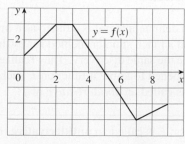

$y = f(x)$

20–23 ▪ Evaluate the integral by interpreting it in terms of areas.

20. $\int_0^3 \left(\tfrac{1}{2}x - 1\right) dx$

21. $\int_{-1}^4 (4 - 2x)\, dx$

22. $\int_{-2}^2 \sqrt{4 - x^2}\, dx$

23. $\int_{-1}^2 |x|\, dx$

24. Find $\int_0^5 f(x)\, dx$ if

$$f(x) = \begin{cases} 3 & \text{for } x < 3 \\ x & \text{for } x \geq 3 \end{cases}$$

25–28 ▪ Use the Midpoint Rule with the given value of n to approximate the integral. Round the answer to four decimal places.

25. $\int_2^{10} \sqrt{x^3 + 1}\, dx$, $n = 4$

26. $\int_0^3 \ln(x^2 + 3)\, dx$, $n = 6$

27. $\int_1^2 4e^{-0.6t}\, dt$, $n = 5$

28. $\int_1^5 x^2 e^{-x}\, dx$, $n = 4$

29. A table of values of an increasing function f is shown. Use the table to find lower and upper estimates for $\int_{10}^{30} f(x)\, dx$.

x	10	14	18	22	26	30
$f(x)$	-12	-6	-2	1	3	8

30. The table gives the values of a function obtained from an experiment. Use them to estimate $\int_3^9 f(x)\, dx$ using three equal subintervals with (a) right endpoints, (b) left endpoints, and (c) midpoints. If the function is known to be an increasing function, can you say whether your estimates are less than or greater than the exact value of the integral?

x	3	4	5	6	7	8	9
$f(x)$	-3.4	-2.1	-0.6	0.3	0.9	1.4	1.8

31. The graph of a function f is given. Estimate $\int_0^8 f(x)\, dx$ using four subintervals with **(a)** right endpoints, **(b)** left endpoints, and **(c)** midpoints.

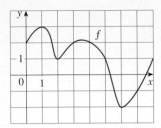

32. The graph of g is shown. Estimate $\int_{-3}^3 g(x)\, dx$ with six subintervals using **(a)** right endpoints, **(b)** left endpoints, and **(c)** midpoints.

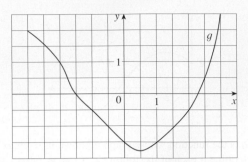

33. **(a)** If $f(x) > 0$ for $2 \leq x \leq 6$, can we say that $\int_2^6 f(x)\, dx > 0$? Explain or give a counterexample.
    (b) If $\int_2^6 f(x)\, dx > 0$, can we say that $f(x) > 0$ for $2 \leq x \leq 6$? Explain or give a counterexample.

34. If f is an odd function (see Section 1.1), explain why $\int_{-a}^a f(x)\, dx = 0$ for any positive constant a.

▪ Challenge Yourself

35. For the function f whose graph is shown, list the following quantities in increasing order, from smallest to largest, and explain your reasoning.

    (A) $\displaystyle\int_0^8 f(x)\, dx$ **(B)** $\displaystyle\int_0^3 f(x)\, dx$

    (C) $\displaystyle\int_3^8 f(x)\, dx$ **(D)** $\displaystyle\int_4^8 f(x)\, dx$

    (E) $f'(1)$

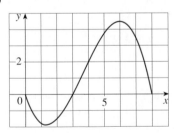

36. If $F(x) = \int_2^x f(t)\, dt$, where f is the function whose graph is given, which of the following values is largest?
    (A) $F(0)$ **(B)** $F(1)$ **(C)** $F(2)$

    (D) $F(3)$ **(E)** $F(4)$

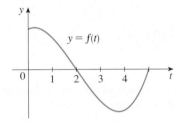

37. **(a)** Use the fact that $1 + 2 + 3 + \cdots + n = \frac{1}{2}n(n+1)$ to find the value of $\int_0^4 x\, dx$ using Definition 5.
    [*Hint:* First show that $x_i = i \cdot (4/n)$ here.]
    (b) Verify your answer in part (a) by evaluating $\int_0^4 x\, dx$ geometrically.

38. Use Definition 5 and the fact that
    $$1^2 + 2^2 + 3^2 + \cdots + n^2 = \frac{1}{6}n(n+1)(2n+1)$$
    to show that the area described in Example 2 is exactly $\frac{1}{3}$.

5.2 The Fundamental Theorem of Calculus

In Section 5.1 we saw that a definite integral is defined as a limit of Riemann sums. Sir Isaac Newton discovered a much simpler method for evaluating integrals and a few years later Leibniz made the same discovery. They realized that they could calculate $\int_a^b f(x)\, dx$ if they happened to know a function F such that $F'(x) = f(x)$. If such a function F exists, it is called an *antiderivative* of f.

> ▪ **Definition**   A function F is called an **antiderivative** of f if $F'(x) = f(x)$.

The discovery of Newton and Leibniz is called the *Fundamental Theorem of Calculus*.

> ■ **The Fundamental Theorem of Calculus**   If f is continuous on the interval $[a, b]$, then
>
> $$\int_a^b f(x)\, dx = F(b) - F(a)$$
>
> where F is any antiderivative of f, that is, $F' = f$.

This theorem states that if we know an antiderivative F of f, then we can evaluate $\int_a^b f(x)\, dx$ simply by subtracting the values of F at the endpoints of the interval $[a, b]$. It is very surprising that $\int_a^b f(x)\, dx$, which was defined by a complicated procedure involving all of the values of $f(x)$ for $a \le x \le b$, can be found by knowing the values of $F(x)$ at only two points, a and b.

For instance, because the derivative of e^{3x} is $3e^{3x}$, an antiderivative of $g(x) = 3e^{3x}$ is $G(x) = e^{3x}$, so according to the Fundamental Theorem of Calculus we can write

$$\int_0^2 3e^{3x}\, dx = G(2) - G(0) = e^{3(2)} - e^{3(0)} = e^6 - 1$$

Comparing this calculation with Definition 5 in Section 5.1, which requires computing a limit of sums, we see that the theorem provides us with a simple and powerful method.

■ **EXAMPLE 1**   Using the Fundamental Theorem of Calculus

Evaluate $\int_0^1 x^2\, dx$.

SOLUTION

An antiderivative of $f(x) = x^2$ is $F(x) = \frac{1}{3}x^3$. This is easily verified by differentiating F:

$$F'(x) = \frac{1}{3} \cdot 3x^2 = x^2 = f(x)$$

The Fundamental Theorem of Calculus then tells us that

$$\int_0^1 x^2\, dx = F(1) - F(0) = \frac{1}{3} \cdot 1^3 - \frac{1}{3} \cdot 0^3 = \frac{1}{3}$$ ■

Although the Fundamental Theorem of Calculus may be surprising at first glance, it becomes plausible if we interpret it in physical terms. If $M(q)$ is the marginal cost of a good after q units have been produced and $C(q)$ is the total cost of producing q units, then $M(q) = C'(q)$, so C is an antiderivative of M. In Section 5.1 we made the guess that the area under the marginal cost curve is equal to the additional cost incurred when raising production levels. In symbols:

$$\int_a^b M(q)\, dq = C(b) - C(a)$$

That is exactly what the Fundamental Theorem says in this context.

At the end of the section we will provide additional insight as to why the Fundamental Theorem of Calculus is true.

▪ Antiderivatives

In light of the Fundamental Theorem, the key to evaluating definite integrals is identifying an antiderivative for a function. In Example 1 we showed that an antiderivative of $f(x) = x^2$ is $F(x) = \frac{1}{3}x^3$. But is it the only one? Notice that $G(x) = \frac{1}{3}x^3 + 5$ also satisfies $G'(x) = x^2$, so both F and G are antiderivatives of f. In fact, any function of the form $H(x) = \frac{1}{3}x^3 + C$, where C is a constant, is an anti-derivative of f. The following theorem says that f has no other antiderivative.

---

(1) ▪ Theorem   If F is an antiderivative of f on an interval, then the most general antiderivative of f on that interval is

$$F(x) + C$$

where C is an arbitrary constant.

Going back to the function $f(x) = x^2$, we see that the general antiderivative of f is $\frac{1}{3}x^3 + C$. By assigning specific values to the constant C, we obtain a family of functions whose graphs are vertical translates of one another (see Figure 1). This makes sense because each curve must have the same slope at any given value of x, so the shape of each curve must be the same, but the vertical location can vary.

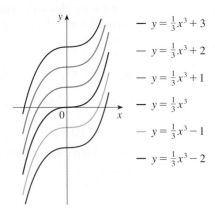

FIGURE 1

Members of the family of antiderivatives of $f(x) = x^2$

$-\; y = \frac{1}{3}x^3 + 3$

$-\; y = \frac{1}{3}x^3 + 2$

$-\; y = \frac{1}{3}x^3 + 1$

$-\; y = \frac{1}{3}x^3$

$-\; y = \frac{1}{3}x^3 - 1$

$-\; y = \frac{1}{3}x^3 - 2$

▪ EXAMPLE 2   Finding Antiderivatives

Find the most general antiderivative of each of the following functions.

(a) $g(t) = t^4$ **(b)** $y = 1/x$ **(c)** $f(x) = x^n, \quad n \neq -1$

SOLUTION

(a) We know that the derivative of t^5 is $5t^4$, so if we let $G(t) = \frac{1}{5}t^5$, we have

$$G'(t) = \frac{1}{5} \cdot 5t^4 = t^4$$

Thus an antiderivative of t^4 is $\frac{1}{5}t^5$. By Theorem 1, the most general antiderivative is $G(t) = \frac{1}{5}t^5 + C$.

(b) Recall that

$$\frac{d}{dx}(\ln x) = \frac{1}{x}$$

So for $x > 0$ (the domain of $\ln x$) the general antiderivative of $1/x$ is

$\ln x + C$. In fact, if $x < 0$ then $\ln|x| = \ln(-x)$ and

$$\frac{d}{dx}\ln|x| = \frac{d}{dx}\ln(-x) = \frac{1}{-x}(-1) = \frac{1}{x}$$

Thus

$$\frac{d}{dx}\left(\ln|x|\right) = \frac{1}{x}$$

for all $x \neq 0$. Theorem 1 then tells us that the general antiderivative of $y = 1/x$ is $\ln|x| + C$ on any interval that doesn't contain 0. So in general we will say that the antiderivative of $1/x$ is $\ln|x| + C$.

(c) We use the Power Rule

$$\frac{d}{dx}(x^n) = n \cdot x^{n-1}$$

to discover an antiderivative of x^n. In fact, if $n \neq -1$, then

$$\frac{d}{dx}\left(\frac{x^{n+1}}{n+1}\right) = \frac{1}{n+1} \cdot (n+1)x^n = x^n$$

Thus the general antiderivative of $f(x) = x^n$ is

$$F(x) = \frac{x^{n+1}}{n+1} + C$$

This is valid for any positive value of n. If n is negative (but $n \neq -1$), it is valid on any interval that doesn't contain 0. ■

As in Example 2, every differentiation formula, when read from right to left, gives rise to an antidifferentiation formula. In Table 2 we summarize differentiation formulas we have seen and list the corresponding antiderivatives. Each antiderivative formula in the table is true because the derivative of the function in the right column appears in the column to its left. In particular, the first formula says that the antiderivative of a constant times a function is the constant times the antiderivative of the function. The second formula says that the antiderivative of a sum (or difference) is the sum (or difference) of the antiderivatives. (We use the notation $F' = f$, $G' = g$.)

(2) Table of Derivative and Antiderivative Formulas

Function	Derivative	Function	General Antiderivative
$cf(x)$	$cf'(x)$	$cf(x)$	$cF(x) + C$
$f(x) \pm g(x)$	$f'(x) \pm g'(x)$	$f(x) \pm g(x)$	$F(x) \pm G(x) + C$
cx	c	c	$cx + C$
$x^n\ (n \neq 0)$	nx^{n-1}	$x^n\ (n \neq -1)$	$\dfrac{x^{n+1}}{n+1} + C$
$\ln x$	$1/x$	$1/x$	$\ln|x| + C$
e^x	e^x	e^x	$e^x + C$
e^{kx}	ke^{kx}	e^{kx}	$\frac{1}{k}e^{kx} + C$
a^x	$a^x \ln a$	a^x	$\frac{1}{\ln a}a^x + C$

■ **EXAMPLE 3**   **Finding a Function, Given its Derivative**

(a) Find all functions g such that

$$g'(x) = 6\sqrt{x} - \frac{4}{x^2}$$

(b) Of the functions g from part (a), which one satisfies $g(1) = 5$?

SOLUTION

(a) Rewriting the function, we see that we need to find an antiderivative of

$$g'(x) = 6x^{1/2} - 4x^{-2}$$

Using the formulas in Table 2 together with Theorem 1, we obtain

$$g(x) = 6 \cdot \frac{x^{3/2}}{3/2} - 4 \cdot \frac{x^{-1}}{-1} + C$$

You can check the answer to Example 3 by differentiating g.

$$= 6 \cdot \frac{2}{3}x^{3/2} - 4\left(-\frac{1}{x}\right) + C = 4x^{3/2} + \frac{4}{x} + C$$

(b) We have

$$g(1) = 4(1)^{3/2} + \frac{4}{1} + C = 5$$

so $4 + 4 + C = 5$ or $C = -3$. Thus the desired function is

$$g(x) = 4x^{3/2} + \frac{4}{x} - 3$$

■

■ **EXAMPLE 4**   **Finding General Antiderivatives**

Find the most general antiderivative of each of the following functions.

(a) $f(t) = 6e^{-0.3t} + 8$ **(b)** $w(p) = 0.4p^3 - \frac{2.9}{p}$

SOLUTION

(a)
$$F(t) = 6 \cdot \frac{1}{-0.3}e^{-0.3t} + 8t + C$$

$$= -20e^{-0.3t} + 8t + C$$

(b) First we rewrite $w(p)$ slightly:

$$w(p) = 0.4p^3 - 2.9\left(\frac{1}{p}\right)$$

Then

$$W(p) = 0.4\left(\frac{p^4}{4}\right) - 2.9 \ln |p| + C$$

$$= 0.1p^4 - 2.9 \ln |p| + C$$

■

■ Using the Fundamental Theorem

Once you are comfortable finding antiderivatives, you can take advantage of the Fundamental Theorem to evaluate definite integrals. When applying the Fundamental Theorem we use the notation

$$F(x)\Big]_a^b = F(b) - F(a)$$

and so we can write

$$\int_a^b f(x)\, dx = F(x)\Big]_a^b \qquad \text{where} \qquad F' = f$$

Other common notations are $F(x)\big|_a^b$ and $[F(x)]_a^b$.

■ EXAMPLE 5

Evaluating a Definite Integral with the Fundamental Theorem

Evaluate $\displaystyle\int_1^3 \frac{1}{t}\, dt$.

The Fundamental Theorem says we can use *any* antiderivative F of f. So in Example 5 we may as well use the simplest one, namely $F(t) = \ln|t|$. It is not necessary to use the most general antiderivative $(\ln|t| + C)$.

SOLUTION

An antiderivative of $f(t) = 1/t$ is $F(t) = \ln|t|$, so we use the Fundamental Theorem as follows:

$$\int_1^3 \frac{1}{t}\, dt = \ln|t|\,\Big]_1^3 = \ln|3| - \ln|1| = \ln 3 - 0 \approx 1.0986$$

■ EXAMPLE 6 Area Under a Curve

Find the area under the curve $y = e^x$ from 1 to 3.

SOLUTION

Since $e^x > 0$ and an antiderivative of $f(x) = e^x$ is $F(x) = e^x$, we have

$$\text{area} = \int_1^3 e^x\, dx = e^x\,\Big]_1^3 = e^3 - e^1 = e^3 - e$$

Thus the area under the graph of the exponential function for $1 \leq x \leq 3$ is $e^3 - e \approx 17.367$. (See Figure 2.)

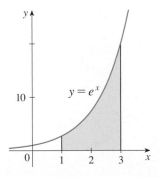

FIGURE 2

■ Indefinite Integrals

We need a convenient notation for antiderivatives that makes them easy to work with. Because of the relation between antiderivatives and integrals given by the Fundamental Theorem, the notation $\int f(x)\, dx$ is traditionally used for an antiderivative of f and is called an **indefinite integral**. Thus

$$\int f(x)\, dx = F(x) \qquad \text{means} \qquad F'(x) = f(x)$$

⊘   You should distinguish carefully between definite and indefinite integrals. A definite integral $\int_a^b f(x)\,dx$ is a *number,* whereas an indefinite integral $\int f(x)\,dx$ is a *function* (or family of functions). The connection between them is given by the Fundamental Theorem: If f is continuous on $[a, b]$, then

$$\int_a^b f(x)\,dx = \int f(x)\,dx \Big]_a^b$$

Note that the antiderivatives in Table 2 can be rewritten in the notation of indefinite integrals. For instance,

$$\int e^{kx}\,dx = \frac{1}{k}e^{kx} + C \qquad \text{because} \qquad \frac{d}{dx}\left(\frac{1}{k}e^{kx} + C\right) = e^{kx}$$

The properties listed in the first two rows of Table 2 can be written as

(3) $$\int cf(x)\,dx = c\int f(x)\,dx$$

(4) $$\int [f(x) \pm g(x)]\,dx = \int f(x)\,dx \pm \int g(x)\,dx$$

Property 3 says that the integral of a constant times a function is the constant times the integral of the function. In other words, a constant (but *only* a constant) can be taken in front of an integral sign. Property 4 says that the integral of a sum (or difference) is the sum (or difference) of the integrals.

▪ **EXAMPLE 7   Finding an Indefinite Integral**

Find the general indefinite integral

$$\int (10x^4 - 3e^x)\,dx$$

SOLUTION
Using Properties 3 and 4 with Table 2, we have

$$\int (10x^4 - 3e^x)\,dx = 10\int x^4\,dx - 3\int e^x\,dx$$

$$= 10 \cdot \frac{x^5}{5} - 3e^x + C$$

$$= 2x^5 - 3e^x + C$$

You should check this answer by differentiating it.   ▪

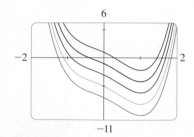

FIGURE 3

The indefinite integral in Example 7 is graphed in Figure 3 for several values of C.

▪ **EXAMPLE 8   An Indefinite and a Definite Integral**

Find   **(a)** $\int (x^3 - 6x)\,dx$ and   **(b)** $\int_0^3 (x^3 - 6x)\,dx$.

SOLUTION

(a) Using Properties 3 and 4 with Table 2, we have

$$\int (x^3 - 6x)\, dx = \frac{x^4}{4} - 6 \cdot \frac{x^2}{2} + C = \tfrac{1}{4}x^4 - 3x^2 + C$$

(b) Using the Fundamental Theorem, we have

$$\int_0^3 (x^3 - 6x)\, dx = \tfrac{1}{4}x^4 - 3x^2 \Big]_0^3$$

$$= \left(\tfrac{1}{4} \cdot 3^4 - 3 \cdot 3^2 \right) - \left(\tfrac{1}{4} \cdot 0^4 - 3 \cdot 0^2 \right)$$

$$= \tfrac{81}{4} - 27 - 0 + 0 = -6.75$$

This answer is consistent with our calculation in Example 5 of Section 5.1. ▪

▪ **EXAMPLE 9**   **An Integral Interpreted as a Net Area**

Find $\int_0^3 (x^3 - 3x^2 + 2)\, dx$ and interpret the result in terms of areas.

SOLUTION

The Fundamental Theorem gives

$$\int_0^3 (x^3 - 3x^2 + 2)\, dx = \frac{x^4}{4} - 3 \cdot \frac{x^3}{3} + 2x \Big]_0^3$$

$$= \tfrac{1}{4}x^4 - x^3 + 2x \Big]_0^3$$

$$= \tfrac{1}{4}(3^4) - 3^3 + 2(3) - 0$$

$$= \tfrac{81}{4} - 27 + 6 = -0.75$$

Figure 4 shows the graph of the integrand. We know from Section 5.1 that the value of the integral can be interpreted as the sum of the areas labeled with a plus sign minus the area labeled with a minus sign.

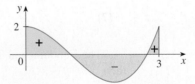

FIGURE 4

▪ **EXAMPLE 10**   **Simplifying before Integrating**

Evaluate $\int_1^8 \dfrac{0.7r^2 + 4.6r + 9}{r}\, dr$.

SOLUTION

First we need to write the integrand in a simpler form by carrying out the division:

$$\int_1^8 \frac{0.7r^2 + 4.6r + 9}{r} \, dr = \int_1^8 \left(\frac{0.7r^2}{r} + \frac{4.6r}{r} + \frac{9}{r} \right) dr$$

$$= \int_1^8 \left(0.7r + 4.6 + \frac{9}{r} \right) dr$$

$$= 0.7 \left(\frac{r^2}{2} \right) + 4.6r + 9 \ln |r| \bigg]_1^8 = 0.35r^2 + 4.6r + 9 \ln |r| \bigg]_1^8$$

$$= [0.35(8^2) + 4.6(8) + 9 \ln 8] - [0.35(1^2) + 4.6(1) + 9 \ln 1]$$

$$= 22.4 + 36.8 + 9 \ln 8 - 0.35 - 4.6 - 0 = 54.25 + 9 \ln 8$$

This is the exact value of the integral. If a decimal approximation is desired, we can use a calculator to approximate ln 8. Doing so, we get

$$\int_1^8 \frac{0.7r^2 + 4.6r + 9}{r} \, dr \approx 72.965$$

■

■ Properties of the Definite Integral

The properties of indefinite integrals given in (3) and (4) carry over to definite integrals:

(5) $$\int_a^b cf(x) \, dx = c \int_a^b f(x) \, dx, \quad \text{where } c \text{ is any constant}$$

(6) $$\int_a^b [f(x) \pm g(x)] \, dx = \int_a^b f(x) \, dx \pm \int_a^b g(x) \, dx$$

When we defined the definite integral $\int_a^b f(x) \, dx$, we implicitly assumed that $a < b$. But the definition as a limit of Riemann sums makes sense even if $a > b$. Notice that if we reverse a and b in Definition 5 of Section 5.1, then Δx changes from $(b - a)/n$ to $(a - b)/n = -(b - a)/n$. Therefore

(7) $$\int_b^a f(x) \, dx = -\int_a^b f(x) \, dx$$

The Fundamental Theorem confirms this result: If $\int_a^b f(x) \, dx = F(b) - F(a)$, then

$$\int_b^a f(x) \, dx = F(a) - F(b) = -[F(b) - F(a)] = -\int_a^b f(x) \, dx$$

Finally, the next property tells us how to combine integrals of the same function over adjacent intervals:

(8) $$\int_a^c f(x) \, dx + \int_c^b f(x) \, dx = \int_a^b f(x) \, dx$$

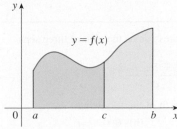

FIGURE 5

This is not easy to prove in general, but for the case where $f(x) \geqslant 0$ and $a < c < b$, Property 8 can be seen from the geometric interpretation in Figure 5: The area under $y = f(x)$ from a to c plus the area from c to b is equal to the total area from a to b.

▪ **EXAMPLE 11**   **Integrating over Adjacent Intervals**

If it is known that $\int_0^{10} f(x)\,dx = 17$ and $\int_0^8 f(x)\,dx = 12$, find $\int_8^{10} f(x)\,dx$.

SOLUTION

By Property 8, we have

$$\int_0^8 f(x)\,dx + \int_8^{10} f(x)\,dx = \int_0^{10} f(x)\,dx$$

so

$$\int_8^{10} f(x)\,dx = \int_0^{10} f(x)\,dx - \int_0^8 f(x)\,dx = 17 - 12 = 5$$

▪

Why Is the Fundamental Theorem True?

Consider $\int_a^b f(x)\,dx$, where f is continuous on the interval $[a, b]$. We divide the interval $[a, b]$ into n subintervals with endpoints $x_0\ (= a),\ x_1,\ x_2,\ \ldots,\ x_n\ (= b)$ and with length $\Delta x = (b - a)/n$. Let F be any antiderivative of f. By subtracting and adding like terms, we can express the total difference in the F values as the sum of the differences over the subintervals:

$$F(b) - F(a) = F(x_n) - F(x_0)$$

$$= F(x_n) + [-F(x_{n-1}) + F(x_{n-1})] + [-F(x_{n-2}) + F(x_{n-2})] + \cdots + [-F(x_1) + F(x_1)] - F(x_0)$$

$$= [F(x_n) - F(x_{n-1})] + [F(x_{n-1}) - F(x_{n-2})] + \cdots + [F(x_2) - F(x_1)] + [F(x_1) - F(x_0)]$$

Now F is differentiable and so, by the definition of the derivative, when Δx is small, we can write

$$F'(x_i) \approx \frac{F(x_i + \Delta x) - F(x_i)}{\Delta x} = \frac{F(x_{i+1}) - F(x_i)}{\Delta x}$$

or equivalently

$$F(x_{i+1}) - F(x_i) \approx F'(x_i)\,\Delta x = f(x_i)\,\Delta x$$

Therefore

$$F(b) - F(a) \approx f(x_{n-1})\,\Delta x + f(x_{n-2})\,\Delta x + \cdots + f(x_1)\,\Delta x + f(x_0)\,\Delta x$$

Now we take the limit of each side of this equation as $n \to \infty$. The left side is a constant and the right side is a Riemann sum for the function f, so

$$F(b) - F(a) = \lim_{n \to \infty} [f(x_0)\,\Delta x + f(x_1)\,\Delta x + \cdots + f(x_{n-1})\,\Delta x] = \int_a^b f(x)\,dx$$

▪ Prepare Yourself

1. Expand and simplify the expression $(4t + 3)^2$.

2. Distribute and simplify: $\sqrt{x}\,(2x^2 - 3)$.

3. Express each of the following as a power of x.

(a) $\sqrt{x}$

(b) $(\sqrt[3]{x})^2$

(c) $1/x^2$

(d) $\dfrac{1}{x\sqrt{x}}$

4. Write the following expression as the sum of three separate powers of x:

$$\frac{2x^4 + 5x^3 + 2}{x^2}$$

5. Find the derivative of each of the following.

(a) $y = \frac{1}{4}x^4$

(b) $B(t) = \frac{2}{3}t^{3/2}$

(c) $L(u) = \ln|u|$

(d) $P = 7.3e^t$

(e) $g(t) = -\dfrac{1}{0.2}\,e^{-0.2t}$

(f) $f(v) = 2\sqrt{v}$

(g) $h(x) = 1/x^3$

(h) $A = 5^t$

▪ Exercises 5.2

1–11 ▪ Find the most general antiderivative of the function. (Check your answer by differentiation.)

1. $f(x) = 6x^2 - 8x + 3$

2. $f(x) = 1 - x^3 + 12x^5$

3. $f(x) = 5x^{1/4} - 7x^{3/4}$

4. $g(u) = 2u + \sqrt[3]{u}$

5. $f(x) = 3\sqrt{x} + \dfrac{5}{x^6}$

6. $p(t) = 9.2t^2 - 4/t$

7. $q(s) = 5e^s + 3.7$

8. $h(t) = 8e^{2t}$

9. $f(q) = 1 + 2e^{0.8q}$

10. $f(x) = \dfrac{x^2 + x + 1}{x}$

11. $g(x) = \dfrac{5 - 4x^3 + 2x^6}{x^6}$

12. Find the antiderivative F of $f(x) = 5x^4 - 2x^5$ that satisfies $F(0) = 4$. Check your answer by comparing the graphs of f and F.

13. Find all functions f so that $f'(x) = 8x^2 - 3/x$.

14. Find all functions g so that $g'(t) = 4.1e^{-0.3t}$.

15–16 ▪ Find f.

15. $f'(x) = \sqrt{x}\,(6 + 5x)$, $f(1) = 10$

16. $f'(x) = 2x - 3/x^4$, $x > 0$, $f(1) = 3$

17–18 ▪ Find (a) f' and (b) f.

17. $f''(x) = 24x^2 + 2x + 10$, $f(1) = 5$, $f'(1) = -3$

18. $f''(x) = 4 - 6x - 40x^3$, $f(0) = 2$, $f'(0) = 1$

19. (a) Use a graphing device to graph $f(x) = e^x - 2x$.

(b) Starting with the graph in part (a), sketch a rough graph of the antiderivative F that satisfies $F(0) = 1$.

(c) Use the rules of this section to find an expression for $F(x)$.

(d) Graph F using the expression in part (c). Compare with your sketch in part (b).

20. Motion   An object moves with acceleration function

$$a(t) = 5 + 4t - 2t^2$$

Its initial velocity is $v(0) = 3$ m/s and its initial position is $s(0) = 10$ m. Find its position after t seconds.

21–40 ▪ Evaluate the integral.

21. $\displaystyle\int_0^4 (6x - 5)\,dx$

22. $\displaystyle\int_3^9 (t^2 + 1)\,dt$

23. $\displaystyle\int_{-1}^3 x^5\,dx$

24. $\displaystyle\int_1^3 (1 + 2x - 4x^3)\,dx$

25. $\displaystyle\int_0^2 (6x^2 - 4x + 5)\,dx$

26. $\displaystyle\int_{-2}^0 (u^5 - u^3 + u^2)\,du$

27. $\displaystyle\int_1^9 4\sqrt{z}\,dz$

28. $\displaystyle\int_1^8 \sqrt[3]{x}\,dx$

29. $\displaystyle\int_{-1}^0 (2x - e^x)\,dx$

30. $\displaystyle\int_0^{2.6} 8.4e^{0.4t}\,dt$

31. $\displaystyle\int_0^4 (1 + 2e^{-0.6q})\,dq$

32. $\displaystyle\int_{-2}^2 (3u + 1)^2\,du$

[*Hint:* First expand the integrand.]

33. $\int_0^4 (2v + 5)(3v - 1)\, dv$

34. $\int_{-2}^{-1} \left(4y^3 + \dfrac{2}{y^3} \right) dy$

35. $\int_1^2 \dfrac{v^3 + 3v^6}{v^4}\, dv$

36. $\int_1^9 \dfrac{3x - 2}{\sqrt{x}}\, dx$

37. $\int_1^9 \dfrac{1}{2x}\, dx$

38. $\int_1^2 \left(x + \dfrac{1}{x} \right) dx$

39. $\int_0^1 5(2^z)\, dz$

40. $\int_0^6 1425(1.12^t)\, dt$

41–44 ■ Find the area under the graph of the function for the given interval.

41. $y = 4x + 0.3e^x, \quad 1 \le x \le 4$

42. $y = 3/x + 2x^3, \quad 2 \le x \le 3$

43. $p = 5\sqrt{t}, \quad 0 \le t \le 4$

44. $T = e^{1.1v} + 2/v^2, \quad 2 \le v \le 5$

---

45. Use a graph to estimate the x-intercepts of the curve $y = x + x^2 - x^4$. Then use this information to estimate the area of the region that lies under the curve and above the x-axis.

46. Repeat Exercise 45 for the curve $y = 2x + 3x^4 - 2x^6$.

47–50 ■ Evaluate the integral and interpret it as a difference of areas. Illustrate with a sketch.

47. $\int_{-1}^2 x^3\, dx$

48. $\int_{-1}^1 (e^x - 1)\, dx$

49. $\int_0^3 (t^2 + 2t - 3)\, dt$

50. $\int_{-2}^2 (u^3 - u^2 - 2u)\, du$

51–62 ■ Find the general indefinite integral.

51. $\int 12x^3\, dx$

52. $\int (8x - 3)\, dx$

53. $\int (t^2 + 3t + 4)\, dt$

54. $\int (9.3 - 1.5v^2)\, dv$

55. $\int (0.01q^3 + 0.6q^2 + 3.5q + 14.9)\, dq$

56. $\int (x + e^x)\, dx$

57. $\int 5e^{2t}\, dt$

58. $\int (1 + 0.4e^{-0.22r})\, dr$

59. $\int (6u^2 - 3\sqrt{u})\, du$

60. $\int (3/s - 2/s^2)\, ds$

61. $\int (1 - t)(2 + t^2)\, dt$

  [*Hint:* First expand the integrand.]

62. $\int x(1 + 2x^4)\, dx$

63–64 ■ Find the general indefinite integral. Illustrate by graphing several members of the family on the same screen.

63. $\int x\sqrt{x}\, dx$

64. $\int (x - 1/x)\, dx$

---

65. For the function $f(x) = \frac{1}{2}x^3 + 4x - 1$ find

  (a) $\int f(x)\, dx$ **(b)** $\int_0^2 f(x)\, dx$

66. For the function $f(x) = 3e^{2x}$ find

  (a) $\int f(x)\, dx$ **(b)** $\int_0^3 f(x)\, dx$

67–68 ■ Verify by differentiation that the formula is correct.

67. $\int \dfrac{x}{\sqrt{x^2 + 1}}\, dx = \sqrt{x^2 + 1} + C$

68. $\int \ln x\, dx = x \ln x - x + C$

---

69. If $\int_0^9 f(x)\, dx = 37$ and $\int_0^9 g(x)\, dx = 16$, find $\int_0^9 [2f(x) + 3g(x)]\, dx$.

70. Given that $\int_1^3 e^x\, dx = e^3 - e$, use the properties of integrals to evaluate $\int_1^3 (2e^x - 1)\, dx$.

71. Write as a single integral in the form $\int_a^b f(x)\, dx$:

$$\int_{-2}^2 f(x)\, dx + \int_2^5 f(x)\, dx - \int_{-2}^{-1} f(x)\, dx$$

72. If $\int_1^5 f(x)\, dx = 12$ and $\int_4^5 f(x)\, dx = 3.6$, find $\int_1^4 f(x)\, dx$.

■ Challenge Yourself

73. The area of the region that lies to the right of the y-axis and to the left of the parabola $x = 2y - y^2$ (the shaded region in the figure) is given by the integral $\int_0^2 (2y - y^2)\, dy$. (Turn your head clockwise and think of the region as lying below the curve $x = 2y - y^2$ from $y = 0$ to $y = 2$.) Find the area of the region.

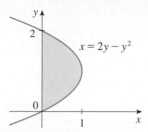

74. The boundaries of the shaded region are the y-axis, the line $y = 1$, and the curve $y = \sqrt[4]{x}$. Find the area of this region by writing x as a function of y and integrating with respect to y (as in Exercise 73).

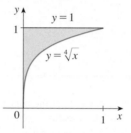

75. If

$$g(x) = \begin{cases} 0.2x^2 + 4 & \text{if } x < 5 \\ 3x - 6 & \text{if } x \geq 5 \end{cases}$$

use Property 8 to evaluate $\int_0^8 g(x)\, dx$.

76. The area labeled B is three times the area labeled A. Express b in terms of a.

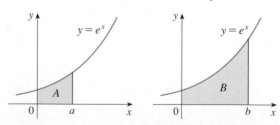

77. Suppose h is a function such that $h(1) = -2$, $h'(1) = 2$, $h''(1) = 3$, $h(2) = 6$, $h'(2) = 5$, $h''(2) = 13$, and h'' is continuous everywhere. Evaluate $\int_1^2 h''(u)\, du$.

5.3 The Net Change Theorem and Average Value

The Fundamental Theorem of Calculus says that if f is continuous on $[a, b]$, then

$$\int_a^b f(x)\, dx = F(b) - F(a)$$

where F is any antiderivative of f. This means that $F' = f$, so the equation can be rewritten as

$$\int_a^b F'(x)\,dx = F(b) - F(a)$$

We know that $F'(x)$ represents the rate of change of $y = F(x)$ with respect to x and $F(b) - F(a)$ is the change in y when x changes from a to b. [Note that y could, for instance, increase, then decrease, then increase again. Although y might change in both directions, $F(b) - F(a)$ represents the *net* change in y.] So we can reformulate the Fundamental Theorem in words as follows.

■ **Net Change Theorem**   The integral of a rate of change is the net change:

$$\int_a^b F'(x)\,dx = F(b) - F(a)$$

■ | Applications of the Net Change Theorem

The Net Change Theorem can be applied to all of the rates of change in the social and natural sciences that we discussed in earlier chapters. Here are a few instances of this idea:

■ If the rate of growth of a population is dP/dt, then

$$\int_{t_1}^{t_2} \frac{dP}{dt}\,dt = P(t_2) - P(t_1)$$

is the net change in population during the time period from t_1 to t_2. (The population increases when births happen and decreases when deaths occur. The net change takes into account both births and deaths.)

■ If $C(q)$ is the cost of producing q units of a commodity, then the marginal cost is the derivative $C'(q)$. So

$$\int_{q_1}^{q_2} C'(q)\,dq = C(q_2) - C(q_1)$$

is the change in cost when production is increased from q_1 units to q_2 units.

■ If $V(t)$ is the volume of water in a reservoir at time t, then its derivative $V'(t)$ is the rate at which water flows into the reservoir at time t. So

$$\int_{t_1}^{t_2} V'(t)\,dt = V(t_2) - V(t_1)$$

is the change in the amount of water in the reservoir between time t_1 and time t_2.

■ If an object moves along a straight line with position function $s(t)$, then its velocity is $v(t) = s'(t)$, so

(1) $$\int_{t_1}^{t_2} v(t)\,dt = s(t_2) - s(t_1)$$

is the net change of position, or *displacement*, of the object during the time period from t_1 to t_2.

■ If want to calculate the distance traveled during the time interval, we have to consider the intervals when $v(t) \geqslant 0$ (the object moves to the right) and also the

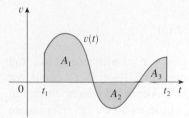

$$\text{displacement} = \int_{t_1}^{t_2} v(t)\, dt$$
$$= A_1 - A_2 + A_3$$

$$\text{distance} = \int_{t_1}^{t_2} |v(t)|\, dt$$
$$= A_1 + A_2 + A_3$$

FIGURE 1

intervals when $v(t) \leq 0$ (the object moves to the left). In both cases the distance is computed by integrating $|v(t)|$, the speed. Therefore

(2) $$\int_{t_1}^{t_2} |v(t)|\, dt = \text{total distance traveled}$$

Figure 1 shows how both displacement and distance traveled can be interpreted in terms of areas under a velocity curve.

▪ The acceleration of an object is the rate of change of velocity, that is, $a(t) = v'(t)$, so

$$\int_{t_1}^{t_2} a(t)\, dt = v(t_2) - v(t_1)$$

is the change in velocity from time t_1 to time t_2.

▪ EXAMPLE 1   Integrating Marginal Revenue

Recall that marginal revenue is the rate of change of revenue with respect to the number of units produced, measured in dollars per unit. Suppose the marginal revenue when a company sells q units of a new product is given by $R'(q) = 26 - 0.04q$. Compute and interpret

$$\int_{200}^{300} R'(q)\, dq$$

SOLUTION

$$\int_{200}^{300} R'(q)\, dq = \int_{200}^{300} (26 - 0.04q)\, dq$$

$$= 26q - 0.04\left(\frac{q^2}{2}\right)\bigg]_{200}^{300} = 26q - 0.02q^2\big]_{200}^{300}$$

$$= 26(300) - 0.02(300^2) - 26(200) + 0.02(200^2)$$

$$= 7800 - 1800 - 5200 + 800 = 1600$$

This is the net change in revenue for $200 \leq q \leq 300$. Our answer says that revenue increases by $1600 when production is increased from 200 units to 300 units.   ▪

▪ EXAMPLE 2   Displacement versus Distance

An object moves along a line so that its velocity at time t is $v(t) = t^2 - t - 6$ (measured in meters per second).

(a) Find the displacement of the object during the time period $1 \leq t \leq 4$.

(b) Find the distance traveled during this time period.

SOLUTION

(a) By Equation 1, the displacement is

$$s(4) - s(1) = \int_1^4 v(t)\, dt = \int_1^4 (t^2 - t - 6)\, dt = \left[\frac{t^3}{3} - \frac{t^2}{2} - 6t\right]_1^4 = -\frac{9}{2}$$

This means that the object's position at time $t = 4$ is 4.5 m to the left of its position at the start of the time period.

(b) We first need to determine when $v(t) \geq 0$ and when $v(t) \leq 0$. Note that $v(t) = t^2 - t - 6 = (t - 3)(t + 2)$ and so $v(t) = 0$ when $t = 3$ or $t = -2$. Here $v(t) \leq 0$ for $-2 \leq t \leq 3$, and $v(t) \geq 0$ for $t \leq -2$ and $t \geq 3$, so for $1 \leq t \leq 4$ we have $v(t) \leq 0$ on the interval $[1, 3]$ and $v(t) \geq 0$ on $[3, 4]$. Recall that if $v(t) \leq 0$, then $|v(t)| = -v(t)$. Thus, from Equation 2, the distance traveled is

To integrate the absolute value of $v(t)$, we use Property 8 of integrals (Section 5.2) to split the integral into two parts, one where $v(t) \leq 0$ and one where $v(t) \geq 0$.

$$\int_1^4 |v(t)| \, dt = \int_1^3 [-v(t)] \, dt + \int_3^4 v(t) \, dt$$

$$= \int_1^3 (-t^2 + t + 6) \, dt + \int_3^4 (t^2 - t - 6) \, dt$$

$$= \left[-\frac{t^3}{3} + \frac{t^2}{2} + 6t \right]_1^3 + \left[\frac{t^3}{3} - \frac{t^2}{2} - 6t \right]_3^4$$

$$= \left(\frac{27}{2} - \frac{37}{6} \right) + \left(-\frac{32}{3} + \frac{27}{2} \right) = \frac{61}{6} \approx 10.17 \text{ m}$$ ▪

▪ EXAMPLE 3   Computing Energy by Integrating Power

Figure 2 shows the power consumption in the city of San Francisco for a day in September (P is measured in megawatts; t is measured in hours starting at midnight). Estimate the energy used on that day.

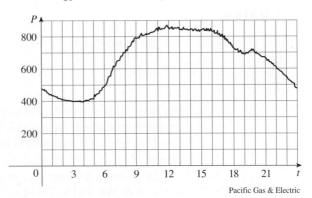

FIGURE 2

Pacific Gas & Electric

SOLUTION

Power is the rate of change of energy: $P(t) = E'(t)$. So, by the Net Change Theorem,

$$\int_0^{24} P(t) \, dt = \int_0^{24} E'(t) \, dt = E(24) - E(0)$$

is the total amount of energy used that day. We approximate the value of the integral using the Midpoint Rule with 12 subintervals and $\Delta t = 2$:

$$\int_0^{24} P(t) \, dt \approx [P(1) + P(3) + P(5) + \cdots + P(21) + P(23)]\Delta t$$

$$\approx (440 + 400 + 420 + 620 + 790 + 840 + 850$$

$$+ 840 + 810 + 690 + 670 + 550)(2)$$

$$\approx 15,840$$

The energy used was approximately 15,840 megawatt-hours. ▪

■ A Note on Units

How did we know what units to use for energy in Example 3? The integral $\int_0^{24} P(t)\,dt$ is defined as the limit of sums of terms of the form $P(t_i)\,\Delta t$. Now $P(t_i)$ is measured in megawatts and Δt is measured in hours, so their product is measured in megawatt-hours. The same is true of the limit. In general, the unit of measurement for $\int_a^b f(x)\,dx$ is the product of the unit for $f(x)$ and the unit for x.

■ EXAMPLE 4   Determining Units in a Definite Integral

Suppose $a(t)$ is the absorption rate in a patient's bloodstream of a drug t hours after the drug is administered. If $a(t)$ is measured in $\mu g/mL$ per hour and t is measured in hours, what units does $\int_1^3 a(t)\,dt$ have? What does the integral represent?

SOLUTION

The units for $\int_1^3 a(t)\,dt$ are $\mu g/mL$ per hour [the unit for $a(t)$] multiplied by hours (the unit for t), giving $\mu g/mL$. The integral represents the net change in drug concentration in the patient's bloodstream from the end of the first hour after administration to the end of the third hour.   ■

■ Average Value

It is easy to calculate the average value of n numbers $y_1, y_2, \ldots, y_n$:

$$y_{\text{ave}} = \frac{y_1 + y_2 + \cdots + y_n}{n}$$

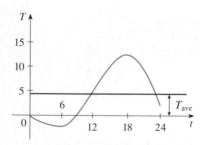

FIGURE 3

But how do we compute the average temperature during a day if infinitely many temperature readings are possible? Figure 3 shows the graph of a temperature function $T(t)$, where t is measured in hours and T in °C, and a guess at the average temperature, T_{ave}.

In general, let's try to compute the average value of a function $y = f(x)$, $a \leq x \leq b$. We start by dividing the interval $[a, b]$ into n equal subintervals, each with length $\Delta x = (b - a)/n$. We can use the right endpoint x_i of each subinterval to find a representative output value $f(x_i)$ for that interval and then calculate the average of these values:

$$\frac{f(x_1) + \cdots + f(x_n)}{n}$$

(For example, if f represents a temperature function and $n = 24$, this means that we take temperature readings at the end of every hour and then average them.) Since $\Delta x = (b - a)/n$, we can write $n = (b - a)/\Delta x$ and the average value becomes

$$\frac{f(x_1) + \cdots + f(x_n)}{(b - a)/\Delta x} = \frac{\Delta x}{b - a}\left[f(x_1) + \cdots + f(x_n)\right]$$

$$= \frac{1}{b - a}\left[f(x_1)\,\Delta x + \cdots + f(x_n)\,\Delta x\right]$$

If we let n increase, we would be computing the average value of a large number of closely spaced values. (For example, we would be averaging temperature readings taken every minute or even every second.) The expression in brackets is a Riemann sum, so if we allow $n \to \infty$ we have

$$\lim_{n \to \infty} \frac{1}{b-a} [f(x_1) \Delta x + \cdots + f(x_n) \Delta x] = \frac{1}{b-a} \int_a^b f(x)\, dx$$

Therefore we define the **average value of f** on the interval $[a, b]$ as

For a positive function, we can think of this definition as saying

$$\frac{\text{area}}{\text{width}} = \text{average height}$$

$$\boxed{f_{\text{ave}} = \frac{1}{b-a} \int_a^b f(x)\, dx}$$

■ **EXAMPLE 5** **Computing the Average Value of a Function**

Find the average value of the function $f(x) = 1 + x^2$ on the interval $[-1, 2]$.

SOLUTION

With $a = -1$ and $b = 2$ we have

$$f_{\text{ave}} = \frac{1}{b-a} \int_a^b f(x)\, dx = \frac{1}{2 - (-1)} \int_{-1}^2 (1 + x^2)\, dx$$

$$= \frac{1}{3} \left[x + \frac{x^3}{3} \right]_{-1}^2 = \frac{1}{3} \left[2 + \frac{8}{3} - (-1) - \left(-\frac{1}{3} \right) \right] = 2$$ ■

■ **EXAMPLE 6** **Finding an Average Temperature**

The temperature (in °F) in Mexico City t hours after midnight during a day in April was modeled by the function

$$T(t) = -0.017t^3 + 0.53t^2 - 2.9t + 65$$

Find the average temperature during that day from 8 AM to 6 PM.

SOLUTION

The indicated hours correspond to the interval $[8, 18]$, so the average temperature was

$$T_{\text{ave}} = \frac{1}{18 - 8} \int_8^{18} (-0.017t^3 + 0.53t^2 - 2.9t + 65)\, dt$$

$$= \frac{1}{10} \left[-0.017 \frac{t^4}{4} + 0.53 \frac{t^3}{3} - 2.9 \frac{t^2}{2} + 65t \right]_8^{18}$$

$$\approx \frac{1}{10} (1284.372 - 500.245) \approx 78.4°F$$ ■

■ Exercises 5.3

1. **Integrating marginal cost**   If $C'(q)$ is the marginal cost, in thousands of dollars per unit, of producing q guitars, what does $\int_{300}^{500} C'(q)\,dq$ represent?

2. **Integrating marginal revenue**   If $R(q)$ is the revenue collected, measured in millions of dollars, after producing q items, then $R'(q)$ is the marginal revenue. What does $\int_{1000}^{5000} R'(q)\,dq$ represent?

3. **Human growth rate**   If $w'(t)$ is the rate of growth of a child, in pounds per year, when the child is t years old, what does $\int_{5}^{10} w'(t)\,dt$ represent?

4. **Integrating marginal profit**   Marginal profit $P'(q)$ is the derivative of the profit function $P(q)$, where q is the number of units sold. What does $\int_{75}^{90} P'(q)\,dq$ represent?

5. **Oil spill**   If oil leaks from a tank at a rate of $r(t)$ gallons per minute at time t, what does $\int_{0}^{120} r(t)\,dt$ represent?

6. **Vehicle value**   If $V'(t)$ is the rate at which an automobile's value declines, in dollars per year, t years after it is first sold, what does $\int_{4}^{6} V'(t)\,dt$ represent?

7. **Trail slope**   If $f(x)$ is the slope of a trail at a (horizontal) distance of x miles from the start of the trail, what does $\int_{3}^{5} f(x)\,dx$ represent?

8. **Insect population**   A honeybee population starts with 100 bees and increases at a rate of $n'(t)$ bees per week. What does $100 + \int_{0}^{15} n'(t)\,dt$ represent?

9. **Data flow**   Data is flowing through a router on a company's network at a rate of $f(t)$ megabytes per second (mB/s) t seconds after midnight. Interpret the statement $\int_{21,600}^{28,800} f(t)\,dt = 18{,}350$ in this context.

10. **Snowfall**   Suppose $f(t)$ is the rate at which snow is falling at a particular location, in inches per hour, t hours after the start of a storm. Interpret the statement $\int_{2}^{5} f(t)\,dt = 8.3$ in this context.

11. **Fabric production**   The marginal cost of manufacturing y yards of a certain fabric is

$$C'(y) = 3 - 0.01y + 0.000006y^2$$

in dollars per yard. Find the increase in cost if the production level is raised from 2000 yards to 4000 yards.

12. **Total cost**   A company estimates that the marginal cost, in dollars per item, of producing q items is $1.92 - 0.002q$. If the fixed costs are \$562, find the cost of producing the first 100 items.

13. **Newspaper subscriptions**   A city's major newspaper has been losing subscribers. If $s(t) = -15{,}000e^{-0.04t}$ is the rate at which the number of subscribers is changing (measured in subscriptions per month) t months after

January 1, 2010, how many subscribers did the paper lose during 2010?

14. **Water flow**   Water flows from the bottom of a storage tank at a rate of $r(t) = 200 - 4t$ liters per minute, where $0 \le t \le 50$. Find the amount of water that flows from the tank during the first 10 minutes.

15. **Bacteria population**   A bacteria culture is growing at a rate of $r(t) = 6e^{0.3t}$ thousand bacteria per hour after t hours. How much did the bacteria population increase during the first two hours?

16. **Bacteria population**   A bacteria population starts with 400 bacteria and grows at a rate of $r(t) = (450.268)e^{1.12567t}$ bacteria per hour. How many bacteria will there be after three hours?

17. **Oil spill**   An oil storage tank ruptures at time $t = 0$ and oil leaks from the tank at a rate of $r(t) = 100e^{-0.01t}$ liters per minute. How much oil leaks out during the first hour?

18. **Spread of disease**   The number of people in a community who have contracted a particular flu virus is increasing at a rate of $g(t) = 9\sqrt{t}$ people per day, where t is the number of days after the start of the month. How many people contracted the flu from the 10th to the 20th day of the month?

19. If the units for t are years and the units for $P(t)$ are thousands of people per year, what are the units for $\int_{a}^{b} P(t)\,dt$?

20. If the units for x are feet and the units for $a(x)$ are pounds per foot, what are the units for da/dx? What units does $\int_{2}^{8} a(x)\,dx$ have?

21. **Energy consumption**   The electric power usage in kilowatts (kW) of a residence during a particular day was modeled by

$$p(t) = -0.016t^3 + 0.44t^2 - 1.4t + 12.1$$

where t is the number of hours after midnight, $0 \le t \le 24$. How much electric energy was consumed during that day? (Use the fact that power is the rate of change of energy.)

22. Suppose that a volcano is erupting and readings of the rate $r(t)$ at which solid materials are spewed into the atmosphere are given in the table. The time t is measured in seconds and the units for $r(t)$ are tonnes (metric tons) per second. Use the Midpoint Rule to estimate the total quantity of erupted materials after six seconds.

t	0	1	2	3	4	5	6
$r(t)$	2	10	24	36	46	54	60

23. Bird velocity The acceleration after t seconds of a hawk flying along a straight path is $a(t) = 0.4 + 0.12t$ ft/s^2. How much did the hawk's speed increase from $t = 5$ to $t = 8$?

24. Motion An object is moving along a straight path that runs north and south. Its velocity, in feet per minute, is given by

$$v(t) = 18 - 8t + 0.3t^2$$

where north is the positive direction. After ten minutes, is the object north of its starting position, or south?

25–26 ■ **Distance vs. displacement** The velocity function, in meters per second, is given for an object moving along a line. Find **(a)** the displacement and **(b)** the distance traveled by the object during the given time interval.

25. $v(t) = 3t - 5$, $0 \leqslant t \leqslant 3$

26. $v(t) = t^2 - 2t - 8$, $1 \leqslant t \leqslant 6$

27–28 ■ **Motion** The acceleration function, in m/s^2, and the initial velocity are given for an object moving along a line. Find **(a)** the velocity at time t and **(b)** the distance traveled during the given time interval.

27. $a(t) = t + 4$, $v(0) = 5$, $0 \leqslant t \leqslant 10$

28. $a(t) = 2t + 3$, $v(0) = -4$, $0 \leqslant t \leqslant 3$

29–32 ■ Find the average value of the function on the given interval.

29. $f(x) = 4x - x^2$, $[0, 4]$

30. $H(t) = 3.2e^{0.5t}$, $[0, 6]$

31. $g(x) = \sqrt[3]{x}$, $[1, 8]$

32. $A(v) = 4v - \dfrac{1}{v}$, $[1, 5]$

33. Average temperature In a certain city the temperature (in °F) t hours after 9 AM was modeled by the function $T(t) = 57 - 2.4t + 0.43t^2 - 0.014t^3$. Find the average temperature during the period from 9 AM to 9 PM.

34. Average temperature If a cup of coffee has temperature 95°C in a room where the temperature is 20°C, then, according to Newton's Law of Cooling, the temperature of the coffee after t minutes is $T(t) = 20 + 75e^{-t/50}$. What is the average temperature of the coffee during the first half hour?

35. Drug concentration The concentration of a drug in a patient's bloodstream, measured in mg/L, t minutes after being injected is given by

$$C(t) = 8(e^{-0.05t} - e^{-0.4t})$$

Find the average concentration of the drug in the bloodstream during the first 20 minutes.

36. Investment balance The balance of an investment account t years after it is opened is $P(t) = 48{,}000e^{0.062t}$ dollars. Compute the average balance of the account during the first five years.

37. Water storage Water flows into and out of a storage tank. A graph of the rate of change $r(t)$ of the volume of water in the tank, in liters per day, is shown. If the amount of water in the tank at time $t = 0$ is 25,000 L, use the Midpoint Rule to estimate the amount of water four days later.

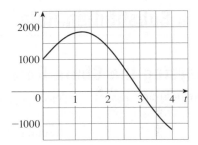

38. Space shuttle altitude On May 7, 1992, the space shuttle *Endeavour* was launched on mission STS-49, the purpose of which was to install a new perigee kick motor in an Intelsat communications satellite. The table gives the velocity data for the shuttle between liftoff and the jettisoning of the solid rocket boosters.

(a) Use a graphing calculator or computer to model these data by a third-degree polynomial.

(b) Use the model in part (a) to estimate the height reached by the *Endeavour*, 125 seconds after liftoff.

Event	Time (s)	Velocity (ft/s)
Launch	0	0
Begin roll maneuver	10	185
End roll maneuver	15	319
Throttle to 89%	20	447
Throttle to 67%	32	742
Throttle to 104%	59	1325
Maximum dynamic pressure	62	1445
Solid rocket booster separation	125	4151

39. The table gives values of a continuous function. Use the Midpoint Rule to estimate the average value of f on $[20, 50]$.

x	20	25	30	35	40	45	50
$f(x)$	42	38	31	29	35	48	60

40. Car velocity The velocity graph of an accelerating car is shown.

(a) Estimate the average velocity of the car during the first 12 seconds.

(b) At what time was the instantaneous velocity equal to the average velocity?

Challenge Yourself

41. Marginal cost If $C'(q)$ is the marginal cost after producing q units of a good, show that the average value of the marginal cost for $a \leqslant q \leqslant b$ is equal to the average rate of change of the total cost for $a \leqslant q \leqslant b$.

42. Find the numbers b such that the average value of

$$f(x) = 2 + 6x - 3x^2$$

on the interval $[0, b]$ is equal to 3.

5.4 The Substitution Rule

Reversing the Chain Rule

The Fundamental Theorem gives us an easy way to evaluate a definite integral of a function if we know an antiderivative for the function. But our antidifferentiation formulas don't tell us how to evaluate integrals such as

(1)
$$\int 2x\sqrt{1 + x^2}\, dx$$

The function $2x\sqrt{1 + x^2}$ does have an antiderivative; in fact,

$$\frac{d}{dx}\left[\tfrac{2}{3}(1 + x^2)^{3/2}\right] = \tfrac{2}{3} \cdot \tfrac{3}{2}(1 + x^2)^{1/2} \cdot 2x = 2x\sqrt{1 + x^2}$$

so $\tfrac{2}{3}(1 + x^2)^{3/2}$ is an antiderivative of $2x\sqrt{1 + x^2}$. But in general, how do we *find* such an antiderivative? Notice that $\tfrac{2}{3}(1 + x^2)^{3/2}$ is a composition of functions, so its derivative requires the Chain Rule. Thus the key to evaluating the integral in (1) is to use the Chain Rule in reverse.

In general, consider the composition of functions $F(g(x))$. By the Chain Rule we have

$$\frac{d}{dx}[F(g(x))] = F'(g(x))\, g'(x)$$

and so, if $F' = f$,

(2)
$$\int f(g(x))\, g'(x)\, dx = F(g(x)) + C$$

■ **EXAMPLE 1**   **Reversing the Chain Rule**

Evaluate the indefinite integral $\int 2x\sqrt{1 + x^2}\, dx$.

SOLUTION

Notice that if we let $g(x) = 1 + x^2$, the expression inside the square root, then $g'(x) = 2x$ and so we can write $\int 2x\sqrt{1 + x^2}\, dx = \int \sqrt{g(x)}\, g'(x)\, dx$. Letting $f(u) = \sqrt{u}$, an antiderivative for f is $F(u) = \frac{2}{3}u^{3/2}$, so by Equation 2 we have

$$\int \underbrace{\sqrt{\underbrace{1 + x^2}_{g(x)}}\ \underbrace{2x}_{g'(x)}}_{f(g(x))} dx = F(g(x)) + C = F(1 + x^2) + C$$

$$= \tfrac{2}{3}(1 + x^2)^{3/2} + C \qquad\blacksquare$$

■ Substitution in Indefinite Integrals

In practice, it is convenient to make a "change of variable" or "substitution" where we express the integrand in terms of a new variable. If we let $u = g(x)$, then $du/dx = g'(x)$. Although du/dx is one symbol, it is a common convention to think of du/dx as a ratio and write $du = g'(x)\, dx$. When regarded as individual quantities, dx and du are called *differentials*, and they are related by the equation $du = g'(x)\, dx$. With this notation, Equation 2 becomes

$$\int f(\underbrace{g(x)}_{u})\ \underbrace{g'(x)\, dx}_{du} = \int f(u)\, du = F(u) + C$$

where $F' = f$. Thus replacing $g(x)$ by u and $g'(x)\, dx$ by du converts an integral in x to an equivalent one in u, and we have established the following rule.

---

(3) ■ The Substitution Rule   If $u = g(x)$ is a differentiable function and f is continuous on the range of g, then

$$\int f(g(x))\, g'(x)\, dx = \int f(u)\, du$$

After evaluating the integral in u, we replace u by $g(x)$. Looking again at Example 1, we could make the substitution $u = 1 + x^2$ and then $du = 2x\, dx$. Thus

$$\int 2x\sqrt{1 + x^2}\, dx = \int \sqrt{u}\, du = \tfrac{2}{3}u^{3/2} + C = \tfrac{2}{3}(1 + x^2)^{3/2} + C$$

■ EXAMPLE 2   Using the Substitution Rule

Find $\int x^3(x^4 + 2)^6\, dx$.

SOLUTION

We make the substitution $u = x^4 + 2$ because its differential is $du = 4x^3\, dx$, which, apart from the constant factor 4, occurs in the integral. Thus $x^3\, dx = \frac{1}{4}\, du$, and so the Substitution Rule gives

$$\int x^3(x^4 + 2)^6 \, dx = \int u^6 \cdot \tfrac{1}{4} \, du = \tfrac{1}{4} \int u^6 \, du$$

$$= \tfrac{1}{4} \cdot \tfrac{1}{7} u^7 + C = \tfrac{1}{28}(x^4 + 2)^7 + C$$

Check the answer by differentiating it.

Notice that at the final stage we had to return to the original variable x.   ▪

The idea behind the Substitution Rule is to replace a relatively complicated integral by a simpler integral. This is accomplished by changing from the original variable x to a new variable u that is a function of x. Thus in Example 2 we replaced the integral $\int x^3(x^4 + 2)^6 \, dx$ by the simpler integral $\tfrac{1}{4}\int u^6 \, du$.

The main challenge in using the Substitution Rule is to think of an appropriate substitution. You should try to choose u to be some function in the integrand whose derivative also occurs (except for a constant factor). This was the case in Example 2. If that is not possible, try choosing u to be some complicated part of the integrand, such as the inner function in a composition of functions. Finding the right substitution is a bit of an art. It's not unusual to guess wrong; if your first guess doesn't work, try another substitution.

▪ EXAMPLE 3   Choosing a Substitution

Evaluate $\displaystyle\int \frac{t^2}{t^3 + 1} \, dt$.

SOLUTION

The expression $t^3 + 1$ in the denominator is the more complicated part of the integrand, and its derivative is $3t^2$, which appears in the numerator (without the constant multiple 3). Thus we let $u = t^3 + 1$. Then $du = 3t^2 \, dt$, so $t^2 \, dt = \tfrac{1}{3} \, du$ and the Substitution Rule gives

In general, when you see an integral of the form

$$\int \frac{g'(x)}{g(x)} \, dx$$

the substitution $u = g(x)$ leads to the antiderivative $\ln|g(x)| + C$.

$$\int \frac{t^2}{t^3 + 1} \, dt = \int \frac{1}{u} \cdot \frac{1}{3} \, du = \frac{1}{3} \int \frac{1}{u} \, du$$

$$= \tfrac{1}{3} \ln|u| + C = \tfrac{1}{3} \ln|t^3 + 1| + C$$ ▪

▪ EXAMPLE 4   Choosing a Substitution

Find $\displaystyle\int \frac{x}{\sqrt{1 - 4x^2}} \, dx$.

SOLUTION

In the denominator we have a composition of functions. Let $u = 1 - 4x^2$, the inner function in the composition. Then $du = -8x \, dx$, so $x \, dx = -\tfrac{1}{8} \, du$ and

$$\int \frac{x}{\sqrt{1 - 4x^2}} \, dx = \int \frac{1}{\sqrt{u}}\left(-\frac{1}{8}\right) du = -\frac{1}{8}\int u^{-1/2} \, du = -\frac{1}{8}\left[\frac{u^{1/2}}{1/2}\right] du$$

$$= -\tfrac{1}{8}\left(2\sqrt{u}\right) + C = -\tfrac{1}{4}\sqrt{1 - 4x^2} + C$$ ▪

The answer to Example 4 could be checked by differentiation, but instead let's check it with a graph. In Figure 1 we have used a computer to graph both the integrand $f(x) = x/\sqrt{1 - 4x^2}$ and its indefinite integral $g(x) = -\frac{1}{4}\sqrt{1 - 4x^2}$ (we take the case $C = 0$). Notice that $g(x)$ decreases when $f(x)$ is negative, increases when $f(x)$ is positive, and has its minimum value when $f(x) = 0$. So it seems reasonable, from the graphical evidence, that g is an antiderivative of f.

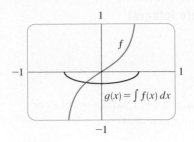

FIGURE 1

$$f(x) = \frac{x}{\sqrt{1 - 4x^2}}$$

$$g(x) = \int f(x)\, dx = -\frac{1}{4}\sqrt{1 - 4x^2}$$

When you see an integral of the form $\int g'(x) e^{g(x)}\, dx$, substituting $u = g(x)$ results in the antiderivative $e^{g(x)} + C$.

■ **EXAMPLE 5   The Substitution Rule with an Exponential Function**

Calculate $\int r e^{5r^2}\, dr$.

SOLUTION

If we let $u = 5r^2$, then the differential $du = 10r\, dr$ appears in the integral (without the constant multiple). Thus $r\, dr = \frac{1}{10}\, du$ and

$$\int r e^{5r^2}\, dr = \int e^u \cdot \frac{1}{10}\, du = \frac{1}{10}\int e^u\, du = \frac{1}{10} e^u + C = \frac{1}{10} e^{5r^2} + C \qquad ■$$

■ Substitution in Definite Integrals

When evaluating a *definite* integral by substitution, two methods are possible. One method is to evaluate the indefinite integral first and then use the Fundamental Theorem. For instance, using the result of Example 3, we have

$$\int_0^1 \frac{t^2}{t^3 + 1}\, dt = \frac{1}{3}\ln|t^3 + 1|\Big]_0^1 = \frac{1}{3}\ln(1 + 1) - \frac{1}{3}\ln(0 + 1) = \frac{1}{3}\ln 2$$

Another method, which is usually preferable, is to change the limits of integration when the variable is changed. This allows us to evaluate the integral without returning to the original variable.

This rule says that when using a substitution in a definite integral, we must put everything in terms of the new variable u, not only x and dx but also the limits of integration. The new limits of integration are the values of u that correspond to $x = a$ and $x = b$.

(4) ■ **The Substitution Rule for Definite Integrals**   If g' is continuous on $[a, b]$ and f is continuous on the range of $u = g(x)$, then

$$\int_a^b f(g(x))\, g'(x)\, dx = \int_{g(a)}^{g(b)} f(u)\, du$$

PROOF Let F be an antiderivative of f. Then, by (2), $F(g(x))$ is an antiderivative of $f(g(x))\, g'(x)$, so by the Fundamental Theorem, we have

$$\int_a^b f(g(x))\, g'(x)\, dx = F(g(x))\Big]_a^b = F(g(b)) - F(g(a))$$

But, applying the Fundamental Theorem a second time, we also have

$$\int_{g(a)}^{g(b)} f(u)\, du = F(u)\Big]_{g(a)}^{g(b)} = F(g(b)) - F(g(a)) \qquad ■$$

▪ **EXAMPLE 6**   **Substitution in a Definite Integral**

Evaluate $\int_0^1 \dfrac{t^2}{t^3 + 1}\, dt$ using (4).

SOLUTION

Using the substitution from Example 3, we have $u = t^3 + 1$, so $du = 3t^2\, dt$ or $\frac{1}{3}\, du = t^2\, dt$. To find the new limits of integration we note that

$$\text{when } t = 0, \quad u = (0)^3 + 1 = 1 \qquad \text{and} \qquad \text{when } t = 1, \quad u = (1)^3 + 1 = 2$$

Therefore

$$\int_0^1 \frac{t^2}{t^3 + 1}\, dt = \int_1^2 \frac{1}{u} \cdot \frac{1}{3}\, du = \tfrac{1}{3} \ln |u|\, \Big]_1^2$$

$$= \tfrac{1}{3} \ln 2 - \tfrac{1}{3} \ln 1 = \tfrac{1}{3} \ln 2 \qquad ▪$$

The integral given in Example 7 is an abbreviation for

$$\int_1^2 \frac{1}{(3 - 5x)^2}\, dx$$

▪ **EXAMPLE 7**   **Substitution in a Definite Integral**

Evaluate $\int_1^2 \dfrac{dx}{(3 - 5x)^2}$.

SOLUTION

Let $u = 3 - 5x$. Then $du = -5\, dx$, so $dx = -\frac{1}{5}\, du$. When $x = 1$, $u = -2$ and when $x = 2$, $u = -7$. Thus

$$\int_1^2 \frac{dx}{(3 - 5x)^2} = \int_{-2}^{-7} -\frac{1}{5} \cdot \frac{1}{u^2}\, du = -\frac{1}{5} \int_{-2}^{-7} u^{-2}\, du$$

$$= -\frac{1}{5} \left[\frac{u^{-1}}{-1} \right]_{-2}^{-7} = -\frac{1}{5} \left[-\frac{1}{u} \right]_{-2}^{-7} = \frac{1}{5} \left[\frac{1}{u} \right]_{-2}^{-7}$$

$$= \frac{1}{5} \left[-\frac{1}{7} - \left(-\frac{1}{2} \right) \right] = \frac{1}{5} \cdot \frac{5}{14} = \frac{1}{14} \qquad ▪$$

Since the function $f(x) = (\ln x)/x$ in Example 8 is positive for $x > 1$, the integral represents the area of the shaded region in Figure 2.

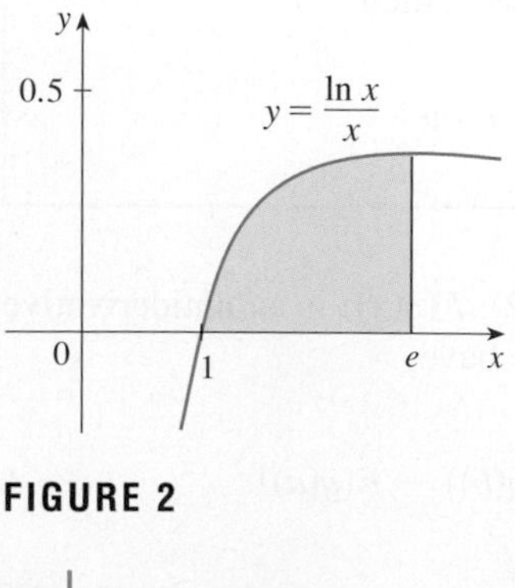

FIGURE 2

▪ **EXAMPLE 8**   **Substitution in a Definite Integral**

Calculate $\int_1^e \dfrac{\ln x}{x}\, dx$.

SOLUTION

We let $u = \ln x$ because its differential $du = (1/x)\, dx$ occurs in the integral. When $x = 1$, $u = \ln 1 = 0$; when $x = e$, $u = \ln e = 1$. Thus

$$\int_1^e \frac{\ln x}{x}\, dx = \int_0^1 u\, du = \frac{u^2}{2} \Bigg]_0^1 = \frac{1}{2} - 0 = \frac{1}{2} \qquad ▪$$

▪ | **Symmetry**

If a function possesses symmetry properties, the calculation of definite integrals can be simplified under certain circumstances. Recall from Section 1.1 that the graph of an even function [where $f(-x) = f(x)$] is symmetric about the y-axis, and the graph of an odd function [where $f(-x) = -f(x)$] is symmetric about the origin.

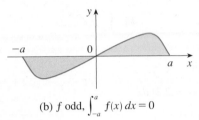

(a) f even, $\int_{-a}^{a} f(x)\, dx = 2\int_{0}^{a} f(x)\, dx$

(b) f odd, $\int_{-a}^{a} f(x)\, dx = 0$

FIGURE 3

Figure 3(a) shows a positive, even function for $-a \leqslant x \leqslant a$. Because of symmetry, the area below the curve on the left side is the same as that on the right side, so the area from $-a$ to a is twice the area from 0 to a. If f is an odd function, the area between the curve and the x-axis is the same for $-a$ to 0 as 0 to a, but one region is above the x-axis and the other is below, so the areas cancel and the integral from $-a$ to a is 0. [See Figure 3(b).]

We summarize these results in the following theorem. You are asked in Exercise 60 to use the Substitution Rule for Definite Integrals (4) to prove it.

---

(5) ■ **Integrals of Symmetric Functions**   Suppose f is continuous on $[-a, a]$.

(a) If f is even $[f(-x) = f(x)]$, then $\int_{-a}^{a} f(x)\, dx = 2\int_{0}^{a} f(x)\, dx$.

(b) If f is odd $[f(-x) = -f(x)]$, then $\int_{-a}^{a} f(x)\, dx = 0$.

■ **EXAMPLE 9**   **Integrating an Even Function**

Since $f(x) = x^6 + 1$ satisfies $f(-x) = f(x)$, it is even and so

$$\int_{-2}^{2} (x^6 + 1)\, dx = 2\int_{0}^{2} (x^6 + 1)\, dx$$

$$= 2\left[\tfrac{1}{7}x^7 + x\right]_{0}^{2} = 2\left(\tfrac{128}{7} + 2\right) = \tfrac{284}{7} \qquad ■$$

■ **EXAMPLE 10**   **Integrating an Odd Function**

Since $f(x) = x^3/(1 + x^2 + x^4)$ satisfies $f(-x) = -f(x)$, it is odd and so

$$\int_{-1}^{1} \frac{x^3}{1 + x^2 + x^4}\, dx = 0 \qquad ■$$

■ **Prepare Yourself**

1. Find the derivative of the function.

   (a) $y = e^{x^3 + 1}$ **(b)** $Q(t) = \ln(3t + t^2)$

   (c) $f(x) = (2x^2 + 3)^4$ **(d)** $g(z) = \sqrt{e^z + 5z}$

   (e) $r = 3^{2t+2}$

2. Evaluate the indefinite integral.

   (a) $\int (3x^5 + 4x - 1)\, dx$ **(b)** $\int 8\sqrt{t}\, dt$

   (c) $\int (5/v)\, dv$ **(d)** $\int (5/v^2)\, dv$

   (e) $\int 4^x\, dx$

3. Find functions f and g so that $h(x) = f(g(x))$.

   (a) $h(x) = (3x^2 + 2)^4$ **(b)** $h(x) = \sqrt{x^3 + 8}$

   (c) $h(x) = \dfrac{1}{x^3 - 2}$ **(d)** $h(x) = e^{x^2 + 1}$

■ **Exercises 5.4**

1–6 ■ Evaluate the integral by making the given substitution.

1. $\int e^{-x}\, dx, \quad u = -x$

2. $\int x^3(2 + x^4)^5\, dx, \quad u = 2 + x^4$

3. $\int x^2\sqrt{x^3 + 1}\, dx, \quad u = x^3 + 1$

4. $\int \dfrac{dt}{(1 - 6t)^4}, \quad u = 1 - 6t$

5. $\int \dfrac{p}{1 + 4p^2}\, dp, \quad u = 1 + 4p^2$

6. $\int \dfrac{r^2}{\sqrt{r^3 + 2}}\, dr, \quad u = r^3 + 2$

7–30 ▪ Evaluate the indefinite integral

7. $\int t(3 - t^2)^4 \, dt$

8. $\int x^2(x^3 + 5)^9 \, dx$

9. $\int (3x - 2)^{20} \, dx$

10. $\int (3t + 2)^{2.4} \, dt$

11. $\int q\sqrt{q^2 + 3.1} \, dq$

12. $\int w^2\sqrt{6 + 2w^3} \, dw$

13. $\int \dfrac{x^2}{\sqrt{0.4x^3 + 2.2}} \, dx$

14. $\int \dfrac{t}{\sqrt{2.7t^2 + 8}} \, dt$

15. $\int \dfrac{5z^2}{(z^3 + 2)^3} \, dz$

16. $\int \dfrac{x}{(x^2 + 1)^2} \, dx$

17. $\int e^x\sqrt{1 + e^x} \, dx$

18. $\int \dfrac{e^v}{(2e^v + 5)^2} \, dv$

19. $\int \dfrac{(\ln x)^2}{x} \, dx$

20. $\int \dfrac{4}{x(\ln x)^2} \, dx$

21. $\int t e^{2t^2} \, dt$

22. $\int p^2 e^{p^3+2} \, dp$

23. $\int \dfrac{dx}{5 - 3x}$

24. $\int \dfrac{z^2}{2z^3 - 10} \, dz$

25. $\int \dfrac{e^{\sqrt{t}+1}}{\sqrt{t}} \, dt$

26. $\int \dfrac{e^x}{e^x + 1} \, dx$

27. $\int 2^{3-4t} \, dt$

28. $\int 3^{x^2+5} x \, dx$

29. $\int (x^2 + 1)(x^3 + 3x)^4 \, dx$

30. $\int \dfrac{a + bx^2}{\sqrt{3ax + bx^3}} \, dx$

▱ **31–32** ▪ Evaluate the indefinite integral. Illustrate and check that your answer is reasonable by graphing both the function and its antiderivative (take $C = 0$).

31. $\int x(x^2 - 1)^3 \, dx$

32. $\int \dfrac{1 + \ln x}{x} \, dx$

33–44 ▪ Evaluate the definite integral.

33. $\int_0^1 \sqrt[3]{1 + 7x} \, dx$

34. $\int_0^1 (3t - 1)^{50} \, dt$

35. $\int_0^1 x^2(1 + 2x^3)^5 \, dx$

36. $\int_0^2 t^2\sqrt{8 - t^3} \, dt$

37. $\int_0^1 \dfrac{1}{(3v + 1)^2} \, dv$

38. $\int_0^2 \dfrac{x}{\sqrt{2x^2 + 1}} \, dx$

39. $\int_1^3 4z e^{z^2-1} \, dz$

40. $\int_0^1 \dfrac{x^3}{3x^4 + 1} \, dx$

41. $\int_1^4 \dfrac{e^{\sqrt{x}}}{\sqrt{x}} \, dx$

42. $\int_0^a x\sqrt{a^2 - x^2} \, dx$

43. $\int_e^{e^4} \dfrac{dx}{x\sqrt{\ln x}}$

44. $\int_1^2 x\left(3^{x^2-1}\right) dx$

45–48 ▪ Use symmetry properties to evaluate the definite integral.

45. $\int_{-1}^1 (3x^8 + x^4) \, dx$

46. $\int_{-3}^3 x^3 e^{-x^2} \, dx$

47. $\int_{-2}^2 \dfrac{t^3}{t^6 + 1} \, dt$

48. $\int_{-10}^{10} (0.2x^4 + 1.5x^2 + 2.2) \, dx$

---

49. Music sales   A recording artist released a new song as a digital download. If the track is downloaded at a rate of

$$50\left(\dfrac{t}{2t^2 + 5}\right) \text{ thousand downloads/week}$$

t weeks after its release, how many copies were downloaded during the first six weeks?

50. Blog readership   A new blog has added readers at a rate of

$$te^{-0.03t^2} \text{ thousand readers/week}$$

t weeks after its start. How many readers were added from the beginning of the second week to the end of the fifth week?

51. Gasoline storage   Gasoline is being added to a supply in a storage tank at the rate of

$$8te^{-0.026t^2} \text{ gallons/minute}$$

How many gallons are added to the tank during the first ten minutes?

52. Calculator production   Alabama Instruments Company has set up a production line to manufacture a new calculator. The rate of production of these calculators after t weeks is

$$\dfrac{dx}{dt} = 5000\left(1 - \dfrac{100}{(t + 10)^2}\right) \text{ calculators/week}$$

(Notice that production approaches 5000 per week as time goes on, but the initial production is lower because of the workers' unfamiliarity with the new techniques.) Find the number of calculators produced from the beginning of the third week to the end of the fourth week.

■ Challenge Yourself

53–54 ■ Evaluate the indefinite integral.

53. $\int x(2x + 5)^8 \, dx$ **54.** $\int x^2\sqrt{2 + x} \, dx$

55. If f is continuous and $\int_0^4 f(x) \, dx = 10$, find $\int_0^2 f(2x) \, dx$.

56. If f is continuous and $\int_0^9 f(x) \, dx = 4$, find $\int_0^3 x f(x^2) \, dx$.

57. Evaluate $\int_{-2}^2 (x + 3)\sqrt{4 - x^2} \, dx$ by writing it as a sum of two integrals and interpreting one of those integrals in terms of an area.

58. Evaluate $\int_0^1 x\sqrt{1 - x^4} \, dx$ by making a substitution and interpreting the resulting integral in terms of an area.

59. Are the following areas equal? Why or why not?

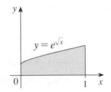
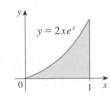

60. In this exercise we prove Theorem 5. First we split the integral $\int_{-a}^a f(x) \, dx$ in two to write

$$\int_{-a}^a f(x) \, dx = \int_{-a}^0 f(x) \, dx + \int_0^a f(x) \, dx$$

$$= -\int_0^{-a} f(x) \, dx + \int_0^a f(x) \, dx$$

(a) Make a substitution to show that

$$-\int_0^{-a} f(x) \, dx = \int_0^a f(-u) \, du$$

(b) Use part (a) to show that, if f is even,

$$\int_{-a}^a f(x) \, dx = 2\int_0^a f(x) \, dx$$

(c) Use part (a) to show that, if f is odd,

$$\int_{-a}^a f(x) \, dx = 0$$

61. If a and b are positive numbers, show that

$$\int_0^1 x^a(1 - x)^b \, dx = \int_0^1 x^b(1 - x)^a \, dx$$

5.5 Integration by Parts

■ Reversing the Product Rule

Every differentiation rule has a corresponding integration rule. For instance, the Substitution Rule for integration corresponds to the Chain Rule for differentiation. The rule that corresponds to the Product Rule for differentiation is called the rule for *integration by parts*.

The Product Rule states that if f and g are differentiable functions, then

$$\frac{d}{dx}\left[f(x)\,g(x)\right] = f(x)\,g'(x) + g(x)\,f'(x)$$

In the notation for indefinite integrals this equation becomes

$$\int \left[f(x)\,g'(x) + g(x)\,f'(x)\right] dx = f(x)\,g(x)$$

or $$\int f(x)\,g'(x) \, dx + \int g(x)\,f'(x) \, dx = f(x)\,g(x)$$

We can rearrange this equation as

(1) $$\int f(x)\,g'(x) \, dx = f(x)\,g(x) - \int g(x)\,f'(x) \, dx$$

■ Indefinite Integration by Parts

Formula 1 is called the **formula for integration by parts**. It is perhaps easier to remember in the following notation. Let $u = f(x)$ and $v = g(x)$. Then the differentials are $du = f'(x)\, dx$ and $dv = g'(x)\, dx$, so, by the Substitution Rule, the formula for integration by parts becomes

(2)
$$\int u\, dv = uv - \int v\, du$$

Notice that the formula for integration by parts converts an integral into an expression containing a different integral, which we still must evaluate.

■ EXAMPLE 1 Integrating by Parts

Find $\int xe^{2x}\, dx$.

SOLUTION USING FORMULA 1

Suppose we choose $f(x) = x$ and $g'(x) = e^{2x}$. Then $f'(x) = 1$ and $g(x) = \frac{1}{2}e^{2x}$. (For g we can choose *any* antiderivative of g', not necessarily the most general one.) Thus, using Formula 1, we have

$$\int xe^{2x}\, dx = f(x)\, g(x) - \int g(x)\, f'(x)\, dx$$

$$= x\left(\tfrac{1}{2}e^{2x}\right) - \int \tfrac{1}{2}e^{2x}(1)\, dx$$

$$= \tfrac{1}{2}xe^{2x} - \tfrac{1}{2}\int e^{2x}\, dx = \tfrac{1}{2}xe^{2x} - \tfrac{1}{4}e^{2x} + C$$

It's wise to check the answer by differentiating it. If we do so, we get xe^{2x}, as expected.

SOLUTION USING FORMULA 2

It is helpful to use the pattern:
$$u = \square \qquad dv = \square$$
$$du = \square \qquad v = \square$$

Let
$$u = x \qquad dv = e^{2x}\, dx$$

Then
$$du = dx \qquad v = \tfrac{1}{2}e^{2x}$$

and so

$$\int \overset{u}{x}\ \overset{dv}{e^{2x}\, dx} = \overset{u}{x} \cdot \overset{v}{\tfrac{1}{2}e^{2x}} - \int \overset{v}{\tfrac{1}{2}e^{2x}}\ \overset{du}{dx}$$

$$= \tfrac{1}{2}xe^{2x} - \tfrac{1}{2} \cdot \tfrac{1}{2}e^{2x} + C$$

$$= \tfrac{1}{2}xe^{2x} - \tfrac{1}{4}e^{2x} + C \qquad ■$$

NOTE: Our aim in using integration by parts is to obtain a simpler integral than the one we started with. Thus in Example 1 we started with $\int xe^{2x}\, dx$ and expressed it in terms of the simpler integral $\int e^{2x}\, dx$. If we had instead chosen $u = e^{2x}$ and $dv = x\, dx$, then $du = 2e^{2x}\, dx$ and $v = x^2/2$, so integration by parts gives

$$\int xe^{2x}\, dx = (e^{2x}) \frac{x^2}{2} - \int x^2 e^{2x}\, dx$$

Although this is true, $\int x^2 e^{2x}\, dx$ is a more difficult integral than the one we started with. In general, when deciding on a choice for u and dv, we usually try to choose $u = f(x)$ to be a function that becomes simpler when differentiated (or at least not more complicated) as long as $dv = g'(x)\, dx$ can be readily integrated to give v.

■ EXAMPLE 2 Integrating by Parts Twice

Find $\int t^2 e^t\, dt$.

SOLUTION

Notice that t^2 becomes simpler when differentiated (whereas e^t is unchanged when differentiated or integrated), so we choose

$$u = t^2 \qquad dv = e^t\, dt$$

Then

$$du = 2t\, dt \qquad v = e^t$$

Integration by parts gives

$$\textbf{(3)} \qquad \int t^2 e^t\, dt = t^2 e^t - 2\int t e^t\, dt$$

The integral that we obtained, $\int te^t\, dt$, is simpler than the original integral but requires integration by parts again. Choosing $u = t$ and $dv = e^t\, dt$ (similar to the choice we made in Example 1), we get $du = dt$, $v = e^t$, and

$$\int te^t\, dt = te^t - \int e^t\, dt = te^t - e^t + C$$

Putting this in Equation 3, we have

$$\int t^2 e^t\, dt = t^2 e^t - 2\int te^t\, dt$$
$$= t^2 e^t - 2(te^t - e^t + C)$$
$$= t^2 e^t - 2te^t + 2e^t - 2C$$

Because C is an arbitrary constant, multiplying C by -2 gives another arbitrary constant, so we could replace $-2C$ by a constant C_1 to make the expression simpler:

$$\int t^2 e^t\, dt = t^2 e^t - 2te^t + 2e^t + C_1 \qquad ■$$

Figure 1 illustrates Example 2 by showing the graphs of $f(t) = t^2 e^t$ and $F(t) = t^2 e^t - 2te^t + 2e^t$. As a visual check on our work, notice that $f(t) = 0$ when F has a horizontal tangent.

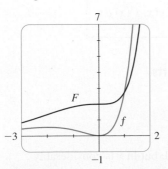

FIGURE 1

NOTE: How do we know when to try integration by parts? The technique is often effective when the integrand is the product of a power function and an exponential or logarithmic function. If you see a composition of functions, the Substitution Rule is likely to be a better choice.

■ **EXAMPLE 3**   **Integrating the Natural Logarithmic Function**

Evaluate $\int \ln x \, dx$.

SOLUTION

Although we don't have an apparent product of two functions, the natural logarithmic function can be integrated using integration by parts. Here we don't have much choice for u and dv. Let

$$u = \ln x \qquad dv = dx$$

Then

$$du = \frac{1}{x} \, dx \qquad v = x$$

Integrating by parts, we get

$$\int \ln x \, dx = x \ln x - \int x \cdot \frac{1}{x} \, dx$$

It's customary to write $\int 1 \, dx$ as $\int dx$.

$$= x \ln x - \int dx$$

$$= x \ln x - x + C$$

Check the answer by differentiating it.

Integration by parts is effective in this example because the derivative of the function $f(x) = \ln x$ is simpler than f.   ■

■ | Definite Integration by Parts

If we combine the formula for integration by parts with the Fundamental Theorem, we can evaluate definite integrals by parts. Evaluating both sides of Formula 1 between a and b, assuming f' and g' are continuous, and using the Fundamental Theorem, we obtain

(4) $$\int_a^b f(x) \, g'(x) \, dx = f(x) \, g(x) \Big]_a^b - \int_a^b g(x) \, f'(x) \, dx$$

■ **EXAMPLE 4**   **Definite Integration by Parts**

Calculate $\int_1^2 x^2 \ln x \, dx$.

SOLUTION

Both x^2 and $\ln x$ become simpler when differentiated, but $\ln x$ is considerably more complicated when integrated (see Example 3), so we choose

$$u = \ln x \qquad dv = x^2 \, dx$$

Then

$$du = \frac{1}{x} \, dx \qquad v = \tfrac{1}{3}x^3$$

Since $x^2 \ln x \geq 0$ for $x \geq 1$, the integral in Example 4 can be interpreted as the area of the region shown in Figure 2.

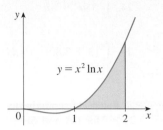

$y = x^2 \ln x$

FIGURE 2

Formula 4 gives

$$\int_1^2 x^2 \ln x\, dx = (\ln x)\tfrac{1}{3}x^3\Big]_1^2 - \int_1^2 \tfrac{1}{3}x^3 \cdot \frac{1}{x}\, dx$$

$$= \tfrac{1}{3}x^3 \ln x\Big]_1^2 - \tfrac{1}{3}\int_1^2 x^2\, dx$$

$$= \tfrac{8}{3}\ln 2 - \tfrac{1}{3}\ln 1 - \tfrac{1}{3}\big[\tfrac{1}{3}x^3\big]_1^2$$

$$= \tfrac{8}{3}\ln 2 - 0 - \tfrac{1}{3}\big[\tfrac{8}{3} - \tfrac{1}{3}\big]$$

$$= \tfrac{8}{3}\ln 2 - \tfrac{7}{9} \approx 1.0706$$

■ Can We Integrate All Continuous Functions?

In addition to the Substitution Rule and integration by parts, there are many more advanced techniques for evaluating integrals that we don't describe here. Lengthy tables of formulas have also been established to help. Mathematical software packages such as *Mathematica* and *Maple*, and some more advanced calculators like the Texas Instruments TI-89, are able to evaluate many integrals using these formulas and techniques.

The question arises: With all the known formulas and techniques, along with computers and calculators, can we find the integral of any continuous function? The answer is No, at least not in terms of the functions we are familiar with. For example, none of the following innocent-looking functions have antiderivatives that can be expressed in terms of familiar functions:

$$\int e^{x^2}\, dx \qquad\qquad \int \frac{e^x}{x}\, dx$$

$$\int \sqrt{x^3 + 1}\, dx \qquad\qquad \int \frac{1}{\ln x}\, dx$$

In fact, the majority of functions fall into this category!

If we cannot find an antiderivative for a function, then we are unable to use the Fundamental Theorem to evaluate a definite integral and our only option is to estimate its value. We have already done this using Riemann sums; in particular the Midpoint Rule (Section 5.1) can be used to approximate a definite integral. More advanced techniques can be found in books with a more complete treatment of calculus. In Exercises 23–27 you are asked to use a calculator or computer to estimate the values of several definite integrals.

In Appendix C we discuss additional techniques for estimating definite integrals.

■ Exercises 5.5

1–2 ■ Evaluate the integral using integration by parts with the indicated choices of u and dv.

1. $\int x \ln x\, dx; \quad u = \ln x,\ dv = x\, dx$

2. $\int te^{-t}\, dt; \quad u = t,\ dv = e^{-t}\, dt$

3–12 ■ Evaluate the indefinite integral.

3. $\int re^{r/2}\, dr$

4. $\int 3te^{0.4t}\, dt$

5. $\int x^3 \ln 2x\, dx$

6. $\int p^5 \ln p\, dp$

7. $\int (1 - 2z)e^{-z}\, dz$

8. $\int (x^2 + 1)\ln x\, dx$

9. $\int \ln \sqrt[3]{x}\ dx$

10. $\int \dfrac{\ln q}{q^3}\ dq$

11. $\int r^2 e^{-3r}\ dr$

12. $\int 3x^2 e^{2x}\ dx$

13–20 ■ Evaluate the definite integral

13. $\int_0^1 x e^{4x}\ dx$

14. $\int_1^2 t^4 \ln t\ dt$

15. $\int_0^1 \dfrac{y}{e^{2y}}\ dy$

16. $\int_0^1 (x^2 + 1)e^{-x}\ dx$

17. $\int_1^2 \dfrac{\ln x}{x^2}\ dx$

18. $\int_4^9 \dfrac{\ln y}{\sqrt{y}}\ dy$

19. $\int_1^2 (\ln x)^2\ dx$ *Hint:* Let $u = (\ln x)^2$, $dv = dx$

20. $\int_0^1 \dfrac{r^3}{\sqrt{4 + r^2}}\ dr$ *Hint:* Let $u = r^2$, $dv = \dfrac{r}{\sqrt{4 + r^2}}\ dr$

21–22 ■ Evaluate the indefinite integral. Illustrate, and check that your answer is reasonable, by graphing both the function and its antiderivative (take $C = 0$).

21. $\int x e^{-2x}\ dx$

22. $\int x^{3/2} \ln x\ dx$

23–26 ■ Use a calculator (or computer software) to estimate the value of the definite integral. Round to three decimal places.

23. $\int_0^4 \sqrt{r^3 + r}\ dr$

24. $\int_2^3 3.9 e^{-0.2t^2}\ dt$

25. $\int_1^5 \dfrac{8}{1 + 2e^{0.8t}}\ dt$

26. $\int_3^8 [3 + 2.8(\ln x)^3]\ dx$

27. Use a calculator (or computer software) to approximate the area under the curve $y = 2x^3 e^{-x}$ for $1 \le x \le 5$. Round your answer to three decimal places.

28. Blood alcohol content   Suppose the function $B'(t) = e^{-1.6t}(1 - 3t)$ gives the rate, measured in mg/mL per hour, at which a patient's blood alcohol concentration is changing during an experiment t hours after the patient finishes drinking an alcoholic beverage. Compute $\int_0^2 B'(t)\ dt$ and interpret your result.

29. Motion   An object that moves along a straight line has velocity $v(t) = t^2 e^{-t}$ meters per second after t seconds.

(a) How far will it travel during the first t seconds?

(b) How far will it travel during the first ten seconds?

Challenge Yourself

30. Rocket altitude   A rocket accelerates by burning its onboard fuel, so its mass decreases with time. Suppose the initial mass of the rocket at liftoff (including its fuel) is m, the fuel is consumed at rate r, and the exhaust gases are ejected with constant velocity v_e (relative to the rocket). A model for the velocity of the rocket at time t is given by the equation

$$v(t) = -gt - v_e \ln \dfrac{m - rt}{m}$$

where g is the acceleration due to gravity and t is not too large. If $g = 9.8$ m/s^2, $m = 30{,}000$ kg, $r = 160$ kg/s, and $v_e = 3000$ m/s, find the height of the rocket one minute after liftoff.

31–32 ■ First make a substitution and then use integration by parts to evaluate the integral.

31. $\int x \ln(1 + x)\ dx$

32. $\int t^3 e^{-t^2}\ dt$

33–34 ■ Use integration by parts to prove the given *reduction formula*.

33. $\int (\ln x)^n\ dx = x(\ln x)^n - n \int (\ln x)^{n-1}\ dx$

34. $\int x^n e^x\ dx = x^n e^x - n \int x^{n-1} e^x\ dx$

35. Use Exercise 33 to find $\int (\ln x)^3\ dx$.

36. Use Exercise 34 to find $\int x^4 e^x\ dx$.

37. Suppose that $f(1) = 2$, $f(4) = 7$, $f'(1) = 5$, $f'(4) = 3$, and f'' is continuous. Find the value of $\int_1^4 x f''(x)\ dx$.

38. (a) Evaluate $\int x^3 \sqrt{x^2 + 1}\ dx$ using the Substitution Rule.

(b) Evaluate the same integral using integration by parts.

(c) Show that your answers from parts (a) and (b) are equivalent.

■ CONCEPT CHECK

1. (a) How do we estimate the area under a curve?
   (b) Write an expression for a Riemann sum of a function f.
   (c) If $f(x) \geq 0$, what is the geometric interpretation of a Riemann sum? Illustrate with a diagram.

2. (a) Write the definition of the definite integral of a continuous function from a to b.
   (b) What is the geometric interpretation of $\int_a^b f(x)\, dx$ if $f(x) \geq 0$ on $[a, b]$?
   (c) What is the geometric interpretation of $\int_a^b f(x)\, dx$ if $f(x)$ takes on both positive and negative values on $[a, b]$? Illustrate with a diagram.

3. What does the Midpoint Rule say?

4. (a) What is an antiderivative of a function f?
   (b) How do we find the most general antiderivative of f?

5. (a) State the Fundamental Theorem of Calculus.
   (b) State the Net Change Theorem.

6. What does the area from a to b under a marginal cost curve represent?

7. If $r(t)$ is the rate at which water flows into a reservoir, what does $\int_{t_1}^{t_2} r(t)\, dt$ represent?

8. Suppose an object moves back and forth along a straight line with velocity $v(t)$, measured in feet per second, and acceleration $a(t)$.
   (a) What is the meaning of $\int_{60}^{120} v(t)\, dt$?
   (b) What is the meaning of $\int_{60}^{120} |v(t)|\, dt$?
   (c) What is the meaning of $\int_{60}^{120} a(t)\, dt$?

9. (a) Explain the meaning of the indefinite integral $\int f(x)\, dx$.
   (b) What is the connection between the definite integral $\int_a^b f(x)\, dx$ and the indefinite integral $\int f(x)\, dx$?

10. If t is measured in minutes and $g(t)$ is measured in gallons per minute, what are the units for $\int_{10}^{30} g(t)\, dt$?

11. How do we find the average value of a function f on an interval $[a, b]$?

12. (a) State the Substitution Rule. In practice, how do you use it?
    (b) State the rule for integration by parts. In practice, how do you use it?

13. Explain exactly what is meant by the statement that "differentiation and integration are inverse processes."

Answers to the Concept Check can be found on the back endpapers.

■ Exercises

1. **Total cost** The table shows the marginal cost for a tool manufacturer's power drill at various production levels.

Units	Marginal cost ($/unit)	Units	Marginal cost ($/unit)
0	17.48	35,000	11.62
5,000	15.69	40,000	11.56
10,000	15.01	45,000	11.68
15,000	14.71	50,000	11.89
20,000	13.84	55,000	12.65
25,000	13.05	60,000	14.18
30,000	12.24		

(a) Estimate the cost (not including fixed costs) to produce the first 60,000 units by considering batches of 10,000 units. Use the initial marginal cost for each batch in your calculations.

(b) Use batches of 5,000 units and the ending marginal cost for each batch to estimate the cost of raising the production level from 20,000 to 40,000 units.

2. **Whale migration** A researcher has installed a transponder on a gray whale and is tracking its movement during migration. The table shows the speeds recorded every two hours during one day. Estimate the distance the whale traveled using the velocities at the beginning of the time intervals.

t (h)	v (ft/s)	t (h)	v (ft/s)
0	3.8	14	7.0
2	4.1	16	9.4
4	4.0	18	5.7
6	5.3	20	5.3
8	6.8	22	4.9
10	6.6	24	4.2
12	11.1		

3. Use the given graph of f to find the Riemann sum with six subintervals using **(a)** left endpoints and **(b)** midpoints. In each case draw a diagram and explain what the Riemann sum represents.

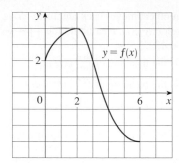

4. Braking distance   The velocity graph of a braking car is shown. Use it to estimate the distance traveled by the car while the brakes are applied.

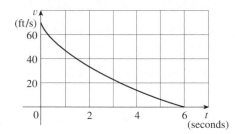

5. Evaluate

$$\int_0^1 \left(x + \sqrt{1 - x^2} \right) dx$$

by writing it as two separate integrals and interpreting them in terms of areas.

6. Use the Midpoint Rule with $n = 6$ subintervals to estimate the area under the curve $y = e^x/x$ from $x = 1$ to $x = 4$.

7. Use the Midpoint Rule with $n = 6$ subintervals to approximate $\int_0^{12} \ln(x^3 + 1) \, dx$.

8. Find f if $f'(x) = -6x^2 + 4x + 3$ and $f(1) = 5$.

9–12 ▪ Find the most general antiderivative of the function.

9. $f(x) = 2x^3 + 6x - 7$ **10.** $g(t) = \dfrac{1}{t^2} - 4\sqrt{t}$

11. $p(r) = 4 + \dfrac{5}{r}$ **12.** $h(z) = 7.6e^{-0.4z}$

13–18 ▪ Evaluate the integral.

13. $\displaystyle\int_1^2 (8x^3 + 3x^2) \, dx$ **14.** $\displaystyle\int_0^1 (x^4 - 8x + 7) \, dx$

15. $\displaystyle\int_0^1 (1 - x^9) \, dx$ **16.** $\displaystyle\int_1^4 12\sqrt{w} \, dw$

17. $\displaystyle\int_0^2 5e^{2t} \, dt$ **18.** $\displaystyle\int_1^3 \left(2 + \dfrac{3}{x} \right) dx$

19. Find the area under the graph of $f(x) = 2\sqrt[3]{x}$ for $1 \le x \le 8$.

20. For the function $h(t) = 2 + 4e^{-0.5t}$ find

(a) $\displaystyle\int h(t) \, dt$ **(b)** $\displaystyle\int_0^4 h(t) \, dt$

21–24 ▪ Find the general indefinite integral.

21. $\displaystyle\int (7.2t^2 - 4.6t + 18.1) \, dt$ **22.** $\displaystyle\int \left(1.8\sqrt{u} + 2.1 \right) du$

23. $\displaystyle\int \left(\dfrac{6}{x} + x \right) dx$ **24.** $\displaystyle\int \left(\dfrac{3x - x^2 + 2}{x} \right) dx$

25. Oil consumption   Let $r(t)$ be the rate at which the world's oil is consumed, where t is measured in years starting at $t = 0$ on January 1, 2000, and $r(t)$ is measured in barrels per year. What does $\int_0^8 r(t) \, dt$ represent?

26. Unemployment rate   Suppose $g(t)$ is the rate at which the percentage of Americans who are unemployed changes, where t is measured in months starting at $t = 0$ on January 1, 2010, and $g(t)$ is measured in percentage points per month. What does $\int_4^6 g(t) \, dt$ represent?

27. Integrating marginal cost   If $C'(q)$ is the marginal cost, in dollars per unit, of producing q laptop computers, what does $\int_{500}^{1000} C'(q) \, dq$ represent?

28. Water flow   Suppose $w(t)$ measures the rate at which water flows through a dam, measured in thousands of gallons per hour. If t is measured in hours and $t = 0$ corresponds to 12:00 AM today, interpret the statement $\int_8^{10} w(t) \, dt = 1450$.

29. Integrating marginal revenue   A company estimates that the marginal revenue after producing q pianos is $4.3 - 0.002q$ thousand dollars per piano. Find the increase in revenue if production is raised from 1200 to 1800 pianos.

30. Bacteria population   A population of bacteria is growing at a rate of $r(t) = 3.4e^{0.24t}$ thousand bacteria per hour after t hours. How much did the bacteria population increase during the first four hours?

31. Leaking gasoline   Gasoline is leaking from a car tank at a rate of $r(t) = 2.1e^{-0.3t}$ gallons per minute t minutes after the tank developed the leak. How much gasoline is lost during the first ten minutes?

32. Insect population   A population of honeybees increased at a rate of $r(t)$ bees per week, where the graph of r is as shown. Use the Midpoint Rule with six subintervals to estimate the increase in the bee population during the first 24 weeks.

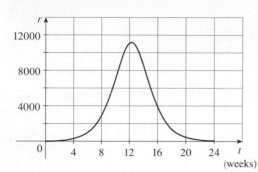

33. Motion   A particle moves along a line with velocity function $v(t) = t^2 - t$, where v is measured in meters per second. Find (a) the displacement and (b) the distance traveled by the particle during the time interval $[0, 5]$.

34. Find the average value of the function $f(x) = x^2\sqrt{1 + x^3}$ on the interval $[0, 2]$.

35. Radioactive decay   The amount of a radioactive material, measured in ounces, remaining t hours after the start of an experiment is given by $A(t) = 7.4e^{-0.12t}$. What is the average amount of material present during the first five hours?

36–45 ■ Use the Substitution Rule to evaluate the integral.

36. $\int t(t^2 - 4)^5\, dt$

37. $\int x^2(1 + x^3)^6\, dx$

38. $\int v\sqrt{3v^2 + 2}\, dv$

39. $\int_0^1 \dfrac{x}{x^2 + 1}\, dx$

40. $\int_0^1 (1 - x)^9\, dx$

41. $\int_0^2 we^{4-w^2}\, dw$

42. $\int_1^2 \dfrac{1}{2 - 3x}\, dx$

43. $\int e^x\sqrt{e^x + 2}\, dx$

44. $\int_1^4 \dfrac{dt}{(2t + 1)^3}$

45. $\int x5^{x^2}\, dx$

46–50 ■ Use integration by parts to evaluate the integral.

46. $\int 4xe^{2x}\, dx$

47. $\int_1^4 x^{3/2}\ln x\, dx$

48. $\int_0^5 ye^{-0.6y}\, dy$

49. $\int \dfrac{t^2}{e^{4t}}\, dt$

50. $\int \dfrac{\ln t}{\sqrt{t}}\, dt$

51–54 ■ Evaluate the integral.

51. $\int_0^3 3^t\, dt$

52. $\int_1^2 x^3\ln x\, dx$

53. $\int \dfrac{x + 2}{\sqrt{x^2 + 4x}}\, dx$

54. $\int \dfrac{3p}{2p^2 + 5}\, dp$

55–56 ■ Evaluate the indefinite integral. Illustrate and check that your answer is reasonable by graphing both the function and its antiderivative (take $C = 0$).

55. $\int 8xe^{-x^2}\, dx$

56. $\int \dfrac{4x}{\sqrt{x^2 + 1}}\, dx$

---

57. Use a graphing calculator (or computer software) to estimate the value of

$$\int_0^{12} \dfrac{23}{1 + 3e^{-0.2t}}\, dt$$

Round to two decimal places.

58. Use symmetry to evaluate $\int_{-3}^3 \dfrac{x}{2x^4 + 5}\, dx$.

59. If $\int_0^6 f(x)\, dx = 10$ and $\int_0^4 f(x)\, dx = 7$, find $\int_4^6 f(x)\, dx$.

6

If you raise the price of tickets at a movie theater, you may get fewer people purchasing them but you collect more money per person. The techniques of this chapter will enable us to analyze the tradeoff between price and attendance and determine the price that produces the most revenue.   © Wernher Krutein / photovault.com

Applications of Integration

In this chapter we explore some of the applications of the definite integral. We see that the area between two curves has meaning in a wide variety of contexts and we learn how to compute it. We investigate several different applications to economics and biology, including measures of the equity of economic markets and predictions of future populations given information about the rates of survival and reproduction of its members. We look at differential equations and how integrals are used to solve them. Finally, we see that definite integrals can be defined over infinite intervals and how these integrals are used to evaluate probabilities.

The common theme in many of these applications is the following general method, which is similar to the one we used to find areas under curves: We break up a quantity Q into a large number of small parts. We next approximate each small part by a quantity of the form $f(x_i)\,\Delta x$ and thus approximate Q by a Riemann sum. Then we take the limit and express Q as an integral.

In Chapter 5 we defined and calculated areas of regions that lie under the graphs of functions. The area of a region that lies between the graphs of two functions can also have meaningful applications. In this section we use integrals to calculate the areas of such regions.

The Area Between Two Curves

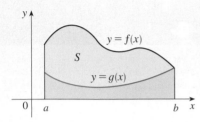

FIGURE 1

$$A = \int_a^b f(x)\,dx - \int_a^b g(x)\,dx$$

Consider the region S that lies between two curves $y = f(x)$ and $y = g(x)$ and between the vertical lines $x = a$ and $x = b$, where f and g are continuous functions and $f(x) \geqslant g(x)$ for all x in $[a, b]$. If f and g are both positive functions, as in Figure 1, then intuitively we can see that the area A of the region S is

$$A = [\text{area under } y = f(x)] - [\text{area under } y = g(x)]$$

$$= \int_a^b f(x)\,dx - \int_a^b g(x)\,dx = \int_a^b [f(x) - g(x)]\,dx$$

If it happens that one or both of the curves go below the x-axis, as in Figure 2, we can shift both graphs upward a distance c so that the curves lie entirely above the x-axis. (See Figure 3.)

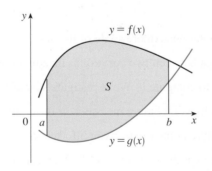

FIGURE 2

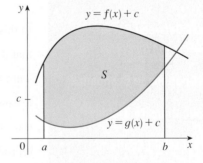

FIGURE 3

The shape of the region between the curves remains unchanged, so the area between the curves is

$$A = [\text{area under } y = f(x) + c] - [\text{area under } y = g(x) + c]$$

$$= \int_a^b [f(x) + c]\,dx - \int_a^b [g(x) + c]\,dx$$

$$= \int_a^b ([f(x) + c] - [g(x) + c])\,dx$$

$$= \int_a^b [f(x) + c - g(x) - c]\,dx = \int_a^b [f(x) - g(x)]\,dx$$

Thus, regardless of whether f and g are positive functions, we have the following formula for the area between two curves.

> **(1)** ▪ The area A of the region bounded by the curves $y = f(x)$, $y = g(x)$, and the lines $x = a$, $x = b$, where f and g are continuous and $f(x) \geqslant g(x)$ for all x in $[a, b]$, is
>
> $$A = \int_a^b [f(x) - g(x)]\, dx$$

▪ EXAMPLE 1 Area Between Two Curves

Find the area of the region bounded above by $y = e^x$, bounded below by $y = x$, and bounded on the sides by $x = 0$ and $x = 1$.

SOLUTION

The region is shown in Figure 4. The upper boundary curve is $y = e^x$ and the lower boundary curve is $y = x$. So we use the area formula (1) with $f(x) = e^x$, $g(x) = x$, $a = 0$, and $b = 1$:

$$A = \int_0^1 (e^x - x)\, dx = e^x - \tfrac{1}{2}x^2 \Big]_0^1$$

$$= e - \tfrac{1}{2} - 1 = e - 1.5 \approx 1.2183$$ ▪

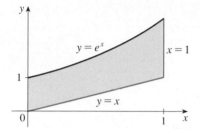

FIGURE 4

In the next example both of the side boundaries reduce to a point, so we must first identify the values a and b.

▪ EXAMPLE 2 Finding the Area Between Intersecting Curves

Find the area of the region enclosed by the parabolas $y = x^2$ and $y = 2x - x^2$.

SOLUTION

We first find the points of intersection of the parabolas by solving their equations simultaneously. This gives $x^2 = 2x - x^2$, or $2x^2 - 2x = 0$. Thus $2x(x - 1) = 0$, so $x = 0$ or 1. The points of intersection are $(0, 0)$ and $(1, 1)$.

We see from Figure 5 that the top and bottom boundaries are

$$y_T = 2x - x^2 \qquad \text{and} \qquad y_B = x^2$$

and the region lies between $x = 0$ and $x = 1$. So the total area is

$$A = \int_0^1 (2x - x^2 - x^2)\, dx = \int_0^1 (2x - 2x^2)\, dx$$

$$= \left[x^2 - \tfrac{2}{3}x^3 \right]_0^1 = 1 - \tfrac{2}{3} - 0 = \tfrac{1}{3}$$ ▪

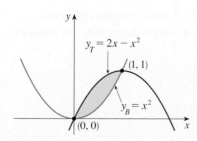

FIGURE 5

Sometimes it's difficult, or even impossible, to find the points of intersection of two curves exactly. In Exercises 23–26 we ask you to use a graphing calculator (or computer) to find approximate values for the intersection points and then proceed as before.

▪ Applications

In Section 3.2 we noted that profit is revenue minus cost. The next example shows that we can interpret profit as the area between the graphs of the marginal cost and revenue functions.

▪ EXAMPLE 3 Area Between Marginal Cost and Revenue Curves

Figure 6 shows graphs of the marginal cost and marginal revenue functions for a bakery's energy snack bar. Estimate the increase in profit the bakery earns by raising production from 1000 to 4000 bars.

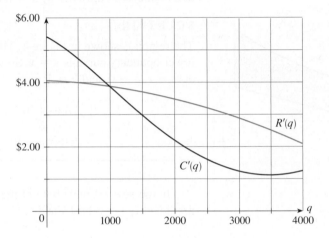

FIGURE 6

SOLUTION

We know from Section 5.1 that the area under a marginal cost curve represents the total increase in cost; similarly, the area under the marginal revenue curve corresponds to the total increase in revenue. Because profit is revenue minus cost, we can interpret the area between the curves as the total increase in profit. (Notice that for $1000 \le q \le 4000$ marginal revenue is greater than marginal cost, so profit is positive.)

  We can estimate this area by using the Midpoint Rule with the *difference* between the function values. Taking subintervals of length $\Delta q = 500$, we estimate the midpoint values as shown in Figure 7.

Notice in Figure 6 that $R'(q) < C'(q)$ for $0 \le q \le 1000$, so there the area between the curves represents a *decrease* in profit. If we were to estimate the value of $\int_0^{4000} [R'(q) - C'(q)] \, dq$, the area between the curves from 0 to 1000 would be *subtracted* from the area between the curves for $1000 \le q \le 4000$.

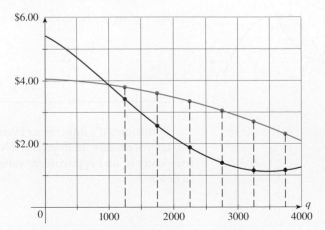

FIGURE 7

We then compile this information in a table:

Midpoint	1250	1750	2250	2750	3250	3750
$R'(q)$	3.8	3.6	3.3	3.1	2.7	2.3
$C'(q)$	3.5	2.6	1.9	1.4	1.1	1.1
$R'(q) - C'(q)$	0.3	1.0	1.4	1.7	1.6	1.2

Then

$$\int_{1000}^{4000} [R'(q) - C'(q)]\, dq \approx [0.3 + 1.0 + 1.4 + 1.7 + 1.6 + 1.2]\, \Delta q$$

$$= 7.2(500) = 3600$$

Thus the increase in profit is approximately $3600.   ▪

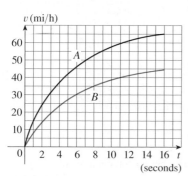

v (mi/h)

FIGURE 8

▪ EXAMPLE 4   Area Between Velocity Curves

Figure 8 shows velocity curves for two cars, A and B, that start side by side and move along the same road. What does the area between the curves represent? Use the Midpoint Rule to estimate it.

SOLUTION

We know from Section 5.3 that the area under the velocity curve A represents the distance traveled by car A during the first 16 seconds. Similarly, the area under curve B is the distance traveled by car B during that time period. So the area between these curves, which is the difference of the areas under the curves, is the distance between the cars after 16 seconds. We read the velocities from the graph and convert them to feet per second $\left(1 \text{ mi/h} = \frac{5280}{3600} \text{ ft/s}\right)$.

t	0	2	4	6	8	10	12	14	16
v_A	0	34	54	67	76	84	89	92	95
v_B	0	21	34	44	51	56	60	63	65
$v_A - v_B$	0	13	20	23	25	28	29	29	30

We use the Midpoint Rule with $n = 4$ intervals, so that $\Delta t = 4$. The midpoints of the intervals are $\bar{t}_1 = 2$, $\bar{t}_2 = 6$, $\bar{t}_3 = 10$, and $\bar{t}_4 = 14$. We estimate the distance between the cars after 16 seconds as follows:

$$\int_0^{16} (v_A - v_B)\, dt \approx \Delta t[13 + 23 + 28 + 29]$$

$$= 4(93) = 372 \text{ ft}$$ ▪

▪ Prepare Yourself

1. Evaluate the definite integral.

(a) $\displaystyle\int_0^8 (0.4x^2 - 6x + 1.8)\, dx$

(b) $\displaystyle\int_0^2 (e^t - t)\, dt$

(c) $\displaystyle\int_1^4 \left(1/x - 2\sqrt{x}\right) dx$

(d) $\displaystyle\int_0^1 (2.5^x - 3x)\, dx$

(e) $\displaystyle\int_1^2 \left(\frac{1}{q} - \frac{4}{q^2}\right) dq$

2. Use the Midpoint Rule with four subintervals to estimate the value of

$$\int_0^{12} \ln(2x^2 + 5)\, dx$$

3. Find the points where the curves $y = 2x^2 - 8$ and $y = x^2 - x + 4$ intersect.

4. If $C'(q)$ is the marginal cost, in thousands of dollars per unit, of producing q units of a commodity, what does $\int_0^{1500} C'(q)\, dq$ measure?

5. If $r(t)$ is the rate at which oil drains from a tank after t minutes, measured in quarts per minute, what does $\int_5^{20} r(t)\, dt$ represent?

▪ Exercises 6.1

1–4 ▪ Find the area of the shaded region.

1.

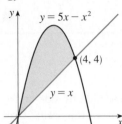

2.

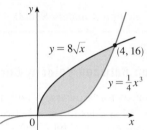

3.

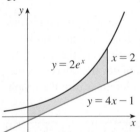

4.

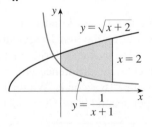

5–18 ▪ Sketch the region enclosed by the given curves and find its area.

5. $y = x + 1$, $y = 9 - x^2$, $x = -1$, $x = 2$

6. $y = x^4$, $y = -x - 1$, $x = 0$, $x = 1$

7. $y = e^x$, $y = x^2 - 1$, $x = -1$, $x = 1$

8. $y = \sqrt{x}$, $y = 3x^2 + 1$, $x = 1$, $x = 4$

9. $y = x$, $y = x^2$

10. $y = x^2 - 4x$, $y = 2x$

11. $y = 1/x$, $y = 1/x^2$, $x = 2$

12. $y = x^2 - 2x$, $y = x + 4$

13. $y = x^2$, $y = \sqrt{x}$

14. $y = x^2$, $y = 4x - x^2$

15. $y = 12 - x^2$, $y = x^2 - 6$

16. $y = x^3 - x$, $y = 3x$, $0 \le x \le 2$

17. $y = e^x$, $y = xe^x$, $x = 0$

18. $y = |x|$, $y = x^2 - 2$

19. The graphs of two functions are shown with the areas of the regions between the curves indicated.

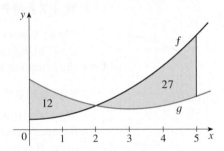

(a) What is the total area between the curves for $0 \le x \le 5$?

(b) What is the value of $\int_0^5 [f(x) - g(x)]\, dx$?

20. Sketch the region enclosed by the curves $y = 8 - x^2$, $y = x^2$, $x = 0$, $x = 3$ and observe that the region between them consists of two separate parts. Find the area of this region.

21–22 ▪ Use the Midpoint Rule with $n = 4$ subintervals to approximate the area of the region bounded by the given curves.

21. $y = \sqrt{x^2 - 1}$, $y = (\ln x)^2$, $x = 1$, $x = 5$

22. $y = \sqrt[3]{16 - x^3}$, $y = x$, $x = 0$

23–26 ▪ Use a graphing calculator (or computer) to find approximate x-coordinates of the points of intersection of the given curves. Then find (approximately) the area of the region enclosed by the curves.

23. $y = 3x^2 - 2x$, $y = x^3 - 3x + 4$

24. $y = e^x$, $y = 2 - x^2$

25. $y = 1/x^2$, $y = 3x - x^2$

26. $y = 1.3^x$, $y = 2\sqrt{x}$

27. Marginal cost and revenue The marginal cost, in dollars per unit, for a company's portable heater is

$$C'(q) = 48 - 0.03q + 0.00002q^2$$

and the marginal revenue is $R'(q) = 44 - 0.007q$. Find the area between the graphs of these functions for $0 \leq q \leq 200$ and interpret your result. [Note that $C'(q) > R'(q)$ for $0 \leq q \leq 200$.]

28. Birth and death rates If the birth rate of a population is $b(t) = 2200e^{0.024t}$ people per year and the death rate is $d(t) = 1460e^{0.018t}$ people per year, find the area between these curves for $0 \leq t \leq 10$. What does this area represent?

29. Hard drive production In 2010 a particular factory produced computer hard drives at a rate of

$$f(t) = 4.3e^{0.0172t} \text{ thousand drives per month}$$

where t is the time in months after the start of that year. In 2011 the production rate was

$$g(t) = 5.3e^{0.0164t} \text{ thousand drives per month}$$

t months after the start of that year. Compute the area between the graphs of these functions for $0 \leq t \leq 12$ and interpret your result in this context.

30. Lake depth The water depth in a lake has been declining. The rate at which the water level changed t weeks after the start of last year was given by

$$r_2(t) = -0.14t - 3.2 \text{ inches per week}$$

The previous year the level changed at a rate of

$$r_1(t) = 0.005t^2 - 0.3t - 1.5 \text{ inches per week}$$

where t is the number of weeks after the start of that year. Compute the area between the graphs of these functions for $0 \leq t \leq 52$. What does your answer represent?

31. Property value An investor purchased two commercial buildings for the same price. The value of one changed at a rate of

$$r_1(t) = 0.063(1.041^t) \text{ million dollars per year}$$

t years after after the purchase date while the value of the other property changed at a rate of

$$r_2(t) = 0.047(1.038^t) \text{ million dollars per year}$$

What is the area between the graphs for $0 \leq t \leq 10$? What does this represent?

32. Rainfall The rates at which rain fell, in inches per hour, in two different locations t hours after the start of a storm are given by

$$f(t) = 0.73t^3 - 2t^2 + t + 0.6$$

and $\quad g(t) = 0.17t^2 - 0.5t + 1.1$

Compute the area between the graphs for $0 \leq t \leq 2$ and interpret your result in this context.

33. Driving distance Two cars, A and B, start side by side and accelerate from rest. The figure shows the graphs of their velocity functions.

  (a) Which car is ahead after one minute? Explain.

  (b) What is the meaning of the area of the shaded region?

  (c) Which car is ahead after two minutes? Explain.

  (d) Estimate the time at which the cars are again side by side.

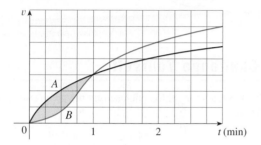

34. Marginal cost and revenue The figure shows graphs of the marginal revenue function R' and the marginal cost function C' for a manufacturer. [Recall from Section 4.7 that $R(x)$ and $C(x)$ represent the revenue and cost when x units are manufactured. Assume that R and C are measured in thousands of dollars.] What is the meaning of the area of the shaded region? Use the Midpoint Rule to estimate the value of this quantity.

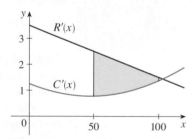

35. Driving distance Racing cars driven by Chris and Kelly are side by side at the start of a race. The table shows the velocities of each car, in miles per hour, during the first ten seconds of the race. Use the Midpoint Rule to estimate how much farther Kelly travels than Chris does during the first ten seconds.

t	v_C	v_K	t	v_C	v_K
0	0	0	6	69	80
1	20	22	7	75	86
2	32	37	8	81	93
3	46	52	9	86	98
4	54	61	10	90	102
5	62	71			

36. Pool area The widths, in meters, of a kidney-shaped swimming pool were measured at 2-meter intervals as indicated in the figure. Use the Midpoint Rule to estimate the area of the pool.

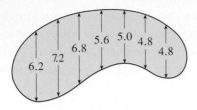

37. Cross-sectional area   A cross-section of an airplane wing is shown. Measurements of the thickness of the wing, in centimeters, at 20-centimeter intervals are 5.8, 20.3, 26.7, 29.0, 27.6, 27.3, 23.8, 20.5, 15.1, 8.7, and 2.8. Use the Midpoint Rule to estimate the area of the wing's cross-section.

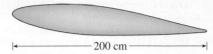

← 200 cm →

▪ Challenge Yourself

38–39 ▪ Sketch the region enclosed by the given curves and find its area.

38. $y = 3x^2$, $y = 8x^2$, $4x + y = 4$, $x \geq 0$

39. $y = 1/x$, $y = x$, $y = \frac{1}{4}x$, $x > 0$

40. Find the values of c such that the area of the region bounded by the parabolas $y = x^2 - c^2$ and $y = c^2 - x^2$ is 576.

41. Find the number b such that the line $y = b$ divides the region bounded by the curves $y = x^2$ and $y = 4$ into two regions with equal area.

42. Find the area of the region bounded by the parabola $y = x^2$, the tangent line to this parabola at $(1, 1)$, and the x-axis.

P R O J E C T   ▪ The Gini Index

How is it possible to measure the distribution of income among the inhabitants of a given country? One such measure is the *Gini index*, named after the Italian economist Corrado Gini, who first devised it in 1912.

We first rank all households in a country by income and then we compute the percentage of households whose income is at most a given percentage of the country's total income. We define a **Lorenz curve** $y = L(x)$ on the interval $[0, 1]$ by plotting the point $(a/100, b/100)$ on the curve if the bottom $a\%$ of households receive at most $b\%$ of the total income. For instance, in Figure 1 the point $(0.4, 0.12)$ is on the Lorenz curve for the United States in 2008 because the poorest 40% of the population received just 12% of the total income. Likewise, the bottom 80% of the population received 50% of the total income, so the point $(0.8, 0.5)$ lies on the Lorenz curve. (The Lorenz curve is named after the American economist Max Lorenz.)

Figure 2 shows some typical Lorenz curves. They all pass through the points $(0, 0)$ and $(1, 1)$ and are concave upward. In the extreme case $L(x) = x$, society is perfectly egalitarian: The poorest $a\%$ of the population receives $a\%$ of the total income and so everybody receives the same income. The area between a Lorenz curve $y = L(x)$ and the line $y = x$ measures how much the income distribution differs from absolute equality. The **Gini index** (sometimes called the **Gini coefficient** or the **coefficient of inequality**) is the area between the Lorenz curve and the line $y = x$ (shaded in Figure 3) divided by the area under $y = x$.

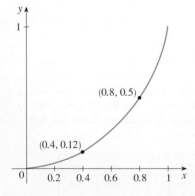

FIGURE 1
Lorenz curve for the US in 2008

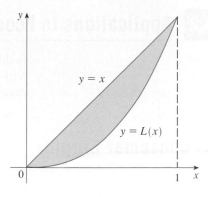

FIGURE 2   **FIGURE 3**

1. **(a)** Show that the Gini index G is twice the area between the Lorenz curve and the line $y = x$, that is,

$$G = 2 \int_0^1 [x - L(x)] \, dx$$

(b) What is the value of G for a perfectly egalitarian society (everybody has the same income)? What is the value of G for a perfectly totalitarian society (a single person receives all the income)?

2. The following table (derived from data supplied by the US Census Bureau) shows values of the Lorenz function for income distribution in the United States for the year 2008.

x	0.0	0.2	0.4	0.6	0.8	1.0
$L(x)$	0.000	0.034	0.120	0.267	0.500	1.000

(a) What percentage of the total US income was received by the richest 20% of the population in 2008?

(b) Use a calculator or computer to fit a quadratic function to the data in the table. Graph the data points and the quadratic function. Is the quadratic model a reasonable fit?

(c) Use the quadratic model for the Lorenz function to estimate the Gini index for the United States in 2008.

3. The following table gives values for the Lorenz function in the years 1970, 1980, 1990, and 2000. Use the method of Problem 2 to estimate the Gini index for the United States for those years and compare with your answer to Problem 2(c). Do you notice a trend?

x	0.0	0.2	0.4	0.6	0.8	1.0
1970	0.000	0.041	0.149	0.323	0.568	1.000
1980	0.000	0.042	0.144	0.312	0.559	1.000
1990	0.000	0.038	0.134	0.293	0.530	1.000
2000	0.000	0.036	0.125	0.273	0.503	1.000

6.2 Applications to Economics

In this section we discuss two prominent topics in economics where integrals naturally arise: consumer and producer surplus, and future and present value of income streams. Others are described in the exercises.

▪ Consumer Surplus

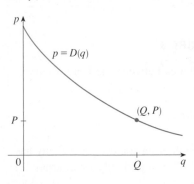

FIGURE 1
A typical demand curve

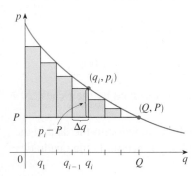

FIGURE 2

Recall from Section 3.2 that the demand function D relates the price p of a commodity to the number of units q that can sell at that price: $p = D(q)$. Usually, selling larger quantities requires lowering prices, so the demand function is a decreasing function. The graph of a typical demand function (the demand curve) is shown in Figure 1. If Q is the amount of the commodity that can currently be sold, then $P = D(Q)$ is the current selling price.

At a given price, some consumers who buy a good would be willing to pay more; they benefit by not having to. The difference between what a consumer is willing to pay and what the consumer actually pays for a good is called the *consumer surplus*. By finding the total consumer surplus among all purchasers of a good, economists can assess the overall benefit of a market to society.

To determine the total consumer surplus, we look at the demand curve and divide the interval $[0, Q]$ into n subintervals, each of length $\Delta q = Q/n$, and let q_i be the right endpoint of the ith subinterval, as in Figure 2. According to the demand curve, q_{i-1} units would be purchased at a price of $p_{i-1} = D(q_{i-1})$ dollars. To increase sales to q_i units, the price would have to be lowered to $p_i = D(q_i)$. In this case, an additional Δq units would be sold (but no more). The consumers who would pay p_i dollars place a high value on the product; they are willing to pay what it is worth to them. So, in paying only P dollars they save an amount of

$$(\text{savings per unit})(\text{number of units}) = [p_i - P]\,\Delta q = [D(q_i) - P]\,\Delta q$$

Considering similar groups of willing consumers for each of the subintervals and adding the savings, we get the total savings:

$$[D(q_1) - P]\,\Delta q + [D(q_2) - P]\,\Delta q + \cdots + [D(q_n) - P]\,\Delta q$$

(This sum corresponds to the area enclosed by the rectangles in Figure 2.) If we let $n \to \infty$, this Riemann sum approaches the integral

$$\int_0^Q [D(q) - P]\,dq$$

This is the total consumer surplus for the commodity.

The consumer surplus is the difference between what consumers are willing to pay for a good and what they actually pay.

---

(1) ▪ The **consumer surplus** of a good when Q units are sold at a price $P = D(Q)$ is given by

$$\int_0^Q [D(q) - P]\,dq$$

The consumer surplus represents the total amount of money saved by consumers in purchasing the commodity at price P, corresponding to an amount

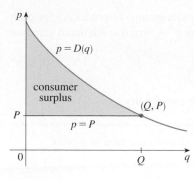

FIGURE 3

demanded of Q. Figure 3 shows the interpretation of the consumer surplus as the area under the demand curve and above the line $p = P$. Note that the total amount actually spent on the goods is the area of the rectangular region below the line $p = P$ for $0 \leq q \leq Q$.

■ **EXAMPLE 1 Consumer Surplus**

The demand for a product, in dollars, is

$$p = 1200 - 0.2q - 0.0001q^2$$

Find the consumer surplus when the sales level is 500.

SOLUTION

Since the number of products sold is $Q = 500$, the corresponding price is

$$P = 1200 - (0.2)(500) - (0.0001)(500)^2 = 1075$$

Therefore, from Definition 1, the consumer surplus is

$$\int_0^{500} [D(q) - P]\,dq = \int_0^{500} (1200 - 0.2q - 0.0001q^2 - 1075)\,dq$$

$$= \int_0^{500} (125 - 0.2q - 0.0001q^2)\,dq$$

$$= 125q - 0.1q^2 - (0.0001)\left(\frac{q^3}{3}\right)\Bigg]_0^{500}$$

$$= (125)(500) - (0.1)(500)^2 - \frac{(0.0001)(500)^3}{3}$$

$$= \$33{,}333.33 \qquad ■$$

■ **Producer Surplus**

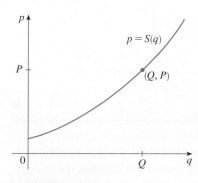

FIGURE 4
A typical supply curve

We now look at the relationship between number of goods sold and price per unit from the producers' point of view. A **supply function** S gives the price per unit p at which producers are willing (and able) to sell q units of a good: $p = S(q)$. The graph of a supply function is called a **supply curve**.

Typically, if a good can be sold at a higher price, a manufacturer has incentive to produce more units of the good. (If more revenue can be collected, producers are more inclined to purchase more machinery and hire more employees, for instance.) Thus we expect S to be an increasing function. A typical supply curve is shown in Figure 4.

If a producer can sell a good for more than its minimum acceptable price, the producer benefits. Economists call the difference between these amounts the *producer surplus*. We can determine the total producer surplus by following the procedure we used for consumer surplus. Let $P = S(Q)$ be the current market price and quantity sold. We again divide the interval $[0, Q]$ into n equal subintervals. If, after the first q_{i-1} units were sold, the price had increased to $p_i = S(q_i)$, an additional Δq units would be supplied. These goods sell for p_i dollars, and the producers receive more revenue than they were counting on; they have collected an excess of

(extra revenue per unit)(number of units) $= [P - p_i]\,\Delta q = [P - S(q_i)]\,\Delta q$

We get the total excess revenue by considering similar groups of producers for each of the subintervals and computing the sum. Letting $n \to \infty$ in this Riemann sum, we get the integral

$$\int_0^Q [P - S(q)] \, dq$$

which measures the total producer surplus for the commodity.

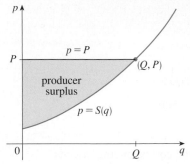

FIGURE 5

---

(2) ▪ The **producer surplus** of a good when Q units are sold at a price $P = S(Q)$ is given by

$$\int_0^Q [P - S(q)] \, dq$$

The total producer surplus can be interpreted as the area under the line $p = P$ and above the supply curve, as shown in Figure 5.

▪ EXAMPLE 2   Producer Surplus

An electronics manufacturer estimates that the supply function for its digital clocks is $S(q) = 5.4 + 0.001q^{1.2}$ dollars. Find the producer surplus when the number of clocks sold is 2000.

SOLUTION

The number of clocks produced is $Q = 2000$ and the corresponding price is

$$P = S(2000) = 5.4 + 0.001(2000)^{1.2} \approx 14.55$$

From Definition 2, the producer surplus is

$$\int_0^{2000} [P - S(Q)] \, dq = \int_0^{2000} (14.55 - 5.4 - 0.001q^{1.2}) \, dq$$

$$= \int_0^{2000} (9.15 - 0.001q^{1.2}) \, dq$$

$$= 9.15q - \frac{0.001}{2.2} q^{2.2} \Big]_0^{2000}$$

$$= 9.15(2000) - \frac{0.001}{2.2} (2000)^{2.2} - 0$$

$$\approx \$9985.36 \qquad \blacksquare$$

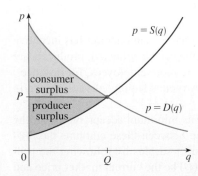

FIGURE 6

Figure 6 shows that the total surplus can be interpreted as the combined areas representing consumer surplus and producer surplus. Total surplus is maximized when P and Q are at the equilibrium values.

It is assumed that in a competitive market, the price of a good is naturally driven to a value where the quantity demanded by consumers matches the quantity producers are willing and able to sell. When this occurs, the market is said to be in *equilibrium*. On the graph, the equilibrium point is where the demand and supply curves intersect.

The sum of consumer surplus and producer surplus is called the **total surplus**, and it is one measure economists use as an indicator of the economic health of a society. We can interpret the total surplus as the area between the supply and demand curves for $0 \le q \le Q$, as shown in Figure 6. This area is maximized when (Q, P) is the point of intersection, so the total surplus is maximized when the market is in equilibrium.

■ **EXAMPLE 3**   **Maximizing Total Surplus**

The demand function for metal thermoses produced by a manufacturer is $p = D(q) = 14e^{-0.15q}$ and the supply function is $p = S(q) = 2e^{0.12q}$ where q is measured in thousands. What price should the thermoses sell for in order to maximize the total surplus? What is the total surplus?

SOLUTION

We know the maximum total surplus occurs when the market is in equilibrium, that is, when the supply and demand functions intersect. Thus we solve

$$14e^{-0.15q} = 2e^{0.12q}$$

$$7e^{-0.15q} \cdot e^{0.15q} = e^{0.12q} \cdot e^{0.15q}$$

$$7 = e^{0.27q}$$

$$\ln 7 = 0.27q$$

$$q = \frac{\ln 7}{0.27} \approx 7.2071$$

The corresponding price for the thermoses is $D(7.2071) = 14e^{-0.15(7.2071)} \approx \$4.75$. To find the total surplus we could compute consumer surplus and producer surplus and add the results. Alternatively, observe that the total surplus is the area between the demand and supply curves for $0 \leq q \leq 7.2071$. Thus

$$\text{Total surplus} = \int_0^{7.2071} [D(q) - S(q)] \, dq = \int_0^{7.2071} [14e^{-0.15q} - 2e^{0.12q}] \, dq$$

$$= 14 \frac{e^{-0.15q}}{-0.15} - 2 \frac{e^{0.12q}}{0.12} \Bigg]_0^{7.2071}$$

$$= \frac{14}{-0.15} e^{-0.15(7.2071)} - \frac{2}{0.12} e^{0.12(7.2071)} - \frac{14}{-0.15} e^0 + \frac{2}{0.12} e^0$$

$$\approx 38.761$$

Because q is measured in thousands, the total surplus is \$38,761.   ■

■ Income Streams

Consider a business that earns income and immediately invests that income, earning interest. For example, suppose you buy an apartment building that generates a profit each month. It is convenient to consider the profit as an income that is earned continuously. If the collected income then earns interest at a constant rate, we can determine the total earned after a specified period of time. This total amount of earnings is called the **future value** of the investment.

If income is continuously earned at a rate of $f(t)$ dollars per year and invested at a constant interest rate r (compounded continuously) for a period of T years, we can approximate the total earnings by dividing the total time T into n subintervals of equal length Δt. (So $\Delta t = T/n$.) If t_i is the time at the end of the ith subinterval, we can assume that the income rate doesn't change much in the interval and say that the income earned during that time interval is approximately $f(t_i) \, \Delta t$ dollars (rate × time). This money will earn interest for $T - t_i$ years, so by using the formula $A = Pe^{rt}$, we see that the value of the income at the end of the time period is

$$f(t_i) \, \Delta t \, e^{r(T-t_i)}$$

If we add these terms for all the time subintervals, we get an estimate for the total amount of earnings at the end of the time period:

$$f(t_1) \, \Delta t \, e^{r(T-t_1)} + f(t_2) \, \Delta t \, e^{r(T-t_2)} + \cdots + f(t_n) \, \Delta t \, e^{r(T-t_n)}$$

But this is just a Riemann sum for the function $f(t)e^{r(T-t)}$ over the interval $[0, T]$. If we take the limit of such sums as $n \to \infty$, we get the following formula.

---

(3) ▪ The future value FV of income earned at a rate of $f(t)$ dollars per year and invested at a constant annual interest rate r (compounded continuously) for a period of T years is given by

$$FV = \int_0^T f(t) e^{r(T-t)} \, dt$$

We can simplify the integral in (3) as follows.

$$\int_0^T f(t) e^{r(T-t)} \, dt = \int_0^T f(t) e^{rT} e^{-rt} \, dt$$

We can take e^{rT} outside the integral sign because it does not depend on t.

$$= e^{rT} \int_0^T f(t) e^{-rt} \, dt$$

Thus

(4)

$$FV = e^{rT} \int_0^T f(t) e^{-rt} \, dt$$

▪ EXAMPLE 4 Computing the Future Value of an Investment

Brad inherits a small apartment building that generates income at an estimated rate of $22{,}000e^{0.02t}$ dollars per year t years from now. He invests this income in a bank account that earns 4.5% interest compounded continuously. What is the value of the income earned from the building after six years?

SOLUTION

If we assume that the income is earned as a continuous stream, we can use Formula 4 to compute the future value of the investment:

$$FV = e^{rT} \int_0^T f(t) e^{-rt} \, dt = e^{0.045(6)} \int_0^6 22{,}000e^{0.02t} \, e^{-0.045t} \, dt$$

$$= 22{,}000e^{0.27} \int_0^6 e^{-0.025t} \, dt = 22{,}000e^{0.27} \left(\frac{1}{-0.025} \right) e^{-0.025t} \bigg]_0^6$$

$$= -880{,}000e^{0.27}(e^{-0.15} - 1) \approx \$160{,}571 \qquad \blacksquare$$

We can also look at the income stream in Example 4 from the reverse direction: What amount would need to be invested now, earning interest at the given rate, to reach the same sum at the end of the time period? This amount is called the **pres-**

ent value of the income stream; it gives us a sense of the value today of the long-term investment.

■ **EXAMPLE 5** **Finding the Present Value of an Income Stream**

Compute the present value of Brad's income stream in Example 4.

SOLUTION

We want to find the amount P that would need to be invested to reach $160,571 in 6 years at a continuously compounded rate of 4.5%. Using the formula $A = Pe^{rt}$ we have

$$Pe^{0.045(6)} = 160{,}571$$

$$P = \frac{160{,}571}{e^{0.27}} \approx 122{,}577$$

Thus the present value of Brad's income stream is $122,577.   ■

In general, the present value of an income stream is determined by finding the principal amount P that, after being invested for T years at a continuously compounded rate r, matches the future value of the income stream. The value of this comparison investment after T years is Pe^{rT}, and from Formula 4 we have

$$Pe^{rT} = e^{rT} \int_0^T f(t) e^{-rt}\, dt$$

$$P = \int_0^T f(t) e^{-rt}\, dt$$

Thus we have the following formula for present value.

---

(5) ■ The present value PV of income earned at a rate of $f(t)$ dollars per year and invested at a constant annual interest rate r (compounded continuously) for a period of T years is given by

$$\text{PV} = \int_0^T f(t) e^{-rt}\, dt$$

■ **EXAMPLE 6** **Finding a Present Value**

A contest winner is awarded $500 per week for 10 years. What is the present value of the prize if the current available interest rate is 5% compounded continuously?

SOLUTION

We assume that there are 52 weeks in a year and treat the prize money as income earned continuously at the rate of $26,000 per year. Using Formula 5 we have

$$\text{PV} = \int_0^{10} 26{,}000 e^{-0.05t}\, dt = \frac{26{,}000}{-0.05} e^{-0.05t} \Big]_0^{10}$$

$$= -520{,}000(e^{-0.5} - 1) \approx 204{,}604$$

Thus the present value of the prize money is $204,604.   ■

▪ Exercises 6.2

1–8 ▪ Consumer surplus Find the consumer surplus for the given demand function and sales level.

1. $p = 1450 - 0.2q$, 2000

2. $p = 6 - q/3500$, 12,000

3. $p = 760 - 0.1q - 0.0002q^2$, 800

4. $p = 0.001q^2 - q + 225$, 250

5. $p = 15,000e^{-0.03q}$, 45

6. $p = 31,000/(q + 100)$, 400

7. $p = 40 - 3.2\sqrt{q}$, 100

8. $p = 85,000(q + 50)^{-0.8}$, 1000

9. **Consumer surplus** The demand function for a particular vacation package is $D(q) = 2000 - 46\sqrt{q}$. Find the consumer surplus when the sales level for the package is 800. Illustrate by drawing the graph and shading an area that represents the consumer surplus.

10. **Consumer surplus** The demand function for a certain commodity is $p = 20 - 0.05q$. Find the consumer surplus when the sales level is 300. Illustrate by drawing the demand curve and identifying the consumer surplus as an area.

11. **Consumer surplus** A demand curve is given by $p = 450/(q + 8)$. Find the consumer surplus when the selling price is $10.

12. **Consumer surplus** The number q, measured in thousands, of a particular model of LCD television that can be sold is related to the sale price p, in dollars, by the equation $p = 18,000e^{-0.2q}$. Compute the consumer surplus if the televisions sell for $800.

13. **Demand and consumer surplus** A concert promoter has been selling 210 T-shirts on average at performances for $18 each. They estimate that for each dollar they lower the price, they will sell an additional 30 shirts. Find the demand function for the shirts and calculate the consumer surplus if the shirts are sold for $15 each.

14. **Demand and consumer surplus** A movie theater has been charging $9.50 per person and selling about 400 tickets on a typical weeknight. After surveying their customers, the theater estimates that for every 50 cents that they lower their price, the number of moviegoers will increase by 35 per night. Find the demand function and calculate the consumer surplus when the tickets are priced at $8.00.

15–18 ▪ Producer surplus Find the producer surplus for the given supply function and sales level.

15. $p = 16 + 0.03q$, 500

16. $p = 0.001q^2 + 0.2q + 8$, 200

17. $p = 22e^{0.002q}$, 600

18. $p = 4.4 + 0.005q^{1.3}$, 2500

19. **Producer surplus** If a supply curve is modeled by the equation $p = 200 + 0.2q^{3/2}$, find the producer surplus when the selling price is $400.

20. **Producer surplus** Calculate the producer surplus for the supply function $S(q) = 3 + 0.01q^2$ at the sales level $q = 10$. Illustrate by drawing the supply curve and identifying the producer surplus as an area.

21. **Consumer surplus** A company modeled the demand curve for its product, in dollars, by the equation

$$p = \frac{800,000e^{-q/5000}}{q + 20,000}$$

Use a graph to estimate the sales level when the selling price is $16. Then find (approximately) the consumer surplus for this sales level.

22. **Producer surplus** A manufacturer estimates that the supply function for its product, in dollars, is

$$p = \sqrt{30 + 0.01qe^{0.001q}}$$

Find (approximately) the producer surplus when the selling price is $30.

23. **Total surplus** The demand function for an electronics company's car stereos is $D(q) = 228.4 - 18q$ and the supply function is $S(q) = 27q + 57.4$, where q is measured in thousands.

  (a) At what price is the market for the stereos in equilibrium?

  (b) Compute the total surplus corresponding to the price from part (a).

  (c) What is the maximum total surplus?

24. **Total surplus** Given the demand curve $p = 50 - \frac{1}{20}x$ and the supply curve $p = 20 + \frac{1}{10}x$, find the consumer surplus and the producer surplus at market equilibrium. Illustrate by sketching the supply and demand curves and identifying the surpluses as areas. What is the total surplus?

25. **Total surplus** A camera company estimates that the demand function for its new consumer digital camera is $p = 312e^{-0.14q}$ and the supply function is estimated to be $p = 26e^{0.2q}$ where q is measured in thousands. What price for the camera maximizes the total surplus?

26. **Total surplus** If the demand function for a good is $D(q) = 30(0.95)^q$ and the supply function is $S(q) = 10(1.01)^q$, where q is measured in thousands, what

price for the good maximizes the total surplus? What is the total surplus?

27. Future value   A retiree is paid $1500 per month by an annuity. If the income is invested in an account that earns 5% interest compounded continuously, what is the future value of the income after ten years?

28. Future value   A scholarship fund receives $20,000 annually from a family trust. Assuming that the income can be invested at a 6% annual rate (compounded continuously), what is the future value of the income after eight years?

29. Present and future value   An heiress receives an income stream from a will at a rate of

$$f(t) = 30,000e^{0.025t} \text{ dollars per year}$$

She invests this income and earns 4.8% interest (compounded continuously).

(a) What is the future value of the income after ten years?

(b) Compute the present value of the income over a ten-year period.

30. Present and future value   Income received at the rate of $f(t) = 8000e^{0.04t}$ dollars per year is invested at a 6.2% interest rate (compounded continuously).

(a) Find the future value of the income for a six-year time period.

(b) What is the present value of the income over a six-year time period?

31. Present value   The winner of a lottery is awarded $1,000,000 to be paid in annual installments of $50,000 for 20 years. Alternatively, the winner can accept a "cash value" one-time payment of $450,000. The winner esti-

mates he can earn 8% annually on the winnings. What is the present value of the installment plan? Should he choose the one-time payment instead?

32. Present value   Compute the present value of income received at the rate $f(t) = 5000e^{0.01t}$ dollars per year for a period of eight years if the income is invested at a 3.5% interest rate (compounded continuously).

33. Present and future value   A charity receives donations from a family trust at the rate of

$$f(t) = 12,000 + 500t \text{ dollars per year}$$

for the next 15 years. If the charity invests the income at an annual rate of 5.2% (compounded continuously), what is the future value of the income? What is the present value?

34. Present and future value   An office park generates income at the rate $f(t) = 650,000e^{0.04t}$ dollars per year. Assuming the income can be invested at a continuously compounded rate of 6%, what is the future value of the income over a five-year time span? What is the present value?

35. Net investment flow   If the amount of capital that a company has at time t is $f(t)$, then the derivative, $f'(t)$, is called the *net investment flow*. Suppose that the net investment flow is $\sqrt{t}$ million dollars per year (where t is measured in years). Find the increase in capital (the *capital formation*) from the fourth year to the eighth year.

36. Net investment flow   If revenue flows into a company at a rate of $f(t) = 9000\sqrt{1 + 2t}$, where t is measured in years and $f(t)$ is measured in dollars per year, find the total revenue obtained in the first four years.

■ Challenge Yourself

37. Pareto's Law of Income   *Pareto's Law of Income* states that the number of people with incomes between $x = a$ and $x = b$ is $N = \int_a^b Ax^{-k} \, dx$, where A and k are constants with $A > 0$ and $k > 1$. The average income of these people is

$$\bar{x} = \frac{1}{N} \int_a^b Ax^{1-k} \, dx$$

Calculate $\bar{x}$.

6.3 Applications to Biology

In this section we consider some of the applications of integration to biology: survival and renewal of populations, blood flow in veins and arteries, and cardiac output. Other applications are explored in the exercises.

■ Survival and Renewal

The strategy we used to analyze income streams in Section 6.2 can be adapted to other contexts. For instance, a population may be continually adding members while some of the existing members die. If we can model how these changes occur with suitable functions, we can predict the population size at any time in the future.

Suppose we start with an initial population P_0 that is adding new members at the rate $R(t)$, where t is the number of years from now. We call R a **renewal function**. In addition, the proportion of the population that survives at least t years from now is given by a **survival function** $S(t)$. [So if $S(5) = 0.8$, then 80% of the current population remains after five years.]

To predict the population in T years, we first note that $S(T) \cdot P_0$ members of the current population survive. To account for the newly added members, we divide the time interval $[0, T]$ into n subintervals, each of length $\Delta t = T/n$, and let t_i be the right endpoint of the ith subinterval. During this time interval approximately $R(t_i) \, \Delta t$ members are added, and the proportion of them that survive until time T is given by $S(T - t_i)$. Thus the remaining members of those added during this time interval is

$$\text{(proportion surviving)}\text{(number of members)} = S(T - t_i) \, R(t_i) \, \Delta t$$

Then the total number of new members to the population who survive after T years is approximately

$$S(T - t_1) \, R(t_1) \, \Delta t + S(T - t_2) \, R(t_2) \, \Delta t + \cdots + S(T - t_n) \, R(t_n) \, \Delta t$$

If we let $n \to \infty$, this Riemann sum approaches the integral

$$\int_0^T S(T - t) \, R(t) \, dt$$

By adding this integral to the number of initial members who survived, we get the total population after T years.

■ A population begins with P_0 members and members are added at a rate given by the renewal function $R(t)$, where t is measured in years. The proportion of the population that remains after t years is given by the survival function $S(t)$. Then the population T years from now is given by

(1) $$P(T) = S(T) \cdot P_0 + \int_0^T S(T - t) \, R(t) \, dt$$

Equation 1 is also valid if t represents any other unit of time such as weeks or months.

■ **EXAMPLE 1**   **Predicting a Future Population**

There are currently 5600 trout in a lake and the trout are reproducing at the rate $R(t) = 720e^{0.1t}$ fish per year. However, pollution is killing many of the trout; the proportion that survives after t years is given by $S(t) = e^{-0.2t}$. How many trout will there be in the lake in ten years?

SOLUTION

We have $P_0 = 5600$ and $T = 10$, so by Formula 1, the population in ten years is

$$P(10) = S(10) \cdot 5600 + \int_0^{10} S(10 - t) \, R(t) \, dt$$

$$= 5600e^{-0.2(10)} + \int_0^{10} e^{-0.2(10-t)} \cdot 720e^{0.1t} \, dt$$

$$= 5600e^{-2} + 720 \int_0^{10} e^{0.3t-2} \, dt$$

Writing $e^{0.3t-2}$ as $e^{0.3t}e^{-2}$ gives

$$P(10) = 5600e^{-2} + 720e^{-2} \int_0^{10} e^{0.3t}\, dt$$

$$= 5600e^{-2} + 720e^{-2} \left. \frac{e^{0.3t}}{0.3} \right]_0^{10}$$

$$= 5600e^{-2} + \frac{720}{0.3}\, e^{-2}(e^3 - e^0)$$

$$= 5600e^{-2} + 2400(e - e^{-2}) \approx 6956.95$$

Thus we predict that there will be about 6960 trout in the lake ten years from now. ◼

Although we presented Formula 1 in the context of populations, it applies to other settings as well, such as administering a drug over time as the body works to eliminate the drug. Additional applications are investigated in the exercises.

◼ | Blood Flow

When we consider the flow of blood through a blood vessel, such as a vein or artery, we can model the shape of the blood vessel by a cylindrical tube with radius R and length l as illustrated in Figure 1.

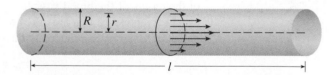

FIGURE 1

Blood flow in an artery

For more detailed information, see W. Nichols and M. O'Rourke (eds.), *McDonald's Blood Flow in Arteries: Theoretic, Experimental, and Clinical Principles*, 5th ed. (New York, 2005).

Because of friction at the walls of the tube, the velocity v of the blood is greatest along the central axis of the tube and decreases as the distance r from the axis increases until v becomes 0 at the wall. The relationship between v and r is given by the **law of laminar flow** discovered by the French physician Jean-Louis-Marie Poiseuille in 1840. This law states that

$$(2) \qquad v(r) = \frac{P}{4\eta l}\,(R^2 - r^2)$$

where η is the viscosity of the blood and P is the pressure difference between the ends of the tube. If P and l are constant, then v is a function of r with domain $[0, R]$.

We can compute the rate of blood flow, or *flux* (volume per unit time), by considering smaller, equally spaced radii $r_1, r_2, \ldots$. The approximate area of the ring with inner radius r_{i-1} and outer radius r_i is

$$2\pi r_i\, \Delta r \qquad \text{where} \qquad \Delta r = r_i - r_{i-1}$$

(See Figure 2.) If Δr is small, then the velocity is almost constant throughout this ring and can be approximated by $v(r_i)$. Thus the volume of blood per unit time that flows across the ring is approximately

$$(2\pi r_i\, \Delta r)\, v(r_i) = 2\pi r_i\, v(r_i)\, \Delta r$$

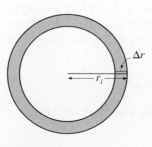

FIGURE 2

and the total volume of blood that flows across a cross-section per unit time is about

$$2\pi r_1\, v(r_1)\, \Delta r + 2\pi r_2\, v(r_2)\, \Delta r + \cdots + 2\pi r_n\, v(r_n)\, \Delta r$$

This approximation is illustrated in Figure 3. Notice that the velocity (and hence the volume per unit time) increases toward the center of the blood vessel. The approximation gets better as n increases. When we take the limit as $n \to \infty$, we get an integral that gives the exact value of the **flux** (or *discharge*), which is the volume of blood that passes a cross-section per unit time:

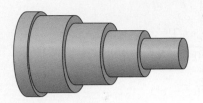

FIGURE 3

$$F = \int_0^R 2\pi r\, v(r)\, dr$$

$$= \int_0^R 2\pi r\, \frac{P}{4\eta l}\, (R^2 - r^2)\, dr$$

$$= \frac{\pi P}{2\eta l} \int_0^R (R^2 r - r^3)\, dr = \frac{\pi P}{2\eta l} \left[R^2 \frac{r^2}{2} - \frac{r^4}{4} \right]_{r=0}^{r=R}$$

$$= \frac{\pi P}{2\eta l} \left[\frac{R^4}{2} - \frac{R^4}{4} \right] = \frac{\pi P R^4}{8\eta l}$$

The resulting equation

(3)
$$F = \frac{\pi P R^4}{8\eta l}$$

is called **Poiseuille's Law**; it shows that the flux is proportional to the fourth power of the radius of the blood vessel.

In Exercise 12 you are asked to investigate the effect on blood pressure if the radius of an artery is reduced to three-fourths of its normal value.

■ Cardiac Output

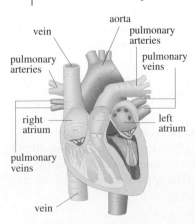

FIGURE 4

Figure 4 shows the human cardiovascular system. Blood returns from the body through the veins, enters the right atrium of the heart, and is pumped to the lungs through the pulmonary arteries for oxygenation. It then flows back into the left atrium through the pulmonary veins and then out to the rest of the body through the aorta. The **cardiac output** of the heart is the volume of blood pumped by the heart per unit time, that is, the rate of flow into the aorta.

The *dye dilution method* is used to measure the cardiac output. Dye is injected into the right atrium and flows through the heart into the aorta. A probe inserted into the aorta measures the concentration of the dye leaving the heart at equally spaced times over a time interval $[0, T]$ until the dye has cleared. Let $c(t)$ be the concentration of the dye at time t. If we divide $[0, T]$ into subintervals of equal length Δt, then the amount of dye that flows past the measuring point during the subinterval from $t = t_{i-1}$ to $t = t_i$ is approximately

$$(\text{concentration})(\text{volume}) = c(t_i)(F\, \Delta t)$$

where F is the rate of flow that we are trying to determine. Thus the total amount of dye is approximately

$$c(t_1) F\, \Delta t + c(t_2) F\, \Delta t + \cdots + c(t_n) F\, \Delta t$$

and, letting $n \to \infty$, we find that the amount of dye is

$$A = \int_0^T c(t) F \, dt = F \int_0^T c(t) \, dt$$

Thus the cardiac output is given by

(4)
$$F = \frac{A}{\displaystyle\int_0^T c(t) \, dt}$$

where the amount of dye A is known and the integral can be approximated from the concentration readings.

■ EXAMPLE 2 Cardiac Output

A 5-mg dose (called a bolus) of dye is injected into a right atrium. The concentration of the dye (in milligrams per liter) is measured in the aorta at one-second intervals as shown in the chart. Estimate the cardiac output.

t	$c(t)$	t	$c(t)$
0	0	6	6.1
1	0.4	7	4.0
2	2.8	8	2.3
3	6.5	9	1.1
4	9.8	10	0
5	8.9		

SOLUTION

Here $A = 5$ and $T = 10$. We can use the Midpoint Rule with $n = 5$ subdivisions to approximate the integral of the concentration. Then $\Delta t = 2$ and

$$\int_0^{10} c(t) \, dt \approx [c(1) + c(3) + c(5) + c(7) + c(9)] \Delta t$$

$$= [0.4 + 6.5 + 8.9 + 4.0 + 1.1](2)$$

$$= 41.8$$

Thus Formula 4 gives the cardiac output to be

$$F = \frac{A}{\displaystyle\int_0^{10} c(t) \, dt} \approx \frac{5}{41.8} \approx 0.12 \text{ L/s} = 7.2 \text{ L/min}$$

■

■ Exercises 6.3

1. **Animal survival and renewal** An animal population currently has 7400 members and is reproducing at a rate of $R(t) = 2240 + 60t$ members per year. The proportion of members that survive after t years is given by $S(t) = 1/(t + 1)$.
   (a) How many of the original members survive four years?
   (b) How many new members are added during the next four years?
   (c) Explain why the animal population four years from now is not the same as the sum of your answers from parts (a) and (b).

2. **City population** A city currently has 36,000 residents and is adding new residents steadily at the rate of 1600 per year. If the proportion of residents who remain after t years is given by $S(t) = 1/(t + 1)$, what is the population of the city seven years from now?

3. **Insect survival and renewal** A population of insects currently numbers 22,500 and is increasing at a rate of $R(t) = 1225e^{0.14t}$ insects per week. If the survival function for the insects is $S(t) = e^{-0.2t}$ where t is measured in weeks, how many insects are there after 12 weeks?

4. **Animal survival and renewal** There are currently 3800 birds of a particular species in a national park and their number is increasing at a rate of

$$R(t) = 525e^{0.05t} \text{ birds per year}$$

If the proportion of birds that survive t years is given by $S(t) = e^{-0.1t}$, what do you predict the bird population will be ten years from now?

5. **Magazine subscriptions**   A magazine currently has 8400 subscribers for its online web version. It is adding members at the rate of $R(t) = 180e^{0.04t}$ subscribers per month. If the proportion of members who remain subscribed t months from now is $S(t) = e^{-0.06t}$, how many online subscribers will the magazine have two years from now?

6. **Cable TV subscribers**   A local cable TV provider has 26,800 subscribers and is adding subscribers at a rate of 450 per month. The proportion of subscribers who remain t months from now is $e^{-0.08t}$. How many subscribers will there be in three years?

7. **Drug concentration**   A drug is administered intravenously to a patient at the rate of 12 mg/h. The patient's body eliminates the drug over time so that after t hours the proportion that remains is $e^{-0.25t}$. If the patient currently has 50 mg of the drug in her bloodstream, how much of the drug is present eight hours from now?

8. **Drug concentration**   A patient receives a drug at a constant rate of 30 mg/h. The drug is eliminated from the bloodstream over time so that the fraction $e^{-0.2t}$ remains after t hours. The patient currently has 80 mg of the drug present in the bloodstream; how much will be present in 24 hours?

9. **Water pollution**   A contaminant is leaking into a lake at a rate of $R(t) = 1600e^{0.06t}$ gal/h. Enzymes that neutralize the contaminant have been added to the lake over time so that after t hours the fraction of contaminant that remains is $S(t) = e^{-0.32t}$. If there are currently 10,000 gallons of the contaminant in the lake, how many gallons will be present in the lake 18 hours from now?

10. **Insect survival and renewal**   Sterile fruit flies are used in an experiment where the proportion that survives at least t days is given by $e^{-0.15t}$. If the experiment begins with 200 fruit flies, and flies are added at the rate of 5 per hour, how many flies are present 14 days after the start of the experiment?

11. **Blood flow**   Use Poiseuille's Law to calculate the rate of flow in a small human artery where we can take $\eta = 0.027$, $R = 0.008$ cm, $l = 2$ cm, and $P = 4000$ dynes/cm^2.

12. **Blood flow**   High blood pressure results from constriction of the arteries. To maintain a normal flow rate (flux), the heart has to pump harder, thus increasing the blood pressure. Use Poiseuille's Law to show that if R_0 and P_0 are normal values of the radius and pressure in an artery and the constricted values are R and P, then for the flux to remain constant, P and R are related by the equation

$$\frac{P}{P_0} = \left(\frac{R_0}{R}\right)^4$$

Deduce that if the radius of an artery is reduced to three-fourths of its former value, then the pressure is more than tripled.

13. **Cardiac output**   The dye dilution method is used to measure cardiac output with 6 mg of dye. The dye concentrations, in mg/L, are modeled by $c(t) = 20te^{-0.6t}$, $0 \leq t \leq 10$, where t is measured in seconds. Find the cardiac output. [*Hint*: Integration by parts is required.]

14. **Cardiac output**   After an 8-mg injection of dye, the readings of dye concentration, in mg/L, at two-second intervals are as shown in the table. Use the Midpoint Rule to estimate the cardiac output.

t	$c(t)$	t	$c(t)$
0	0	12	3.9
2	2.4	14	2.3
4	5.1	16	1.6
6	7.8	18	0.7
8	7.6	20	0
10	5.4		

15. **Cardiac output**   The graph of the concentration function $c(t)$ is shown after a 7-mg injection of dye into a heart. Use the Midpoint Rule to estimate the cardiac output.

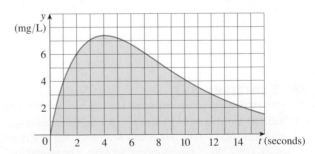

16. **Blood flow**   Use the law of laminar flow given in Equation 2 to find the average velocity (with respect to r) of blood flowing through a blood vessel over the interval $0 \leq r \leq R$. Compare the average velocity with the maximum velocity.

▪ Challenge Yourself

17. **Drug administration**   A patient is continually receiving a drug. If the drug is eliminated from the body over time so that the fraction that remains after t hours is $e^{-0.4t}$, at what constant rate should the drug be administered to maintain a steady level of the drug in the bloodstream?

6.4 Differential Equations

■ Introduction to Differential Equations

Now is a good time to read (or reread) the discussion of mathematical modeling on page 10.

In describing the process of modeling in Section 1.1, we talked about formulating a mathematical model of a real-world problem either through intuitive reasoning about the phenomenon or from a physical law based on evidence from experiments. The mathematical model often takes the form of a **differential equation**, that is, an equation that contains an unknown function and one or more of its derivatives. This is not surprising because in a real-world problem we often notice that changes occur and we want to predict future behavior on the basis of how current values change.

For example, a function may be related to its derivative as follows:

(1) $$f'(x) = xf(x)$$

This is a differential equation, and a function f is a solution if it satisfies the equation for all values of x in some interval. It is often convenient to use y as the unknown function of x, and y' or dy/dx in place of $f'(x)$. Thus the differential equation in (1) can be written as

$$y' = xy$$

■ EXAMPLE 1 Verify a Solution of a Differential Equation

Verify that the function $y = e^{x^2/2}$ is a solution of the differential equation $y' = xy$.

SOLUTION

If $y = e^{x^2/2}$ then $y' = e^{x^2/2} \cdot \frac{1}{2}(2x) = xe^{x^2/2} = xy$ as desired. ■

It turns out that the function $y = e^{x^2/2}$ is not the only solution to the differential equation in Example 1. In fact, any function of the form $y = Ce^{x^2/2}$, where C is any constant, is a solution (and there are no others).

When we are asked to solve a differential equation, we are expected to find all possible solutions of the equation. Generally speaking, solving a differential equation is not an easy matter. There is no systematic technique that will enable us to solve all differential equations. But in this section we examine a certain type of differential equation that *can* be solved explicitly.

■ Separable Differential Equations

A **separable equation** is a differential equation involving an unknown function and its first derivative in which the expression for dy/dx (or y') can be factored as a function of x times a function of y. In other words, it can be written in the form

$$\frac{dy}{dx} = g(x)f(y)$$

The name *separable* comes from the fact that the expression on the right side can be "separated" into a function of x and a function of y. Equivalently, if $f(y) \neq 0$, we could write

(2)
$$\frac{dy}{dx} = \frac{g(x)}{h(y)}$$

where $h(y) = 1/f(y)$. To solve this equation we think of dy and dx as separate differentials and rewrite the equation in the differential form

$$h(y)\, dy = g(x)\, dx$$

so that all y's are on one side of the equation and all x's are on the other side. Then we integrate both sides of the equation:

(3)
$$\int h(y)\, dy = \int g(x)\, dx$$

Equation 3 defines y implicitly as a function of x. In some cases we may be able to solve for y in terms of x.

The technique for solving separable differential equations was first used by James Bernoulli (in 1690) in solving a problem about pendulums and by Leibniz (in a letter to Huygens in 1691). John Bernoulli explained the general method in a paper published in 1694.

We use the Chain Rule to justify this procedure: If h and g satisfy (3), then

$$\frac{d}{dx}\left(\int h(y)\, dy\right) = \frac{d}{dx}\left(\int g(x)\, dx\right)$$

so

$$\frac{d}{dy}\left(\int h(y)\, dy\right)\frac{dy}{dx} = g(x)$$

and

$$h(y)\frac{dy}{dx} = g(x)$$

Thus Equation 2 is satisfied.

▪ EXAMPLE 2   Solving a Separable Differential Equation

Solve the differential equation $\dfrac{dy}{dx} = \dfrac{x^2}{y^2}$.

SOLUTION

Multiply both sides of the equation by dx and by y^2:

$$y^2\, dy = x^2\, dx$$

We now have the equation written in terms of differentials, with y's on one side and x's on the other, so we integrate both sides:

$$\int y^2\, dy = \int x^2\, dx$$

$$\tfrac{1}{3}y^3 = \tfrac{1}{3}x^3 + C$$

where C is an arbitrary constant. (We could have used a constant C_1 on the left side and another constant C_2 on the right side. But then we could combine these constants by writing $C = C_2 - C_1$.)

Solving for y, we get

$$y^3 = x^3 + 3C$$

$$y = \sqrt[3]{x^3 + 3C}$$

We could leave the solution like this or we could write it in the form

$$y = \sqrt[3]{x^3 + K}$$

where $K = 3C$. (Since C is an arbitrary constant, so is $3C$ or K.)   ▪

The solution we found to the differential equation in Example 2 is called the **general solution**; it describes the *family* of all possible solution functions. Each different value of C (or K) gives a different solution function.

In many application problems, however, we need to find one specific solution that satisfies an additional requirement of the form $y(x_0) = y_0$. (Here we use y interchangeably as both the name of the function and the output variable.) This is called an **initial condition**, and the problem of finding a solution of the differential equation that satisfies the initial condition is called an **initial-value problem**. Geometrically, the initial condition $y(x_0) = y_0$ corresponds to looking at the family of solution curves and picking the one that passes through the point (x_0, y_0).

Figure 1 shows graphs of several members of the family of solutions of the differential equation in Example 2. The solution of the initial-value problem in Example 3 is shown in red.

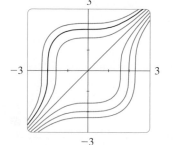

FIGURE 1

■ **EXAMPLE 3** Solving an Initial-Value Problem

Find the solution of the differential equation from Example 2 that satisfies the initial condition $y(0) = 2$.

SOLUTION

If we put $x = 0$ in the general solution in Example 2, we get $y(0) = \sqrt[3]{K}$. To satisfy the initial condition $y(0) = 2$, we must have $\sqrt[3]{K} = 2$ and so $K = 8$. Thus the solution of the initial-value problem is

$$y = \sqrt[3]{x^3 + 8}$$ ▪

■ | **Models of Population Growth**

In Section 3.6 we looked at the situation in which a quantity grows at a rate proportional to its size and showed that functions of the form $A(t) = Ce^{kt}$ follow this description. We are now prepared to show that these are the *only* functions that satisfy the requirement.

As we mentioned in Section 3.6, some populations do grow at a rate proportional to the size of the population. Suppose P is such a population at time t. We can represent this description as the differential equation

$$\frac{dP}{dt} = kP$$

where k is the constant of proportionality. We solve this separable differential equation in the following example.

■ **EXAMPLE 4** Exponential Growth

Solve the differential equation $dP/dt = kP$.

SOLUTION

$$\frac{dP}{dt} = kP$$

$$\frac{1}{P} \, dP = k \, dt$$

$$\int \frac{1}{P} \, dP = \int k \, dt$$

$$\ln|P| = kt + C$$

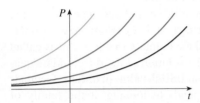

FIGURE 2
The family of solutions of $dP/dt = kP$
where $P > 0$ ($k > 0$)

Because population P is never negative, we can write $|P| = P$ and solve the equation explicitly for P as follows:

$$e^{\ln P} = e^{kt+C}$$

$$P = e^{kt} e^{C}$$

$$P = Ae^{kt}$$

where $A = e^{C}$ is an arbitrary positive constant (e^{C} is positive for any value of C). Several members of the family of solutions are graphed in Figure 2.

To see the significance of the constant A, we observe that

$$P(0) = Ae^{k \cdot 0} = A$$

Therefore A is the initial population and we can write the solution to the differential equation as

$$P = P_0 e^{kt}$$

where $P_0 = P(0)$. ▪

The exponential growth function from Example 4 is appropriate for modeling population growth under ideal conditions, but we have to recognize that a more realistic model must reflect the fact that a given environment has limited resources. Many populations start by increasing in an exponential manner, but the population levels off when it approaches its *carrying capacity M*. For a model to take into account both trends, we make two assumptions:

- $\dfrac{dP}{dt} \approx kP$ if P is small   (Initially, the growth rate is proportional to P.)

- $\dfrac{dP}{dt}$ approaches 0 if the value of P approaches M

  (The growth rate slows if the population nears its carrying capacity.)

A simple expression that incorporates both assumptions is given by the equation

Equation 4 is called the **logistic differential equation** and was proposed by the Belgian mathematical biologist Pierre-François Verhulst in the 1840s as a model for world population growth.

(4)
$$\boxed{\frac{dP}{dt} = kP\left(1 - \frac{P}{M}\right)}$$

Notice that if P is small compared with M, then P/M is close to 0 and therefore $dP/dt \approx kP$. If P is close to M, then P/M is almost 1 and so $1 - P/M$ is almost 0.

We first observe that the constant functions $P(t) = 0$ and $P(t) = M$ are solutions because, in either case, one of the factors on the right side of Equation 4 is zero. (This certainly makes physical sense: If the population is ever either 0 or at the carrying capacity, then it stays that way.) These two constant solutions are called *equilibrium solutions*.

Next observe that if the population approaches the carrying capacity ($P \rightarrow M$), then $dP/dt \rightarrow 0$, which means the population levels off. So we expect that the solutions of the logistic differential equation have graphs that look something like the ones in Figure 3. Notice that the graphs move away from the equilibrium solution $P = 0$ and move toward the equilibrium solution $P = M$.

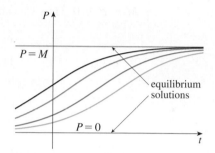

FIGURE 3
Solutions of the logistic equation

In the next example we solve the logistic differential equation to find an explicit family of solution functions.

■ **EXAMPLE 5**   **The Logistic Equation**

Solve the logistic differential equation given in (4).

SOLUTION

The logistic equation is separable:

$$\frac{dP}{dt} = kP\left(1 - \frac{P}{M}\right)$$

$$\frac{1}{P(1 - P/M)}\, dP = k\, dt$$

Then

(5)
$$\int \frac{1}{P(1 - P/M)}\, dP = \int k\, dt$$

To evaluate the integral on the left side, we write

$$\frac{1}{P(1 - P/M)} \cdot \frac{M}{M} = \frac{M}{P(M - P)}$$

It turns out that we can split this expression into two fractions that are easier to integrate. You can verify that

$$\frac{M}{P(M - P)} = \frac{1}{P} + \frac{1}{M - P}$$

by taking the terms on the right-hand side to a common denominator. This allows us to rewrite Equation 5 as

$$\int \left(\frac{1}{P} + \frac{1}{M - P} \right) dP = \int k \, dt$$

Splitting the integral on the left and using the substitution $u = M - P$ on the second integral, we get

$$\ln |P| - \ln |M - P| = kt + C$$

Because P represents a population, we know that $P > 0$, and we have $P < M$ so $M - P > 0$. Thus we can write

$$\ln P - \ln(M - P) = kt + C$$

$$-\ln P + \ln(M - P) = -kt - C$$

$$\ln \frac{M - P}{P} = -kt - C$$

$$\frac{M - P}{P} = e^{-kt-C} = e^{-C}e^{-kt}$$

(6) $$\frac{M - P}{P} = Ae^{-kt}$$

where $A = e^{-C}$. Solving Equation 6 for P, we get

$$\frac{M}{P} - 1 = Ae^{-kt}$$

$$\frac{M}{P} = 1 + Ae^{-kt}$$

$$\frac{P}{M} = \frac{1}{1 + Ae^{-kt}}$$

$$P = \frac{M}{1 + Ae^{-kt}}$$

We find the value of A by putting $t = 0$ in Equation 6. If $t = 0$, then $P = P_0$ (the initial population), so

$$\frac{M - P_0}{P_0} = Ae^0 = A$$

Thus the solution to the logistic differential equation is

(7) $$P(t) = \frac{M}{1 + Ae^{-kt}} \qquad \text{where } A = \frac{M - P_0}{P_0}$$

This is the logistic equation given in Section 3.6. You can verify that the graphs of the functions for different values of A do indeed look like the curves shown in Figure 3. ▪

Differential Equations in Economics

In Section 4.7 we discussed elasticity of demand as a measure of how a change in price affects a change in demand. Recall that if the demand is elastic ($E > 1$), then a relative change in price causes a greater relative change in demand. We have the opposite situation if demand is inelastic ($E < 1$). If $E = 1$ (demand is unit elastic), then changes in price and demand are proportional. In the next example we determine precisely which demand functions have unit elasticity.

■ EXAMPLE 6   For Which Demand Functions Is $E(q) = 1$?

Determine which demand functions $p = D(q)$ have unit elasticity, that is,

$$E(q) = -\frac{p/q}{dp/dq} = 1$$

SOLUTION

We first write the differential equation in differential form:

$$-\frac{p/q}{dp/dq} = 1$$

$$-\frac{p\,dq}{q\,dp} = 1$$

$$q\,dp = -p\,dq$$

$$\frac{1}{p}\,dp = -\frac{1}{q}\,dq$$

Then we integrate both sides.

$$\int \frac{1}{p}\,dp = \int -\frac{1}{q}\,dq$$

$$\ln|p| = -\ln|q| + C$$

Both p and q are positive, so $|p| = p$, $|q| = q$ and we solve for p:

$$\ln p = -\ln q + C$$

$$e^{\ln p} = e^{-\ln q + C}$$

$$p = e^C e^{-\ln q}$$

Since $e^{-\ln q} = e^{\ln q^{-1}} = q^{-1} = 1/q$, we have

$$p = e^C \cdot \frac{1}{q}$$

$$p = \frac{k}{q}$$

where $k = e^C$. Thus all demand functions with unit elasticity are of the form

$$p = D(q) = \frac{k}{q}$$

where k is a positive constant. ■

▪ Prepare Yourself

1. If $y = e^{-3x}$, what is y'? What is $y + y'$?

2. If $y = x^2 e^x$, find $xy' + y$.

3. Evaluate the indefinite integral.

  (a) $\int (1/x)\, dx$ (b) $\int (1/x^2)\, dx$

  (c) $\int 1/(x + 4)\, dx$

  (d) $\int x/(x^2 + 4)\, dx$

  (e) $\int e^{-2t}\, dt$

  (f) $\int \sqrt{t}\, dt$

▪ Exercises 6.4

1–4 ▪ Verify that the given function is a solution of the differential equation.

1. $y = \frac{2}{3}e^x + e^{-2x}, \quad y' + 2y = 2e^x$

2. $y = 1/\sqrt{4 - x^2}, \quad y' = xy^3$

3. $y = xe^x, \quad y'/y = 1 + 1/x$

4. $y = 3/(3 - x^3), \quad y' = x^2 y^2$

5–14 ▪ Solve the separable differential equation.

5. $\dfrac{dy}{dx} = xy^2$

6. $\dfrac{dy}{dx} = xe^{-y}$

7. $\dfrac{dy}{dx} = \dfrac{x + 1}{y - 1}$

8. $e^y y' = x^2$

9. $\dfrac{dy}{dx} = \dfrac{y}{x}$

10. $\dfrac{dy}{dx} = \dfrac{\sqrt{x}}{e^y}$

11. $(x^2 + 1)y' = xy$

12. $(y^2 + xy^2)y' = 1$

13. $\dfrac{du}{dt} = (1 + u)(2 + t)$

14. $\dfrac{dz}{dt} = -e^t e^z$

15–20 ▪ Find the solution of the differential equation that satisfies the given initial condition.

15. $\dfrac{dy}{dx} = \dfrac{x}{y}, \quad y(0) = -3$

16. $\dfrac{dr}{dq} = q^2 r, \quad r(3) = 2$

17. $y' = \dfrac{e^{2x}}{y^2}, \quad y(0) = 3$

18. $\dfrac{dP}{dt} = \sqrt{Pt}, \quad P(1) = 2$

19. $\dfrac{dA}{dt} = \dfrac{1}{At} \ (t > 0, A > 0), \quad A(1) = 4$

20. $y' = \dfrac{e^x}{e^y}, \quad y(0) = 1$

21. (a) What can you say about a solution of the equation $y' = -y^2$ just by looking at the differential equation?

  (b) Verify that all members of the family $y = 1/(x + C)$ are solutions of the equation in part (a).

  (c) Can you think of a solution of the differential equation $y' = -y^2$ that is not a member of the family in part (b)?

  (d) Find a solution of the initial-value problem

$$y' = -y^2 \qquad y(0) = 0.5$$

22. (a) Show that every member of the family of functions $y = (\ln x + C)/x$ is a solution of the differential equation $x^2 y' + xy = 1$.

  (b) Illustrate part (a) by graphing several members of the family of solutions on a common screen.

  (c) Find a solution of the differential equation that satisfies the initial condition $y(1) = 2$.

  (d) Find a solution of the differential equation that satisfies the initial condition $y(2) = 1$.

23. Find an equation of the curve that passes through the point $(0, 1)$ and whose slope at (x, y) is xy.

24. Find the family of functions whose graphs have slope $x^2 y^2$ at every point (x, y). Then identify the one function whose graph passes through the point $(3, 10)$.

25. Suppose B is a function of time t and B always increases at a rate inversely proportional to its value. Write a differential equation that describes this relationship.

26. Spread of disease A cattle rancher discovers a viral disease spreading among his 4000 cattle. Suppose the rate at which the disease spreads is proportional to the product of the number of cattle that have the disease and the number that do not. Write a differential equation that describes this scenario.

27. Population growth Suppose a population always grows at a rate equal to $1/100$ of the population size (measured in members per year). Write a differential equation that represents this relationship. If the population is currently 20,000, find a solution to the differential equation that gives the population t years from now.

28. Radioactive decay A radioactive material decays at a rate equal to $1/5000$ of the amount of the material present (measured in grams per year). An experiment starts with 20 g of the material. Write a differential equation that describes this situation and solve it to find a function that gives the amount of material present after t years.

29. Insect population   Suppose at any time t the number of ants in a colony is increasing at a rate of 500 more than $1/50$ of the population, measured in ants per month. Write a differential equation that describes this scenario. If there are currently 300,000 ants in the colony, how many will there be a year from now?

30. Coffee temperature   Suppose you have just poured a cup of freshly brewed coffee with temperature 95°C in a room where the temperature is 20°C.

(a) When do you think the coffee cools most quickly? What happens to the rate of cooling as time goes by? Explain.

(b) **Newton's Law of Cooling** states that the rate of cooling of an object is proportional to the temperature difference between the object and its surroundings, provided that this difference is not too large. If it is known that the coffee cools at a rate of 1°C per minute when its temperature is 70°C, write a differential equation that expresses Newton's Law of Cooling for this particular situation. What is the initial condition?

(c) Solve the differential equation to find an expression for the temperature of the coffee at time t.

31. Learning rate   Psychologists interested in learning theory study **learning curves**. A learning curve is the graph of a function $P(t)$, the performance of someone learning a skill as a function of the training time t. The derivative dP/dt represents the rate at which performance improves.

(a) When do you think P increases most rapidly? What happens to dP/dt as t increases? Explain.

(b) If M is the maximum level of performance of which the learner is capable, explain why the differential equation

$$\frac{dP}{dt} = k(M - P) \qquad k \text{ a positive constant}$$

is a reasonable model for learning.

(c) Make a rough sketch of a possible solution of this differential equation.

(d) Solve this differential equation to find an expression for $P(t)$. What is the limit of this expression as $t \to \infty$?

32. Homeostasis   *Homeostasis* refers to a state in which the nutrient content of a consumer is independent of the nutrient content of its food. In the absence of homeostasis, a model proposed by Sterner and Elser is given by

$$\frac{dy}{dx} = \frac{1}{\theta}\frac{y}{x}$$

where x and y represent the nutrient content of the food and the consumer, respectively, and θ is a constant with $\theta \geqslant 1$.

(a) Solve the differential equation.

(b) What happens when $\theta = 1$? What happens when $\theta \to \infty$?

■ **Challenge Yourself**

33. Solve the differential equation $\dfrac{dy}{dt} = \dfrac{te^t}{y\sqrt{1 + y^2}}$.

34. Solve the differential equation $\dfrac{dy}{dx} = \dfrac{\ln x}{xy}$ given the initial condition $y(1) = 2$.

35. Solve the differential equation $y' = x + y$ by making the change of variable $u = x + y$.

36. The function with the given graph is a solution of one of the following differential equations. Decide which is the correct equation and justify your answer.

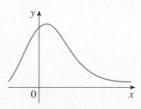

A. $y' = 1 + xy$ B. $y' = -2xy$ C. $y' = 1 - 2xy$

37. Glucose level   A glucose solution is administered intravenously into the bloodstream at a constant rate r. As the glucose is added, it is converted into other substances and removed from the bloodstream at a rate that is proportional to the concentration at that time. Thus a model for the concentration $C = C(t)$ of the glucose solution in the bloodstream is

$$\frac{dC}{dt} = r - kC$$

where k is a positive constant.

(a) Suppose that the concentration at time $t = 0$ is C_0. Determine the concentration at any time t by solving the differential equation.

(b) Assuming that $C_0 < r/k$, find $\lim_{t \to \infty} C(t)$ and interpret your answer.

38. Allometric growth   *Allometric growth* in biology refers to relationships between sizes of parts of an organism (skull length and body length, for instance). If $L_1(t)$ and $L_2(t)$ are the sizes of two organs in an organism of age t, then L_1 and L_2 satisfy an allometric law if their specific growth rates are proportional:

$$\frac{1}{L_1}\frac{dL_1}{dt} = k\frac{1}{L_2}\frac{dL_2}{dt}$$

where k is a constant.

(a) Use the allometric law to write a differential equation relating L_1 and L_2 and solve it to express L_1 as a function of L_2.

(b) In a study of several species of unicellular algae, the proportionality constant in the allometric law relating B (cell biomass) and V (cell volume) was found to be $k = 0.0794$. Write B as a function of V.

39. Tissue growth   Let $A(t)$ be the area of a tissue culture at time t and let M be the final area of the tissue when growth is complete. Most cell divisions occur on the periphery of the tissue and the number of cells on the periphery is proportional to $\sqrt{A(t)}$. So a reasonable model for the growth of tissue is obtained by assuming that the rate of growth of the area is jointly proportional to $\sqrt{A(t)}$ and $M - A(t)$. Formulate a differential equation and use it to show that the tissue grows fastest when $A(t) = \frac{1}{3}M$.
[*Hint:* Use the Chain Rule to differentiate the expression for dA/dt, and then substitute for dA/dt from the differential equation where appropriate.]

6.5 Improper Integrals

In defining a definite integral $\int_a^b f(x)\,dx$ we dealt with a function f defined on a finite interval $[a, b]$. In this section we extend the concept of a definite integral to the case where the interval is infinite. We call such an integral an *improper* integral. One of the most important applications of this idea, probability distributions, will be studied in the next section.

▪ Integrating on Infinite Intervals

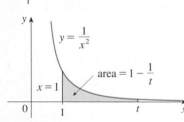

FIGURE 1

Consider the infinite region S that lies under the curve $y = 1/x^2$, above the x-axis, and to the right of the line $x = 1$. You might think that, since S is infinite in extent, its area must be infinite, but let's take a closer look. The area of the part of S that lies to the left of the line $x = t$ (shaded in Figure 1) is

$$A(t) = \int_1^t \frac{1}{x^2}\,dx = -\frac{1}{x}\Big]_1^t = 1 - \frac{1}{t}$$

Notice that $A(t) < 1$ no matter how large t is chosen.
   We also observe that

$$\lim_{t\to\infty} A(t) = \lim_{t\to\infty}\left(1 - \frac{1}{t}\right) = 1$$

The area of the shaded region approaches 1 as $t \to \infty$ (see Figure 2), so we say that the area of the infinite region S is equal to 1 and we write

$$\int_1^\infty \frac{1}{x^2}\,dx = \lim_{t\to\infty}\int_1^t \frac{1}{x^2}\,dx = 1$$

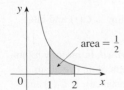 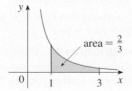 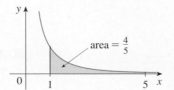 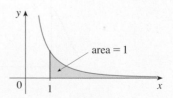

FIGURE 2

Using this example as a guide, we define the integral of f (not necessarily a positive function) over an infinite interval as the limit of integrals over finite intervals.

---

(1) ▪ **Integrating from a to ∞**   If $\int_a^t f(x)\,dx$ exists for every number $t \geq a$, then

$$\int_a^\infty f(x)\,dx = \lim_{t \to \infty} \int_a^t f(x)\,dx$$

provided this limit exists (as a finite number). An improper integral is called **convergent** if the corresponding limit exists and **divergent** if the limit does not exist.

The improper integral in Definition 1 can be interpreted as an area provided that f is a positive function. For instance, if $f(x) \geq 0$ and the integral $\int_a^\infty f(x)\,dx$ is convergent, then we define the area of the region below the graph of f, above the x-axis, and to the right of the line $x = a$ (as in Figure 3) to be

$$\text{area} = \int_a^\infty f(x)\,dx$$

This is appropriate because $\int_a^\infty f(x)\,dx$ is the limit as $t \to \infty$ of the area under the graph of f from a to t.

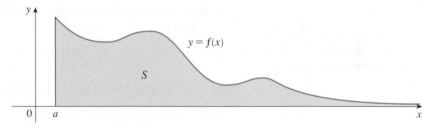

FIGURE 3

▪ EXAMPLE 1   A Divergent Improper Integral

Determine whether the integral $\int_1^\infty (1/x)\,dx$ is convergent or divergent.

SOLUTION

According to Definition 1, we have

$$\int_1^\infty \frac{1}{x}\,dx = \lim_{t \to \infty} \int_1^t \frac{1}{x}\,dx = \lim_{t \to \infty} \ln|x| \Big]_1^t$$

$$= \lim_{t \to \infty} (\ln t - \ln 1) = \lim_{t \to \infty} \ln t = \infty$$

The limit does not exist as a finite number and so the improper integral $\int_1^\infty (1/x)\,dx$ is divergent. ▪

Let's compare the result of Example 1 with the example given at the beginning of this section:

$$\int_1^\infty \frac{1}{x^2}\,dx \text{ converges} \qquad \int_1^\infty \frac{1}{x}\,dx \text{ diverges}$$

Geometrically, this says that although the curves $y = 1/x^2$ and $y = 1/x$ look very similar for $x > 0$, the region under $y = 1/x^2$ to the right of $x = 1$ (the shaded region in Figure 4) has finite area whereas the corresponding region under $y = 1/x$ (in Figure 5) has infinite area. Note that both $1/x^2$ and $1/x$ approach 0 as $x \to \infty$ but $1/x^2$ approaches 0 faster than $1/x$. The values of $1/x$ don't decrease fast enough for its integral to have a finite value.

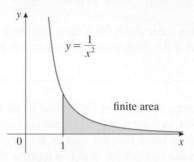

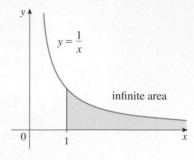

FIGURE 4 $\int_1^{\infty} (1/x^2)\, dx$ converges

FIGURE 5 $\int_1^{\infty} (1/x)\, dx$ diverges

TEC In Module 6.5 you can investigate visually and numerically whether several improper integrals are convergent or divergent.

We next define an improper integral over an interval that extends infinitely in the negative direction.

Again we call the improper integral defined here convergent if the limit exists and divergent otherwise.

> **(2)** ▪ **Integrating From $-\infty$ to b**   If $\int_t^b f(x)\, dx$ exists for every number $t \le b$, then
>
> $$\int_{-\infty}^b f(x)\, dx = \lim_{t \to -\infty} \int_t^b f(x)\, dx$$
>
> provided this limit exists (as a finite number).

▪ **EXAMPLE 2   A Convergent Improper Integral**

Evaluate $\displaystyle\int_{-\infty}^0 e^x\, dx$.

SOLUTION

Using Definition 2, we have

$$\int_{-\infty}^0 e^x\, dx = \lim_{t \to -\infty} \int_t^0 e^x\, dx = \lim_{t \to -\infty} e^x \Big]_t^0$$

$$= \lim_{t \to -\infty} (1 - e^t)$$

As $t \to -\infty$, $e^t \to 0$, and so

$$\int_{-\infty}^0 e^x\, dx = \lim_{t \to -\infty} (1 - e^t) = 1 - 0 = 1$$

▪

Finally, we can integrate a function over the entire real line by splitting the real line into two separate parts.

> **(3) ■ Integrating Over the Entire Real Line**   If both $\int_a^\infty f(x)\, dx$ and $\int_{-\infty}^a f(x)\, dx$ are convergent, then we define
>
> $$\int_{-\infty}^\infty f(x)\, dx = \int_{-\infty}^a f(x)\, dx + \int_a^\infty f(x)\, dx$$
>
> where a is any real number.

■ EXAMPLE 3   An Improper Integral over the Entire Real Line

Evaluate $\displaystyle\int_{-\infty}^\infty e^x\, dx.$

SOLUTION

According to Definition 3, we must split the integral into two parts. If we split at $x = 0$, then

$$\int_{-\infty}^\infty e^x\, dx = \int_{-\infty}^0 e^x\, dx + \int_0^\infty e^x\, dx$$

We evaluated the first integral in Example 2. For the second integral we have

$$\int_0^\infty e^x\, dx = \lim_{t \to \infty} \int_0^t e^x\, dx = \lim_{t \to \infty} e^x\Big]_0^t$$

$$= \lim_{t \to \infty} (e^t - 1)$$

But $e^t \to \infty$ as $t \to \infty$, so $\lim_{t \to \infty} e^t = \infty$ and

$$\lim_{t \to \infty} (e^t - 1) = \infty$$

Thus $\int_0^\infty e^x\, dx$ is divergent. Although $\int_{-\infty}^0 e^x\, dx$ is convergent, the sum of divergent and convergent integrals is divergent, and so $\int_{-\infty}^\infty e^x\, dx$ diverges.   ■

■ | Applications of Improper Integrals

In Section 6.2 we discussed the present value of an income stream. Recall that if income flows in at a rate of $f(t)$ dollars per year and earns interest at a constant rate r (compounded continuously) for a period of T years, then the present value is

$$\text{PV} = \int_0^T f(t) e^{-rt}\, dt$$

A **perpetuity** is an income stream without end; its present value is computed by using an improper integral.

> ■ The **present value of a perpetuity** where income is received at a rate of $f(t)$ dollars per year and earns interest at a constant rate r (compounded continuously) is
>
> **(4)** $$\text{PV} = \int_0^\infty f(t) e^{-rt}\, dt$$

▪ EXAMPLE 4 Present Value of a Perpetuity

The British government has in the past issued perpetual bonds called *consols* that pay $2\frac{1}{2}\%$ annually to the bond holder forever. Bond holders are paid quarterly, but if we consider this income as a continuous stream, we can compute the present value using Formula 4. Suppose a £5000 bond is purchased, and the payments received are invested at 4% interest (compounded continuously). What is the present value of the consol?

SOLUTION

The annual payment on the consol is $2\frac{1}{2}\%$ of £5000, or $5000(0.025) = £125$. If we consider this a continuous income stream, then $f(t) = 125$ pounds per year in Formula 4. We have $r = 0.04$, so the present value is

$$\text{PV} = \int_0^\infty f(t)e^{-rt}\,dt = \lim_{T\to\infty} \int_0^T 125e^{-0.04t}\,dt$$

$$= \lim_{T\to\infty}\left[\frac{125}{-0.04}e^{-0.04t}\right]_0^T = \lim_{T\to\infty}\left[-3125(e^{-0.04T}-1)\right]$$

$$= -3125\left[\left(\lim_{T\to\infty}\frac{1}{e^{0.04T}}\right)-1\right]$$

$$= -3125(0-1) = 3125$$

It may seem counterintuitive that the present value of a perpetuity that pays income indefinitely is finite, but remember that the present value of money is less and less as the time period grows.

Thus the present value of the consol is £3125. Since this is less than the cost of the bond, perhaps a consol is not the wisest investment! ▪

▪ Prepare Yourself

1. Evaluate each of the following limits.

(a) $\lim_{t\to\infty} e^{-t}$ (b) $\lim_{x\to\infty}(x+e^{-x})$

(c) $\lim_{x\to-\infty}(4+e^{-x})$ (d) $\lim_{t\to\infty}\left(1+5\sqrt{t}\right)$

(e) $\lim_{t\to\infty}\left(1+5/\sqrt{t}\right)$ (f) $\lim_{x\to\infty}(1/\ln x)$

(g) $\lim_{x\to-\infty}\ln(1+x^2)$

2. Evaluate the integral.

(a) $\int_2^5 \frac{1}{(x+1)^{5/2}}\,dx$ (b) $\int_1^w e^{-x/3}\,dx$

(c) $\int te^{t^2}\,dt$ (d) $\int \frac{1}{x\ln x}\,dx$

(e) $\int 1/t^{1.4}\,dt$

▪ Exercises 6.5

1. Find the area under the curve $y = 1/x^3$ from $x=1$ to $x=t$ and evaluate it for $t = 10$, 100, and 1000. Then find the total area under this curve for $x \ge 1$.

▦ **2.** (a) Graph the functions $f(x) = 1/x^{1.1}$ and $g(x) = 1/x^{0.9}$ in the viewing rectangles $[0, 10]$ by $[0, 1]$ and $[0, 100]$ by $[0, 1]$.

(b) Find the areas under the graphs of f and g from $x=1$ to $x=t$ and evaluate for $t = 10$, 100, 10^4, 10^6, 10^{10}, and 10^{20}.

(c) Find the total area under each curve for $x \ge 1$, if it exists.

3–16 ▪ Determine whether the integral is convergent or divergent. Evaluate those that are convergent.

3. $\int_1^\infty \frac{2}{x^3}\,dx$ **4.** $\int_0^\infty e^{-3x}\,dx$

5. $\int_3^\infty \frac{1}{(x-2)^{3/2}}\,dx$ **6.** $\int_0^\infty \frac{1}{\sqrt[4]{1+x}}\,dx$

7. $\int_{-\infty}^{-1} \dfrac{1}{\sqrt{2-w}}\, dw$

8. $\int_{0}^{\infty} \dfrac{x}{(x^2+2)^2}\, dx$

9. $\int_{4}^{\infty} e^{-y/2}\, dy$

10. $\int_{-\infty}^{-1} e^{-2t}\, dt$

11. $\int_{-\infty}^{\infty} x e^{-x^2}\, dx$

12. $\int_{-\infty}^{\infty} (y^3 - 3y^2)\, dy$

13. $\int_{1}^{\infty} \dfrac{\ln x}{x}\, dx$

14. $\int_{1}^{\infty} \dfrac{e^{-\sqrt{x}}}{\sqrt{x}}\, dx$

15. $\int_{e}^{\infty} \dfrac{1}{x(\ln x)^3}\, dx$

16. $\int_{-\infty}^{\infty} x^3 e^{-x^4}\, dx$

17–20 ▪ Sketch the region and find its area (if the area is finite).

17. $x \leq 1, \quad 0 \leq y \leq e^x$

18. $x \geq -2, \quad 0 \leq y \leq e^{-x/2}$

19. $x \geq 1, \quad 0 \leq y \leq 1/\sqrt{x}$

20. $x \geq 0, \quad 0 \leq y \leq 2x/(x^2+1)$

---

21. Present value   A certain perpetuity pays the holder $400 per month. If the money is invested at a rate of 5% annually (compounded continuously), what is the present value of the perpetuity?

22. Present value   Find the present value of an indefinite income stream that pays $2000 annually if the income received earns 6% interest (compounded continuously).

23. Present value   A government bond is offered that pays 3% annually to the holder indefinitely. If the income received is invested at 4.2% interest (compounded continuously), what is the present value of a $10,000 bond?

24. Perpetuity value   An insurance company offers a perpetuity for $100,000 that pays $600 per month indefinitely. If we assume an interest rate of 5% (compounded continuously) for the received income, is the perpetuity a wise purchase?

25. Present value   A charitable organization receives donations from a corporate trust continually at the rate of $f(t) = 8000e^{0.01t}$ dollars per year t years from now. If the money is invested at 5.5% interest (compounded continuously) and the donations continue indefinitely, what is the present value of the income?

26. Present value   A family owns an apartment building that is estimated to generate income at a rate of $f(t) = 250,000e^{0.025t}$ dollars per year t years from now. If we assume a 6.3% interest rate on the income, and the building stays within the family indefinitely, what is the present value of the income?

27. Lightbulb lifetime   A manufacturer of lightbulbs wants to produce bulbs that last about 700 hours but, of course, some bulbs burn out faster than others. Let $F(t)$ be the fraction of the company's bulbs that burn out before t hours, so $F(t)$ always lies between 0 and 1.

(a) Make a rough sketch of what you think the graph of F might look like.

(b) What is the meaning of the derivative $r(t) = F'(t)$?

(c) What is the value of $\int_{0}^{\infty} r(t)\, dt$? Why?

▪ Challenge Yourself

28. Find the values of p for which the integral $\int_{1}^{\infty} \dfrac{1}{x^p}\, dx$ converges and evaluate the integral for those values of p. [*Hint:* Consider separately the cases where $p < 0$, $p = 0$, $0 < p < 1$, $p = 1$, and $p > 1$.]

6.6 Probability

Calculus plays a role in the analysis of random behavior. Suppose we consider the cholesterol level of a person chosen at random from a certain age group, or the height of an adult female chosen at random, or the lifetime of a randomly chosen battery of a certain type. Such quantities are called **continuous random variables** because their values actually range over an interval of real numbers, although they might be measured or recorded only to the nearest integer. We might want to know the probability that a blood cholesterol level is greater than 250, or the probability that the height of an adult female is between 60 and 70 inches, or the probability that the battery we are buying lasts between 100 and 200 hours. If X represents the lifetime of that type of battery, we denote this last probability as follows:

$$P(100 \leqslant X \leqslant 200)$$

According to the frequency interpretation of probability, this number is the long-run proportion of all batteries of the specified type whose lifetimes are between 100 and 200 hours. Since it represents a proportion, the probability naturally falls between 0 and 1.

Probability Density Functions

Every continuous random variable X has a **probability density function** f. This means that the probability that X lies between a and b is found by integrating f from a to b:

(1)
$$P(a \leqslant X \leqslant b) = \int_a^b f(x)\, dx$$

For example, Figure 1 shows the graph of a model for the probability density function f for a random variable X defined to be the height in inches of an adult female in the United States (according to data from the National Health Survey). The probability that the height of a woman chosen at random from this population is between 60 and 70 inches is equal to the area under the graph of f from 60 to 70.

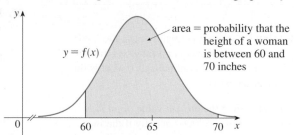

FIGURE 1
Probability density function
for the height of an adult female

In general, the probability density function f of a random variable X satisfies the condition $f(x) \geqslant 0$ for all x. Because probabilities are measured on a scale from 0 to 1, it follows that

(2)
$$\int_{-\infty}^{\infty} f(x)\, dx = 1$$

▪ EXAMPLE 1 A Probability Density Function

Let $f(x) = 0.006x(10 - x)$ for $0 \leqslant x \leqslant 10$ and $f(x) = 0$ for all other values of x.
(a) Verify that f is a probability density function.
(b) Find $P(4 \leqslant X \leqslant 8)$.

SOLUTION

(a) For $0 \leqslant x \leqslant 10$ we have $0.006x(10 - x) \geqslant 0$, so $f(x) \geqslant 0$ for all x. We also need to check that Equation 2 is satisfied:

We need only integrate from 0 to 10 because $f(x) = 0$ outside these values.

$$\int_{-\infty}^{\infty} f(x)\, dx = \int_0^{10} 0.006x(10 - x)\, dx = 0.006 \int_0^{10} (10x - x^2)\, dx$$

$$= 0.006\left[5x^2 - \tfrac{1}{3}x^3\right]_0^{10} = 0.006\left(500 - \tfrac{1000}{3}\right) = 1$$

Therefore f is a probability density function.

(b) The probability that X lies between 4 and 8 is

$$P(4 \leqslant X \leqslant 8) = \int_4^8 f(x)\,dx = 0.006 \int_4^8 (10x - x^2)\,dx$$

$$= 0.006\big[5x^2 - \tfrac{1}{3}x^3\big]_4^8 = 0.544$$ ▪

▪ EXAMPLE 2 Probability Density Function for Waiting Times

Phenomena such as waiting times and equipment failure times are commonly modeled by exponentially decreasing probability density functions. Find the exact form of such a function.

SOLUTION

Think of the random variable as being the time you wait on hold before an agent of a company you're telephoning answers your call. So instead of x, let's use t to represent time, in minutes. If f is the probability density function and you call at time $t = 0$, then, from Definition 1, $\int_0^2 f(t)\,dt$ represents the probability that an agent answers within the first two minutes and $\int_4^5 f(t)\,dt$ is the probability that your call is answered during the fifth minute.

It's clear that $f(t) = 0$ for $t < 0$ (the agent can't answer before you place the call). For $t > 0$ we are told to use an exponentially decreasing function, that is, a function of the form $f(t) = Ae^{-ct}$ where A and c are positive constants. Thus

$$f(t) = \begin{cases} 0 & \text{if } t < 0 \\ Ae^{-ct} & \text{if } t \geqslant 0 \end{cases}$$

We use Equation 2 to determine the value of A. First we split the integral:

$$\int_{-\infty}^{\infty} f(t)\,dt = \int_{-\infty}^{0} f(t)\,dt + \int_0^{\infty} f(t)\,dt$$

Because $f(t) = 0$ for $t < 0$, $\int_{-\infty}^0 f(t)\,dt = 0$. Thus

$$\int_{-\infty}^{\infty} f(t)\,dt = \int_0^{\infty} f(t)\,dt = \int_0^{\infty} Ae^{-ct}\,dt$$

$$= \lim_{x \to \infty} \int_0^x Ae^{-ct}\,dt = \lim_{x \to \infty} \left[-\frac{A}{c}e^{-ct} \right]_0^x$$

$$= \lim_{x \to \infty} \left[-\frac{A}{c}e^{-cx} + \frac{A}{c}e^0 \right] = \lim_{x \to \infty} \frac{A}{c}(-e^{-cx} + 1)$$

$$= \lim_{x \to \infty} \frac{A}{c}\left(1 - \frac{1}{e^{cx}} \right) = \frac{A}{c}(1 - 0) = \frac{A}{c}$$

By Equation 2, $\int_{-\infty}^{\infty} f(t)\,dt = 1$, therefore $A/c = 1$ and so $A = c$. Thus every exponential density function has the form

$$f(t) = \begin{cases} 0 & \text{if } t < 0 \\ ce^{-ct} & \text{if } t \geqslant 0 \end{cases}$$

A typical graph is shown in Figure 2. ▪

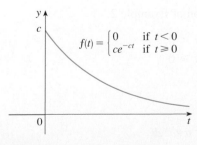

$$f(t) = \begin{cases} 0 & \text{if } t < 0 \\ ce^{-ct} & \text{if } t \geqslant 0 \end{cases}$$

FIGURE 2

An exponential density function

■ Average Values

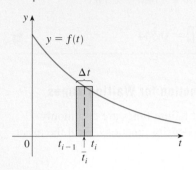

FIGURE 3

Suppose you're waiting for a company to answer your phone call and you wonder how long, on average, you can expect to wait. Let $f(t)$ be the corresponding density function, where t is measured in minutes, and think of a sample of N people who have called this company. Most likely, none of them had to wait more than an hour, so let's restrict our attention to the interval $0 \leqslant t \leqslant 60$. Let's divide that interval into n subintervals of length Δt and endpoints $0, t_1, t_2, \ldots, t_n = 60$. (Think of Δt as lasting a minute, or half a minute, or 10 seconds, or even a second.) The probability that somebody's call gets answered during the time period from t_{i-1} to t_i is the area under the curve $y = f(t)$ from t_{i-1} to t_i, which is approximately equal to $f(\bar{t}_i) \Delta t$, where $\bar{t}_i$ is the midpoint of the interval. (This is the area of the approximating rectangle in Figure 3.)

Since the long-run proportion of calls that get answered in the time period from t_{i-1} to t_i is $f(\bar{t}_i) \Delta t$, we expect that, out of our sample of N callers, the number whose call was answered in that time period is approximately $Nf(\bar{t}_i) \Delta t$ and the time that each waited is about $\bar{t}_i$. Therefore the total time they waited is the product of these numbers: approximately $\bar{t}_i[Nf(\bar{t}_i) \Delta t]$. Adding over all such intervals, we get the approximate total of everybody's waiting times:

$$N \bar{t}_1 f(\bar{t}_1) \Delta t + N \bar{t}_2 f(\bar{t}_2) \Delta t + \cdots + N \bar{t}_n f(\bar{t}_n) \Delta t$$

If we now divide by the number of callers N, we get the approximate *average* waiting time:

$$\bar{t}_1 f(\bar{t}_1) \Delta t + \bar{t}_2 f(\bar{t}_2) \Delta t + \cdots + \bar{t}_n f(\bar{t}_n) \Delta t$$

We recognize this as a Riemann sum for the function $tf(t)$. As the time interval shrinks (that is, $\Delta t \to 0$ and $n \to \infty$), this Riemann sum approaches the integral

$$\int_0^{60} tf(t) \, dt$$

This integral is called the *mean waiting time*.

In general, the **mean** of any probability density function f is defined to be

It is traditional to denote the mean by the Greek letter μ (mu).

$$\mu = \int_{-\infty}^{\infty} x f(x) \, dx$$

The mean can be interpreted as the long-run average value of the random variable X. It can also be interpreted as a measure of centrality of the probability density function.

■ EXAMPLE 3   Computing a Mean

Find the mean of the exponential density function of Example 2:

$$f(t) = \begin{cases} 0 & \text{if } t < 0 \\ ce^{-ct} & \text{if } t \geqslant 0 \end{cases}$$

SOLUTION

According to the definition of a mean, we have

$$\mu = \int_{-\infty}^{\infty} tf(t) \, dt = \int_0^{\infty} tce^{-ct} \, dt$$

To evaluate $\int tce^{-ct} \, dt$ we use integration by parts with $u = t$ and $dv = ce^{-ct} \, dt$.

Then $du = dt$ and $v = -e^{-ct}$, and

$$\int tce^{-ct}\, dt = -te^{-ct} - \int -e^{-ct}\, dt = -te^{-ct} - \frac{e^{-ct}}{c} + C$$

Then

$$\int_0^\infty tce^{-ct}\, dt = \lim_{x \to \infty} \int_0^x tce^{-ct}\, dt = \lim_{x \to \infty} \left[-te^{-ct} - \frac{e^{-ct}}{c} \right]_0^x$$

$$= \lim_{x \to \infty} \left(-xe^{-cx} - \frac{e^{-cx}}{c} + 0 + \frac{e^0}{c} \right)$$

$$= \lim_{x \to \infty} \left(-\frac{x}{e^{cx}} - \frac{1}{ce^{cx}} + \frac{1}{c} \right)$$

To evaluate $\lim_{x \to \infty} x/e^{cx}$, note that e^{cx} increases at a much higher rate than x, so the ratio x/e^{cx} becomes smaller and smaller as x increases. Thus

$$\lim_{x \to \infty} \frac{x}{e^{cx}} = 0$$

and

$$\int_0^\infty tce^{-ct}\, dt = \lim_{x \to \infty} \left(-\frac{x}{e^{cx}} - \frac{1}{ce^{cx}} + \frac{1}{c} \right) = 0 - 0 + \frac{1}{c} = \frac{1}{c}$$

The mean is $\mu = 1/c$, so $c = 1/\mu$ and we can rewrite the probability density function as

$$f(t) = \begin{cases} 0 & \text{if } t < 0 \\ \mu^{-1}e^{-t/\mu} & \text{if } t \geq 0 \end{cases}$$ ■

▪ EXAMPLE 4   Find a Probability Given the Mean

Suppose the average waiting time for a customer's call to be answered by a company representative is five minutes.

(a) Find the probability that a call is answered during the first minute.

(b) Find the probability that a customer waits more than five minutes to be answered.

SOLUTION

(a) We are given that the mean of the exponential distribution is $\mu = 5$ min and so, from the result of Example 3, we know that the probability density function is

$$f(t) = \begin{cases} 0 & \text{if } t < 0 \\ \frac{1}{5}e^{-t/5} & \text{if } t \geq 0 \end{cases}$$

Thus the probability that a call is answered during the first minute is

$$P(0 \leq T \leq 1) = \int_0^1 f(t)\, dt$$

$$= \int_0^1 \frac{1}{5}e^{-t/5}\, dt$$

$$= \frac{1}{5}(-5)e^{-t/5} \Big]_0^1 = -1(e^{-1/5} - e^0)$$

$$= 1 - e^{-1/5} \approx 0.1813$$

So about 18% of customers' calls are answered during the first minute.

(b) The probability that a customer waits more than five minutes is

$$P(T > 5) = \int_5^{\infty} f(t)\, dt = \int_5^{\infty} \tfrac{1}{5} e^{-t/5}\, dt$$

$$= \lim_{x \to \infty} \int_5^{x} \tfrac{1}{5} e^{-t/5}\, dt = \lim_{x \to \infty} \left[-e^{-t/5} \right]_5^{x}$$

$$= \lim_{x \to \infty} \left(-e^{-x/5} + e^{-1} \right) = 0 + e^{-1}$$

$$= \frac{1}{e} \approx 0.368$$

About 37% of customers wait more than five minutes before their calls are answered. ■

Notice the result of Example 4(b): Even though the mean waiting time is 5 minutes, only 37% of callers wait more than 5 minutes. The reason is that some callers have to wait much longer (maybe 10 or 15 minutes), and this brings up the average.

Another measure of centrality of a probability density function is the *median*, which is a number *m* such that half the callers have a waiting time less than *m* and the other callers have a waiting time longer than *m*. In general, the **median** of a probability density function is the number *m* such that

$$\int_m^{\infty} f(x)\, dx = \tfrac{1}{2}$$

This means that half the area under the graph of *f* lies to the right of *m*. In Exercise 9 you are asked to show that the median waiting time for the company described in Example 4 is approximately 3.5 minutes.

■ Normal Distributions

Many important random phenomena—such as test scores on aptitude tests, heights and weights of individuals from a homogeneous population, annual rainfall in a given location—are modeled by a **normal distribution**. This means that the probability density function of the random variable *X* is a member of the family of functions

(3) $$f(x) = \frac{1}{\sigma \sqrt{2\pi}} e^{-(x-\mu)^2/(2\sigma^2)}$$

You can verify that the mean for this function is μ. The positive constant σ is called the **standard deviation**; it measures how spread out the values of *X* are. From the bell-shaped graphs of members of the family in Figure 4, we see that for small val-

The standard deviation is denoted by the lowercase Greek letter σ (sigma).

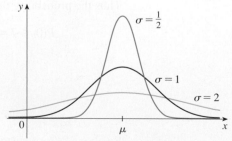

FIGURE 4
Normal distributions

ues of σ the values of X are clustered about the mean, whereas for larger values of σ the values of X are more spread out. Statisticians have methods for using sets of data to estimate μ and σ.

The factor $1/(\sigma\sqrt{2\pi}\,)$ is needed to make f a probability density function. In fact, it can be verified using the methods of multivariable calculus that

$$\int_{-\infty}^{\infty} \frac{1}{\sigma\sqrt{2\pi}} e^{-(x-\mu)^2/(2\sigma^2)} \, dx = 1$$

■ EXAMPLE 5   A Normal Distribution

Intelligence Quotient (IQ) scores are distributed normally with mean 100 and standard deviation 15. (Figure 5 shows the corresponding probability density function.)

(a) What percentage of the population has an IQ score between 85 and 115?

(b) What percentage of the population has an IQ above 140?

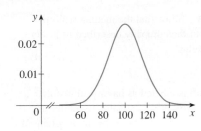

FIGURE 5
Distribution of IQ scores

SOLUTION

(a) Since IQ scores are normally distributed, we use the probability density function given by Equation 3 with $\mu = 100$ and $\sigma = 15$:

$$P(85 \leqslant X \leqslant 115) = \int_{85}^{115} \frac{1}{15\sqrt{2\pi}} e^{-(x-100)^2/(2\cdot15^2)} \, dx$$

As we mentioned in Section 5.5, the function $y = e^{-x^2}$ doesn't have an anti-derivative in the usual sense, so we can't evaluate the integral exactly. But we can use the numerical integration capability of a calculator or computer (or the Midpoint Rule) to estimate the integral. Doing so, we find that

$$P(85 \leqslant X \leqslant 115) \approx 0.68$$

So about 68% of the population has an IQ between 85 and 115, that is, within one standard deviation of the mean.

(b) The probability that the IQ score of a person chosen at random is more than 140 is

$$P(X > 140) = \int_{140}^{\infty} \frac{1}{15\sqrt{2\pi}} e^{-(x-100)^2/450} \, dx$$

To avoid the improper integral we could approximate it by the integral from 140 to 200. (It's quite safe to say that people with an IQ over 200 are extremely rare.) Then

$$P(X > 140) \approx \int_{140}^{200} \frac{1}{15\sqrt{2\pi}} e^{-(x-100)^2/450} \, dx \approx 0.0038$$

Therefore about 0.4% of the population has an IQ over 140.   ■

Exercises 6.6

1. Tire lifetime   Let $f(x)$ be the probability density function for the lifetime of a manufacturer's highest quality car tire, where x is measured in miles. Explain the meaning of each integral.

(a) $\displaystyle\int_{30,000}^{40,000} f(x)\, dx$ (b) $\displaystyle\int_{25,000}^{\infty} f(x)\, dx$

2. Commuting time   Let $f(t)$ be the probability density function for the time it takes you to drive to school in the morning, where t is measured in minutes. Express the following probabilities as integrals.

(a) The probability that you drive to school in less than 15 minutes

(b) The probability that it takes you more than half an hour to get to school

3. Let $f(x) = \frac{3}{64}x\sqrt{16 - x^2}$ for $0 \le x \le 4$ and $f(x) = 0$ for all other values of x.

(a) Verify that f is a probability density function.

(b) Find $P(X < 2)$.

4. Let $f(x) = xe^{-x}$ if $x \ge 0$ and $f(x) = 0$ if $x < 0$.

(a) Verify that f is a probability density function. [*Hint:* Use integration by parts.]

(b) Find $P(1 \le X \le 2)$.

5. Let $f(x) = cxe^{-x^2}$ if $x \ge 0$ and $f(x) = 0$ if $x < 0$.

(a) For what value of c is f a probability density function?

(b) For that value of c, find $P(1 < X < 4)$.

6. Let $f(x) = kx^2(1 - x)$ if $0 \le x \le 1$ and $f(x) = 0$ if $x < 0$ or $x > 1$.

(a) For what value of k is f a probability density function?

(b) For that value of k, find $P\!\left(X \ge \tfrac{1}{2}\right)$.

(c) Find the mean.

7. Game of chance   A spinner from a board game randomly indicates a real number between 0 and 10. The spinner is fair in the sense that it indicates a number in a given interval with the same probability as it indicates a number in any other interval of the same length.

(a) Explain why the function

$$f(x) = \begin{cases} 0.1 & \text{if } 0 \le x \le 10 \\ 0 & \text{if } x < 0 \text{ or } x > 10 \end{cases}$$

is a probability density function for the spinner's values.

(b) What does your intuition tell you about the value of the mean? Check your guess by evaluating an integral.

8. (a) Explain why the function whose graph is shown is a probability density function.

(b) Use the graph to find the following probabilities:

    (i) $P(X < 3)$ (ii) $P(3 \le X \le 8)$

(c) Calculate the mean.

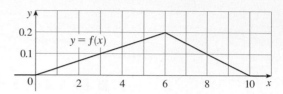

9. Phone waiting time   Show that the median waiting time for a phone call to the company described in Example 4 is about 3.5 minutes.

10. Lightbulb lifetime

(a) A type of lightbulb is labeled as having an average lifetime of 1000 hours. It's reasonable to model the probability of failure of these bulbs by an exponential density function with mean $\mu = 1000$. Use this model to find the probability that a bulb

    (i) fails within the first 200 hours.

    (ii) burns for more than 800 hours.

(b) What is the median lifetime of these lightbulbs?

11. Restaurant waiting times   The manager of a fast-food restaurant determines that the average time that her customers wait for service is 2.5 minutes.

(a) Find the probability that a customer has to wait more than four minutes.

(b) Find the probability that a customer is served within the first two minutes.

(c) The manager wants to advertise that anybody who isn't served within a certain number of minutes gets a free hamburger. But she doesn't want to give away free hamburgers to more than 2% of her customers. What should the advertisement say?

12. Human heights   According to the National Health Survey, the heights of adult males in the United States are normally distributed with mean 69.0 inches and standard deviation 2.8 inches.

(a) What is the probability that an adult male chosen at random is between 65 inches and 73 inches tall?

(b) What percentage of the adult male population is more than 6 feet tall?

13. The Garbage Project   The "Garbage Project" at the University of Arizona reports that the amount of paper discarded by households per week is normally distributed with mean 9.4 lb and standard deviation 4.2 lb. What percentage of households throw out at least 10 lb of paper a week?

14. Food packaging   Boxes are labeled as containing 500 g of cereal. The machine filling the boxes produces weights that are normally distributed with standard deviation 12 g.

   (a) If the target weight is 500 g, what is the probability that the machine produces a box with less than 480 g of cereal?

   (b) Suppose a law states that no more than 5% of a manufacturer's cereal boxes can contain less than the stated weight of 500 g. At what target weight should the manufacturer set its filling machine?

15. Driving speeds   The speeds of vehicles on a highway with speed limit 100 km/h are normally distributed with mean 112 km/h and standard deviation 8 km/h.

   (a) What is the probability that a randomly chosen vehicle is traveling at a legal speed?

   (b) If police are instructed to ticket motorists driving 125 km/h or more, what percentage of motorists are targeted?

■ Challenge Yourself

16. Show that the probability density function for a normally distributed random variable has inflection points at $x = \mu \pm \sigma$.

17. For any normal distribution, find the probability that the random variable lies within two standard deviations of the mean.

18. The standard deviation for a random variable with probability density function f and mean μ is defined by

$$\sigma = \left[\int_{-\infty}^{\infty} (x - \mu)^2 f(x)\, dx \right]^{1/2}$$

Find the standard deviation for an exponential density function with mean μ. [*Hint:* Integrate by parts twice and use the fact that $\lim_{x \to \infty} x^n e^{-x} = 0$ for any integer $n \geq 0$.]

CHAPTER 6 REVIEW

■ CONCEPT CHECK

1. How do you find the area between two curves? Does it make a difference if one of the curves lies below the x-axis?

2. What does the area between marginal revenue and marginal cost curves represent?

3. Suppose that Sue runs faster than Kathy throughout a 1500-meter race. What is the physical meaning of the area between their velocity curves for the first minute of the race?

4. (a) Given a demand function $p = D(q)$, explain what is meant by the consumer surplus when the amount of a commodity currently available is Q and the current selling price is P. Illustrate with a sketch. How do we compute it?

   (b) Given a supply function $p = S(q)$, explain what is meant by the producer surplus when the amount of a commodity currently available is Q and the current selling price is P. How is it computed?

   (c) What is the total surplus? When is it maximized?

5. (a) What is the future value of a continuous income stream? How is it computed?

   (b) How do we find the present value of a continuous income stream? What does this mean?

6. If we have survival and renewal functions for a population, how do we predict the size of the population T years from now?

7. (a) What is the cardiac output of the heart?

   (b) Explain how the cardiac output can be measured by the dye dilution method.

8. What is a differential equation? What is an initial-value problem?

9. What is a separable differential equation? How do you solve it?

10. (a) Write a differential equation for a quantity that grows at a rate proportional to its size.

   (b) Under what circumstances is this an appropriate model for population growth?

   (c) What is the solution of this equation?

11. (a) Write the logistic equation.

   (b) Under what circumstances is this an appropriate model for population growth?

12. Define the following improper integrals.

   (a) $\int_a^{\infty} f(x)\, dx$ (b) $\int_{-\infty}^{b} f(x)\, dx$ (c) $\int_{-\infty}^{\infty} f(x)\, dx$

13. What is a probability density function? What properties does such a function have?

14. Suppose $f(x)$ is the probability density function for the weight of a female college student, where x is measured in pounds.

(a) What is the meaning of the integral $\int_0^{130} f(x)\, dx$?

(b) Write an expression for the mean of this density function.

(c) How can we find the median of this density function?

15. What is a normal distribution? What is the significance of the standard deviation?

Answers to the Concept Check can be found on the back endpapers.

▪ **Exercises**

1–4 ▪ Sketch the region enclosed by the given curves and find its area.

1. $y = x^2 + 1$, $\quad y = 5 + 2x$, $\quad x = 0$, $\quad x = 2$

2. $y = 2e^x$, $\quad y = \sqrt{x}$, $\quad x = 0$, $\quad x = 1$

3. $y = \frac{1}{2}x^2$, $\quad y = 4\sqrt{x}$

4. $y = 2^x$, $\quad y = x + 5$, $\quad x = 0$

5–8 ▪ Find the area of the region bounded by the given curves.

5. $y = x^2$, $\quad y = 4x - x^2$

6. $y = 1/x$, $\quad y = x^2$, $\quad x = e$

7. $y = 1 - 2x^2$, $\quad y = |x|$

8. $y = x^3$, $\quad y = \sqrt{x}$

---

9. Marginal cost and revenue The marginal cost for a manufacturer's hiking backpacks is given by $C'(q) = 16 - 0.02q + 0.00004q^2$ and the marginal revenue is $R'(q) = 22 - 0.005q$. Find the area between the graphs of these functions for $0 \le q \le 500$ and interpret your result.

10. Income and spending Money is being contributed to a trust fund at a rate of $2.8e^{0.02t}$ million dollars per year but funds are being spent from the account at a rate of $1.3e^{0.01t}$ million dollars per year. Find the area between the graphs of these functions for $0 \le t \le 5$ and interpret your result in this context.

11. Consumer surplus The demand function for a commodity is given by

$$p = 2000 - 0.1q - 0.01q^2$$

Find the consumer surplus when the sales level is 100.

12. Demand and consumer surplus A publisher currently has 34,000 magazine subscribers paying $26 per year. It estimates that for each dollar the price is lowered, an additional 2500 people will subscribe. Write a demand function for the subscriptions and find the consumer surplus if the subscription is priced at $14.

13. Producer surplus The supply function for a good is given by

$$p = 18e^{0.003q}$$

Find the producer surplus when the sales level is 500.

14. Total surplus The demand function for a distributor's gourmet coffee beans is

$$D(q) = 8.8 - 0.02q \text{ dollars per pound}$$

and the supply function is $S(q) = 2.5 + 0.01q$, where q is measured in thousands.

(a) At what price is the market for the coffee in equilibrium?

(b) Compute the maximum total surplus.

15. Future value An annuity pays $2000 per month, and the income is invested at 6.2% interest (compounded continuously). What is the future value of the income after six years?

16. Present and future value If income is received at the rate of $f(t) = 150,000e^{0.03t}$ dollars per year for a period of ten years and the income is invested at an annual interest rate of 4.8% (compounded continuously), what is the future value of the income? What is the present value?

17. Survival and renewal Suppose a city's population is currently 75,000 and the renewal function is $R(t) = 3200e^{0.05t}$. If the survival function is $S(t) = e^{-0.1t}$, predict the population in ten years.

18. Animal survival and renewal The fish population in a lake is currently 3400 and is increasing at a rate of $R(t) = 650e^{0.04t}$ fish per month. If the proportion of fish that remain after t months is $S(t) = e^{-0.09t}$, how many fish will be in the lake in three years?

19. Organization membership The membership of a particular organization is currently 26,500 and is increasing at a rate of $R(t) = 1720$ members per year, and the proportion of members remaining after t years is given by

$S(t) = 2/(t + 2)$. How many people will belong to the organization eight years from now?

20. Cardiac output After a 6-mg injection of dye into a heart, the readings of dye concentration at two-second intervals are as shown in the table. Use the Midpoint Rule to estimate the cardiac output.

t	$c(t)$	t	$c(t)$
0	0	14	4.7
2	1.9	16	3.3
4	3.3	18	2.1
6	5.1	20	1.1
8	7.6	22	0.5
10	7.1	24	0
12	5.8		

21. If A is a function of time t and A increases at a rate inversely proportional to its square, write a differential equation describing this relationship.

22. Verify that the function $y = (1 + \ln x)/x$ is a solution of the differential equation $x^2 y' = 1 - xy$.

23. Solve the separable differential equation $dB/dt = e^{2B}/\sqrt{t}$.

24. Solve the separable differential equation $xyy' = y^2 + 1$, $x > 0$.

25. Find the solution of the differential equation $y' = xy^3$ that satisfies the initial condition $y(\frac{1}{4}) = 4$.

26. Find the family of functions whose graphs have slope x^2/y at every point (x, y). Then identify the one function whose graph passes through the point $(0, 5)$.

27–29 ■ Determine whether the improper integral converges or diverges. It it converges, evaluate it.

27. $\int_1^\infty \dfrac{1}{(2x + 1)^3}\, dx$

28. $\int_4^\infty \dfrac{1}{x^{3/2}}\, dx$

29. $\int_{-\infty}^0 e^{-x/2}\, dx$

30. Perpetuity value Suppose income is received indefinitely at a rate of $f(t) = 5000e^{0.02t}$ dollars per year t years from now and the income is invested at an annual interest rate of 6% (compounded continuously). What is the present value of this income?

31. (a) Explain why the function

$$f(x) = \begin{cases} \frac{1}{288}(12x - x^2) & \text{if } 0 \le x \le 12 \\ 0 & \text{if } x < 0 \text{ or } x > 12 \end{cases}$$

is a probability density function.
(b) Find $P(X < 4)$.
(c) Calculate the mean. Is the value what you would expect?

32. Gestation period Lengths of human pregnancies are normally distributed with mean 268 days and standard deviation 15 days. What percentage of pregnancies last between 250 days and 280 days?

33. Waiting times The length of time spent waiting in line at a certain bank is modeled by an exponential density function with mean eight minutes.
(a) What is the probability that a customer is served in the first three minutes?
(b) What is the probability that a customer has to wait more than ten minutes?
(c) What is the median waiting time?

34. Use a graph to find approximate x-coordinates where the curves $y = e^{x/2}$ and $y = x^3 - 3x^2 + 1$ intersect. Then find (approximately) the area between the curves.

7

Graphs of functions of two variables are surfaces that can take a variety of shapes, including that of a saddle or mountain pass. At this location in southern Utah (Phipps Arch), you can see a point that is a minimum in one direction but a maximum in another direction. Such surfaces are discussed in Section 7.3.
Photo by Stan Wagon, Macalester College

Functions of Several Variables

Functions with more than one input are called functions of two or more variables. Many quantities depend on more than one variable: a manufacturer's yearly production cost depends on the number of items produced, the cost of raw materials, labor, and so on; the temperature at a given location depends on the latitude, longitude, and time; wave heights in the open sea depend on the wind speed and the amount of time the wind has been blowing at that speed.

In this chapter we extend some of the ideas of calculus (derivatives, maximum and minimum values) to functions of two or more variables.

7.1 Functions of Several Variables

We begin by considering functions with just two inputs.

▪ Functions of Two Variables

The temperature T at a point on the surface of the earth at any given time depends on the longitude x and latitude y of the point. We can think of T as being a function of the two variables x and y, or as a function of the pair (x, y). We indicate this functional dependence by writing $T = f(x, y)$.

> ▪ A **function f of two variables** is a rule that assigns to each input pair (x, y) exactly one output number $f(x, y)$.

The **domain** of f is the set of all allowable input pairs. In other words, it's the set of all ordered pairs (x, y) in the coordinate plane such that $f(x, y)$ is defined.

We often write $z = f(x, y)$ to make explicit the value taken on by f at the general point (x, y). The variables x and y are **independent variables** and z is the **dependent variable**. [Compare this with the notation $y = f(x)$ for functions of a single variable.]

▪ EXAMPLE 1   A Cost Function with Two Inputs

A company makes two kinds of chocolate bars: plain, and with almonds. Fixed production costs are $10,000 and it costs $1.10 to make a plain chocolate bar and $1.25 to make one with almonds.

(a) Express the cost of making x plain bars and y bars with almonds as a function of two variables $C = f(x, y)$.

(b) Find $f(2000, 1000)$ and interpret it.

(c) What is the domain of f?

SOLUTION

(a) It costs $1.10 to make a plain bar and there are x such bars, so the total cost of making the plain bars is $1.1x$. Likewise, the cost of making y bars with almonds is $1.25y$. Including the fixed costs, we see that the total cost is

$$C = f(x, y) = 10,000 + 1.1x + 1.25y$$

(b) When $x = 2000$ and $y = 1000$, we have

$$f(2000, 1000) = 10,000 + 1.1(2000) + 1.25(1000)$$
$$= 10,000 + 2,200 + 1,250 = 13,450$$

This means that the cost of making 2000 plain chocolate bars and 1000 bars with almonds is $13,450.

(c) The number of bars can't be negative, so $x \geq 0$ and $y \geq 0$. The domain of f is therefore

$$D = \{(x, y) \mid x \geq 0, y \geq 0\}$$

which is the first quadrant of the coordinate plane. (See Figure 1.) ▪

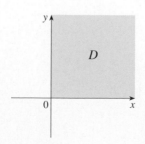

FIGURE 1

The domain of the cost function in Example 1.

■ EXAMPLE 2 Domain of a Function of Two Variables

Let $g(x, y) = 3x^2y - 2xy^3 + \sqrt{y - x}$.

(a) Find $g(2, 6)$.   **(b)** Find the domain of g.

SOLUTION

(a) If we put $x = 2$ and $y = 6$, we get

$$g(2, 6) = 3(2^2)(6) - 2(2)(6^3) + \sqrt{6 - 2}$$
$$= 72 - 864 + \sqrt{4} = -790$$

(b) The expression for $g(x, y)$ is defined whenever $y - x \geq 0$, that is, $y \geq x$. So the domain is

$$D = \{(x, y) \mid y \geq x\}$$

which is the set of points that lie on or above the line $y = x$. (See Figure 2.) ■

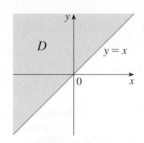

FIGURE 2
The domain of the function g
in Example 2.

■ EXAMPLE 3 The Cobb-Douglas Production Function

In 1928 Charles Cobb and Paul Douglas published a study in which they modeled the growth of the American economy during the period 1899–1922. They considered a simplified view of the economy in which production output is determined by the amount of labor involved and the amount of capital invested. While there are many other factors affecting economic performance, their model proved to be remarkably accurate. The function they used to model production was of the form

(1) $$P(L, K) = bL^aK^{1-a}$$

where P is the total production (the monetary value of all goods produced in a year), L is the amount of labor (the total number of person-hours worked in a year), and K is the amount of capital invested (the monetary worth of all machinery, equipment, and buildings). In Equation 1, b is a positive constant and a is a number between 0 and 1.

Cobb and Douglas used economic data published by the government to obtain Table 1. They took the year 1899 as a baseline and P, L, and K for 1899 were each assigned the value 100. The values for other years were expressed as percentages of the 1899 figures.

TABLE 1

Year	P	L	K	Year	P	L	K
1899	100	100	100	1911	153	148	216
1900	101	105	107	1912	177	155	226
1901	112	110	114	1913	184	156	236
1902	122	117	122	1914	169	152	244
1903	124	122	131	1915	189	156	266
1904	122	121	138	1916	225	183	298
1905	143	125	149	1917	227	198	335
1906	152	134	163	1918	223	201	366
1907	151	140	176	1919	218	196	387
1908	126	123	185	1920	231	194	407
1909	155	143	198	1921	179	146	417
1910	159	147	208	1922	240	161	431

Cobb and Douglas used the method of least squares to fit the data of Table 1 to the function

(2) $$P(L, K) = 1.01L^{0.75}K^{0.25}$$

If we use the model given by the function in Equation 2 to compute the production in the years 1910 and 1920, we get the values

$$P(147, 208) = 1.01(147)^{0.75}(208)^{0.25} \approx 161.9$$
$$P(194, 407) = 1.01(194)^{0.75}(407)^{0.25} \approx 235.8$$

which are quite close to the actual values, 159 and 231.

The production function (1) has subsequently been used in many settings, ranging from individual firms to global economic questions. It has become known as the **Cobb-Douglas production function**. Its domain is $\{(L, K) \mid L \geq 0, K \geq 0\}$, because L and K represent labor and capital and are therefore never negative. ▪

▪ Three-Dimensional Coordinates and Graphs

To graph a function of two variables, we need a three-dimensional coordinate system because when we write $z = f(x, y)$, we have three variables x, y, and z, where x and y are the independent variables and z is the dependent variable.

In order to represent points in space, we first choose a fixed point O (the origin) and three directed lines through O that are perpendicular to each other, called the **coordinate axes** and labeled the x-axis, y-axis, and z-axis. Usually we think of the x- and y-axes as being horizontal and the z-axis as being vertical, and we draw the orientation of the axes as in Figure 3.

The three coordinate axes determine the three **coordinate planes** illustrated in Figure 4(a). The xy-plane is the plane that contains the x- and y-axes; the yz-plane contains the y- and z-axes; the xz-plane contains the x- and z-axes. These three coordinate planes divide space into eight parts, called **octants**. The **first octant**, in the foreground, is determined by the positive axes.

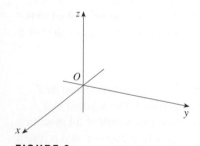

FIGURE 3
Coordinate axes

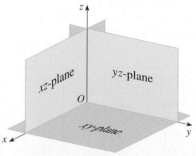
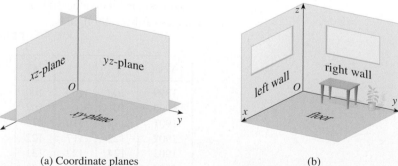

FIGURE 4   (a) Coordinate planes   (b)

Because many people have some difficulty visualizing diagrams of three-dimensional figures, you may find it helpful to do the following [see Figure 4(b)]. Look at any bottom corner of a room and call the corner the origin. The wall on your left is in the xz-plane, the wall on your right is in the yz-plane, and the floor is in the xy-plane. The x-axis runs along the intersection of the floor and the left wall. The y-axis runs along the intersection of the floor and the right wall. The z-axis runs

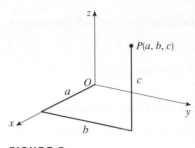

FIGURE 5

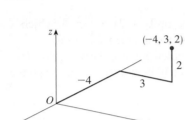

FIGURE 6

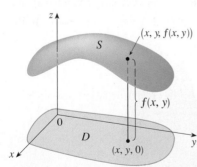

FIGURE 7

up from the floor toward the ceiling along the intersection of the two walls. You are situated in the first octant, and you can now imagine seven other rooms situated in the other seven octants (three on the same floor and four on the floor below), all connected by the common corner point O.

Now if P is any point in space, let a be the (directed) distance from the yz-plane to P, let b be the distance from the xz-plane to P, and let c be the distance from the xy-plane to P. We represent the point P by the ordered triple (a, b, c) of real numbers and we call a, b, and c the **coordinates** of P; a is the x-coordinate, b is the y-coordinate, and c is the z-coordinate. Thus, to locate the point (a, b, c), we can start at the origin O and move a units along the x-axis, then b units parallel to the y-axis, and then c units parallel to the z-axis as in Figure 5. Notice that the first octant can be described as the set of points whose coordinates are all positive.

■ **EXAMPLE 4** **Plotting in Three Dimensions**

Plot the point $(-4, 3, 2)$.

SOLUTION

Starting at the origin, we move 4 units along the x-axis in the negative direction, then 3 units in the direction of the positive y-axis, and then 2 units upward (see Figure 6). ■

In two-dimensional analytic geometry, the graph of an equation involving x and y is a curve in the plane. In three-dimensional analytic geometry, an equation in x, y, and z represents a **surface** in three-dimensional space.

> ■ The **graph** of a function f of two variables is the surface consisting of all points (x, y, z) in space such that $z = f(x, y)$.

Just as the graph of a function f of one variable is a curve C with equation $y = f(x)$, so the graph of a function f of two variables is a surface S with equation $z = f(x, y)$. We can visualize the graph S of f as lying directly above or below its domain D in the xy-plane (see Figure 7).

■ **EXAMPLE 5** **Graphing a Constant Function in Three Dimensions**

Sketch the graph of the constant function $f(x, y) = 3$.

SOLUTION

The equation of the graph of f is $z = 3$. So the graph is the surface consisting of all points (x, y, z) whose z-coordinate is 3. This is the horizontal plane that is parallel to the xy-plane and three units above it, as shown in Figure 8.

FIGURE 8
The graph of $f(x, y) = 3$
is a horizontal plane.

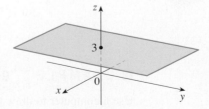

■

A function of the form

$$f(x, y) = ax + by + c$$

where a, b, and c are constants, is called a **linear function**. Its graph has the equation

$$z = ax + by + c \qquad \text{or} \qquad ax + by - z + c = 0$$

and it can be shown that such an equation always represents a plane.

▪ **EXAMPLE 6** **Graphing a Linear Function of Two Variables**

Sketch the graph of the function $f(x, y) = 6 - 3x - 2y$.

SOLUTION

The graph of f has the equation $z = 6 - 3x - 2y$, or $3x + 2y + z = 6$, which represents a plane. To graph the plane we first find the intercepts. Putting $y = z = 0$ in the equation, we get $x = 2$ as the x-intercept. Similarly, the y-intercept is 3 and the z-intercept is 6. This helps us sketch the portion of the graph that lies in the first octant (see Figure 9). ▪

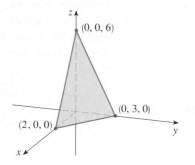

FIGURE 9

Although we were able to sketch the graphs of the simple functions in Examples 5 and 6, it's quite difficult to draw the graphs of most functions of two variables by hand. But computer programs are available for graphing these functions. In most such programs, the computer depicts the surface by drawing the curves of intersection of the surface with the vertical planes $x = k$ (parallel to the yz-plane) and $y = k$ (parallel to the xz-plane). These curves are called **vertical traces** and show vertical cross-sections of the surface.

Figure 10 shows computer-generated graphs of a function of two variables. We get an especially good picture of a function when rotation is used to give views from different vantage points. Notice that the graph of f is very flat and close to the xy-plane except near the origin; this is because $e^{-x^2-y^2}$ is very small when x or y is large.

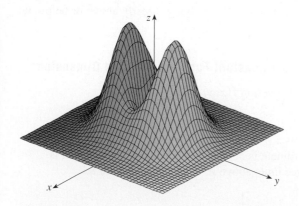
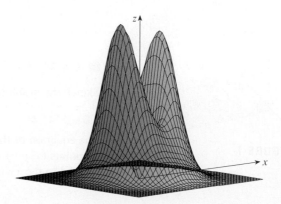

FIGURE 10

$f(x, y) = (x^2 + 3y^2)e^{-x^2-y^2}$

▪ **EXAMPLE 7** **Graphing a Cobb-Douglas Production Function**

Use a computer to draw the graph of the Cobb-Douglas production function $P(L, K) = 1.01L^{0.75}K^{0.25}$.

SOLUTION

Figure 11 shows the graph of P for values of the labor L and capital K that lie between 0 and 300. The computer has drawn the surface by plotting vertical traces. We see from these traces that the value of the production P increases as either L or K increases, as is to be expected.

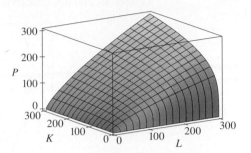

FIGURE 11

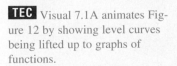

Level Curves

One method for visualizing a function of two variables is to look at its graph. Another method, borrowed from mapmakers, is a **contour map** on which points of constant elevation are joined to form *contour curves*, or *level curves*.

> ▪ The **level curves** of a function f of two variables are the curves with equations $f(x, y) = k$, where k is a constant (in the range of f).

A level curve $f(x, y) = k$ is the set of all points in the domain of f at which f takes on a given value k. In other words, it shows where the graph of f has height k.

You can see from Figure 12 the relation between level curves and **horizontal traces**, which are obtained by slicing the graph of f with horizontal planes. The level curves $f(x, y) = k$ are just the traces of the graph of f in the horizontal plane $z = k$ projected down to the xy-plane. So if you draw the level curves of a function and visualize them being lifted up to the surface at the indicated height, then you

TEC Visual 7.1A animates Figure 12 by showing level curves being lifted up to graphs of functions.

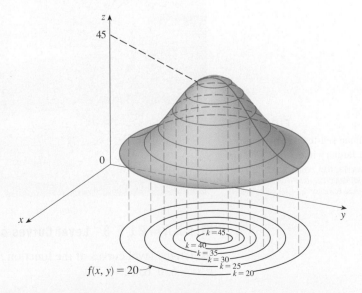

FIGURE 12

can mentally piece together a picture of the graph. The surface is steep where the level curves are close together. It is somewhat flatter where they are farther apart.

One common example of level curves occurs in topographic maps of mountainous regions, such as the map in Figure 13. The level curves are curves of constant elevation above sea level. If you walk along one of these contour lines you neither ascend nor descend.

FIGURE 13

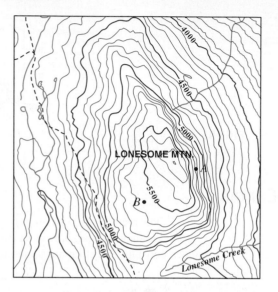

Another common example is the temperature function introduced in the opening paragraph of this section. Here the level curves are called **isothermals** and join locations with the same temperature. Figure 14 shows a weather map of the world indicating the average January temperatures. The isothermals are the curves that separate the colored bands.

FIGURE 14
World mean sea-level temperatures
in January in degrees Celsius

■ **EXAMPLE 8**   **Level Curves of a Linear Function**

Sketch the level curves of the function $f(x, y) = 6 - 3x - 2y$ for the values $k = -6, 0, 6, 12$.

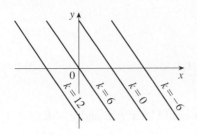

FIGURE 15
Contour map of
$f(x, y) = 6 - 3x - 2y$

SOLUTION
The level curves are

$$6 - 3x - 2y = k \qquad \text{or} \qquad 3x + 2y + (k - 6) = 0$$

This is a family of lines with slope $-\frac{3}{2}$. The four particular level curves with $k = -6, 0, 6,$ and 12 are $3x + 2y - 12 = 0$, $3x + 2y - 6 = 0$, $3x + 2y = 0$, and $3x + 2y + 6 = 0$. They are sketched in Figure 15. The level curves are equally spaced parallel lines because the graph of f is a plane (compare with Figure 9). ▪

▪ **EXAMPLE 9   Drawing a Contour Map**

Sketch a contour map of the function $h(x, y) = x^2 + y^2 + 2$.

SOLUTION
Recall that a contour map consists of a collection of level curves. Each level curve has the equation $x^2 + y^2 + 2 = k$ or $x^2 + y^2 = k - 2$. If $k > 2$, we recognize this as the equation of a circle with center the origin and radius $\sqrt{k - 2}$. We sketch these circles for several values of k in Figure 16(a). Then Figure 16(b) shows these level curves lifted up to form the graph of h, which has the equation $z = x^2 + y^2 + 2$. This surface is called a **paraboloid** because vertical traces have the equations $z = x^2 + k^2 + 2$ and $z = k^2 + y^2 + 2$, which are parabolas.

TEC Visual 7.1B demonstrates the connection between surfaces and their contour maps.

FIGURE 16
The graph of $h(x, y) = x^2 + y^2 + 2$ is formed by lifting the level curves.

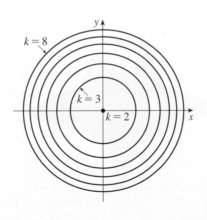

(a) Contour map of $h(x, y) = x^2 + y^2 + 2$

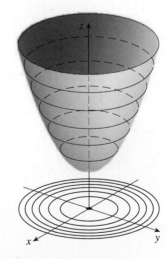

(b) Graph of h ▪

▪ **EXAMPLE 10   Contour Map of the Cobb-Douglas Function**

Plot level curves for the Cobb-Douglas production function of Example 3.

SOLUTION
To plot the level curve with $k = 140$, for instance, we solve the equation $P = 140$ for K. Writing $P(L, K) = 1.01L^{3/4}K^{1/4}$, we get

$$1.01L^{3/4}K^{1/4} = 140$$

So
$$K^{1/4} = \frac{140}{1.01} \frac{1}{L^{3/4}}$$

$$K = \left(\frac{140}{1.01}\right)^4 \frac{1}{L^3}$$

To write this equation in a more familiar way, we could let $x = L$ and $y = K$:

$$y = \left(\frac{140}{1.01}\right)^4 \frac{1}{x^3}$$

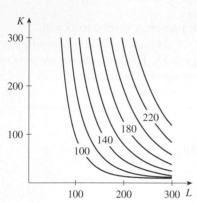

FIGURE 17

This level curve and several others are graphed in Figure 17. Level curves are labeled with the value of the production P. For instance, the level curve labeled 140 shows all values of the labor L and capital investment K that result in a production of $P = 140$. We see that, for a fixed value of P, as L increases K decreases, and vice versa.

Economists call the level curves in Figure 17 **isoquants**. (*Iso* means same and *quant* is short for quantity.) For some purposes, a contour map is more useful than a graph. That is certainly true in Example 10. (Compare Figure 17 with Figure 11.)

Figure 18 shows some computer-generated level curves together with the corresponding computer-generated graphs. Notice that the level curves in part (c) crowd together near the origin. That corresponds to the fact that the graph in part (d) is very steep near the origin.

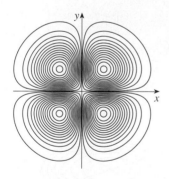

(a) Level curves of $f(x, y) = -xye^{-x^2-y^2}$

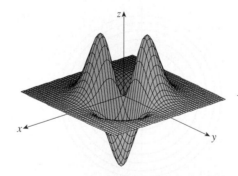

(b) Two views of $f(x, y) = -xye^{-x^2-y^2}$

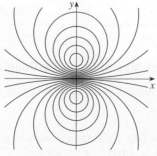

(c) Level curves of $f(x, y) = \dfrac{-3y}{x^2 + y^2 + 1}$

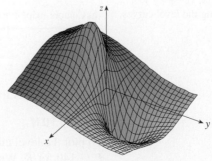

(d) $f(x, y) = \dfrac{-3y}{x^2 + y^2 + 1}$

FIGURE 18

■ Functions of Three or More Variables

A **function of three variables**, f, is a rule that assigns to each input triple (x, y, z) exactly one output number $f(x, y, z)$. For instance, the temperature T at a point on the surface of the earth depends on the longitude x and latitude y of the point and on the time t, so we could write $T = f(x, y, t)$. The domain of a function of three variables is a set in three-dimensional space. It's very difficult to visualize such a function by its graph because that graph would lie in a four-dimensional space.

■ **EXAMPLE 11**   **A Function of Three Variables**

Let $f(x, y, z) = e^{x-y+z}(\sqrt{z-2} + x^2 y)$.

(a) Evaluate $f(1, 4, 3)$.   **(b)** Find the domain of f.

SOLUTION

(a) We substitute $x = 1$, $y = 4$, and $z = 3$ in the formula for f:

$$f(1, 4, 3) = e^{1-4+3}(\sqrt{3-2} + 1^2 \cdot 4) = e^0(1 + 4) = 5$$

(b) The expression for $f(x, y, z)$ is defined when x and y take on any real values and $z - 2 \geqslant 0$ so the domain of f is

$$\{(x, y, z) \mid z \geqslant 2\}$$

This is the set of all points that lie on or above the horizontal plane $z = 2$.   ■

Functions of any number of variables can also be considered. A **function of n variables** is a function with n inputs. For example, if a company uses n different ingredients in making a food product, c_i is the cost per unit of the ith ingredient, and x_i units of the ith ingredient are used, then the total cost C of the ingredients is a function of the n variables $x_1, x_2, \ldots, x_n$:

$$C = f(x_1, x_2, \ldots, x_n) = c_1 x_1 + c_2 x_2 + \cdots + c_n x_n$$

■ Prepare Yourself

1. Find the domain of the function.

  (a) $f(x) = \sqrt{x + 5}$ **(b)** $g(x) = \sqrt{4 - x^2}$

  (c) $F(a) = \dfrac{a^2}{a - 2}$ **(d)** $L(t) = \ln(1 - 2t)$

2. Sketch a graph of the equation.

  (a) $3x + y = 6$ **(b)** $x^2 + y = -1$

  (c) $xy = 4$

  (d) $\sqrt{8 - x^2 - y^2} = 2$

■ Exercises 7.1

1–8 ■ Find the values of the function.

1. $f(x, y) = 1 + 4xy - 3y^2$

  (a) $f(6, 2)$ **(b)** $f(-1, 4)$

  (c) $f(0, -3)$ **(d)** $f(x, 2)$

2. $f(x, y) = x^2/(1 + y^2)$

  (a) $f(-2, 0)$ **(b)** $f(10, 1)$

  (c) $f\left(\frac{1}{2}, -\frac{1}{2}\right)$ **(d)** $f(-6, y)$

3. $g(x, y) = x^2 e^{3y}$

  (a) $g(-3, 0)$ **(b)** $g\left(3, \frac{1}{3}\right)$

  (c) $g(1, -1)$ **(d)** $g(-2, y)$

4. $f(s, t) = \sqrt{s^2 + 4t^2}$

  (a) $f(0, -1)$ **(b)** $f(-2, 2)$

  (c) $f(1, \sqrt{2})$ **(d)** $f(s, 3)$

5. $f(x, y, z) = \dfrac{x}{y - z}$

(a) $f(12, 2, -2)$ (b) $f(6, 5, 1)$
(c) $f\left(\frac{1}{6}, \frac{1}{2}, \frac{1}{3}\right)$ (d) $f(x, 2, 3)$

6. $g(x, y, z) = e^{x-2y} \ln z$

(a) $g(2, 1, 2)$ (b) $g(2, -1, 1)$
(c) $g(-6, 1, e)$ (d) $g(1, 2, 3)$

7. $f(x, y) = e^{xy}$

(a) $f(x + h, y) - f(x, y)$ (b) $f(x, y + h) - f(x, y)$

8. $f(x, y) = x^2 - 2xy$,

(a) $\dfrac{f(x + h, y) - f(x, y)}{h}$ (b) $\dfrac{f(x, y + h) - f(x, y)}{h}$

9. Let $f(x, y) = x^2 e^{3xy}$.

(a) Evaluate $f(2, 0)$.   (b) Find the domain of f.

10. Let $f(x, y) = \ln(x + y - 1)$.

(a) Evaluate $f(1, 1)$.   (b) Evaluate $f(e, 1)$.
(c) Find and sketch the domain of f.

11–16 ▪ Find and sketch the domain of the function.

11. $f(x, y) = \sqrt{x + y}$ **12.** $f(x, y) = \sqrt{x} + \sqrt{y}$

13. $f(x, y) = \dfrac{xy}{x - y}$ **14.** $f(x, y) = x \ln y$

15. $g(x, y) = \ln(4 - x^2 - y^2)$ **16.** $h(x, y) = \dfrac{e^{xy}}{y - x^2}$

17. Suppose you start at the origin, move along the x-axis a distance of 4 units in the positive direction, and then move downward a distance of 3 units. What are the coordinates of your position?

18. Plot the points $(0, 5, 2)$, $(4, 0, -1)$, $(2, 4, 6)$, and $(1, -1, 2)$ on a single set of coordinate axes.

19. Plot the points $(0, 0, 3)$, $(1, 2, 0)$, $(5, 1, -2)$, and $(-1, 2, 3)$ on a single set of coordinate axes.

20–22 ▪ Sketch the graph of the function.

20. $f(x, y) = 8$

21. $f(x, y) = 1 - x - y$

22. $f(x, y) = 4 - 2x - \frac{4}{3}y$

23–28 ▪ Draw a contour map of the function showing several level curves.

23. $f(x, y) = 2x - y$ **24.** $f(x, y) = y - x^2$

25. $f(x, y) = xy$

26. $f(x, y) = \sqrt{36 - x^2 - y^2}$

27. $f(x, y) = ye^x$ **28.** $f(x, y) = \dfrac{y}{x + 1}$

29–32 ▪ Match the function **(a)** with its graph (labeled A–D on page 393) and **(b)** with its contour map (labeled I–IV). Give reasons for your choices.

29. $z = (1 - x^2)(1 - y^2)$ **30.** $z = \dfrac{x - y}{1 + x^2 + y^2}$

31. $z = \dfrac{1}{x^2 + 4y^2}$ **32.** $z = x^2 y^2 e^{-x^2 - y^2}$

33. Cobb-Douglas production function   A manufacturer has modeled its yearly production function P (the monetary value of its entire production) as a Cobb-Douglas function

$$P(L, K) = 1.47 L^{0.65} K^{0.35}$$

where L is the number of labor hours (in thousands) and K is the invested capital (in millions of dollars). Find $P(120, 20)$ and interpret it.

34. Cobb-Douglas production function   Verify for the Cobb-Douglas production function

$$P(L, K) = 1.01 L^{0.75} K^{0.25}$$

discussed in Example 3 that the production will be doubled if both the amount of labor and the amount of capital are doubled. Is this also true for the general production function $P(L, K) = bL^a K^{1-a}$?

35. Joint cost function   A company makes three sizes of cardboard boxes: small, medium, and large. It costs $2.50 to make a small box, $4.00 for a medium box, and $4.50 for a large box. Fixed costs are $8000.

(a) Express the cost of making x small boxes, y medium boxes, and z large boxes as a function of three variables: $C = f(x, y, z)$.

(b) Find $f(3000, 5000, 4000)$ and interpret it.

(c) What is the domain of f?

36. Wind-chill index   The wind-chill index measures how cold it feels at a given temperature when the wind is blowing at a certain speed. It is modeled by the function

$$W(T, v) = 13.12 + 0.6215T - 11.37v^{0.16} + 0.3965Tv^{0.16}$$

where T is the temperature in °C, and v is the wind speed in km/h. Evaluate $W(-15, 30)$ and interpret it.

GRAPHS AND CONTOUR MAPS FOR EXERCISES 29–32

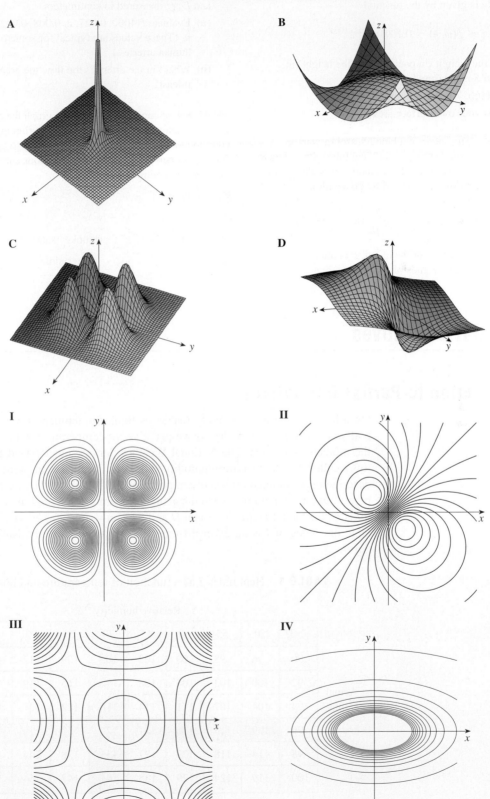

37. Human body area   A model for the surface area of a human body is given by the function

$$S = f(w, h) = 0.1091 w^{0.425} h^{0.725}$$

where w is the weight (in pounds), h is the height (in inches), and S is measured in square feet.

(a) Find $f(160, 70)$ and interpret it.

(b) What is your own surface area?

38. Blood flow   The shape of a blood vessel (a vein or artery) can be modeled by a cylindrical tube with radius R and length L. The velocity of the blood is modeled by the **law of laminar flow** (discovered by Poiseuille):

$$v = f(P, \eta, L, R, r) = \frac{P}{4\eta L}(R^2 - r^2)$$

where η is the viscosity of the blood, P is the pressure difference between the ends of the tube (in dynes/cm^2),

r is the distance from the central axis of the tube, and r, R, and L are measured in centimeters.

(a) Evaluate $f(4000, 0.027, 2, 0.008, 0.002)$ and interpret it. (These values are typical for some of the smaller human arteries.)

(b) Where in the artery is the flow the greatest? Where is it least?

39–42 ▪ Use computer software to graph the function using various domains and viewpoints. Get a printout of one that, in your opinion, gives a good view. If your software also produces level curves, then plot some contour curves of the same function and compare with the graph.

39. $f(x, y) = e^{-x^2} + e^{-2y^2}$

40. $f(x, y) = (1 - 3x^2 + y^2)e^{1-x^2-y^2}$

41. $f(x, y) = xy^2 - x^3$ (monkey saddle)

42. $f(x, y) = xy^3 - yx^3$ (dog saddle)

7.2 Partial Derivatives

▪ Introduction to Partial Derivatives

On a hot day, extreme humidity makes us think the temperature is higher than it really is, whereas in very dry air we perceive the temperature to be lower than the thermometer indicates. The National Weather Service has devised the *heat index* (also called the temperature-humidity index, or humidex, in some countries) to describe the combined effects of temperature and humidity. The heat index I is the perceived air temperature when the actual temperature is T and the relative humidity is H. So I is a function of T and H and we can write $I = f(T, H)$. The following table of values of I is an excerpt from a table compiled by the National Weather Service.

TABLE 1   Heat index I as a function of temperature and humidity

Relative humidity (%)

T \ H	50	55	60	65	70	75	80	85	90
90	96	98	100	103	106	109	112	115	119
92	100	103	105	108	112	115	119	123	128
94	104	107	111	114	118	122	127	132	137
96	109	113	116	121	125	130	135	141	146
98	114	118	123	127	133	138	144	150	157
100	119	124	129	135	141	147	154	161	168

Actual temperature (°F)

If we concentrate on the highlighted column of the table, which corresponds to a relative humidity of $H = 70\%$, we are considering the heat index as a function of the single variable T for a fixed value of H. Let's write $g(T) = f(T, 70)$. Then $g(T)$ describes how the heat index I increases as the actual temperature T increases when the relative humidity is 70%. The derivative of g when $T = 96°F$ is the rate of change of I with respect to T when $T = 96°F$:

$$g'(96) = \lim_{h \to 0} \frac{g(96 + h) - g(96)}{h} = \lim_{h \to 0} \frac{f(96 + h, 70) - f(96, 70)}{h}$$

We can approximate it using the values in Table 1 by taking $h = 2$ and -2:

$$g'(96) \approx \frac{g(98) - g(96)}{2} = \frac{f(98, 70) - f(96, 70)}{2} = \frac{133 - 125}{2} = 4$$

$$g'(96) \approx \frac{g(94) - g(96)}{-2} = \frac{f(94, 70) - f(96, 70)}{-2} = \frac{118 - 125}{-2} = 3.5$$

Averaging these values, we can say that the derivative $g'(96)$ is approximately 3.75. This means that, when the actual temperature is 96°F and the relative humidity is 70%, the apparent temperature (heat index) rises by about 3.75°F for every degree that the actual temperature rises!

Now let's look at the highlighted row in Table 1, which corresponds to a fixed temperature of $T = 96°F$. The numbers in this row are values of the function $G(H) = f(96, H)$, which describes how the heat index increases as the relative humidity H increases when the actual temperature is $T = 96°F$. The derivative of this function when $H = 70\%$ is the rate of change of I with respect to H when $H = 70\%$:

$$G'(70) = \lim_{h \to 0} \frac{G(70 + h) - G(70)}{h} = \lim_{h \to 0} \frac{f(96, 70 + h) - f(96, 70)}{h}$$

By taking $h = 5$ and -5, we approximate $G'(70)$ using the tabular values:

$$G'(70) \approx \frac{G(75) - G(70)}{5} = \frac{f(96, 75) - f(96, 70)}{5} = \frac{130 - 125}{5} = 1$$

$$G'(70) \approx \frac{G(65) - G(70)}{-5} = \frac{f(96, 65) - f(96, 70)}{-5} = \frac{121 - 125}{-5} = 0.8$$

By averaging these values we get the estimate $G'(70) \approx 0.9$. This says that, when the temperature is 96°F and the relative humidity is 70%, the heat index rises about 0.9°F for every percent that the relative humidity rises.

In general, if f is a function of two variables x and y, suppose we let only x vary while keeping y fixed, say $y = b$, where b is a constant. Then we are really considering a function of a single variable x, namely, $g(x) = f(x, b)$. If g has a derivative at a, then we call it the **partial derivative of f with respect to x at (a, b)** and denote it by $f_x(a, b)$. Thus

(1) $$\boxed{f_x(a, b) = g'(a) \quad \text{where} \quad g(x) = f(x, b)}$$

By the definition of a derivative, we have

$$g'(a) = \lim_{h \to 0} \frac{g(a + h) - g(a)}{h}$$

and so Equation 1 becomes

(2)

$$f_x(a, b) = \lim_{h \to 0} \frac{f(a + h, b) - f(a, b)}{h}$$

Similarly, the **partial derivative of f with respect to y at (a, b)**, denoted by $f_y(a, b)$, is obtained by keeping x fixed ($x = a$) and finding the ordinary derivative at b of the function $G(y) = f(a, y)$:

(3)

$$f_y(a, b) = \lim_{h \to 0} \frac{f(a, b + h) - f(a, b)}{h}$$

With this notation for partial derivatives, we can write the rates of change of the heat index I with respect to the actual temperature T and relative humidity H when $T = 96°F$ and $H = 70\%$ as follows:

$$f_T(96, 70) \approx 3.75 \qquad f_H(96, 70) \approx 0.9$$

■ **EXAMPLE 1**   **Computing Values of Partial Derivatives**

If $f(x, y) = x^3 + x^2y^3 - 2y^2$, find $f_x(2, 1)$ and $f_y(2, 1)$.

SOLUTION

According to Definition 1 we find $f_x(x, 1)$ by considering the function of one variable that we get by letting $y = 1$ in the expression for $f(x, y)$:

$$f(x, 1) = x^3 + x^2 - 2$$

Then we differentiate with respect to x:

$$f_x(x, 1) = 3x^2 + 2x$$

Now we put $x = 2$:

$$f_x(2, 1) = 3 \cdot 4 + 2 \cdot 2 = 16$$

Similarly, we find $f_y(2, y)$ by putting $x = 2$ in the expression for $f(x, y)$:

$$f(2, y) = 8 + 4y^3 - 2y^2$$
$$f_y(2, y) = 12y^2 - 4y$$
$$f_y(2, 1) = 12 \cdot 1 - 4 \cdot 1 = 8 \qquad ■$$

If we now let the point (a, b) vary in Equations 2 and 3, f_x and f_y become functions of two variables.

(4) ▪ If f is a function of two variables, its **partial derivatives** are the functions f_x and f_y defined by

$$f_x(x, y) = \lim_{h \to 0} \frac{f(x + h, y) - f(x, y)}{h}$$

$$f_y(x, y) = \lim_{h \to 0} \frac{f(x, y + h) - f(x, y)}{h}$$

The partial derivative symbol ∂ is used in place of d for functions of more than one variable. The curly d shape comes from the Cyrillic alphabet.

There are many alternative notations for partial derivatives. For instance, instead of f_x we can write $\partial f / \partial x$.

▪ **Notations for Partial Derivatives**   If $z = f(x, y)$, we write

$$f_x(x, y) = f_x = \frac{\partial f}{\partial x} = \frac{\partial}{\partial x} f(x, y) = \frac{\partial z}{\partial x}$$

$$f_y(x, y) = f_y = \frac{\partial f}{\partial y} = \frac{\partial}{\partial y} f(x, y) = \frac{\partial z}{\partial y}$$

To compute partial derivatives, all we have to do is remember from Equation 1 that the partial derivative with respect to x is just the *ordinary* derivative of the function g of a single variable that we get by keeping y fixed. Thus we have the following rule.

▪ **Rules for Finding Partial Derivatives of** $z = f(x, y)$

1. To find f_x, regard y as a constant and differentiate $f(x, y)$ with respect to x.

2. To find f_y, regard x as a constant and differentiate $f(x, y)$ with respect to y.

The following example shows an alternative method for solving Example 1.

▪ **EXAMPLE 2**   **Revisiting Example 1**

If $f(x, y) = x^3 + x^2y^3 - 2y^2$, find $f_x(2, 1)$ and $f_y(2, 1)$.

SOLUTION

Holding y constant and differentiating with respect to x, we get

$$f_x(x, y) = 3x^2 + 2xy^3$$

and so $$f_x(2, 1) = 3 \cdot 2^2 + 2 \cdot 2 \cdot 1^3 = 16$$

Holding x constant and differentiating with respect to y, we get

$$f_y(x, y) = 3x^2y^2 - 4y$$

$$f_y(2, 1) = 3 \cdot 2^2 \cdot 1^2 - 4 \cdot 1 = 8$$ ▪

▪ **EXAMPLE 3**   **Finding Partial Derivatives**

If $f(x, y) = \left(\dfrac{x}{1 + y}\right)^5$, calculate $\dfrac{\partial f}{\partial x}$ and $\dfrac{\partial f}{\partial y}$.

SOLUTION

Using the Chain Rule for functions of one variable, we have

$$\frac{\partial f}{\partial x} = 5\left(\frac{x}{1 + y}\right)^4 \cdot \frac{\partial}{\partial x}\left(\frac{x}{1 + y}\right) = 5\left(\frac{x}{1 + y}\right)^4 \cdot \frac{\partial}{\partial x}\left(\frac{1}{1 + y} \cdot x\right)$$

$$= 5\left(\frac{x}{1 + y}\right)^4 \cdot \frac{1}{1 + y} = \frac{5x^4}{(1 + y)^5}$$

$$\frac{\partial f}{\partial y} = 5\left(\frac{x}{1 + y}\right)^4 \cdot \frac{\partial}{\partial y}\left(\frac{x}{1 + y}\right) = 5\left(\frac{x}{1 + y}\right)^4 \cdot \frac{\partial}{\partial y}\left(x \cdot \frac{1}{1 + y}\right)$$

$$= 5\left(\frac{x}{1 + y}\right)^4 \cdot \frac{-x}{(1 + y)^2} = -\frac{5x^5}{(1 + y)^6}$$ ▪

▪ **EXAMPLE 4**   **A Partial Derivative Requiring the Product Rule**

If $g(v, w) = we^{vw}$, find g_v and g_w.

SOLUTION

When computing g_v, we treat w as a constant and v as a variable:

$$g_v(v, w) = w\left[e^{vw} \cdot \frac{\partial}{\partial v}(vw)\right] = w(e^{vw} \cdot w) = w^2 e^{vw}$$

For g_w, we treat v as a constant and w as a variable, so the Product Rule is required:

$$g_w(v, w) = w \cdot \frac{\partial}{\partial w}(e^{vw}) + e^{vw} \cdot \frac{\partial}{\partial w}(w)$$

$$= w \cdot (e^{vw} \cdot v) + e^{vw} \cdot 1 = (vw + 1)e^{vw}$$ ▪

▪ | Interpretations of Partial Derivatives

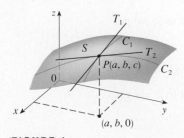

FIGURE 1

The partial derivatives of f at (a, b) are the slopes of the tangents to C_1 and C_2.

To give a geometric interpretation of partial derivatives, we recall that the equation $z = f(x, y)$ represents a surface S (the graph of f). If $f(a, b) = c$, then the point $P(a, b, c)$ lies on S. By fixing $y = b$, we are restricting our attention to the curve C_1 in which the vertical plane $y = b$ intersects S. (In other words, C_1 is the trace of S in the plane $y = b$.) Likewise, the vertical plane $x = a$ intersects S in a curve C_2. Both of the curves C_1 and C_2 pass through the point P. (See Figure 1.)

Notice that the curve C_1 is the graph of the function $g(x) = f(x, b)$, so the slope of its tangent T_1 at P is $g'(a) = f_x(a, b)$. The curve C_2 is the graph of the function $G(y) = f(a, y)$, so the slope of its tangent T_2 at P is $G'(b) = f_y(a, b)$.

Thus the partial derivatives $f_x(a, b)$ and $f_y(a, b)$ can be interpreted geometrically as the slopes of the tangent lines at $P(a, b, c)$ to the traces C_1 and C_2 of S in the planes $y = b$ and $x = a$.

■ **EXAMPLE 5**   **Partial Derivatives as Slopes**

If $f(x, y) = 4 - x^2 - 2y^2$, find $f_x(1, 1)$ and $f_y(1, 1)$ and interpret these numbers as slopes.

SOLUTION

We have

$$f_x(x, y) = -2x \qquad\qquad f_y(x, y) = -4y$$

$$f_x(1, 1) = -2 \qquad\qquad f_y(1, 1) = -4$$

The graph of f is the paraboloid $z = 4 - x^2 - 2y^2$ and the vertical plane $y = 1$ intersects it in the parabola $z = 2 - x^2$, $y = 1$. (As in the preceding discussion, we label it C_1 in Figure 2.) The slope of the tangent line to this parabola at the point $(1, 1, 1)$ is $f_x(1, 1) = -2$. (Look in the positive x-direction.) Similarly, the curve C_2 in which the plane $x = 1$ intersects the paraboloid is the parabola $z = 3 - 2y^2$, $x = 1$, and the slope of the tangent line at $(1, 1, 1)$ is $f_y(1, 1) = -4$. (See Figure 3.)

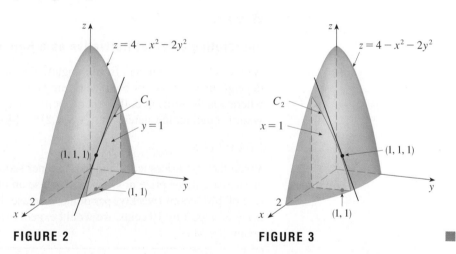

FIGURE 2   **FIGURE 3**   ■

Figure 4 is a computer-drawn counterpart to Figure 2. Part (a) shows the plane $y = 1$ intersecting the surface to form the curve C_1 and part (b) shows C_1 and T_1.

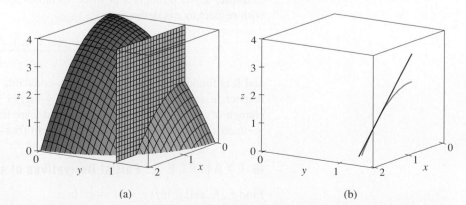

FIGURE 4   (a)   (b)

Similarly, Figure 5 corresponds to Figure 3.

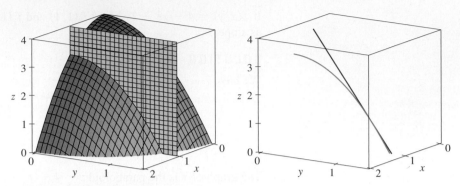

FIGURE 5

As we have seen in the case of the heat index function, partial derivatives can also be interpreted as *rates of change*. If $z = f(x, y)$, then $\partial z/\partial x$ represents the rate of change of z with respect to x when y is fixed. Similarly, $\partial z/\partial y$ represents the rate of change of z with respect to y when x is fixed.

▪ **EXAMPLE 6**

Interpreting a Partial Derivative as a Rate of Change

A cafe sells ice cream and frozen yogurt. The quantity q of ice cream sold weekly depends on the prices of both ice cream and yogurt, so we can say $q = f(x, y)$, where x is the price in dollars (per scoop) of ice cream and y is the price of yogurt. Interpret the statement $f_y(1.8, 1.2) = 140$ in this context.

SOLUTION

When the cafe prices ice cream at $1.80 per scoop and yogurt at $1.20 per scoop, an increase in the price of yogurt will cause an increase in ice cream sales at a rate of 140 scoops (weekly) per dollar increase. For instance, if the cafe raises the price of yogurt by 10 cents, we would expect to sell about 14 more scoops of ice cream per week.   ▪

▪ | Functions of More Than Two Variables

Partial derivatives can also be defined for functions of three or more variables. For example, if f is a function of three variables x, y, and z, then its partial derivative with respect to x is defined as

$$f_x(x, y, z) = \lim_{h \to 0} \frac{f(x + h, y, z) - f(x, y, z)}{h}$$

and it is found by regarding y and z as constants and differentiating $f(x, y, z)$ with respect to x. If $w = f(x, y, z)$, then $f_x = \partial w/\partial x$ can be interpreted as the rate of change of w with respect to x when y and z are held fixed. But we can't interpret it geometrically because the graph of f lies in four-dimensional space.

▪ **EXAMPLE 7**   **Partial Derivatives of a Function of Three Variables**

Find f_x, f_y, and f_z if $f(x, y, z) = e^{xy} \ln z$.

SOLUTION

Holding y and z constant and differentiating with respect to x, we have

$$f_x = ye^{xy} \ln z$$

Similarly, $\qquad f_y = xe^{xy} \ln z \qquad$ and $\qquad f_z = \dfrac{e^{xy}}{z}$ ■

■ Second Derivatives

If f is a function of two variables, then its partial derivatives f_x and f_y are also functions of two variables, so we can consider their partial derivatives $(f_x)_x$, $(f_x)_y$, $(f_y)_x$, and $(f_y)_y$, which are called the **second partial derivatives** of f. If $z = f(x, y)$, we use the following notation:

$$(f_x)_x = f_{xx} = \frac{\partial}{\partial x}\left(\frac{\partial f}{\partial x}\right) = \frac{\partial^2 f}{\partial x^2} = \frac{\partial^2 z}{\partial x^2}$$

$$(f_x)_y = f_{xy} = \frac{\partial}{\partial y}\left(\frac{\partial f}{\partial x}\right) = \frac{\partial^2 f}{\partial y\, \partial x} = \frac{\partial^2 z}{\partial y\, \partial x}$$

$$(f_y)_x = f_{yx} = \frac{\partial}{\partial x}\left(\frac{\partial f}{\partial y}\right) = \frac{\partial^2 f}{\partial x\, \partial y} = \frac{\partial^2 z}{\partial x\, \partial y}$$

$$(f_y)_y = f_{yy} = \frac{\partial}{\partial y}\left(\frac{\partial f}{\partial y}\right) = \frac{\partial^2 f}{\partial y^2} = \frac{\partial^2 z}{\partial y^2}$$

Thus the notation f_{xy} (or $\partial^2 f/\partial y\, \partial x$) means that we first differentiate with respect to x and then with respect to y, whereas in computing f_{yx} the order is reversed.

■ EXAMPLE 8 Second Partial Derivatives

Find the second partial derivatives of

$$f(x, y) = x^3 + x^2 y^3 - 2y^2$$

SOLUTION

In Example 2 we found that

$$f_x(x, y) = 3x^2 + 2xy^3 \qquad\qquad f_y(x, y) = 3x^2 y^2 - 4y$$

Therefore

$$f_{xx} = \frac{\partial}{\partial x}\,(3x^2 + 2xy^3) = 6x + 2y^3 \qquad f_{xy} = \frac{\partial}{\partial y}\,(3x^2 + 2xy^3) = 6xy^2$$

$$f_{yx} = \frac{\partial}{\partial x}\,(3x^2 y^2 - 4y) = 6xy^2 \qquad f_{yy} = \frac{\partial}{\partial y}\,(3x^2 y^2 - 4y) = 6x^2 y - 4$$ ■

Notice that $f_{xy} = f_{yx}$ in Example 8. This is not just a coincidence. It turns out that the mixed partial derivatives f_{xy} and f_{yx} are equal for most functions that one meets in practice.

■ The Cobb-Douglas Production Function

In Example 3 in Section 7.1 we described the work of Cobb and Douglas in modeling the total production P of an economic system as a function of the amount of labor L and the capital investment K. The particular form of their model follows from certain assumptions they made about the economy.

If the production function is denoted by $P = P(L, K)$, then the partial derivative $\partial P/\partial L$ is the rate at which production changes with respect to the amount of labor. Economists call it the marginal production with respect to labor or the **marginal productivity of labor**. Likewise, the partial derivative $\partial P/\partial K$ is the rate of change of production with respect to capital and is called the **marginal productivity of capital**. In these terms, the assumptions made by Cobb and Douglas can be stated as follows.

 (i) If either labor or capital vanishes, then so will production.

 (ii) The marginal productivity of labor is proportional to the amount of production per unit of labor.

(iii) The marginal productivity of capital is proportional to the amount of production per unit of capital.

It can be shown from these assumptions that

(5) $$P(L, K) = bL^a K^{1-a}$$

where b is a positive constant, $0 < a < 1$, L is the amount of labor, and K is the amount of capital invested.

■ EXAMPLE 9   Marginal Productivities

Using the data in Table 1 on page 383, Cobb and Douglas found that $b = 1.01$ and $a = 0.75$ in Equation 5. So in this case the Cobb-Douglas function is

$$P(L, K) = 1.01L^{0.75}K^{0.25}$$

(a) Calculate P_L and P_K.

(b) Find the marginal productivity of labor and the marginal productivity of capital in the year 1920, when $L = 194$ and $K = 407$ (compared with the assigned values of $L = 100$ and $K = 100$ in 1899). Interpret the results.

(c) In the year 1920 which would have benefited production more, an increase in capital investment or an increase in spending on labor?

SOLUTION

(a) $$P_L = \frac{\partial}{\partial L}(1.01L^{0.75}K^{0.25}) = 1.01(0.75L^{-0.25})K^{0.25} = 0.7575\left(\frac{K}{L}\right)^{0.25}$$

$$P_K = \frac{\partial}{\partial K}(1.01L^{0.75}K^{0.25}) = 1.01L^{0.75}(0.25K^{-0.75}) = 0.2525\left(\frac{L}{K}\right)^{0.75}$$

(b) The marginal productivity of labor in 1920 was

$$P_L(194, 407) = 0.7575\left(\frac{407}{194}\right)^{0.25} \approx 0.91$$

This means that if capital investment were held constant at 407 units, then production would increase at a rate of 0.91 units per unit increase in labor. The marginal productivity of capital in 1920 was

$$P_K(194, 407) = 0.2525\left(\frac{194}{407}\right)^{0.75} \approx 0.14$$

so if labor were held constant at 194 units, then production would increase at a rate of 0.14 units per unit increase in capital investment.

The conclusion in part (c) is valid because both L and K are measured as percentages of 1899 values.

(c) From part (b) we see that an increase in spending on labor would benefit production more than an increase in capital investment.   ▪

▪ Substitute and Complementary Products

Two quantities are called **substitute products** if an increase in the demand for one product results in a decrease in demand for the other product. (For example, butter and margarine are substitute products. If consumers are buying more margarine, they are probably buying less butter.) If, on the other hand, an increase in the demand for one product results in an increase in demand for the other, then the products are called **complementary**. (Cars and gasoline are examples of complementary products.)

For two related products we will use demand functions of the form

$$q_1 = D_1(p_1, p_2) \qquad q_2 = D_2(p_1, p_2)$$

where q_1 and q_2 are the demands for the first and second products. Because the products are related, the demand for each product depends on the prices of both products.

▪ EXAMPLE 10   Substitute and Complementary Products

Suppose the demand functions for two products A and B are $q_1 = D_1(p_1, p_2)$ and $q_2 = D_2(p_1, p_2)$, where p_1 and p_2 are the prices per unit for A and B.

(a) For substitute products, what can you say about the partial derivatives $\partial q_2/\partial p_1$ and $\partial q_1/\partial p_2$?

(b) What about complementary products?

SOLUTION

(a) Suppose we have two substitute (competitive) products A and B. (Think of A as butter and B as margarine.) Assume that p_2 (the price of B) is fixed. If p_1 increases, then some consumers will switch from A to B, so q_2 increases and therefore $\partial q_2/\partial p_1 > 0$. Similarly, if p_1 is fixed and p_2 increases, there will be a switch from B to A, so q_1 increases and $\partial q_1/\partial p_2 > 0$. So we see that for substitute products, both $\partial q_2/\partial p_1$ and $\partial q_1/\partial p_2$ are positive.

(b) Now we suppose that A and B are complementary products. (Think of A as cars and B as gas.) If p_2 is fixed and p_1 increases, then $\partial q_1/\partial p_1 < 0$ and so $\partial q_2/\partial p_1 < 0$. (If the price of cars increases, then fewer are sold, and so less gas is used.) Similarly, if p_1 is fixed and p_2 increases, then $\partial q_2/\partial p_2 < 0$ and so $\partial q_1/\partial p_2 < 0$. So we see that for complementary products, both $\partial q_2/\partial p_1$ and $\partial q_1/\partial p_2$ are negative. ▪

The reasoning in Example 10 is reversible and so we can determine whether related products are complementary or substitute by computing partial derivatives: If the mixed partial derivatives $\partial q_2/\partial p_1$ and $\partial q_1/\partial p_2$ are both positive, then the products are substitute. If they are both negative, the products are complementary.

▪ Prepare Yourself

1. Suppose $f(t)$ is the total amount of rainfall, in inches, t hours after the start of a storm. Interpret the statement $f'(1.5) = 0.6$ in this context.

2. Find the derivative of the function.

  (a) $g(x) = 5x^3 - 8x^2 + 13x - 4$

  (b) $f(x) = (x + 2)^8$

  (c) $K(v) = 3^v$ **(d)** $B(u) = u^3 e^u$

  (e) $H(t) = \dfrac{e^t}{t^3}$ **(f)** $f(x) = \dfrac{7x}{x^2 + 1}$

  (g) $g(y) = \sqrt{y} + \ln y + 1$ **(h)** $y = e^{x^2+2}$

  (i) $y = \ln(t^2 - 5t)$ **(j)** $A(t) = t\sqrt{t^3 - 1}$

3. Find dz/dx.

  (a) $z = a^2 + x^3$ **(b)** $z = ae^x$

  (c) $z = \ln(x + ax^2 + b)$ **(d)** $z = \dfrac{x}{2x - c}$

4. Find the second derivative of the function.

  (a) $y = \dfrac{x}{x - 1}$ **(b)** $y = \sqrt{x^2 + 1}$

▪ Exercises 7.2

1. Package delivery   Suppose $C(x, w)$ is the amount, in dollars, a courier service charges to deliver a package that weighs w pounds a distance of x miles. What does $C_w(150, 80)$ represent in this context? What are the units?

2. Food temperature   Suppose $f(w, t)$ gives the temperature, in °F, of a turkey that weighs w pounds after it has been in the oven for t hours. Interpret the statements $f_w(20, 2) = -12$ and $f_t(20, 2) = 18$ in this context.

3. Temperature   The temperature T, in °C, at a location in the Northern Hemisphere depends on the longitude x, latitude y, and time t, so we can write $T = f(x, y, t)$. Let's measure time in hours from the beginning of January.

  (a) What are the meanings of the partial derivatives $\partial T/\partial x$, $\partial T/\partial y$, and $\partial T/\partial t$?

  (b) Honolulu has longitude 158°W and latitude 21°N. Suppose that at 9:00 AM on January 1 the wind is blowing hot air to the northeast, so the air to the west and south is warm and the air to the north and east is cooler. Would you expect $f_x(158, 21, 9)$, $f_y(158, 21, 9)$, and $f_t(158, 21, 9)$ to be positive or negative? Explain.

4. Heat index   At the beginning of this section we discussed the function $I = f(T, H)$, where I is the heat index, T is the temperature, and H is the relative humidity. Use

Table 1 to estimate $f_T(92, 60)$ and $f_H(92, 60)$. What are the practical interpretations of these values?

5. Wind-chill index   The wind-chill index W is the perceived temperature when the actual temperature is T and the wind speed is v, so we can write $W = f(T, v)$. The following table of values was compiled by the NOAA Weather Service of the United States and the Meteorological Service of Canada.

Wind speed (km/h)

$\quad\quad v$ T	20	30	40	50	60	70
-10	-18	-20	-21	-22	-23	-23
-15	-24	-26	-27	-29	-30	-30
-20	-30	-33	-34	-35	-36	-37
-25	-37	-39	-41	-42	-43	-44

Actual temperature (°C)

  (a) Estimate the values of $f_T(-15, 30)$ and $f_v(-15, 30)$. What are the practical interpretations of these values?

  (b) In general, what can you say about the signs of $\partial W/\partial T$ and $\partial W/\partial v$?

  (c) What appears to be the value of the following limit?

$$\lim_{v \to \infty} \frac{\partial W}{\partial v}$$

6. Wave heights   The wave heights h in the open sea depend on the speed v of the wind and the length of time t that the wind has been blowing at that speed. Values of the function $h = f(v, t)$ are recorded in feet in the following table.

Duration (hours)

v＼t	5	10	15	20	30	40	50
10	2	2	2	2	2	2	2
15	4	4	5	5	5	5	5
20	5	7	8	8	9	9	9
30	9	13	16	17	18	19	19
40	14	21	25	28	31	33	33
50	19	29	36	40	45	48	50
60	24	37	47	54	62	67	69

Wind speed (knots)

(a) What are the meanings of the partial derivatives $\partial h / \partial v$ and $\partial h / \partial t$?

(b) Estimate the values of $f_v(40, 15)$ and $f_t(40, 15)$. What are the practical interpretations of these values?

(c) What appears to be the value of the following limit?

$$\lim_{t \to \infty} \frac{\partial h}{\partial t}$$

7. Determine the signs of the partial derivatives for the function f whose graph is shown.

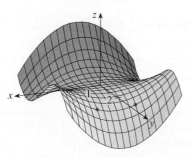

(a) $f_x(1, 2)$ (b) $f_y(1, 2)$
(c) $f_x(-1, 2)$ (d) $f_y(-1, 2)$

8. A contour map is given for a function f.

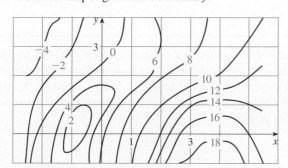

(a) Is $f_x(1, 1)$ positive, or negative? What about $f_y(1, 1)$?
(b) Which is larger, $f_x(2, 1)$ or $f_x(3, 1)$?
(c) Estimate the value of $f_y(2, 1)$.

9. If $f(x, y) = 16 - 4x^2 - y^2$, find $f_x(1, 2)$ and $f_y(1, 2)$ and interpret these numbers as slopes. Illustrate with either hand-drawn sketches or computer plots.

10. If $f(x, y) = \sqrt{4 - x^2 - 4y^2}$, find $f_x(1, 0)$ and $f_y(1, 0)$ and interpret these numbers as slopes. Illustrate with either hand-drawn sketches or computer plots.

11–12 ▪
(a) Find $f(x, -2)$ and use it to calculate $f_x(x, -2)$.
(b) Find $f(3, y)$ and use it to calculate $f_y(3, y)$.

11. $f(x, y) = x^2 y - 5xy^2$ **12.** $f(x, y) = xe^{2y} + y \ln x$

13–36 ▪ Find the first partial derivatives of the function.

13. $f(x, y) = y^5 - 3xy$ **14.** $f(x, y) = x^4 y^3 + 8x^2 y$

15. $f(x, y) = x^4 + x^2 y^2 + y^4$ **16.** $z = x^2 y^2 (x^4 + y^4)$

17. $z = (2x + 3y)^{10}$ **18.** $w = e^v/(u + v^2)$

19. $f(x, y) = \dfrac{x - y}{x + y}$ **20.** $f(x, t) = \sqrt{x} \ln t$

21. $f(r, s) = r \ln(r^2 + s^2)$ **22.** $g(x, y) = \ln(x + \ln y)$

23. $u = te^{w/t}$ **24.** $f(x, y) = x^y$

25. $f(s, t) = \sqrt{2 - 3s^2 - 5t^2}$ **26.** $f(s, t) = s/\sqrt{s^2 + t^2}$

27. $f(x, y, z) = xz - 5x^2 y^3 z^4$

28. $f(x, y, z) = x^2 y z^3 + xy - z$

29. $w = \ln(x + 2y + 3z)$ **30.** $w = ze^{xyz}$

31. $f(x, y, z) = \dfrac{x}{y + z}$ **32.** $f(x, y, z) = x\sqrt{yz}$

33. $f(x, y, z) = x^{yz}$ **34.** $f(x, y, z, t) = \dfrac{xy^2}{t + 2z}$

35. $f(x, y, z, t) = xy^2 z^3 t^4$ **36.** $f(x, y, z, t) = \dfrac{x - y}{z - t}$

37–42 ▪ Find the indicated partial derivative.

37. $f(x, y) = x^3 y^5$; $f_x(3, -1)$

38. $f(x, y) = \sqrt{2x + 3y}$; $f_y(2, 4)$

39. $f(x, y) = \ln(x + \sqrt{x^2 + y^2})$; $f_x(3, 4)$

40. $f(x, y) = xe^{-y} + 3y$; $\dfrac{\partial f}{\partial y}(1, 0)$

41. $f(x, y, z) = \dfrac{y}{x + y + z}$; $f_y(2, 1, -1)$

42. $g(K, L, M) = \sqrt{K^2 + L^2 + M^2}$; $g_L(2, -2, 1)$

43–46 ▪ Find all the second partial derivatives.

43. $f(x, y) = x^3y^5 + 2x^4y$ **44.** $u = x^2y^3z^4$

45. $w = \sqrt{u^2 + v^2}$ **46.** $v = \dfrac{xy}{x - y}$

47–50 ▪ Verify that $u_{xy} = u_{yx}$.

47. $u = 3x^2y - 8x^3 + 2y^2$ **48.** $u = (2x + 3y)^5$

49. $u = \ln \sqrt{x^2 + y^2}$ **50.** $u = xye^y$

---

51. Cobb-Douglas production function   A manufacturer has modeled its yearly production function P (the monetary value of its entire production) as a Cobb-Douglas function

$$P(L, K) = 1.47L^{0.65}K^{0.35}$$

where L is the number of labor hours, in thousands, and K is the invested capital, in millions of dollars.
   (a) Calculate P_L and P_K.
   (b) Find the marginal productivity of labor and the marginal productivity of capital when $L = 120$ and $K = 20$.
   (c) Interpret your answers in part (b).

52. Country productivity   The productivity of a certain country is modeled by the function

$$P(L, K) = 140L^{0.712}K^{0.288}$$

with L units of labor and K units of capital.
   (a) The country is now expending amounts of 750 units of labor and 1200 units of capital. What are the current marginal productivities of labor and capital?
   (b) Should the government encourage increased spending on labor or capital investment? Do we have enough information to make such a judgement?

53. Complementary vs. substitute products   The demand functions for two products are

$$q_1 = 8000 - 25p_1 - 10p_2$$
$$q_2 = 15{,}000 - 120p_1 - 50p_2$$

Are the products complementary or substitute?

54. Complementary vs. substitute products   The demand functions for products A and B are

$$q_1 = 300 \frac{\sqrt[3]{p_2}}{p_1} \qquad q_2 = 400 \frac{p_1}{\sqrt[3]{p_2^2}}$$

Are A and B complementary or substitute?

55. Electrical resistance   The total resistance R produced by three conductors with resistances R_1, R_2, R_3 connected in a parallel electrical circuit is given by the formula

$$\frac{1}{R} = \frac{1}{R_1} + \frac{1}{R_2} + \frac{1}{R_3}$$

Find $\partial R / \partial R_1$.

56. Temperature   The temperature at a point (x, y) on a flat metal plate is given by $T(x, y) = 60/(1 + x^2 + y^2)$, where T is measured in °C and x, y in meters. Find the rate of change of temperature with respect to distance at the point $(2, 1)$ in (a) the x-direction and (b) the y-direction.

57. Kinetic energy   The kinetic energy of a body with mass m and velocity v is $K = \frac{1}{2}mv^2$. Show that

$$\frac{\partial K}{\partial m} \frac{\partial^2 K}{\partial v^2} = K$$

58. Wind-chill index   The wind-chill index is modeled by the function

$$W = 13.12 + 0.6215T - 11.37v^{0.16} + 0.3965Tv^{0.16}$$

where T is the temperature, in °C, and v is the wind speed, in km/h. When $T = -15$°C and $v = 30$ km/h, by how much would you expect the apparent temperature to drop if the actual temperature decreases by 1°C? What if the wind speed increases by 1 km/h?

7.3   Maximum and Minimum Values

As we saw in Chapter 4, one of the main uses of ordinary derivatives is in finding maximum and minimum values. In this section we see how to use partial derivatives to locate maximum and minimum values of functions of two variables. In particular, in Example 4 we will see how a company that makes two products can set prices in order to maximize its profit.

Local and Absolute Extreme Values

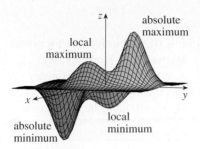

FIGURE 1

When we say that something is true when (x, y) is near (a, b), we mean that it is true when (x, y) lies within some circle with center (a, b).

Look at the hills and valleys in the graph of f shown in Figure 1. There are two points (a, b) where f has a *local maximum*, that is, where $f(a, b)$ is larger than nearby values of $f(x, y)$. The larger of these two values is the *absolute maximum*. Likewise, f has two *local minimum values*, where $f(a, b)$ is smaller than nearby values. The smaller of these two values is the *absolute minimum*.

> ▪ **Definition**   If f is a function of two variables, the number $f(a, b)$ is a
> ▪ **local maximum** value of f if $f(a, b) \geq f(x, y)$ when (x, y) is near (a, b).
> ▪ **local minimum** value of f if $f(a, b) \leq f(x, y)$ when (x, y) is near (a, b).

If the inequalities in this definition hold for *all* points (x, y) in the domain of f, then f has an **absolute maximum** (or **absolute minimum**) at (a, b).

Recall that in finding extreme values of a function of a single variable, we started by using Fermat's Theorem: If $f(x)$ has a local maximum or minimum at c and $f'(c)$ exists, then $f'(c) = 0$. For functions of two variables, we use the following fact.

> ▪ **Fermat's Theorem For Functions of Two Variables**   If f has a local maximum or minimum at (a, b) and f_x and f_y exist there, then $f_x(a, b) = 0$ and $f_y(a, b) = 0$.

This follows from the one-variable version because if $g(x) = f(x, b)$, then g has a local maximum or minimum at a and so $f_x(a, b) = g'(a) = 0$.

A point (a, b) is called a **critical point** (or *stationary point*) of f if $f_x(a, b) = 0$ and $f_y(a, b) = 0$, or if one of these partial derivatives does not exist. Fermat's Theorem says that if f has a local maximum or minimum at (a, b), then (a, b) is a critical point of f. However, as in single-variable calculus, not all critical points give rise to maxima or minima. At a critical point, a function could have a local maximum or a local minimum or neither.

▪ EXAMPLE 1 A Function with an Absolute Minimum

Let $f(x, y) = x^2 + y^2 + 2$. The partial derivatives are easily calculated:

$$f_x(x, y) = 2x \qquad f_y(x, y) = 2y$$

So $f_x = 0$ and $f_y = 0$ when $x = 0$ and $y = 0$ and the only critical point is $(0, 0)$. Because x^2 and y^2 are never negative, we have

$$f(x, y) = x^2 + y^2 + 2 \geq 2 \qquad \text{for all } x \text{ and } y$$

So, in fact, $f(0, 0) = 2$ is both a local minimum value and the absolute minimum value of f. You can also see this from the graph of f in Figure 2. We have met this surface (a paraboloid) before. (See Example 9 in Section 7.1.) ▪

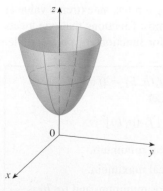

FIGURE 2
$z = x^2 + y^2 + 2$

▪ **EXAMPLE 2**   **A Function With No Extreme Value**

Find the extreme values of $f(x, y) = y^2 - x^2$.

SOLUTION

Since $f_x = -2x$ and $f_y = 2y$, the only critical point is $(0, 0)$. Notice that for points on the x-axis we have $y = 0$, so $f(x, y) = -x^2 < 0$ (if $x \neq 0$). However, for points on the y-axis we have $x = 0$, so $f(x, y) = y^2 > 0$ (if $y \neq 0$). Thus every circle with center $(0, 0)$ encloses points where f takes positive values as well as points where f takes negative values. Therefore $f(0, 0) = 0$ can't be an extreme value for f, so f has no extreme value.   ▪

Example 2 illustrates the fact that a function need not have a maximum or minimum value at a critical point. Figure 3 shows how this is possible. The graph of $f(x, y) = y^2 - x^2$ is a surface with equation $z = y^2 - x^2$ and is called a *hyperbolic paraboloid*. You can see that $f(0, 0) = 0$ is a maximum in the direction of the x-axis but a minimum in the direction of the y-axis. Near the origin the graph has the shape of a saddle and so $(0, 0)$ is called a *saddle point* of f.

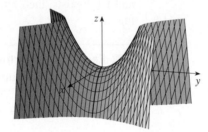

Photo by Stan Wagon, Macalester College

FIGURE 3
$z = y^2 - x^2$

A mountain pass also has the shape of a saddle. As the photograph of the geological formation illustrates, for people hiking in one direction the saddle point is the lowest point on their route, while for those traveling in a different direction the saddle point is the highest point.

▪ | Testing Critical Points

We need to be able to determine whether or not a function has an extreme value at a critical point. The following test, which is true for most functions that one meets in practice, is similar to the Second Derivative Test for functions of one variable.

---

(1) ▪ **Second Derivatives Test**   Suppose that $f_x(a, b) = 0$ and $f_y(a, b) = 0$ [so that (a, b) is a critical point of f]. Let

$$D = D(a, b) = f_{xx}(a, b) f_{yy}(a, b) - [f_{xy}(a, b)]^2$$

(a) If $D > 0$ and $f_{xx}(a, b) > 0$, then $f(a, b)$ is a local minimum.

(b) If $D > 0$ and $f_{xx}(a, b) < 0$, then $f(a, b)$ is a local maximum.

(c) If $D < 0$, then $f(a, b)$ is not a local maximum or minimum and (a, b) is called a **saddle point** of f.

(d) If $D = 0$, the test gives no information: f could have a local maximum or local minimum at (a, b), or (a, b) could be a saddle point of f.

■ **EXAMPLE 3   Classifying Critical Points**

Find the local maximum and minimum values and saddle points of
$f(x, y) = x^4 + y^4 - 4xy + 1$.

SOLUTION

We first locate the critical points:

$$f_x = 4x^3 - 4y \qquad\qquad f_y = 4y^3 - 4x$$

Setting these partial derivatives equal to 0, we obtain the equations

$$x^3 - y = 0 \qquad \text{and} \qquad y^3 - x = 0$$

To solve these equations we substitute $y = x^3$ from the first equation into the second one. This gives

$$x^9 - x = x(x^8 - 1) = 0$$

Factoring, using the difference of squares twice, we get

$$x(x^4 - 1)(x^4 + 1) = 0$$

$$x(x^2 - 1)(x^2 + 1)(x^4 + 1) = 0$$

So there are three real solutions: $x = 0, 1, -1$. The three critical points are $(0, 0)$, $(1, 1)$, and $(-1, -1)$.

Next we calculate the second partial derivatives and $D(x, y)$:

$$f_{xx} = 12x^2 \qquad f_{xy} = -4 \qquad f_{yy} = 12y^2$$

$$D(x, y) = f_{xx}f_{yy} - (f_{xy})^2 = 144x^2y^2 - 16$$

Since $D(0, 0) = -16 < 0$, it follows from case (c) of the Second Derivatives Test that the origin is a saddle point; that is, f has no local maximum or minimum at $(0, 0)$. Since $D(1, 1) = 128 > 0$ and $f_{xx}(1, 1) = 12 > 0$, we see from case (a) of the test that $f(1, 1) = -1$ is a local minimum. Similarly, we have $D(-1, -1) = 128 > 0$ and $f_{xx}(-1, -1) = 12 > 0$, so $f(-1, -1) = -1$ is also a local minimum.

The graph of f is shown in Figure 4.   ■

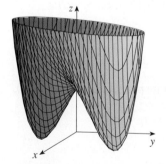

FIGURE 4
$z = x^4 + y^4 - 4xy + 1$

The ideas of this section can be applied to business: minimizing costs and maximizing revenue or profit. On the one hand, if a company makes too many of its products, the excess goods will depress prices and lower profits. On the other hand, if the company makes too few products, lower sales will result in lower profits. Somewhere between these extremes, a profit function will have a maximum, as the next example illustrates.

■ **EXAMPLE 4   Maximizing Profit**

A company makes two kinds of energy bars, bar 1 and bar 2. Production costs are $0.50 and $1.00 per bar for bars 1 and 2, respectively. The quantities q_1 and q_2 of bars 1 and 2 that can be sold each month are modeled by the demand functions

$$q_1 = 600(p_2 - p_1) \qquad q_2 = 600(6 + p_1 - 2p_2)$$

where p_1 and p_2 are the selling prices, in dollars, of the two bars. Find the selling prices that maximize the company's profit.

SOLUTION

The joint cost function is

$$C = 0.50q_1 + 1.00q_2$$

$$= 0.50(600)(p_2 - p_1) + 600(6 + p_1 - 2p_2)$$

$$= 300(p_2 - p_1) + 300 \cdot 2(6 + p_1 - 2p_2)$$

$$= 300(12 + p_1 - 3p_2)$$

The revenue is

$$R = p_1q_1 + p_2q_2$$

$$= p_1(600)(p_2 - p_1) + p_2(600)(6 + p_1 - 2p_2)$$

$$= 600(p_1p_2 - p_1^2) + 600(6p_2 + p_1p_2 - 2p_2^2)$$

$$= 600(-p_1^2 - 2p_2^2 + 2p_1p_2 + 6p_2)$$

So the profit is

$$P = R - C$$

$$= 300 \cdot 2(-p_1^2 - 2p_2^2 + 2p_1p_2 + 6p_2) - 300(12 + p_1 - 3p_2)$$

$$= 300(-2p_1^2 - 4p_2^2 + 4p_1p_2 + 12p_2 - 12 - p_1 + 3p_2)$$

$$= 300(-2p_1^2 - 4p_2^2 + 4p_1p_2 - p_1 + 15p_2 - 12)$$

The critical numbers of the profit function occur when its partial derivatives are 0:

$$\frac{\partial P}{\partial p_1} = 300(-4p_1 + 4p_2 - 1) = 0$$

$$\frac{\partial P}{\partial p_2} = 300(4p_1 - 8p_2 + 15) = 0$$

Solving the system of equations

$$-4p_1 + 4p_2 - 1 = 0$$

$$4p_1 - 8p_2 + 15 = 0$$

we get $p_1 = 3.25$ and $p_2 = 3.50$. Also,

$$\frac{\partial^2 P}{\partial p_1^2} = -1200 \qquad \frac{\partial^2 P}{\partial p_2^2} = -2400 \qquad \frac{\partial^2 P}{\partial p_1 \partial p_2} = 1200$$

so

$$D(3.25, 3.5) = (-1200)(-2400) - (1200)^2 > 0$$

As we commented before Example 4, we are expecting the profit function to have an absolute maximum value.

Since $\partial^2 P/\partial p_1^2$ is negative, the Second Derivatives Test says that these values of p_1 and p_2 give a (local) maximum. So to maximize profit, the company should sell energy bar 1 at $3.25 and energy bar 2 at $3.50. ▪

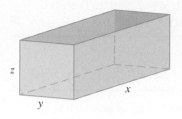

FIGURE 5

■ EXAMPLE 5   Maximizing Volume

A rectangular box without a lid is to be made from 12 m² of cardboard. Find the maximum volume of such a box.

SOLUTION

Let the length, width, and height of the box, in meters, be x, y, and z, as shown in Figure 5. Then the volume of the box is

$$V = xyz$$

We can express V as a function of just two variables x and y by using the fact that the area of the four sides and the bottom of the box is 12 m²:

$$2xz + 2yz + xy = 12$$

We solve this equation for z:

$$2z(x + y) = 12 - xy$$

$$z = \frac{12 - xy}{2(x + y)}$$

Substituting into the expression for V gives

$$V = xy \cdot \frac{12 - xy}{2(x + y)} = \frac{12xy - x^2y^2}{2(x + y)}$$

We compute the partial derivatives using the Quotient Rule:

$$\frac{\partial V}{\partial x} = \frac{2(x + y) \cdot (12y - 2xy^2) - (12xy - x^2y^2) \cdot 2}{[2(x + y)]^2}$$

$$= \frac{2(12xy - 2x^2y^2 + 12y^2 - 2xy^3 - 12xy + x^2y^2)}{4(x + y)^2}$$

$$= \frac{y^2(12 - 2xy - x^2)}{2(x + y)^2}$$

Similarly,

$$\frac{\partial V}{\partial y} = \frac{x^2(12 - 2xy - y^2)}{2(x + y)^2}$$

If V is a maximum, then the partial derivatives must be 0. If $\partial V/\partial x = 0$, then $y^2(12 - 2xy - x^2) = 0$ so $y = 0$ or $12 - 2xy - x^2 = 0$. Similarly, if $\partial V/\partial y = 0$, then $x = 0$ or $12 - 2xy - y^2 = 0$. But $x = 0$ or $y = 0$ gives $V = 0$, so we solve the system of equations

$$12 - 2xy - x^2 = 0 \qquad 12 - 2xy - y^2 = 0$$

Combining these equations, we get $x^2 - y^2 = 0$ or $x^2 = y^2$, and so $x = y$. (Note that x and y must both be positive in this problem.) If we put $x = y$ in either equation, we get $12 - 3x^2 = 0$, which gives $x = 2$, $y = 2$, and $z = (12 - 2 \cdot 2)/[2(2 + 2)] = 1$.

We could use the Second Derivatives Test to show that this gives a local maximum of V, or we could simply argue from the physical nature of this problem that there must be an absolute maximum volume, which has to occur at a critical point of V, so it must occur when $x = 2$, $y = 2$, $z = 1$. Then $V = 2 \cdot 2 \cdot 1 = 4$, so the maximum volume of the box is 4 m^3. ■

■ Exercises 7.3

1. Suppose $(1, 1)$ is a critical point of f. In each case, what can you say about f?
(a) $f_{xx}(1, 1) = 4$, $f_{xy}(1, 1) = 1$, $f_{yy}(1, 1) = 2$
(b) $f_{xx}(1, 1) = 4$, $f_{xy}(1, 1) = 3$, $f_{yy}(1, 1) = 2$

2. Suppose $(0, 2)$ is a critical point of g. In each case, what can you say about g?
(a) $g_{xx}(0, 2) = -1$, $g_{xy}(0, 2) = 6$, $g_{yy}(0, 2) = 1$
(b) $g_{xx}(0, 2) = -1$, $g_{xy}(0, 2) = 2$, $g_{yy}(0, 2) = -8$
(c) $g_{xx}(0, 2) = 4$, $g_{xy}(0, 2) = 6$, $g_{yy}(0, 2) = 9$

3–20 ■ Find the local maximum and minimum values and saddle point(s) of the function. If you have three-dimensional graphing software, graph the function with a domain and viewpoint that reveal all the important aspects of the function.

3. $f(x, y) = 9 - 2x + 4y - x^2 - 4y^2$

4. $f(x, y) = 2x^2 + 3y^2 - 5x + 4y$

5. $f(x, y) = (x + 1)^2 + (y - 2)^2 - 4$

6. $f(x, y) = x^3y + 12x^2 - 8y$

7. $f(x, y) = x^2 + xy + y^2 + y$

8. $f(x, y) = xy - 2x - 2y - x^2 - y^2$

9. $f(x, y) = (x - y)(1 - xy)$

10. $f(x, y) = (1 + xy)(x + y)$

11. $f(x, y) = xy(1 - x - y)$

12. $f(x, y) = 2x^3 + xy^2 + 5x^2 + y^2$

13. $f(x, y) = y^3 + 3x^2y - 6x^2 - 6y^2 + 2$

14. $f(x, y) = 2 - x^4 + 2x^2 - y^2$

15. $f(x, y) = x^3 - 12xy + 8y^3$ **16.** $f(x, y) = xy + \dfrac{1}{x} + \dfrac{1}{y}$

17. $f(x, y) = xe^{-2x^2-2y^2}$ **18.** $f(x, y) = e^{4y-x^2-y^2}$

19. $f(x, y) = (x^2 + y^2)e^{y^2-x^2}$ **20.** $f(x, y) = e^y(y^2 - x^2)$

21. Find three positive numbers whose sum is 100 and whose product is a maximum.

22. Find three positive numbers whose sum is 12 and the sum of whose squares is as small as possible.

23. Maximizing revenue   A store sells two competitive products, A and B. A model for the revenue from selling

q_A units of product A and q_B units of product B is
$$R = 40q_A + 60q_B - 6q_A^2 - 9q_B^2 - 4q_Aq_B$$
What values of q_A and q_B maximize the revenue?

24. Maximizing profit   A company makes two types of cookies and the production costs are 50 cents/lb and 60 cents/lb, respectively. The demand functions for these cookies are
$$q_1 = 10(p_2 - p_1) \qquad q_2 = 600 + 10p_1 - 15p_2$$
where p_1 and p_2 are measured in cents. Determine the selling prices that maximize profit.

25. Maximizing profit   A manufacturer produces two products. The joint cost function is
$$C = 12q_1 + 10q_2 + q_1q_2$$
and the demand functions are
$$q_1 = 60 - 2p_1 \qquad q_2 = 40 - p_2$$
Find the selling prices that maximize profit. What are the corresponding quantities?

26. Different markets and profit   A company makes only one product but engages in price discrimination by selling it for different prices in different markets. Suppose that in markets A and B the demand functions are
$$p_A = 50 - 0.5q_A \qquad p_B = 40 - 0.4q_B$$
where q_A and q_B are the quantities sold per month in markets A and B. If the company's cost function is
$$C = 100 + 6(q_A + q_B)$$
determine the selling prices and quantities sold that maximize profit.

27. Box design   A cardboard box without a lid is to have a volume of 32,000 cm^3. Find the dimensions that minimize the amount of cardboard used.

28. Box design   Find the dimensions of a rectangular box with largest volume if the total surface area is given as 64 cm^2.

29. Box design   Find the dimensions of a rectangular box of maximum volume such that the sum of the lengths of its 12 edges is 30 ft.

30. Aquarium design   The base of an aquarium with given volume V is made of slate and the sides are made of glass. If slate costs five times as much (per unit area) as glass, find the dimensions of the aquarium that minimize the cost of the materials.

P R O J E C T ▪ Designing a Dumpster

For this project we locate a rectangular trash Dumpster in order to study its shape and construction. We then attempt to determine the dimensions of a container of similar design that minimize construction cost.

1. First locate a trash Dumpster in your area. Carefully study and describe all details of its construction, and determine its volume. Include a sketch of the container.

2. While maintaining the general shape and method of construction, determine the dimensions such a container of the same volume should have in order to minimize the cost of construction. Use the following assumptions in your analysis:

   ▪ The sides, back, and front are to be made from 12-gauge (0.1046 inch thick) steel sheets, which cost $0.70 per square foot (including any required cuts or bends).

   ▪ The base is to be made from a 10-gauge (0.1345 inch thick) steel sheet, which costs $0.90 per square foot.

   ▪ Lids cost approximately $50.00 each, regardless of dimensions.

   ▪ Welding costs approximately $0.18 per foot for material and labor combined.

   Give justification of any further assumptions or simplifications made of the details of construction.

3. Describe how any of your assumptions or simplifications may affect the final result.

4. If you were hired as a consultant on this investigation, what would your conclusions be? Would you recommend altering the design of the Dumpster? If so, describe the savings that would result.

7.4 Lagrange Multipliers

In Section 7.3 we learned how to find the maximum or minimum value of a function, but sometimes we need to determine not the *overall* extreme value but rather the extreme value subject to a *constraint*.

Suppose, for instance, that a company has modeled its yearly production function P as a Cobb-Douglas function

$$P(L, K) = 1000L^{0.7}K^{0.3}$$

where L is the number of units of labor and K is the number of units of capital. The company wants to maximize production, subject to the constraint that its total budget for the year is 2 million dollars. If each unit of labor costs $2000 and each unit of capital costs $4000, then the total cost, in thousands of dollars, is

$$C(L, K) = 2L + 4K$$

and so the budget constraint is

$$2L + 4K = 2000$$

We will return to this problem in Example 2.

▪ Lagrange's Method for a Function of Two Variables

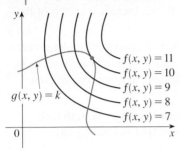

FIGURE 1

In general, suppose we want to maximize or minimize a function $f(x, y)$ subject to a constraint of the form $g(x, y) = k$, where k is a constant. In other words, we seek the extreme values of $f(x, y)$ when the point (x, y) is restricted to lie on the level curve $g(x, y) = k$. Figure 1 shows this curve together with several level curves of f. These have the equations $f(x, y) = c$, where $c = 7, 8, 9, 10, 11$. To maximize $f(x, y)$ subject to $g(x, y) = k$ is to find the largest value of c such that the level curve $f(x, y) = c$ intersects $g(x, y) = k$. It appears from Figure 1 that this happens when these curves just touch each other, that is, when they have a common tangent line. (Otherwise, the value of c could be increased further.) Using this geometric idea, the French mathematician Joseph-Louis Lagrange (1736–1813) devised the following method for finding extreme values subject to a constraint.

The *Lagrangian function*

$$L(x, y, \lambda) = f(x, y) - \lambda[g(x, y) - k]$$

introduces a new variable λ (called a **Lagrange multiplier**). Lagrange's procedure starts by finding the critical numbers of $L(x, y, \lambda)$. Solving the equations

$$L_x = f_x - \lambda g_x = 0 \qquad L_y = f_y - \lambda g_y = 0 \qquad L_\lambda = -[g(x, y) - k] = 0$$

we see that

$$f_x = \lambda g_x \qquad f_y = \lambda g_y \qquad g(x, y) = k$$

The extreme values of $f(x, y)$ subject to the constraint $g(x, y) = k$ occur at the values of x and y that satisfy these three equations.

▪ Method of Lagrange Multipliers for Functions of Two Variables

To find the maximum and minimum values of $f(x, y)$ subject to the constraint $g(x, y) = k$ [assuming that these extreme values exist]:

Step 1 Solve the system of equations

$$f_x(x, y) = \lambda g_x(x, y)$$

$$f_y(x, y) = \lambda g_y(x, y)$$

$$g(x, y) = k$$

for the unknowns x, y, and λ.

Step 2 Evaluate f at all the points (x, y) that result from Step 1.

Step 3 The largest of the values from Step 2 is the maximum value of f; the smallest is the minimum value of f.

■ **EXAMPLE 1** **Extreme Values of a Function of Two Variables**

Find the maximum and minimum values of the function $f(x, y) = x^2 + 2y^2$ subject to the constraint $x^2 + y^2 = 1$.

SOLUTION

Using Lagrange multipliers, we solve the equations

$$f_x = \lambda g_x \qquad f_y = \lambda g_y \qquad g(x, y) = 1$$

where $g(x, y) = x^2 + y^2$. Computing the partial derivatives, we get the system of equations

(1) $$2x = 2x\lambda$$

(2) $$4y = 2y\lambda$$

(3) $$x^2 + y^2 = 1$$

From (1) we have $2x(1 - \lambda) = 0$, so $x = 0$ or $\lambda = 1$. If $x = 0$, then (3) gives $y^2 = 1$, so $y = \pm 1$. If $\lambda = 1$, then (2) gives $y = 0$, and so $x = \pm 1$ from (3). Therefore f has possible extreme values at the points $(0, 1)$, $(0, -1)$, $(1, 0)$, and $(-1, 0)$. Evaluating f at these four points, we find that

$$f(0, 1) = 2 \qquad f(0, -1) = 2 \qquad f(1, 0) = 1 \qquad f(-1, 0) = 1$$

Therefore the maximum value of f subject to the constraint $x^2 + y^2 = 1$ is $f(0, \pm 1) = 2$ and the minimum value is $f(\pm 1, 0) = 1$. ▪

In geometric terms, Example 1 asks for the highest and lowest points on the curve C in Figure 2 that lies on the paraboloid $z = x^2 + 2y^2$ and directly above the constraint curve $x^2 + y^2 = 1$. The *overall* minimum value of f is $f(0, 0) = 0$ but the *constrained* minimum value is $f(1, 0) = f(-1, 0) = 1$.

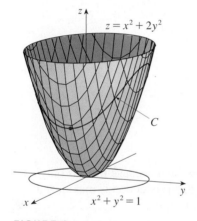

FIGURE 2

■ **EXAMPLE 2** **Maximizing Production**

The yearly production function for a company is

$$P(L, K) = 1000L^{0.7}K^{0.3}$$

where L is the number of units of labor and K is the number of units of invested capital. If a unit of labor costs $2000, a unit of capital is $4000, and the total budget for the year is 2 million dollars, what are the values of L and K that maximize production?

SOLUTION

The partial derivatives of P are

$$P_L(L, K) = 700L^{-0.3}K^{0.3} \qquad P_K(L, K) = 300L^{0.7}K^{-0.7}$$

The cost of L units of labor and K units of capital, in thousands of dollars, is

$$C(L, K) = 2L + 4K$$

and the total budget is 2 million dollars, so the constraint is $C(L, K) = 2000$. According to the Method of Lagrange Multipliers, we need to solve the equations

$$P_L = \lambda C_L \qquad P_K = \lambda C_K \qquad C(L, K) = 2000$$

which become

(4)
$$700L^{-0.3}K^{0.3} = 2\lambda$$

(5)
$$300L^{0.7}K^{-0.7} = 4\lambda$$

(6)
$$2L + 4K = 2000$$

Solving Equations 4 and 5 for λ and equating the results, we get

$$\left(\tfrac{1}{2}\right)700L^{-0.3}K^{0.3} = \left(\tfrac{1}{4}\right)300L^{0.7}K^{-0.7}$$

To eliminate the negative exponents, we multiply both sides of this equation by $2L^{0.3}K^{0.7}$:

$$700K = 150L \qquad \text{so} \qquad K = \tfrac{3}{14}L$$

Now we substitute this expression for K into Equation 6:

$$2L + \tfrac{6}{7}L = 2000$$

$$\tfrac{20}{7}L = 2000$$

$$L = 700$$

$$K = \tfrac{3}{14}L = 150$$

We expect a maximum value (and not a minimum), and because there is only one solution to the Lagrange equations, we conclude that production is maximized when $L = 700$ units of labor and $K = 150$ units of capital. ▪

The Meaning of a Lagrange Multiplier

In certain situations it is possible to give a useful interpretation to the Lagrange multiplier λ. From the Lagrange equations $f_x = \lambda g_x$ and $f_y = \lambda g_y$ we see that, for optimal values of x and y, we have

(7)
$$\lambda = \frac{f_x}{g_x} = \frac{f_y}{g_y}$$

If x varies, then

$$\lambda = \frac{f_x}{g_x} \approx \frac{\Delta f/\Delta x}{\Delta g/\Delta x} = \frac{\Delta f}{\Delta g}$$

and similarly for y. If x and y vary through optimal values of f, then λ is approximately the ratio of the change in the optimal value of f to the change in g. But the value of g is the constraint value k, so we can think of λ as the approximate increase in the optimal value of f resulting from an increase in the constraint value k by one unit.

For instance, in Example 2 we found that the optimal values of labor and capital are $L = 700$ and $K = 150$. From Equation 4 we get the corresponding value of λ:

$$\lambda = \tfrac{1}{2}(700)\left(\frac{K}{L}\right)^{0.3} \approx 220.5$$

This means that if we increase the constraint value ($k = 2000$) by one unit ($1000), we will increase the production level by approximately 220 units.

Recall from Section 7.2 that P_L is called the marginal productivity of labor and P_K is the marginal productivity of capital. Economists call the Lagrange multiplier λ for a production function the *marginal productivity of money*.

Notice that we can rewrite Equation 7 in the form

$$\frac{f_x}{f_y} = \frac{g_x}{g_y}$$

In the context of a production function this becomes

$$\frac{P_L}{P_K} = \frac{C_L}{C_K}$$

which can be stated as a law of economics: If labor and capital are at optimal levels, then the ratio of their marginal productivities is equal to the ratio of their marginal costs.

■ Functions of Three Variables

The method of Lagrange multipliers can be extended to a function of three variables $f(x, y, z)$ as follows.

■ **Method of Lagrange Multipliers for Functions of Three Variables**

To find the maximum and minimum values of $f(x, y, z)$ subject to the constraint $g(x, y, z) = k$ [assuming that these extreme values exist]:

Step 1 Solve the system of equations

$$f_x(x, y, z) = \lambda g_x(x, y, z)$$

$$f_y(x, y, z) = \lambda g_y(x, y, z)$$

$$f_z(x, y, z) = \lambda g_z(x, y, z)$$

$$g(x, y, z) = k$$

for the unknowns x, y, z, and λ.

Step 2 Evaluate f at all the points (x, y, z) that result from Step 1.

Step 3 The largest of the values from Step 2 is the maximum value of f; the smallest is the minimum value of f.

We illustrate the method by reconsidering the problem posed as Example 5 in Section 7.3.

▪ **EXAMPLE 3**   **Maximizing a Function of Three Variables**

A rectangular box without a lid is to be made from 12 m^2 of cardboard. Find the maximum volume of such a box.

SOLUTION

As in Example 5 in Section 7.3, we let x, y, and z be the length, width, and height, respectively, of the box in meters. Then we wish to maximize

$$V = xyz$$

subject to the constraint

$$g(x, y, z) = 2xz + 2yz + xy = 12$$

Using the Method of Lagrange Multipliers, we look for values of x, y, z, and λ such that

$$V_x = \lambda g_x \qquad V_y = \lambda g_y \qquad V_z = \lambda g_z \qquad 2xz + 2yz + xy = 12$$

These equations become

(8) $$yz = \lambda(2z + y)$$

(9) $$xz = \lambda(2z + x)$$

(10) $$xy = \lambda(2x + 2y)$$

(11) $$2xz + 2yz + xy = 12$$

There are no general rules for solving systems of equations. Sometimes some ingenuity is required. In the present example you might notice that if we multiply (8) by x, (9) by y, and (10) by z, then the left sides of these equations will be identical. Doing this, we have

Another method for solving the system of equations (8–11) is to solve each of Equations 8, 9, and 10 for λ and then to equate the resulting expressions.

(12) $$xyz = \lambda(2xz + xy)$$

(13) $$xyz = \lambda(2yz + xy)$$

(14) $$xyz = \lambda(2xz + 2yz)$$

We observe that $\lambda \neq 0$ because $\lambda = 0$ would imply $yz = xz = xy = 0$ from (8), (9), and (10) and this would contradict (11). Therefore, from (12) and (13), we have

$$2xz + xy = 2yz + xy$$

which gives $xz = yz$. But $z \neq 0$ (since $z = 0$ would give $V = 0$), so $x = y$. From (13) and (14) we have

$$2yz + xy = 2xz + 2yz$$

which gives $2xz = xy$ and so (since $x \neq 0$) $y = 2z$. If we now put $x = y = 2z$ in (11), we get

$$4z^2 + 4z^2 + 4z^2 = 12$$

Since x, y, and z are all positive, we therefore have $z = 1$ and so $x = 2$ and $y = 2$. This agrees with our answer in Section 7.3.   ▪

▪ Exercises 7.4

1–12 ▪ Use Lagrange multipliers to find the maximum or minimum values of the function subject to the given constraint.

1. $f(x, y) = xy;$ $x - 2y = 1$

2. $f(x, y) = xy;$ $x^2 + y^2 = 1$

3. $f(x, y) = x^2 + y^2;$ $xy = 1$

4. $f(x, y) = 4x + 6y;$ $x^2 + y^2 = 13$

5. $f(x, y) = x^2 y;$ $x^2 + 2y^2 = 6$

6. $f(x, y) = x^2 + 2y;$ $3x + y = 2$

7. $f(x, y) = x^4 + y^4;$ $x^2 + y^2 = 2$

8. $f(x, y) = x^2 + y^2;$ $x^4 + y^4 = 1$

9. $f(x, y, z) = 2x + 6y + 10z;$ $x^2 + y^2 + z^2 = 35$

10. $f(x, y, z) = 8x - 4z;$ $x^2 + 10y^2 + z^2 = 5$

11. $f(x, y, z) = 2x + 2y + z;$ $x^2 + y^2 + z^2 = 9$

12. $f(x, y, z) = x^2 + y^2 + z^2;$ $x + y + z = 12$

---

13. Find two positive numbers whose sum is 100 and whose product is as large as possible.

14. Find two numbers whose difference is 10 and whose product is as small as possible.

15. Find three positive numbers whose sum is 100 and whose product is a maximum.

16. Find three positive numbers whose sum is 12 and the sum of whose squares is as small as possible.

17. **Production schedule**   A company makes two products, A and B, from the same materials and needs to decide how to allocate limited resources to the production of A and B. If it produces x units of A and y units of B, then x and y satisfy the equation

$$16x^2 + 25y^2 = 40{,}000$$

which is called a *production possibilities curve* and is shown in the figure.

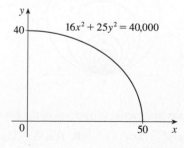

Each point (x, y) on this curve represents a *production schedule* specifying that x units of A and y units of B should be made. If each unit of A yields a profit of $8 and each unit of B yields a profit of $10, then the profit function is $P(x, y) = 8x + 10y$. Determine the production schedule that maximizes profit.

18. **Distributing production**   A manufacturer receives an order for 1000 units of its product and wants to distribute production between its two factories. The joint cost function is

$$C(q_1, q_2) = 2q_1^2 + 3q_2^2 + q_1 q_2 + 10q_1 + 26q_2 + 700$$

where q_1 and q_2 are the quantities made in each factory. In order to minimize costs, how many units should be supplied by each factory?

19. **Maximizing production**   A manufacturer's total production of a new product is

$$P(L, K) = 1200L^{2/3}K^{1/3}$$

where L is the number of units of labor and K is the number of units of capital. A unit of labor costs $100, a unit of capital $400, and the total budget for labor and capital is $360,000.

(a) To maximize production, how much should be spent on labor? How much on capital?

(b) Find the value of the Lagrange multiplier and give an interpretation of it.

20. **Minimizing cost**   A firm's production function is

$$P(x, y) = 500x^{0.6}y^{0.4}$$

where x and y are the number of units of labor and capital, respectively. Labor costs $80 per unit and capital costs $200 per unit. Suppose the firm wants to make 5000 units of their product.

(a) Find the amounts of labor and capital that minimize the cost.

(b) Find the value of the Lagrange multiplier and interpret it.

(c) Show that if cost is minimized, then

$$\frac{\text{marginal productivity of labor}}{\text{marginal productivity of capital}} = \frac{\text{marginal cost of labor}}{\text{marginal cost of capital}}$$

21–24 ▪ Use the Method of Lagrange Multipliers.

21. **Box design**   A cardboard box without a lid is to have a volume of 32,000 cm³. Find the dimensions that minimize the amount of cardboard used.

22. Box design   Find the dimensions of a rectangular box with largest volume if the total surface area is given as 64 cm².

23. Box design   Find the dimensions of a rectangular box of maximum volume such that the sum of the lengths of its 12 edges is 30 ft.

24. Aquarium design   The base of an aquarium with given volume V is made of slate and the sides are made of glass. If slate costs five times as much (per unit area) as glass, find the dimensions of the aquarium that minimize the cost of the materials.

CHAPTER **7** REVIEW

▪ CONCEPT CHECK

1. (a) What is a function of two variables?
   (b) What is the graph of a function of two variables?

2. What is a linear function of two variables? What type of surface is its graph?

3. (a) What is a level curve?
   (b) What is a contour map?

4. What is a function of three variables?

5. (a) Write expressions for the partial derivatives $f_x(a, b)$ and $f_y(a, b)$ as limits.
   (b) How do you interpret $f_x(a, b)$ and $f_y(a, b)$ geometrically? How do you interpret them as rates of change?
   (c) If $f(x, y)$ is given by a formula, how do you calculate f_x and f_y?
   (d) How do you calculate f_{xy}?

6. What do the following statements mean?
   (a) f has a local maximum at (a, b).
   (b) f has an absolute maximum at (a, b).
   (c) f has a local minimum at (a, b).
   (d) f has an absolute minimum at (a, b).
   (e) f has a saddle point at (a, b).

7. (a) If f has a local maximum at (a, b), what can you say about its partial derivatives at (a, b)?
   (b) What is a critical point of f?

8. State the Second Derivatives Test.

9. Explain how the method of Lagrange multipliers works in finding the extreme values of $f(x, y)$ subject to the constraint $g(x, y) = k$.

Answers to the Concept Check can be found on the back endpapers.

▪ Exercises

1. Plot the points $(4, 0, -2)$ and $(1, 2, 3)$ on a single set of coordinate axes.

2. Let $f(x, y) = \sqrt{y - x}$.
   (a) Evaluate $f(1, 5)$.
   (b) Find and sketch the domain of f.
   (c) Evaluate $f_x(1, 5)$ and $f_y(1, 5)$.

3. Let $g(x, y) = \dfrac{e^{xy}}{x - 1}$.
   (a) Evaluate $g(2, 1)$.
   (b) Find and sketch the domain of g.
   (c) Evaluate $g_x(2, 1)$ and $g_y(2, 1)$.

4. Let $f(x, y) = 4 - x - 2y$.
   (a) What is the domain of f?
   (b) Sketch the graph of f.
   (c) Sketch several level curves of f.

5. Draw a contour map of the function $f(x, y) = y - x^2$ showing several level curves.

6. A contour map of a function f is shown. Use it to make a rough sketch of the graph of f.

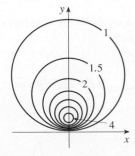

7–14 ▪ Find the first partial derivatives of the function.

7. $f(x, y) = x^2 y^4 - 2xy^5$ **8.** $f(x, y) = (5y^3 + 2x^2 y)^8$

9. $h(x, y) = xe^{2xy}$

10. $g(u, v) = \dfrac{u + 2v}{u^2 + v^2}$

11. $F(\alpha, \beta) = \alpha^2 \ln(\alpha^2 + \beta^2)$

12. $H(s, t) = ste^{s/t}$

13. $G(x, y, z) = \dfrac{x}{y + 2z}$

14. $M(x, y, t) = x^2 y\sqrt{t}$

15–16 ■ Find all second partial derivatives of f.

15. $f(x, y) = 4x^3 - xy^2$

16. $f(x, y, z) = \sqrt{x^2 + 2y^2 + 3z^2}$

---

17. Temperature   A metal plate is situated in the xy-plane and occupies the rectangle $0 \le x \le 10$, $0 \le y \le 8$, where x and y are measured in meters. The temperature at the point (x, y) in the plate is $T(x, y)$, where T is measured in degrees Celsius. Temperatures at equally spaced points were measured and recorded in the table. Estimate the values of the partial derivatives $T_x(6, 4)$ and $T_y(6, 4)$. What are the units?

x \\ y	0	2	4	6	8
0	30	38	45	51	55
2	52	56	60	62	61
4	78	74	72	68	66
6	98	87	80	75	71
8	96	90	86	80	75
10	92	92	91	87	78

18. Speed of sound   The speed of sound traveling through ocean water is a function of temperature, salinity, and pressure. It has been modeled by the function

$$C = 1449.2 + 4.6T - 0.055T^2 + 0.00029T^3$$
$$+ (1.34 - 0.01T)(S - 35) + 0.016D$$

where C is the speed of sound (in meters per second), T is the temperature (in degrees Celsius), S is the salinity (the concentration of salts in parts per thousand, which means the number of grams of dissolved solids per 1000 g of water), and D is the depth below the ocean surface (in meters). Compute $\partial C/\partial T$, $\partial C/\partial S$, and $\partial C/\partial D$ when $T = 10°C$, $S = 35$ parts per thousand, and $D = 100$ m.

Explain the physical significance of these partial derivatives.

19. Cobb-Douglas production function   A company modeled its production function as a Cobb-Douglas function

$$P(L, K) = 600L^{2/3}K^{1/3}$$

with L units of labor and K units of capital.

(a) Calculate P_L and P_K.

(b) Find the marginal productivity of labor and the marginal productivity of capital when $L = 100$ and $K = 80$. Interpret these values.

(c) Which would increase productivity more, an increase in spending on labor or an increase in capital investment?

20. Maximizing profit   A company manufactures two types of nails, for which production costs are 30 cents/lb and 36 cents/lb, respectively. The demand functions for these products are

$$q_1 = 100p_2 - 100p_1 \qquad q_2 = 400 + 100p_1 - 200p_2$$

Find the selling prices that maximize profit.

21–24 ■ Find the local maximum and minimum values and saddle points of the function. If you have three-dimensional graphing software, graph the function with a domain and viewpoint that reveal all the important aspects of the function.

21. $f(x, y) = x^2 - xy + y^2 + 9x - 6y + 10$

22. $f(x, y) = x^3 - 6xy + 8y^3$

23. $f(x, y) = 3xy - x^2y - xy^2$

24. $f(x, y) = (x^2 + y)e^{y/2}$

25–27 ■ Use Lagrange multipliers to find the maximum and minimum values of f subject to the given constraint.

25. $f(x, y) = x^2 y$; $x^2 + y^2 = 1$

26. $f(x, y) = \dfrac{1}{x} + \dfrac{1}{y}$; $\dfrac{1}{x^2} + \dfrac{1}{y^2} = 1$

27. $f(x, y, z) = xyz$; $x^2 + y^2 + z^2 = 3$

---

28. Box design   A package in the shape of a rectangular box can be mailed by the US Postal Service if the sum of its length and girth (the perimeter of a cross-section perpendicular to the length) is at most 108 in. Find the dimensions of the package with largest volume that can be mailed.

Appendixes

A ▪ Review of Algebra

Here we review the basic rules and procedures of algebra that you need to know in order to be successful in calculus.

▪ Arithmetic Operations

The real numbers have the following properties:

$a + b = b + a$ $ab = ba$		(Commutative Law)
$(a + b) + c = a + (b + c)$ $(ab)c = a(bc)$		(Associative Law)
$a(b + c) = ab + ac$		(Distributive law)

In particular, putting $a = -1$ in the Distributive Law, we get

$$-(b + c) = (-1)(b + c) = (-1)b + (-1)c$$

and so

$$-(b + c) = -b - c$$

▪ EXAMPLE 1

(a) $(3xy)(-4x) = 3(-4)x^2y = -12x^2y$

(b) $2t(7x + 2tx - 11) = 14tx + 4t^2x - 22t$

(c) $4 - 3(x - 2) = 4 - 3x + 6 = 10 - 3x$ ▪

If we use the Distributive Law three times, we get

$$(a + b)(c + d) = (a + b)c + (a + b)d = ac + bc + ad + bd$$

The acronym "FOIL" can be used to remember the four products that result from multiplying two binomials, where the letters stand for First, Outer, Inner, Last.

This says that we multiply two factors by multiplying each term in one factor by each term in the other factor and adding the products. Schematically, we have

$$(a + b)(c + d)$$

In the case where $c = a$ and $d = b$, we have

$$(a + b)^2 = a^2 + ba + ab + b^2$$

or

(1)

$$(a + b)^2 = a^2 + 2ab + b^2$$

Similarly, we obtain

(2)

$$(a - b)^2 = a^2 - 2ab + b^2$$

■ **EXAMPLE 2**

(a) $(2x + 1)(3x - 5) = 6x^2 - 10x + 3x - 5 = 6x^2 - 7x - 5$

(b) $(x + 6)^2 = x^2 + 12x + 36$

(c) $3(x - 1)(4x + 3) - 2(x + 6) = 3(4x^2 - x - 3) - 2x - 12$
$$= 12x^2 - 3x - 9 - 2x - 12$$
$$= 12x^2 - 5x - 21$$ ■

■ Fractions

To add two fractions with the same denominator, we add the numerators and keep the same denominator:

$$\frac{a}{b} + \frac{c}{b} = \frac{a + c}{b}$$

Thus it is true that

$$\boxed{\frac{a + c}{b} = \frac{a}{b} + \frac{c}{b}}$$

But remember to avoid the following common error:

$$\frac{a}{b + c} \not= \frac{a}{b} + \frac{a}{c}$$

(For instance, take $a = b = c = 1$ to see the error.)

To add two fractions with different denominators, we use a common denominator:

$$\boxed{\frac{a}{b} + \frac{c}{d} = \frac{a}{b} \cdot \frac{d}{d} + \frac{c}{d} \cdot \frac{b}{b} = \frac{ad + bc}{bd}}$$

We multiply such fractions as follows:

$$\boxed{\frac{a}{b} \cdot \frac{c}{d} = \frac{ac}{bd}}$$

In particular, it is true that

$$\boxed{\frac{-a}{b} = -\frac{a}{b} = \frac{a}{-b}}$$

To divide two fractions, we invert and multiply:

$$\boxed{\frac{\dfrac{a}{b}}{\dfrac{c}{d}} = \frac{a}{b} \div \frac{c}{d} = \frac{a}{b} \times \frac{d}{c} = \frac{ad}{bc}}$$

▪ **EXAMPLE 3**

(a) $\dfrac{x+3}{x} = \dfrac{x}{x} + \dfrac{3}{x} = 1 + \dfrac{3}{x}$

(b) $\dfrac{3}{x-1} + \dfrac{x}{x+2} = \dfrac{3}{x-1} \cdot \dfrac{x+2}{x+2} + \dfrac{x}{x+2} \cdot \dfrac{x-1}{x-1}$

$\qquad\qquad = \dfrac{3(x+2) + x(x-1)}{(x-1)(x+2)} = \dfrac{3x+6+x^2-x}{x^2+x-2}$

$\qquad\qquad = \dfrac{x^2+2x+6}{x^2+x-2}$

(c) $\dfrac{s^2t}{u} \cdot \dfrac{ut}{-2} = \dfrac{s^2t^2u}{-2u} = -\dfrac{s^2t^2}{2}$

(d) $\dfrac{\dfrac{x}{y}+1}{1-\dfrac{y}{x}} = \dfrac{\dfrac{x+y}{y}}{\dfrac{x-y}{x}} = \dfrac{x+y}{y} \times \dfrac{x}{x-y} = \dfrac{x(x+y)}{y(x-y)} = \dfrac{x^2+xy}{xy-y^2}$

▪

▪ Factoring

We have used the Distributive Law to expand certain algebraic expressions. We sometimes need to reverse this process (again using the Distributive Law) by factoring an expression as a product of simpler ones. The easiest situation occurs when the expression has terms with a common factor as follows:

$$\xleftarrow{\text{Expanding} \longrightarrow}$$
$$3x(x-2) = 3x^2 - 6x$$
$$\xleftarrow{\text{Factoring}}$$

To factor a quadratic of the form $x^2 + bx + c$ we note that

$$(x+r)(x+s) = x^2 + (r+s)x + rs$$

so we need to choose numbers r and s so that $r+s=b$ and $rs=c$.

▪ **EXAMPLE 4**

Factor $x^2 + 5x - 24$.

SOLUTION

The two integers that add to give 5 and multiply to give -24 are -3 and 8. Therefore

$$x^2 + 5x - 24 = (x-3)(x+8)$$

▪

▪ **EXAMPLE 5**

Factor $2x^2 - 7x - 4$.

SOLUTION

Even though the coefficient of x^2 is not 1, we can still look for factors of the form $2x + r$ and $x + s$, where $rs = -4$. Experimentation reveals that

$$2x^2 - 7x - 4 = (2x + 1)(x - 4)$$ ▪

Some special quadratics can be factored by using Equations 1 or 2 (from right to left) or by using the formula for a difference of squares:

(3)
$$a^2 - b^2 = (a - b)(a + b)$$

The analogous formula for a difference of cubes is

(4)
$$a^3 - b^3 = (a - b)(a^2 + ab + b^2)$$

which you can verify by expanding the right side. For a sum of cubes we have

(5)
$$a^3 + b^3 = (a + b)(a^2 - ab + b^2)$$

▪ **EXAMPLE 6**

(a) $x^2 - 6x + 9 = (x - 3)^2$ (Equation 2; $a = x, b = 3$)

(b) $4x^2 - 25 = (2x - 5)(2x + 5)$ (Equation 3; $a = 2x, b = 5$)

(c) $x^3 + 8 = (x + 2)(x^2 - 2x + 4)$ (Equation 5; $a = x, b = 2$)   ▪

▪ **EXAMPLE 7**

Simplify $\dfrac{x^2 - 16}{x^2 - 2x - 8}$.

SOLUTION

Factoring numerator and denominator, we have

$$\frac{x^2 - 16}{x^2 - 2x - 8} = \frac{(x - 4)(x + 4)}{(x - 4)(x + 2)} = \frac{x + 4}{x + 2}$$ ▪

▪ Completing the Square

Completing the square is a useful technique for graphing parabolas and circles; it involves rewriting a quadratic $ax^2 + bx + c$ in the form $a(x + p)^2 + q$ and can be accomplished by:

1. Factoring the number a from the terms involving x.

2. Adding and subtracting the square of half the coefficient of x.

In general, we have

$$ax^2 + bx + c = a\left[x^2 + \frac{b}{a}x \right] + c$$

$$= a\left[x^2 + \frac{b}{a}x + \left(\frac{b}{2a}\right)^2 - \left(\frac{b}{2a}\right)^2 \right] + c$$

$$= a\left(x + \frac{b}{2a} \right)^2 + \left(c - \frac{b^2}{4a} \right)$$

▪ EXAMPLE 8

Rewrite $x^2 + x + 1$ by completing the square.

SOLUTION

The square of half the coefficient of x is $\frac{1}{4}$. Thus

$$x^2 + x + 1 = x^2 + x + \tfrac{1}{4} - \tfrac{1}{4} + 1 = \left(x + \tfrac{1}{2}\right)^2 + \tfrac{3}{4} \qquad ▪$$

▪ EXAMPLE 9

$$2x^2 - 12x + 11 = 2[x^2 - 6x] + 11 = 2[x^2 - 6x + 9 - 9] + 11$$

$$= 2[(x - 3)^2 - 9] + 11 = 2(x - 3)^2 - 7 \qquad ▪$$

▪ Quadratic Formula

By completing the square as above we can obtain the following formula for the roots (solutions) of a quadratic equation.

> **(6) ▪ The Quadratic Formula**   The solutions of the quadratic equation $ax^2 + bx + c = 0$ are
>
> $$x = \frac{-b \pm \sqrt{b^2 - 4ac}}{2a}$$

▪ EXAMPLE 10

Solve the equation $5x^2 + 3x - 3 = 0$.

SOLUTION

With $a = 5$, $b = 3$, $c = -3$, the quadratic formula gives the solutions

$$x = \frac{-3 \pm \sqrt{3^2 - 4(5)(-3)}}{2(5)} = \frac{-3 \pm \sqrt{69}}{10} \qquad ▪$$

The quantity $b^2 - 4ac$ that appears in the quadratic formula is called the **discriminant**. There are three possibilities:

1. If $b^2 - 4ac > 0$, the equation has two real-valued solutions.

2. If $b^2 - 4ac = 0$, the solutions are equal.

3. If $b^2 - 4ac < 0$, the equation has no real-valued solutions.

These three cases correspond to the fact that the number of times the parabola $y = ax^2 + bx + c$ crosses the x-axis is 2, 1, or 0. (See Figure 1.) In case (3) the quadratic $ax^2 + bx + c$ can't be factored and is called **irreducible.**

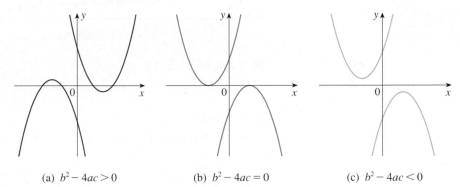

FIGURE 1
Possible graphs of $y = ax^2 + bx + c$

(a) $b^2 - 4ac > 0$ (b) $b^2 - 4ac = 0$ (c) $b^2 - 4ac < 0$

■ EXAMPLE 11

The quadratic $x^2 + x + 2$ is irreducible because its discriminant is negative:

$$b^2 - 4ac = 1^2 - 4(1)(2) = -7 < 0$$

Therefore it is impossible to factor $x^2 + x + 2$.   ■

■ The Binomial Theorem

Recall the binomial expression from Equation 1:

$$(a + b)^2 = a^2 + 2ab + b^2$$

If we multiply both sides by $(a + b)$ and simplify, we get the binomial expansion

(7)
$$(a + b)^3 = a^3 + 3a^2b + 3ab^2 + b^3$$

Repeating this procedure, we get

$$(a + b)^4 = a^4 + 4a^3b + 6a^2b^2 + 4ab^3 + b^4$$

In general, we have the following formula.

(8) ■ **The Binomial Theorem** If k is a positive integer, then

$$(a + b)^k = a^k + ka^{k-1}b + \frac{k(k - 1)}{1 \cdot 2} a^{k-2}b^2$$

$$+ \frac{k(k - 1)(k - 2)}{1 \cdot 2 \cdot 3} a^{k-3}b^3$$

$$+ \cdots + \frac{k(k - 1) \cdots (k - n + 1)}{1 \cdot 2 \cdot 3 \cdot \cdots \cdot n} a^{k-n}b^n$$

$$+ \cdots + kab^{k-1} + b^k$$

■ **EXAMPLE 12**

Expand $(x - 2)^5$.

SOLUTION

Using the Binomial Theorem with $a = x$, $b = -2$, $k = 5$, we have

$$(x - 2)^5 = x^5 + 5x^4(-2) + \frac{5 \cdot 4}{1 \cdot 2} x^3(-2)^2 + \frac{5 \cdot 4 \cdot 3}{1 \cdot 2 \cdot 3} x^2(-2)^3 + 5x(-2)^4 + (-2)^5$$

$$= x^5 - 10x^4 + 40x^3 - 80x^2 + 80x - 32$$ ■

Radicals

The most commonly occurring radicals are square roots. The symbol $\sqrt{}$ means "the positive square root of." Thus

$$x = \sqrt{a} \qquad \text{means} \qquad x^2 = a \quad \text{and} \quad x \geqslant 0$$

Since $a = x^2 \geqslant 0$, the symbol $\sqrt{a}$ makes sense only when $a \geqslant 0$. Here are two rules for working with square roots:

(9)

$$\sqrt{ab} = \sqrt{a} \, \sqrt{b} \qquad\qquad \sqrt{\frac{a}{b}} = \frac{\sqrt{a}}{\sqrt{b}}$$

However, there is no similar rule for the square root of a sum. In fact, you should remember to avoid the following common error:

⊘

$$\sqrt{a + b} \neq \sqrt{a} + \sqrt{b}$$

(For instance, take $a = 9$ and $b = 16$ to see the error.)

▪ **EXAMPLE 13**

(a) $\dfrac{\sqrt{18}}{\sqrt{2}} = \sqrt{\dfrac{18}{2}} = \sqrt{9} = 3$ **(b)** $\sqrt{x^2 y} = \sqrt{x^2}\,\sqrt{y} = |x|\sqrt{y}$

Notice that $\sqrt{x^2} = |x|$ because $\sqrt{}$ indicates the positive square root. (See Absolute Value on page A13.) ▪

In general, if n is a positive integer,

$$\boxed{\begin{array}{c} x = \sqrt[n]{a} \qquad \text{means} \qquad x^n = a \\[4pt] \text{If } n \text{ is even, then } a \geqslant 0 \text{ and } x \geqslant 0. \end{array}}$$

Thus $\sqrt[3]{-8} = -2$ because $(-2)^3 = -8$, but $\sqrt[4]{-8}$ and $\sqrt[6]{-8}$ are not defined. The following rules are valid:

$$\sqrt[n]{ab} = \sqrt[n]{a}\,\sqrt[n]{b} \qquad\qquad \sqrt[n]{\dfrac{a}{b}} = \dfrac{\sqrt[n]{a}}{\sqrt[n]{b}}$$

▪ **EXAMPLE 14**

If $x > 0$, then $\sqrt{x^3} = \sqrt{x^2 x} = \sqrt{x^2}\,\sqrt{x} = x\sqrt{x}$ ▪

To **rationalize** a numerator or denominator that contains an expression such as $\sqrt{a} - \sqrt{b}$, we multiply both the numerator and the denominator by the conjugate radical $\sqrt{a} + \sqrt{b}$. Then we can take advantage of the formula for a difference of squares:

$$(\sqrt{a} - \sqrt{b})(\sqrt{a} + \sqrt{b}) = (\sqrt{a})^2 - (\sqrt{b})^2 = a - b$$

▪ **EXAMPLE 15**

Rationalize the numerator in the expression $\dfrac{\sqrt{x+4} - 2}{x}$.

SOLUTION

We multiply the numerator and the denominator by the conjugate radical $\sqrt{x+4} + 2$:

$$\dfrac{\sqrt{x+4} - 2}{x} = \left(\dfrac{\sqrt{x+4} - 2}{x}\right)\left(\dfrac{\sqrt{x+4} + 2}{\sqrt{x+4} + 2}\right) = \dfrac{(x+4) - 4}{x(\sqrt{x+4} + 2)}$$

$$= \dfrac{x}{x(\sqrt{x+4} + 2)} = \dfrac{1}{\sqrt{x+4} + 2}$$ ▪

▪ Exponents

Let a be any positive number and let n be a positive integer. Then, by definition,

1. $\underbrace{a^n = a \cdot a \cdot \cdots \cdot a}_{n \text{ factors}}$

2. $a^0 = 1$

3. $a^{-n} = \dfrac{1}{a^n}$

4. $a^{1/n} = \sqrt[n]{a}$

$a^{m/n} = \sqrt[n]{a^m} = \left(\sqrt[n]{a}\right)^m$ m is any integer

---

(10) ▪ **Laws of Exponents**   Let a and b be positive numbers and let r and s be any rational numbers (that is, ratios of integers). Then

1. $a^r \times a^s = a^{r+s}$ **2.** $\dfrac{a^r}{a^s} = a^{r-s}$ **3.** $(a^r)^s = a^{rs}$

4. $(ab)^r = a^r b^r$ **5.** $\left(\dfrac{a}{b}\right)^r = \dfrac{a^r}{b^r}$ $b \neq 0$

In words, these five laws can be stated as follows:

1. To multiply two powers of the same number, we add the exponents.

2. To divide two powers of the same number, we subtract the exponents.

3. To raise a power to a new power, we multiply the exponents.

4. To raise a product to a power, we raise each factor to the power.

5. To raise a quotient to a power, we raise both numerator and denominator to the power.

▪ EXAMPLE 16

(a) $2^8 \times 8^2 = 2^8 \times (2^3)^2 = 2^8 \times 2^6 = 2^{14}$

(b) $\dfrac{x^{-2} - y^{-2}}{x^{-1} + y^{-1}} = \dfrac{\dfrac{1}{x^2} - \dfrac{1}{y^2}}{\dfrac{1}{x} + \dfrac{1}{y}} = \dfrac{\dfrac{y^2 - x^2}{x^2 y^2}}{\dfrac{y + x}{xy}} = \dfrac{y^2 - x^2}{x^2 y^2} \cdot \dfrac{xy}{y + x}$

$= \dfrac{(y - x)(y + x)}{xy(y + x)} = \dfrac{y - x}{xy}$

(c) $4^{3/2} = \sqrt{4^3} = \sqrt{64} = 8$ Alternative solution: $4^{3/2} = \left(\sqrt{4}\right)^3 = 2^3 = 8$

(d) $\dfrac{1}{\sqrt[3]{x^4}} = \dfrac{1}{x^{4/3}} = x^{-4/3}$

(e) $\left(\dfrac{x}{y}\right)^3 \left(\dfrac{y^2 x}{z}\right)^4 = \dfrac{x^3}{y^3} \cdot \dfrac{y^8 x^4}{z^4} = x^7 y^5 z^{-4}$

▪

▪ Intervals

Certain sets of real numbers, called **intervals**, occur frequently in calculus and correspond geometrically to line segments. For example, if $a < b$, the **open interval** from a to b consists of all numbers between a and b and is denoted by the symbol (a, b). Using set-builder notation, we can write

$$(a, b) = \{x \mid a < x < b\}$$

Notice that the endpoints of the interval—namely, a and b—are excluded. This is indicated by the round brackets ( ) and by the open dots in Figure 2. The **closed interval** from a to b is the set

$$[a, b] = \{x \mid a \le x \le b\}$$

Here the endpoints of the interval are included. This is indicated by the square brackets [ ] and by the solid dots in Figure 3. It is also possible to include only one endpoint in an interval, as shown in Table 1.

FIGURE 2
Open interval (a, b)

FIGURE 3
Closed interval $[a, b]$

Table 1 lists the nine possible types of intervals. When these intervals are discussed, it is always assumed that $a < b$.

(1) TABLE OF INTERVALS

Notation	Set description	Picture
(a, b)	$\{x \mid a < x < b\}$	
$[a, b]$	$\{x \mid a \le x \le b\}$	
$[a, b)$	$\{x \mid a \le x < b\}$	
$(a, b]$	$\{x \mid a < x \le b\}$	
(a, ∞)	$\{x \mid x > a\}$	
$[a, \infty)$	$\{x \mid x \ge a\}$	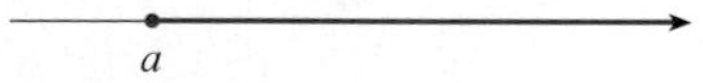
$(-\infty, b)$	$\{x \mid x < b\}$	
$(-\infty, b]$	$\{x \mid x \le b\}$	
$(-\infty, \infty)$	$\mathbb{R}$ (set of all real numbers)	

We also need to consider infinite intervals such as

$$(a, \infty) = \{x \mid x > a\}$$

This does not mean that ∞ ("infinity") is a number. The notation (a, ∞) stands for the set of all numbers that are greater than a, so the symbol ∞ simply indicates that the interval extends indefinitely in the positive direction.

▪ Inequalities

When working with inequalities, note the following rules.

▪ Rules for Inequalities

1. If $a < b$, then $a + c < b + c$.

2. If $a < b$ and $c < d$, then $a + c < b + d$.

3. If $a < b$ and $c > 0$, then $ac < bc$.

4. If $a < b$ and $c < 0$, then $ac > bc$.

5. If $0 < a < b$, then $1/a > 1/b$.

Rule 1 says that we can add any number to both sides of an inequality, and Rule 2 says that two inequalities can be added. However, we have to be careful with multiplication. Rule 3 says that we can multiply both sides of an inequality by a *positive* number, but Rule 4 says that *if we multiply both sides of an inequality by a negative number, then we reverse the direction of the inequality.* For example, if we take the inequality $3 < 5$ and multiply by 2, we get $6 < 10$, but if we multiply by -2, we get $-6 > -10$. Finally, Rule 5 says that if we take reciprocals, then we reverse the direction of an inequality (provided the numbers are positive).

■ EXAMPLE 17

Solve the inequality $1 + x < 7x + 5$.

SOLUTION

The given inequality is satisfied by some values of x but not by others. To *solve* an inequality means to determine the set of numbers x for which the inequality is true. This is called the *solution set.*

First we subtract 1 from each side of the inequality (using Rule 1 with $c = -1$):

$$x < 7x + 4$$

Then we subtract $7x$ from both sides (Rule 1 with $c = -7x$):

$$-6x < 4$$

Now we divide both sides by -6 $\left(\text{Rule 4 with } c = -\frac{1}{6}\right)$:

$$x > -\tfrac{4}{6} = -\tfrac{2}{3}$$

These steps can all be reversed, so the solution set consists of all numbers greater than $-\frac{2}{3}$. In other words, the solution of the inequality is the interval $\left(-\frac{2}{3}, \infty\right)$. ■

■ EXAMPLE 18

Solve the inequality $x^2 - 5x + 6 \leq 0$.

SOLUTION

First we factor the left side:

$$(x - 2)(x - 3) \leq 0$$

We know that the corresponding equation $(x - 2)(x - 3) = 0$ has the solutions 2 and 3. The numbers 2 and 3 divide the real line into three intervals:

$$(-\infty, 2) \qquad (2, 3) \qquad (3, \infty)$$

On each of these intervals we determine the signs of the factors. For instance,

$$\text{if} \quad x < 2 \quad \text{then} \quad x - 2 < 0$$

A visual method for solving Example 18 is to use a graphing device to graph the parabola $y = x^2 - 5x + 6$ (as in Figure 4) and observe that the curve lies on or below the x-axis when $2 \leq x \leq 3$.

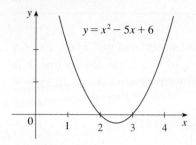

FIGURE 4

FIGURE 5

Then we record these signs in the following chart:

Interval	$x - 2$	$x - 3$	$(x - 2)(x - 3)$
$x < 2$	$-$	$-$	$+$
$2 < x < 3$	$+$	$-$	$-$
$x > 3$	$+$	$+$	$+$

Another method for obtaining the information in the chart is to use *test values*. For instance, if we use the test value $x = 1$ for the interval $(-\infty, 2)$, then substitution in $x^2 - 5x + 6$ gives

$$1^2 - 5(1) + 6 = 2$$

The polynomial $x^2 - 5x + 6$ doesn't change sign inside any of the three intervals, so we conclude that it is positive on $(-\infty, 2)$.

Then we read from the chart that $(x - 2)(x - 3)$ is negative when $2 < x < 3$. Thus the solution of the inequality $(x - 2)(x - 3) \leq 0$ is

$$\{x \mid 2 \leq x \leq 3\} = [2, 3]$$

Notice that we have included the endpoints 2 and 3 because we are looking for values of x such that the product is either negative or zero. The solution is illustrated in Figure 5. ▪

▪ Absolute Value

The **absolute value** of a number a, denoted by $|a|$, is the distance from a to 0 on the real number line. Distances are always positive or 0, so we have

$$|a| \geq 0 \qquad \text{for every number } a$$

For example,

$$|3| = 3 \qquad |-3| = 3 \qquad |0| = 0$$

$$|\sqrt{2} - 1| = \sqrt{2} - 1 \qquad |3 - \pi| = \pi - 3$$

In general, we have

Remember that if a is negative, then $-a$ is positive.

(11)

$$|a| = a \qquad \text{if } a \geq 0$$
$$|a| = -a \qquad \text{if } a < 0$$

▪ EXAMPLE 19

Express $|3x - 2|$ without using the absolute-value symbol.

SOLUTION

$$|3x - 2| = \begin{cases} 3x - 2 & \text{if } 3x - 2 \geq 0 \\ -(3x - 2) & \text{if } 3x - 2 < 0 \end{cases}$$

$$= \begin{cases} 3x - 2 & \text{if } x \geq \frac{2}{3} \\ 2 - 3x & \text{if } x < \frac{2}{3} \end{cases}$$ ▪

⊘ Recall that the symbol $\sqrt{\ }$ means "the positive square root of." Thus, $\sqrt{r} = s$ means $s^2 = r$ and $s \geq 0$. Therefore, the equation $\sqrt{a^2} = a$ is not always true. It is true only when $a \geq 0$. If $a < 0$, then $-a > 0$, so we have $\sqrt{a^2} = -a$. In view of (11), we then have the equation

(12) $$\boxed{\sqrt{a^2} = |a|}$$

which is true for all values of a.

Hints for the proofs of the following properties are given in the exercises.

▪ **Properties of Absolute Values**   Suppose a and b are any real numbers and n is an integer. Then

1. $|ab| = |a||b|$ **2.** $\left|\dfrac{a}{b}\right| = \dfrac{|a|}{|b|}$ $(b \neq 0)$ **3.** $|a^n| = |a|^n$

For solving equations or inequalities involving absolute values, it's often very helpful to use the following statements.

▪ Suppose $a > 0$. Then

4. $|x| = a$ if and only if $x = \pm a$

5. $|x| < a$ if and only if $-a < x < a$

6. $|x| > a$ if and only if $x > a$ or $x < -a$

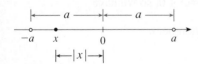

FIGURE 6

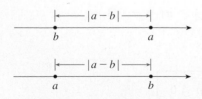

FIGURE 7

Length of a line segment $= |a - b|$

For instance, the inequality $|x| < a$ says that the distance from x to the origin is less than a, and you can see from Figure 6 that this is true if and only if x lies between $-a$ and a.

If a and b are any real numbers, then the distance between a and b is the absolute value of the difference, namely, $|a - b|$, which is also equal to $|b - a|$. (See Figure 7.)

▪ **EXAMPLE 20**

Solve $|2x - 5| = 3$.

SOLUTION

By Property 4 of absolute values, $|2x - 5| = 3$ is equivalent to

$$2x - 5 = 3 \quad \text{or} \quad 2x - 5 = -3$$

So $2x = 8$ or $2x = 2$. Thus $x = 4$ or $x = 1$.   ■

■ Exercises A

1–16 ■ Expand and simplify.

1. $(-6ab)(0.5ac)$ **2.** $-(2x^2y)(-xy^4)$

3. $2x(x - 5)$ **4.** $(4 - 3x)x$

5. $-2(4 - 3a)$ **6.** $8 - (4 + x)$

7. $4(x^2 - x + 2) - 5(x^2 - 2x + 1)$

8. $5(3t - 4) - (t^2 + 2) - 2t(t - 3)$

9. $(4x - 1)(3x + 7)$ **10.** $x(x - 1)(x + 2)$

11. $(2x - 1)^2$ **12.** $(2 + 3x)^2$

13. $y^4(6 - y)(5 + y)$

14. $(t - 5)^2 - 2(t + 3)(8t - 1)$

15. $(1 + 2x)(x^2 - 3x + 1)$ **16.** $(1 + x - x^2)^2$

17–28 ■ Perform the indicated operations and simplify.

17. $\dfrac{2 + 8x}{2}$ **18.** $\dfrac{9b - 6}{3b}$

19. $\dfrac{1}{x + 5} + \dfrac{2}{x - 3}$ **20.** $\dfrac{1}{x + 1} + \dfrac{1}{x - 1}$

21. $u + 1 + \dfrac{u}{u + 1}$ **22.** $\dfrac{2}{a^2} - \dfrac{3}{ab} + \dfrac{4}{b^2}$

23. $\dfrac{x/y}{z}$ **24.** $\dfrac{x}{y/z}$

25. $\left(\dfrac{-2r}{s}\right)\left(\dfrac{s^2}{-6t}\right)$ **26.** $\dfrac{a}{bc} \div \dfrac{b}{ac}$

27. $\dfrac{1 + \dfrac{1}{c - 1}}{1 - \dfrac{1}{c - 1}}$ **28.** $1 + \dfrac{1}{1 + \dfrac{1}{1 + x}}$

29–44 ■ Factor the expression.

29. $2x + 12x^3$ **30.** $5ab - 8abc$

31. $x^2 + 7x + 6$ **32.** $x^2 - x - 6$

33. $x^2 - 2x - 8$ **34.** $2x^2 + 7x - 4$

35. $9x^2 - 36$ **36.** $8x^2 + 10x + 3$

37. $6x^2 - 5x - 6$ **38.** $x^2 + 10x + 25$

39. $t^3 + 1$ **40.** $4t^2 - 9s^2$

41. $4t^2 - 12t + 9$ **42.** $x^3 - 27$

43. $x^3 + 2x^2 + x$ **44.** $3q^3 + q^2 - 10q$

45–50 ■ Simplify the expression.

45. $\dfrac{x^2 + x - 2}{x^2 - 3x + 2}$ **46.** $\dfrac{2x^2 - 3x - 2}{x^2 - 4}$

47. $\dfrac{x^2 - 1}{x^2 - 9x + 8}$ **48.** $\dfrac{x^3 + 5x^2 + 6x}{x^2 - x - 12}$

49. $\dfrac{1}{x + 3} + \dfrac{1}{x^2 - 9}$

50. $\dfrac{x}{x^2 + x - 2} - \dfrac{2}{x^2 - 5x + 4}$

51–56 ■ Complete the square.

51. $x^2 + 2x + 5$ **52.** $x^2 - 16x + 80$

53. $x^2 - 5x + 10$ **54.** $x^2 + 3x + 1$

55. $4x^2 + 4x - 2$ **56.** $3x^2 - 24x + 50$

57–62 ■ Solve the equation.

57. $x^2 + 9x - 10 = 0$ **58.** $x^2 - 2x - 8 = 0$

59. $x^2 + 9x - 1 = 0$ **60.** $x^2 - 2x - 7 = 0$

61. $3x^2 + 5x + 1 = 0$ **62.** $2x^2 + 7x + 2 = 0$

63–66 ■ Which of the quadratics are irreducible?

63. $2x^2 + 3x + 4$ **64.** $2x^2 + 9x + 4$

65. $3x^2 + x - 6$ **66.** $x^2 + 3x + 6$

67–68 ▪ Use the Binomial Theorem to expand the expression.

67. $(a + b)^6$ **68.** $(a + b)^7$

69–74 ▪ Simplify the radicals.

69. $\sqrt{32}\,\sqrt{2}$ **70.** $\dfrac{\sqrt{48}}{\sqrt{2}}$ **71.** $\sqrt{16a^4b^3}$

72. $\sqrt{xy}\,\sqrt{x^3y}$ **73.** $\sqrt{6p^5r}\,\sqrt{15r^3}$ **74.** $\dfrac{4\sqrt{2mn^2}}{\sqrt{8n}}$

75–88 ▪ Use the Laws of Exponents to rewrite and simplify the expression.

75. $3^{10} \times 9^8$ **76.** $2^{16} \times 4^{10} \times 16^6$

77. $\dfrac{x^9(2x)^4}{x^3}$ **78.** $\dfrac{a^n \times a^{2n+1}}{a^{n-2}}$

79. $\dfrac{a^{-3}b^4}{a^{-5}b^5}$ **80.** $\dfrac{x^{-1} + y^{-1}}{(x + y)^{-1}}$

81. $3^{-1/2}$ **82.** $96^{1/5}$

83. $125^{2/3}$ **84.** $64^{-4/3}$

85. $(2x^2y^4)^{3/2}$ **86.** $(x^{-5}y^3z^{10})^{-3/5}$

87. $\dfrac{1}{(\sqrt{t}\,)^5}$ **88.** $(\sqrt[4]{a}\,)^3$

89–94 ▪ Rationalize the expression.

89. $\dfrac{\sqrt{x} - 3}{x - 9}$ **90.** $\dfrac{(1/\sqrt{x}) - 1}{x - 1}$

91. $\dfrac{x\sqrt{x} - 8}{x - 4}$ **92.** $\dfrac{\sqrt{2 + h} + \sqrt{2 - h}}{h}$

93. $\dfrac{2}{3 - \sqrt{5}}$ **94.** $\dfrac{1}{\sqrt{x} - \sqrt{y}}$

95–102 ▪ State whether or not the equation is true for all values of the variable.

95. $\sqrt{x^2} = x$ **96.** $\sqrt{x^2 + 4} = |x| + 2$

97. $\dfrac{16 + a}{16} = 1 + \dfrac{a}{16}$ **98.** $\dfrac{1}{x^{-1} + y^{-1}} = x + y$

99. $\dfrac{x}{x + y} = \dfrac{1}{1 + y}$ **100.** $\dfrac{2}{4 + x} = \dfrac{1}{2} + \dfrac{2}{x}$

101. $(x^3)^4 = x^7$

102. $6 - 4(x + a) = 6 - 4x - 4a$

103–116 ▪ Solve the inequality in terms of intervals and illustrate the solution set on the real number line.

103. $2x + 7 > 3$ **104.** $4 - 3x \geqslant 6$

105. $1 - x \leqslant 2$ **106.** $1 + 5x > 5 - 3x$

107. $0 \leqslant 1 - x < 1$ **108.** $1 < 3x + 4 \leqslant 16$

109. $(x - 1)(x - 2) > 0$ **110.** $x^2 < 2x + 8$

111. $x^2 < 3$ **112.** $x^2 \geqslant 5$

113. $x^3 - x^2 \leqslant 0$

114. $(x + 1)(x - 2)(x + 3) \geqslant 0$

115. $x^3 > x$ **116.** $x^3 + 3x < 4x^2$

117. The relationship between the Celsius and Fahrenheit temperature scales is given by $C = \frac{5}{9}(F - 32)$, where C is the temperature in degrees Celsius and F is the temperature in degrees Fahrenheit. What interval on the Celsius scale corresponds to the temperature range $50 \leqslant F \leqslant 95$?

118. Use the relationship between C and F given in Exercise 117 to find the interval on the Fahrenheit scale corresponding to the temperature range $20 \leqslant C \leqslant 30$.

119. As dry air moves upward, it expands and in so doing cools at a rate of about 1°C for each 100-m rise, up to about 12 km.
   (a) If the ground temperature is 20°C, write a formula for the temperature at height h.
   (b) What range of temperature can be expected if a plane takes off and reaches a maximum height of 5 km?

120. If a ball is thrown upward from the top of a building 128 ft high with an initial velocity of 16 ft/s, then the height h above the ground t seconds later will be

$$h = 128 + 16t - 16t^2$$

During what time interval will the ball be at least 32 ft above the ground?

121–122 ▪ Solve the equation for x.

121. $|x + 3| = 4$ **122.** $|3x + 5| = 1$

123. Prove that $|ab| = |a||b|$. [*Hint:* Use Equation 12.]

124. Show that if $0 < a < b$, then $a^2 < b^2$.

B Coordinate Geometry and Lines

The points in a plane can be identified with ordered pairs of real numbers. We start by drawing two perpendicular coordinate lines that intersect at the origin O on each line. Usually one line is horizontal with positive direction to the right and is called the x-axis; the other line is vertical with positive direction upward and is called the y-axis.

Any point P in the plane can be located by a unique ordered pair of numbers as follows. Draw lines through P perpendicular to the x- and y-axes. These lines intersect the axes in points with coordinates a and b as shown in Figure 1. Then the point P is assigned the ordered pair (a, b). The first number a is called the **x-coordinate** of P; the second number b is called the **y-coordinate** of P. We say that P is the point with coordinates (a, b), and we denote the point by the symbol $P(a, b)$. Several points are labeled with their coordinates in Figure 2.

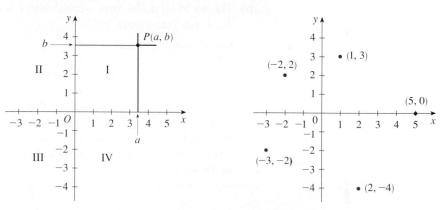

FIGURE 1   **FIGURE 2**

By reversing the preceding process we can start with an ordered pair (a, b) and arrive at the corresponding point P. Often we identify the point P with the ordered pair (a, b) and refer to "the point (a, b)." [Although the notation used for an open interval (a, b) is the same as the notation used for a point (a, b), you will be able to tell from the context which meaning is intended.]

This coordinate system is called the **rectangular coordinate system** or the **Cartesian coordinate system** in honor of the French mathematician René Descartes (1596–1650), even though another Frenchman, Pierre Fermat (1601–1665), invented the principles of analytic geometry at about the same time as Descartes. The plane supplied with this coordinate system is called the **coordinate plane** or the **Cartesian plane** and is denoted by $\mathbb{R}^2$.

The x- and y-axes are called the **coordinate axes** and divide the Cartesian plane into four quadrants, which are labeled I, II, III, and IV in Figure 1. Notice that the first quadrant consists of those points whose x- and y-coordinates are both positive.

▪ **EXAMPLE 1**

Describe and sketch the regions given by the following sets.

(a) $\{(x, y) \mid x \geq 0\}$ **(b)** $\{(x, y) \mid y = 1\}$ **(c)** $\{(x, y) \mid |y| < 1\}$

SOLUTION

(a) The points whose x-coordinates are 0 or positive lie on the y-axis or to the right of it as indicated by the shaded region in Figure 3(a).

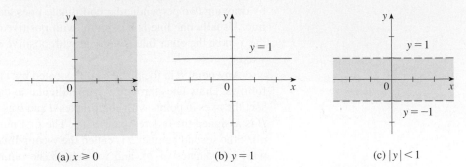

FIGURE 3   (a) $x \geq 0$ (b) $y = 1$ (c) $|y| < 1$

(b) The set of all points with y-coordinate 1 is a horizontal line one unit above the x-axis [see Figure 3(b)].

(c) Recall from Appendix A that

$$|y| < 1 \qquad \text{if and only if} \qquad -1 < y < 1$$

The given region consists of those points in the plane whose y-coordinates lie between -1 and 1. Thus the region consists of all points that lie between (but not on) the horizontal lines $y = 1$ and $y = -1$. [These lines are shown as dashed lines in Figure 3(c) to indicate that the points on these lines don't lie in the set.] ■

Recall from Appendix A that the distance between points a and b on a number line is $|a - b| = |b - a|$. Thus the distance between points $P_1(x_1, y_1)$ and $P_3(x_2, y_1)$ on a horizontal line must be $|x_2 - x_1|$ and the distance between $P_2(x_2, y_2)$ and $P_3(x_2, y_1)$ on a vertical line must be $|y_2 - y_1|$. (See Figure 4.)

To find the distance $|P_1P_2|$ between any two points $P_1(x_1, y_1)$ and $P_2(x_2, y_2)$, we note that triangle $P_1P_2P_3$ in Figure 4 is a right triangle, and so by the Pythagorean Theorem we have

$$|P_1P_2| = \sqrt{|P_1P_3|^2 + |P_2P_3|^2} = \sqrt{|x_2 - x_1|^2 + |y_2 - y_1|^2}$$

$$= \sqrt{(x_2 - x_1)^2 + (y_2 - y_1)^2}$$

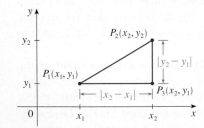

FIGURE 4

■ **Distance Formula**   The distance between the points $P_1(x_1, y_1)$ and $P_2(x_2, y_2)$ is

$$|P_1P_2| = \sqrt{(x_2 - x_1)^2 + (y_2 - y_1)^2}$$

■ **EXAMPLE 2**

The distance between $(1, -2)$ and $(5, 3)$ is

$$\sqrt{(5 - 1)^2 + [3 - (-2)]^2} = \sqrt{4^2 + 5^2} = \sqrt{41}$$ ■

■ | Lines

We want to find an equation of a given line L; such an equation is satisfied by the coordinates of the points on L and by no other point. To find the equation of L we use its *slope,* which is a measure of the steepness of the line.

> **■ Definition**   The **slope** of a nonvertical line that passes through the points $P_1(x_1, y_1)$ and $P_2(x_2, y_2)$ is
>
> $$m = \frac{\Delta y}{\Delta x} = \frac{y_2 - y_1}{x_2 - x_1}$$
>
> The slope of a vertical line is not defined.

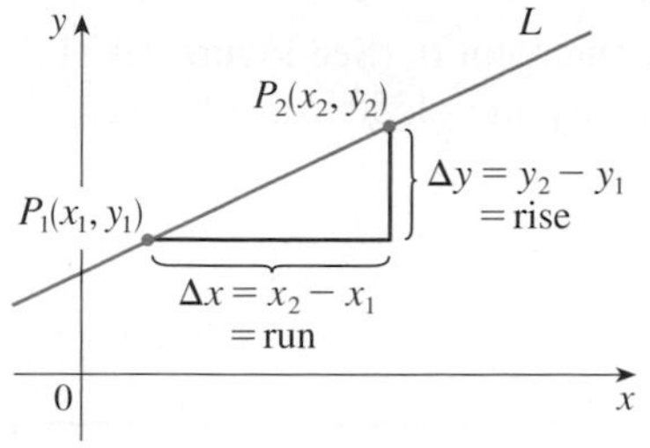

Thus the slope of a line is the ratio of the change in y, Δy, to the change in x, Δx. (See Figure 5.) The slope is therefore the rate of change of y with respect to x. The fact that the line is straight means that the rate of change is constant.

Figure 6 shows several lines labeled with their slopes. Notice that lines with positive slope slant upward to the right, whereas lines with negative slope slant downward to the right. Notice also that the steepest lines are the ones for which the absolute value of the slope is largest, and a horizontal line has slope 0.

Now let's find an equation of the line that passes through a given point $P_1(x_1, y_1)$ and has slope m. A point $P(x, y)$ with $x \neq x_1$ lies on this line if and only if the slope of the line through P_1 and P is equal to m; that is,

$$\frac{y - y_1}{x - x_1} = m$$

This equation can be rewritten in the form

$$y - y_1 = m(x - x_1)$$

and we observe that this equation is also satisfied when $x = x_1$ and $y = y_1$. Therefore it is an equation of the given line.

> **■ Point-Slope Form of the Equation of a Line**   An equation of the line passing through the point $P_1(x_1, y_1)$ and having slope m is
>
> $$y - y_1 = m(x - x_1)$$

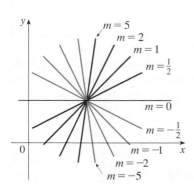

FIGURE 5

FIGURE 6

■ EXAMPLE 3

Find an equation of the line through $(1, -7)$ with slope $-\frac{1}{2}$.

SOLUTION

Using the point-slope form of the equation of a line with $m = -\frac{1}{2}$, $x_1 = 1$, and $y_1 = -7$, we obtain

$$y + 7 = -\tfrac{1}{2}(x - 1)$$

which we can rewrite as

$$2y + 14 = -x + 1 \qquad \text{or} \qquad x + 2y = -13 \qquad ■$$

▪ EXAMPLE 4

Find an equation of the line through the points $(-2, -9)$ and $(4, 6)$.

SOLUTION

The slope of the line is

$$m = \frac{6 - (-9)}{4 - (-2)} = \frac{15}{6} = \frac{5}{2}$$

Using the point-slope form with $x_1 = 4$ and $y_1 = 6$, we obtain

$$y - 6 = \tfrac{5}{2}(x - 4)$$

which simplifies to $5x - 2y = 8$ ▪

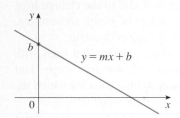

FIGURE 7

Suppose a nonvertical line has slope m and y-intercept b. (See Figure 7.) This means it intersects the y-axis at the point $(0, b)$, so the point-slope form of the equation of the line, with $x_1 = 0$ and $y_1 = b$, becomes

$$y - b = m(x - 0)$$

This simplifies as follows.

> ▪ **Slope-Intercept Form of the Equation of a Line**   An equation of the line with slope m and y-intercept b is
>
> $$y = mx + b$$

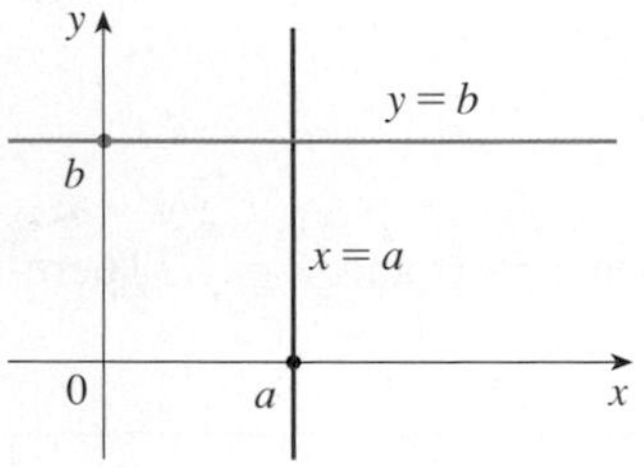

In particular, if a line is horizontal, its slope is $m = 0$, so its equation is $y = b$, where b is the y-intercept (see Figure 8). A vertical line does not have a slope, but we can write its equation as $x = a$, where a is the x-intercept, because the x-coordinate of every point on the line is a.

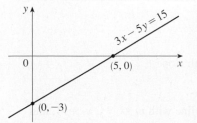

FIGURE 8

▪ EXAMPLE 5

Sketch the graph of the linear equation $3x - 5y = 15$.

SOLUTION 1

Since the equation is linear, its graph is a line. To draw the graph, we can simply find two points on the line. It's easiest to find the intercepts. Substituting $y = 0$ (the equation of the x-axis) in the given equation, we get $3x = 15$, so $x = 5$ is the x-intercept. Substituting $x = 0$ in the equation, we see that the y-intercept is -3. This allows us to sketch the graph as in Figure 9.

SOLUTION 2

We put the equation into slope-intercept form by solving for y:

$$3x - 5y = 15$$

$$-5y = -3x + 15$$

$$y = \frac{-3x}{-5} + \frac{15}{-5} = \frac{3}{5}x - 3$$

FIGURE 9

Thus the slope is $\frac{3}{5}$ and the y-intercept is -3. We plot a point at $(0, -3)$ and arrive at a second point $(5, 0)$ by moving 3 units upward and 5 units to the right.   ▪

▪ EXAMPLE 6

Graph the inequality $x + 2y > 5$.

SOLUTION

We are asked to sketch the graph of the set $\{(x, y) \mid x + 2y > 5\}$ and we begin by solving the inequality for y:

$$x + 2y > 5$$

$$2y > -x + 5$$

$$y > -\tfrac{1}{2}x + \tfrac{5}{2}$$

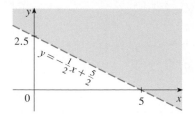

FIGURE 10

Compare this inequality with the equation $y = -\frac{1}{2}x + \frac{5}{2}$, which represents a line with slope $-\frac{1}{2}$ and y-intercept $\frac{5}{2}$. We see that the given graph consists of points whose y-coordinates are *larger* than those on the line $y = -\frac{1}{2}x + \frac{5}{2}$. Thus the graph is the region that lies *above* the line, as illustrated in Figure 10.   ▪

▪ | Parallel and Perpendicular Lines

Slopes can be used to show that lines are parallel or perpendicular. The following facts are proved, for instance, in *Precalculus: Mathematics for Calculus, Sixth Edition* by Stewart, Redlin, and Watson (Belmont, CA, 2012).

▪ Parallel and Perpendicular Lines

1. Two nonvertical lines are parallel if and only if they have the same slope.

2. Two lines with slopes m_1 and m_2 are perpendicular if and only if $m_1 m_2 = -1$; that is, their slopes are negative reciprocals:

$$m_2 = -\frac{1}{m_1}$$

▪ EXAMPLE 7

Find an equation of the line through the point $(5, 2)$ that is parallel to the line $4x + 6y + 5 = 0$.

SOLUTION

The given line can be written in the form

$$y = -\tfrac{2}{3}x - \tfrac{5}{6}$$

which is in slope-intercept form with $m = -\frac{2}{3}$. Parallel lines have the same slope, so the required line has slope $-\frac{2}{3}$ and its equation in point-slope form is

$$y - 2 = -\tfrac{2}{3}(x - 5)$$

We can write this equation as $2x + 3y = 16$. ∎

■ EXAMPLE 8

Show that the lines $2x + 3y = 1$ and $6x - 4y - 1 = 0$ are perpendicular.

SOLUTION

The equations can be written as

$$y = -\tfrac{2}{3}x + \tfrac{1}{3} \qquad \text{and} \qquad y = \tfrac{3}{2}x - \tfrac{1}{4}$$

from which we see that the slopes are

$$m_1 = -\tfrac{2}{3} \qquad \text{and} \qquad m_2 = \tfrac{3}{2}$$

Since $m_1 m_2 = -1$, the lines are perpendicular. ∎

■ Exercises B

1–6 ■ Find the distance between the points.

1. $(1, 1)$, $(4, 5)$

2. $(1, -3)$, $(5, 7)$

3. $(6, -2)$, $(-1, 3)$

4. $(1, -6)$, $(-1, -3)$

5. $(2, 5)$, $(4, -7)$

6. (a, b), (b, a)

7–10 ■ Find the slope of the line through P and Q.

7. $P(1, 5)$, $Q(4, 11)$

8. $P(-1, 6)$, $Q(4, -3)$

9. $P(-3, 3)$, $Q(-1, -6)$

10. $P(-1, -4)$, $Q(6, 0)$

11–12 ■ Sketch the graph of the equation.

11. $x = 3$

12. $y = -2$

13–28 ■ Find an equation of the line that satisfies the given conditions.

13. Through $(1, -8)$, slope 4

14. Through $(-1, 4)$, slope -3

15. Through $(1, 7)$, slope $\frac{2}{3}$

16. Through $(-6, -4)$, slope $-\frac{5}{2}$

17. Through $(4, 3)$ and $(3, 8)$

18. Through $(-2, -3)$ and $(3, 2)$

19. Slope 6, y-intercept -4

20. Slope $\frac{3}{4}$, y-intercept 3

21. x-intercept 2, y-intercept -4

22. x-intercept -7, y-intercept 11

23. Through $(4, 5)$, parallel to the x-axis

24. Through $(4, 5)$, parallel to the y-axis

25. Through $(1, -6)$, parallel to the line $x + 2y = 6$

26. y-intercept 6, parallel to the line $2x + 3y + 4 = 0$

27. Through $(-1, -2)$, perpendicular to the line $2x + 5y + 8 = 0$

28. Through $\left(\frac{1}{2}, -\frac{2}{3}\right)$, perpendicular to the line $4x - 8y = 1$

29–34 ■ Find the slope and y-intercept of the line and draw its graph.

29. $x + 3y = 0$

30. $2x - 5y = 0$

31. $y = -2$

32. $2x - 3y + 6 = 0$

33. $3x - 4y = 12$

34. $4x + 5y = 10$

35–44 ▪ Sketch the region in the *xy*-plane.

35. $\{(x, y) \mid x < 0\}$

36. $\{(x, y) \mid y > 0\}$

37. $\{(x, y) \mid xy < 0\}$

38. $\{(x, y) \mid x \geqslant 1 \text{ and } y < 3\}$

39. $\{(x, y) \mid |x| \leqslant 2\}$

40. $\{(x, y) \mid |x| < 3 \text{ and } |y| < 2\}$

41. $\{(x, y) \mid 0 \leqslant y \leqslant 4 \text{ and } x \leqslant 2\}$

42. $\{(x, y) \mid y > 2x - 1\}$

43. $\{(x, y) \mid 1 + x \leqslant y \leqslant 1 - 2x\}$

44. $\{(x, y) \mid -x \leqslant y < \frac{1}{2}(x + 3)\}$

45. Show that the lines $2x - y = 4$ and $6x - 2y = 10$ are not parallel and find their point of intersection.

46. Show that the lines $3x - 5y + 19 = 0$ and $10x + 6y - 50 = 0$ are perpendicular and find their point of intersection.

C   Approximate Integration

There are two situations in which it is impossible to find the exact value of a definite integral.

The first situation arises from the fact that in order to evaluate $\int_a^b f(x)\, dx$ using the Fundamental Theorem of Calculus we need to know an antiderivative of f. Sometimes, however, it is difficult, or even impossible, to find an antiderivative (see Section 5.5). For example, it is impossible to evaluate the following integrals exactly:

$$\int_0^1 e^{x^2} dx \qquad \int_{-1}^1 \sqrt{1 + x^3}\, dx$$

The second situation arises when the function is determined from a scientific experiment through instrument readings or collected data. There may be no formula for the function (see Example 3).

In both cases we need to find approximate values of definite integrals. We already know one such method. Recall that the definite integral is defined as a limit of Riemann sums, so any Riemann sum could be used as an approximation to the integral: If we divide $[a, b]$ into n subintervals of equal length $\Delta x = (b - a)/n$, then we have

$$\int_a^b f(x)\, dx \approx f(x_1^*)\, \Delta x + f(x_2^*)\, \Delta x + \cdots + f(x_n^*)\, \Delta x$$

where x_i^* is any point in the ith subinterval $[x_{i-1}, x_i]$. If x_i^* is chosen to be the left endpoint of the interval, then $x_i^* = x_{i-1}$ and we have

(1) $$\int_a^b f(x)\, dx \approx L_n = f(x_0)\, \Delta x + f(x_1)\, \Delta x + \cdots + f(x_{n-1})\, \Delta x$$

If $f(x) \geqslant 0$, then the integral represents an area and (1) represents an approximation of this area by the rectangles shown in Figure 1(a). If we choose x_i^* to be the right endpoint, then $x_i^* = x_i$ and we have

(2) $$\int_a^b f(x)\, dx \approx R_n = f(x_1)\, \Delta x + f(x_2)\, \Delta x + \cdots + f(x_n)\, \Delta x$$

[See Figure 1(b).] The approximations L_n and R_n defined by Equations 1 and 2 are called the **left endpoint approximation** and **right endpoint approximation**, respectively.

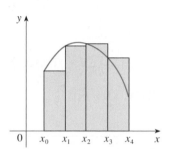

(a) Left endpoint approximation

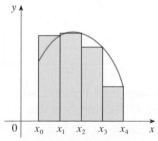

(b) Right endpoint approximation

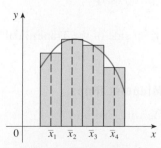

(c) Midpoint approximation

FIGURE 1

In Section 5.1 we also considered the case where x_i^* is chosen to be the midpoint $\bar{x}_i$ of the subinterval $[x_{i-1}, x_i]$. Figure 1(c) shows the midpoint approximation M_n, which appears to be better than either L_n or R_n.

▪ **Midpoint Rule**

$$\int_a^b f(x)\, dx \approx M_n = \Delta x\, [f(\bar{x}_1) + f(\bar{x}_2) + \cdots + f(\bar{x}_n)]$$

where
$$\Delta x = \frac{b - a}{n}$$

and
$$\bar{x}_i = \text{midpoint of } [x_{i-1}, x_i] = \tfrac{1}{2}(x_{i-1} + x_i)$$

Another approximation, called the Trapezoidal Rule, results from averaging the approximations in Equations 1 and 2:

$$\int_a^b f(x)\, dx \approx \tfrac{1}{2}[f(x_0)\, \Delta x + f(x_1)\, \Delta x + f(x_2)\, \Delta x + \cdots + f(x_{n-1})\, \Delta x$$
$$+ f(x_1)\, \Delta x + f(x_2)\, \Delta x + \cdots + f(x_{n-1})\, \Delta x + f(x_n)\, \Delta x]$$

$$= \frac{\Delta x}{2}\, [f(x_0) + 2f(x_1) + 2f(x_2) + \cdots + 2f(x_{n-1}) + f(x_n)]$$

▪ **Trapezoidal Rule**

$$\int_a^b f(x)\, dx \approx T_n = \frac{\Delta x}{2}\, [f(x_0) + 2f(x_1) + 2f(x_2) + \cdots + 2f(x_{n-1}) + f(x_n)]$$

where $\Delta x = (b - a)/n$ and $x_i = a + i\, \Delta x$.

The reason for the name Trapezoidal Rule can be seen from Figure 2, which illustrates the case with $f(x) \geq 0$ and $n = 4$. The area of the trapezoid that lies above the ith subinterval is the base times the average height:

$$\Delta x \left(\frac{f(x_{i-1}) + f(x_i)}{2} \right) = \frac{\Delta x}{2}\, [f(x_{i-1}) + f(x_i)]$$

If we add the areas of all these trapezoids, we get the right side of the Trapezoidal Rule.

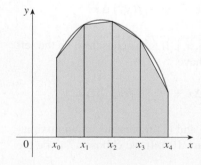

FIGURE 2
Trapezoidal approximation

▪ **EXAMPLE 1 Using the Trapezoidal and Midpoint Rules**

Use **(a)** the Trapezoidal Rule and **(b)** the Midpoint Rule with $n = 5$ to approximate the integral $\int_1^2 (1/x)\, dx$.

SOLUTION

(a) With $n = 5$, $a = 1$, and $b = 2$, we have $\Delta x = (2 - 1)/5 = 0.2$, and so the

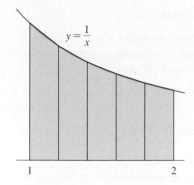

FIGURE 3

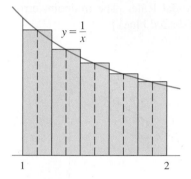

FIGURE 4

$$\int_a^b f(x)\,dx = \text{approximation} + \text{error}$$

TEC Module 5.1 allows you to compare approximation methods.

$$\text{Approximations to } \int_1^2 \frac{1}{x}\,dx$$

Trapezoidal Rule gives

$$\int_1^2 \frac{1}{x}\,dx \approx T_5 = \frac{0.2}{2}[f(1) + 2f(1.2) + 2f(1.4) + 2f(1.6) + 2f(1.8) + f(2)]$$

$$= 0.1\left(\frac{1}{1} + \frac{2}{1.2} + \frac{2}{1.4} + \frac{2}{1.6} + \frac{2}{1.8} + \frac{1}{2}\right) \approx 0.695635$$

This approximation is illustrated in Figure 3.

(b) The midpoints of the five subintervals are 1.1, 1.3, 1.5, 1.7, and 1.9, so the Midpoint Rule gives

$$\int_1^2 \frac{1}{x}\,dx \approx M_5 = \Delta x\,[f(1.1) + f(1.3) + f(1.5) + f(1.7) + f(1.9)]$$

$$= \frac{1}{5}\left(\frac{1}{1.1} + \frac{1}{1.3} + \frac{1}{1.5} + \frac{1}{1.7} + \frac{1}{1.9}\right) \approx 0.691908$$

This approximation is illustrated in Figure 4. ▪

In Example 1 we deliberately chose an integral whose value can be computed explicitly so that we can see how accurate the Trapezoidal and Midpoint Rules are. By the Fundamental Theorem of Calculus,

$$\int_1^2 \frac{1}{x}\,dx = \ln x\big]_1^2 = \ln 2 = 0.693147\ldots$$

The **error** in using an approximation is defined to be the amount that needs to be added to the approximation to make it exact. From the values in Example 1 we see that the errors in the Trapezoidal and Midpoint Rule approximations for $n = 5$ are

$$E_T \approx -0.002488 \quad\text{and}\quad E_M \approx 0.001239$$

In general, we have

$$E_T = \int_a^b f(x)\,dx - T_n \quad\text{and}\quad E_M = \int_a^b f(x)\,dx - M_n$$

The following tables show the results of calculations similar to those in Example 1, but for $n = 5$, 10, and 20 and for the left and right endpoint approximations as well as the Trapezoidal and Midpoint Rules.

n	L_n	R_n	T_n	M_n
5	0.745635	0.645635	0.695635	0.691908
10	0.718771	0.668771	0.693771	0.692835
20	0.705803	0.680803	0.693303	0.693069

Corresponding errors

n	E_L	E_R	E_T	E_M
5	−0.052488	0.047512	−0.002488	0.001239
10	−0.025624	0.024376	−0.000624	0.000312
20	−0.012656	0.012344	−0.000156	0.000078

We can make several observations from these tables:

1. In all of the methods we get more accurate approximations when we increase the value of n. (But very large values of n result in so many arithmetic operations that we have to beware of accumulated round-off error.)

It turns out that these observations are true in most cases.

2. The Trapezoidal and Midpoint Rules are much more accurate than the endpoint approximations.

3. The size of the error in the Midpoint Rule is about half the size of the error in the Trapezoidal Rule.

Figure 5 shows why we can usually expect the Midpoint Rule to be more accurate than the Trapezoidal Rule. The area of a typical rectangle in the Midpoint Rule is the same as the area of the trapezoid $ABCD$ whose upper side is tangent to the graph at P. The area of this trapezoid is closer to the area under the graph than is the area of the trapezoid $AQRD$ used in the Trapezoidal Rule. [The midpoint error (shaded red) is smaller than the trapezoidal error (shaded blue).]

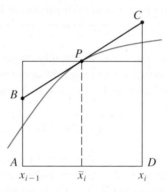 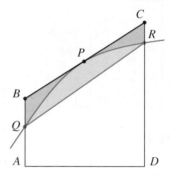

FIGURE 5

◼ Simpson's Rule

Another rule for approximate integration results from using parabolas instead of straight line segments to approximate a curve. As before, we divide $[a, b]$ into n subintervals of equal length $\Delta x = (b - a)/n$, but this time we assume that n is an *even* number. Then on each consecutive pair of intervals we approximate the curve $y = f(x) \geq 0$ by a parabola as shown in Figure 6. If $y_i = f(x_i)$, then $P_i(x_i, y_i)$ is the point on the curve lying above x_i. A typical parabola passes through three consecutive points P_i, P_{i+1}, and P_{i+2}.

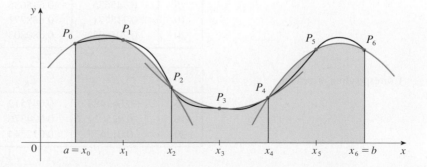

FIGURE 6

It can be shown that the area under the parabola through P_i, P_{i+1}, and P_{i+2} is

(3) $$\frac{\Delta x}{3}(y_i + 4y_{i+1} + y_{i+2})$$

(See Exercise 30.) Thus the area under the parabola through P_0, P_1, and P_2 from $x = x_0$ to $x = x_2$ in Figure 6 is

$$\frac{\Delta x}{3}(y_0 + 4y_1 + y_2)$$

and the area under the parabola through P_2, P_3, and P_4 from $x = x_2$ to $x = x_4$ is

$$\frac{\Delta x}{3}(y_2 + 4y_3 + y_4)$$

If we compute the areas under all the parabolas in this manner and add the results, we get

$$\int_a^b f(x)\, dx \approx \frac{\Delta x}{3}(y_0 + 4y_1 + y_2) + \frac{\Delta x}{3}(y_2 + 4y_3 + y_4)$$
$$+ \cdots + \frac{\Delta x}{3}(y_{n-2} + 4y_{n-1} + y_n)$$
$$= \frac{\Delta x}{3}(y_0 + 4y_1 + 2y_2 + 4y_3 + 2y_4 + \cdots + 2y_{n-2} + 4y_{n-1} + y_n)$$

Although we have derived this approximation for the case in which $f(x) \geqslant 0$, it is a reasonable approximation for any continuous function f and is called Simpson's Rule after the English mathematician Thomas Simpson (1710–1761). Note the pattern of coefficients: 1, 4, 2, 4, 2, 4, 2, ..., 4, 2, 4, 1.

Simpson

Thomas Simpson was a weaver who taught himself mathematics and went on to become one of the best English mathematicians of the 18th century. What we call Simpson's Rule was actually known to Cavalieri and Gregory in the 17th century, but Simpson popularized it in his best-selling calculus textbook, *A New Treatise of Fluxions*.

■ **Simpson's Rule**

$$\int_a^b f(x)\, dx \approx S_n = \frac{\Delta x}{3}[f(x_0) + 4f(x_1) + 2f(x_2) + 4f(x_3) + \cdots$$
$$+ 2f(x_{n-2}) + 4f(x_{n-1}) + f(x_n)]$$

where n is even and $\Delta x = (b - a)/n$.

■ **EXAMPLE 2**   **Using Simpson's Rule**

Use Simpson's Rule with $n = 10$ to approximate $\int_1^2 (1/x)\, dx$.

SOLUTION

Putting $f(x) = 1/x$, $n = 10$, and $\Delta x = 0.1$ in Simpson's Rule, we obtain

$$\int_1^2 \frac{1}{x}\, dx \approx S_{10}$$

$$= \frac{\Delta x}{3}[f(1) + 4f(1.1) + 2f(1.2) + 4f(1.3) + \cdots + 2f(1.8) + 4f(1.9) + f(2)]$$

$$= \frac{0.1}{3}\left(\frac{1}{1} + \frac{4}{1.1} + \frac{2}{1.2} + \frac{4}{1.3} + \frac{2}{1.4} + \frac{4}{1.5} + \frac{2}{1.6} + \frac{4}{1.7} + \frac{2}{1.8} + \frac{4}{1.9} + \frac{1}{2}\right)$$

$$\approx 0.693150 \qquad ■$$

Notice that, in Example 2, Simpson's Rule gives us a *much* better approximation ($S_{10} \approx 0.693150$) to the true value of the integral ($\ln 2 \approx 0.693147\ldots$) than does the Trapezoidal Rule ($T_{10} \approx 0.693771$) or the Midpoint Rule ($M_{10} \approx 0.692835$). It turns out (see Exercise 29) that the approximations in Simpson's Rule are weighted averages of those in the Trapezoidal and Midpoint Rules:

$$S_{2n} = \tfrac{1}{3}T_n + \tfrac{2}{3}M_n$$

In many applications of calculus we need to evaluate an integral even if no explicit formula is known for y as a function of x. A function may be given graphically or as a table of values of collected data. If there is evidence that the values are not changing rapidly, then the Trapezoidal Rule or Simpson's Rule can still be used to find an approximate value for $\int_a^b y \, dx$, the integral of y with respect to x.

▪ EXAMPLE 3

Approximating the Integral of a Function Defined Graphically

Figure 7 shows data traffic on the link from the United States to SWITCH, the Swiss academic and research network, on February 10, 1998. $D(t)$ is the data throughput, measured in megabits per second (Mb/s). Use Simpson's Rule to estimate the total amount of data transmitted on the link from midnight to noon on that day.

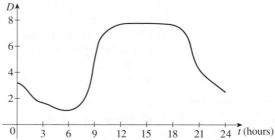

FIGURE 7

SOLUTION

Because we want the units to be consistent and $D(t)$ is measured in megabits per second, we convert the units for t from hours to seconds. If we let $A(t)$ be the amount of data (in megabits) transmitted by time t, where t is measured in seconds, then $A'(t) = D(t)$. So, by the Net Change Theorem (see Section 5.3), the total amount of data transmitted by noon (when $t = 12 \times 60^2 = 43{,}200$) is

$$A(43{,}200) = \int_0^{43{,}200} D(t) \, dt$$

We estimate the values of $D(t)$ at hourly intervals from the graph and compile them in the table.

t (hours)	t (seconds)	$D(t)$	t (hours)	t (seconds)	$D(t)$
0	0	3.2	7	25,200	1.3
1	3,600	2.7	8	28,800	2.8
2	7,200	1.9	9	32,400	5.7
3	10,800	1.7	10	36,000	7.1
4	14,400	1.3	11	39,600	7.7
5	18,000	1.0	12	43,200	7.9
6	21,600	1.1			

Then we use Simpson's Rule with $n = 12$ and $\Delta t = 3600$ to estimate the integral:

$$\int_0^{43,200} A(t)\, dt \approx \frac{\Delta t}{3}[D(0) + 4D(3600) + 2D(7200) + \cdots + 4D(39,600) + D(43,200)]$$

$$\approx \frac{3600}{3}[3.2 + 4(2.7) + 2(1.9) + 4(1.7) + 2(1.3) + 4(1.0)$$

$$+ 2(1.1) + 4(1.3) + 2(2.8) + 4(5.7) + 2(7.1) + 4(7.7) + 7.9]$$

$$= 143,880$$

Thus the total amount of data transmitted from midnight to noon is about 144,000 megabits, or 144 gigabits. ▪

Figure 8 illustrates the calculation using Simpson's Rule in Example 4. Notice that the parabolic arcs are so close to the graph of $y = e^{x^2}$ that they are practically indistinguishable from it.

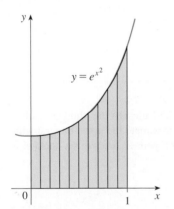

FIGURE 8

Many calculators and computer algebra systems have a built-in algorithm that computes an approximation of a definite integral. Some of these machines use Simpson's Rule; others use more sophisticated techniques such as *adaptive* numerical integration. This means that if a function fluctuates much more on a certain part of the interval than it does elsewhere, then that part gets divided into more subintervals. This strategy reduces the number of calculations required to achieve a prescribed accuracy.

■ **EXAMPLE 4** **Comparing Approximate Integration Techniques**

(a) Use (**i**) Simpson's Rule, (**ii**) the Trapezoidal Rule, amd (**iii**) the Midpoint Rule with $n = 10$ to approximate the integral $\int_0^1 e^{x^2}\, dx$.

(b) Use a graphing calculator or computer software to approximate the integral. How do the approximations from part (a) compare to this value?

SOLUTION

(a) If $n = 10$, then $\Delta x = (1 - 0)/10 = 0.1$.

(**i**) Simpson's Rule gives

$$\int_0^1 e^{x^2}\, dx \approx \frac{\Delta x}{3}[f(0) + 4f(0.1) + 2f(0.2) + \cdots + 2f(0.8) + 4f(0.9) + f(1)]$$

$$= \frac{0.1}{3}[e^0 + 4e^{0.01} + 2e^{0.04} + 4e^{0.09} + 2e^{0.16} + 4e^{0.25} + 2e^{0.36}$$

$$+ 4e^{0.49} + 2e^{0.64} + 4e^{0.81} + e^1]$$

$$\approx 1.462681$$

(**ii**) The Trapezoidal Rule gives

$$\int_0^1 e^{x^2}\, dx \approx \frac{\Delta x}{2}[f(0) + 2f(0.1) + 2f(0.2) + \cdots + 2f(0.9) + f(1)]$$

$$= \frac{0.1}{2}[e^0 + 2e^{0.01} + 2e^{0.04} + 2e^{0.09} + 2e^{0.16} + 2e^{0.25} + 2e^{0.36}$$

$$+ 2e^{0.49} + 2e^{0.64} + 2e^{0.81} + e^1]$$

$$\approx 1.467175$$

(**iii**) The Midpoint Rule gives

$$\int_0^1 e^{x^2}\, dx \approx \Delta x[f(0.05) + f(0.15) + f(0.25) + \cdots + f(0.95)]$$

$$= 0.1[e^{0.0025} + e^{0.0225} + e^{0.0625} + e^{0.1225} + e^{0.2025}$$

$$+ e^{0.3025} + e^{0.4225} + e^{0.5625} + e^{0.7225} + e^{0.9025}]$$

$$\approx 1.460393$$

(b) Using a graphing calculator, we get

$$\int_0^1 e^{x^2}\,dx \approx 1.462652$$

Assuming that this is the most accurate estimate, we see that the approximation using Simpson's Rule is closest (with a difference of 0.000029) followed by the approximation given by the Midpoint Rule (differing by 0.002259). The approximation from the Trapezoidal Rule is the least accurate (a difference of 0.004523). These results are to be expected in light of the observations we made in this section.   ■

■ Exercises C

1. Let $I = \int_0^4 f(x)\,dx$, where f is the function whose graph is shown.

  (a) Use the graph to find L_2, R_2, and M_2.

  (b) Are these underestimates or overestimates of I?

  (c) Use the graph to find T_2. How does it compare with I?

  (d) For any value of n, list the numbers L_n, R_n, M_n, T_n, and I in increasing order.

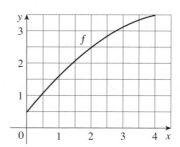

2. The left, right, Trapezoidal, and Midpoint Rule approximations were used to estimate $\int_0^2 f(x)\,dx$, where f is the function whose graph is shown. The estimates were 0.7811, 0.8675, 0.8632, and 0.9540, and the same number of subintervals were used in each case.

  (a) Which rule produced which estimate?

  (b) Between which two approximations does the true value of $\int_0^2 f(x)\,dx$ lie?

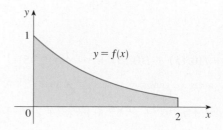

3–4 ■ Use **(a)** the Midpoint Rule and **(b)** Simpson's Rule to approximate the given integral with the specified value of n. (Round your answers to six decimal places.) Compare your results to the actual value to determine the error in each approximation.

3. $\displaystyle\int_0^2 \frac{x}{1+x^2}\,dx, \quad n = 10$

4. $\displaystyle\int_0^1 e^{-2x}\,dx, \quad n = 4$

5–12 ■ Use **(a)** the Trapezoidal Rule, **(b)** the Midpoint Rule, and **(c)** Simpson's Rule to approximate the given integral with the specified value of n. (Round your answers to six decimal places.)

5. $\displaystyle\int_1^2 \sqrt{x^3 - 1}\,dx, \quad n = 10$

6. $\displaystyle\int_0^2 \frac{1}{1+x^6}\,dx, \quad n = 8$

7. $\displaystyle\int_0^2 \frac{e^x}{1+x^2}\,dx, \quad n = 10$

8. $\displaystyle\int_0^3 \frac{dt}{1+t^2+t^4}, \quad n = 6$

9. $\displaystyle\int_1^4 \sqrt{\ln x}\,dx, \quad n = 6$

10. $\displaystyle\int_0^1 \sqrt{z}\,e^{-z}\,dz, \quad n = 10$

11. $\displaystyle\int_{-1}^1 e^{e^x}\,dx, \quad n = 10$

12. $\displaystyle\int_4^6 \ln(x^3 + 2)\,dx, \quad n = 10$

13–14 ▪
(a) Use (**i**) the Midpoint Rule, (**ii**) the Trapezoidal Rule, and (**iii**) Simpson's Rule to approximate the given integral with the specified value of n. (Round your answers to five decimal places.)

(b) Use a graphing calculator (or computer software) to approximate the integral, and compare the value to the results of part (a).

13. $\int_0^2 \sqrt{t^3 + 2}\, dt$, $\quad n = 8$ **14.** $\int_1^2 e^{1/x}\, dx$, $\quad n = 10$

---

15. Estimate the area under the graph in the figure by using (**a**) the Trapezoidal Rule, (**b**) the Midpoint Rule, and (**c**) Simpson's Rule, each with $n = 6$.

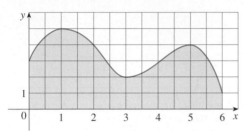

16. The widths (in meters) of a kidney-shaped swimming pool were measured at 2-meter intervals as indicated in the figure. Use Simpson's Rule to estimate the area of the pool.

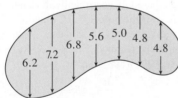

17. Use the Midpoint Rule and the given data to estimate the value of the integral $\int_1^5 f(x)\, dx$.

x	$f(x)$	x	$f(x)$
1.0	2.4	3.5	4.0
1.5	2.9	4.0	4.1
2.0	3.3	4.5	3.9
2.5	3.6	5.0	3.5
3.0	3.8		

18. A table of values of a function g is given. Use Simpson's Rule to estimate $\int_0^{1.6} g(x)\, dx$.

x	$g(x)$	x	$g(x)$
0.0	12.1	1.0	12.2
0.2	11.6	1.2	12.6
0.4	11.3	1.4	13.0
0.6	11.1	1.6	13.2
0.8	11.7		

19. A graph of the temperature in New York City on September 19, 2009 is shown. Use Simpson's Rule with $n = 12$ to estimate the average temperature on that day.

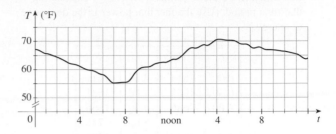

20. A radar gun was used to record the speed of a runner during the first 5 seconds of a race (see the table). Use Simpson's Rule to estimate the distance the runner covered during those 5 seconds.

t (s)	v (m/s)	t (s)	v (m/s)
0	0	3.0	10.51
0.5	4.67	3.5	10.67
1.0	7.34	4.0	10.76
1.5	8.86	4.5	10.81
2.0	9.73	5.0	10.81
2.5	10.22		

21. The graph of the acceleration $a(t)$ of a car measured in ft/s² is shown. Use Simpson's Rule to estimate the increase in the velocity of the car during the 6-second time interval.

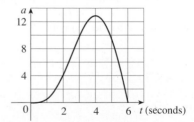

22. Water leaked from a tank at a rate of $r(t)$ liters per hour, where the graph of r is as shown. Use Simpson's Rule to estimate the total amount of water that leaked out during the first six hours.

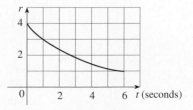

23. The table (supplied by San Diego Gas and Electric) gives the power consumption P in megawatts in San Diego County from midnight to 6:00 AM on a day in December. Use Simpson's Rule to estimate the energy used during that time period. (Use the fact that power is the derivative of energy.)

t	P	t	P
0:00	1814	3:30	1611
0:30	1735	4:00	1621
1:00	1686	4:30	1666
1:30	1646	5:00	1745
2:00	1637	5:30	1886
2:30	1609	6:00	2052
3:00	1604		

24. Shown is the graph of traffic on an Internet service provider's T1 data line from midnight to 8:00 AM. D is the data throughput, measured in megabits per second. Use Simpson's Rule to estimate the total amount of data transmitted during that time period.

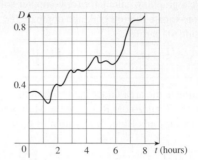

■ Challenge Yourself

25. Sketch the graph of a continuous function on $[0, 2]$ for which the Trapezoidal Rule with $n = 2$ is more accurate than the Midpoint Rule.

26. Sketch the graph of a continuous function on $[0, 2]$ for which the right endpoint approximation with $n = 2$ is more accurate than Simpson's Rule.

27. If f is a positive function and $f''(x) < 0$ for $a \leqslant x \leqslant b$, show that

$$T_n < \int_a^b f(x)\, dx < M_n$$

28. (a) Use Simpson's Rule with $n = 6$ to approximate $\int_0^{12} (x^3 + 5x)\, dx$ and compare to the actual value of the integral. What do you notice?

(b) Show that if f is a polynomial of degree 3 or lower, then Simpson's Rule gives the exact value of $\int_a^b f(x)\, dx$.

29. Show that $\frac{1}{3}T_n + \frac{2}{3}M_n = S_{2n}$.

30. Here we verify the area formula given by (3). Because shifting a parabola horizontally does not change the area under it, shift the points P_i, P_{i+1}, and P_{i+2} so that $x_{i+1} = 0$. Let $h = \Delta x$: then $x_i = -h$ and $x_{i+2} = h$. Show that the area under the parabola $y = Ax^2 + Bx + C$ through these points on $[-h, h]$ is

$$\frac{h}{3}(2Ah^2 + 6C)$$

Then show that $y_i + 4y_{i+1} + y_{i+2} = 2Ah^2 + 6C$.

D ▍ Double Integrals

In Chapter 5 we defined the definite integral of a continuous function f of one variable on an interval $[a, b]$ as

(1) $$\int_a^b f(x)\, dx = \lim_{n \to \infty} [f(x_1)\, \Delta x + f(x_2)\, \Delta x + \cdots + f(x_n)\, \Delta x]$$

where $\Delta x = (b - a)/n$ and $x_1, x_2, \ldots, x_n$ are the endpoints of the subintervals of $[a, b]$ with width Δx. We saw that if $f(x)$ is a positive function, then $\int_a^b f(x)\, dx$ can be interpreted as an area and we used $\int_a^b f(x)\, dx$ to compute average values.

In a similar way we will show here how to define the double integral of a function of two variables $f(x, y)$ on a rectangle. We will see how to interpret it as a volume if $f(x, y)$ is a positive function and how to use it to calculate average values.

■ Double Integrals over Rectangles

We start with a function $f(x, y)$ whose domain is a rectangle

$$R = \{(x, y) \mid a \leq x \leq b, c \leq y \leq d\}$$

If we divide the interval $[a, b]$ into m subintervals of equal width $\Delta x = (b - a)/m$ and we divide the interval $[c, d]$ into n subintervals of equal width $\Delta y = (d - c)/n$, then, as shown in Figure 1, R is divided into mn subrectangles each with area $\Delta A = \Delta x \, \Delta y$. The upper right corner of a typical subrectangle has coordinates (x_i, y_j).

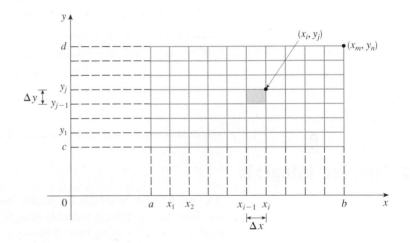

FIGURE 1
Dividing R into subrectangles

By analogy with Equation 1 we define the **double integral** of f over the rectangle R as a limit of double Riemann sums:

(2)
$$\iint\limits_R f(x, y) \, dA$$
$$= \lim_{m,n \to \infty} [f(x_1, y_1) \, \Delta A + f(x_1, y_2) \, \Delta A + \cdots + f(x_m, y_n) \, \Delta A]$$

There are a total of mn terms in the Riemann sum in Definition 2, one for each of the mn subrectangles in Figure 1. If the limit exists, f is called *integrable*.

For a positive function we can interpret $f(x_i, y_j)$ as the height of a thin rectangular column with base area ΔA and volume $f(x_i, y_j) \, \Delta A$. (See Figure 2 on page A34.) So the Riemann sum in Definition 2 can be interpreted as the sum of volumes of columns (see Figure 3) and this sum is an approximation to the volume of the solid that lies under the graph of the surface $z = f(x, y)$ and above the rectangle R.

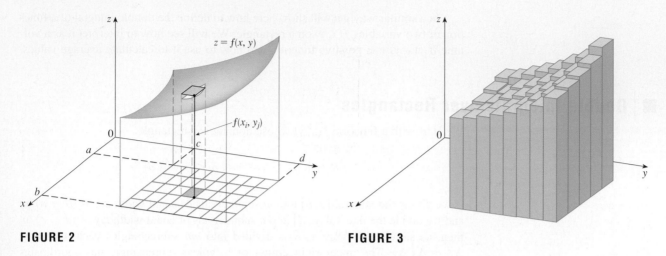

FIGURE 2   **FIGURE 3**

As m and n become large in Definition 2 and Figures 2 and 3, the approximation becomes closer and closer to the actual volume, so we define the *volume* of the solid to be the value of the double integral.

> ▪ If $f(x, y) \geqslant 0$, the **volume** of the solid that lies under the surface $z = f(x, y)$ and above the rectangle R is
>
> $$V = \iint\limits_{R} f(x, y) \, dA$$

▪ | Iterated Integrals

It's very difficult to evaluate a double integral using Definition 2 directly, so now we show how to express a double integral as an iterated integral, which can then be evaluated by calculating two single integrals.

Suppose that f is a function of two variables that is integrable on the rectangle $R = [a, b] \times [c, d]$. We use the notation $\int_c^d f(x, y) \, dy$ to mean that x is held fixed (and treated as a constant) and $f(x, y)$ is integrated with respect to y from $y = c$ to $y = d$. This procedure is called *partial integration with respect to y*. (Notice its similarity to partial differentiation.) Now $\int_c^d f(x, y) \, dy$ is a number that depends on the value of x, so it defines a function of x:

$$A(x) = \int_c^d f(x, y) \, dy$$

If we now integrate the function A with respect to x from $x = a$ to $x = b$, we get

(3) $$\int_a^b A(x) \, dx = \int_a^b \left[\int_c^d f(x, y) \, dy \right] dx$$

The integral on the right side of Equation 3 is called an **iterated integral**. Usually the brackets are omitted. Thus

(4) $$\int_a^b \int_c^d f(x, y) \, dy \, dx = \int_a^b \left[\int_c^d f(x, y) \, dy \right] dx$$

means that we first integrate with respect to y from c to d and then with respect to x from a to b.

Similarly, the iterated integral

(5)
$$\int_c^d \int_a^b f(x, y)\, dx\, dy = \int_c^d \left[\int_a^b f(x, y)\, dx \right] dy$$

means that we first integrate with respect to x (holding y constant) from $x = a$ to $x = b$ and then we integrate the resulting function of y with respect to y from $y = c$ to $y = d$. Notice that in both Equations 4 and 5 we work *from the inside out*.

■ **EXAMPLE 1**

Evaluate the iterated integrals.

(a) $\displaystyle\int_0^3 \int_1^2 x^2 y\, dy\, dx$ **(b)** $\displaystyle\int_1^2 \int_0^3 x^2 y\, dx\, dy$

SOLUTION

(a) Working from the inside out, we first evaluate $\int_1^2 x^2 y\, dy$. Regarding x as a constant, we obtain

$$\int_1^2 x^2 y\, dy = \left[x^2 \frac{y^2}{2} \right]_{y=1}^{y=2} = x^2 \left(\frac{2^2}{2} \right) - x^2 \left(\frac{1^2}{2} \right) = \tfrac{3}{2} x^2$$

Thus the function A in the preceding discussion is given by $A(x) = \tfrac{3}{2} x^2$ in this example. We now integrate this function of x from 0 to 3:

$$\int_0^3 \int_1^2 x^2 y\, dy\, dx = \int_0^3 \left[\int_1^2 x^2 y\, dy \right] dx$$

$$= \int_0^3 \tfrac{3}{2} x^2\, dx = \frac{x^3}{2} \Bigg]_0^3 = \frac{27}{2}$$

(b) Here we first integrate with respect to x:

$$\int_1^2 \int_0^3 x^2 y\, dx\, dy = \int_1^2 \left[\int_0^3 x^2 y\, dx \right] dy = \int_1^2 \left[\frac{x^3}{3} y \right]_{x=0}^{x=3} dy$$

$$= \int_1^2 9y\, dy = 9 \frac{y^2}{2} \Bigg]_1^2 = \frac{27}{2} \qquad ■$$

Notice that in Example 1 we obtained the same answer whether we integrated with respect to y or x first. In general, it turns out (see Theorem 6) that the two iterated integrals in Equations 4 and 5 are always equal; that is, the order of integration does not matter. (This is similar to the equality of mixed partial derivatives.)

The practical method for evaluating a *double* integral is to express it as an *iterated* integral (in either order). The following theorem is true for most functions that one meets in practice. It is proved in courses on advanced calculus.

(6) ▪ If $R = \{(x, y) \mid a \leqslant x \leqslant b, c \leqslant y \leqslant d\}$, then

$$\iint\limits_R f(x, y)\, dA = \int_a^b \int_c^d f(x, y)\, dy\, dx = \int_c^d \int_a^b f(x, y)\, dx\, dy$$

▪ EXAMPLE 2

Find the volume of the solid S that is enclosed by the paraboloid $x^2 + 2y^2 + z = 16$, the planes $x = 2$ and $y = 2$, and the three coordinate planes.

SOLUTION

We first observe that S is the solid that lies under the surface $z = 16 - x^2 - 2y^2$ and above the square $R = \{(x, y) \mid 0 \leqslant x \leqslant 2, 0 \leqslant y \leqslant 2\}$. (See Figure 4.) Using Theorem 6 we express the double integral for the volume as an iterated integral:

$$V = \iint\limits_R (16 - x^2 - 2y^2)\, dA$$

$$= \int_0^2 \int_0^2 (16 - x^2 - 2y^2)\, dx\, dy$$

$$= \int_0^2 \left[16x - \tfrac{1}{3}x^3 - 2y^2 x\right]_{x=0}^{x=2} dy$$

$$= \int_0^2 \left(\tfrac{88}{3} - 4y^2\right) dy = \left[\tfrac{88}{3}y - \tfrac{4}{3}y^3\right]_0^2 = 48 \qquad ▪$$

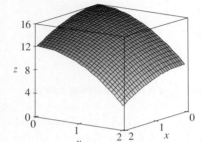

FIGURE 4

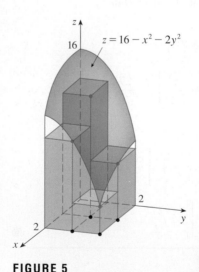

FIGURE 5

NOTE: Figure 5 illustrates the definitions of volume and the double integral by showing how the solid in Example 2 is approximated by the four columns in the Riemann sum with $m = n = 2$. Figure 6 shows how the columns become better approximations to the volume as m and n increase.

FIGURE 6

The Riemann sum approximations to the volume under $z = 16 - x^2 - 2y^2$ become more accurate as m and n increase.

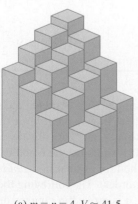

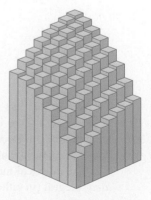

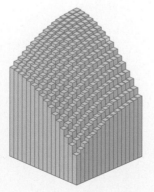

(a) $m = n = 4$, $V \approx 41.5$ (b) $m = n = 8$, $V \approx 44.875$ (c) $m = n = 16$, $V \approx 46.46875$

▪ Double Integrals over More General Regions

What happens if we need to integrate a function $f(x, y)$ over a region D that is not a rectangle? Suppose, for instance, that the domain D of f lies between the graphs of two continuous functions of x:

$$D = \left\{ (x, y) \mid a \leqslant x \leqslant b, \ g_1(x) \leqslant y \leqslant g_2(x) \right\}$$

Figure 7 shows three examples of such regions.

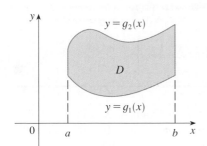
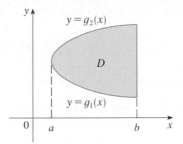
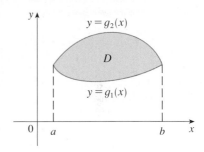

FIGURE 7

The double integral of f over D, $\iint_D f(x, y) \, dA$, can be defined by a limit similar to the one in Definition 2 and it can be evaluated as an iterated integral similar to the one in Theorem 6:

(7)
$$\iint_D f(x, y) \, dA = \int_a^b \int_{g_1(x)}^{g_2(x)} f(x, y) \, dy \, dx$$

Notice that the lower and upper limits of integration in the inner integral in Equation 7 are functions of x: $y = g_1(x)$ and $y = g_2(x)$. This makes sense because for a fixed value of x between a and b, y goes from the lower boundary curve $y = g_1(x)$ to the upper boundary curve $y = g_2(x)$. But in evaluating the inner integral we regard x as being constant not only in $f(x, y)$ but also in the limits of integration, $g_1(x)$ and $g_2(x)$. Notice also that in the special case where $g_1(x) = c$ and $g_2(x) = d$, D is a rectangle and Equation 7 is the same as the first part of Theorem 6.

▪ EXAMPLE 3

Evaluate $\iint_D (x + 2y) \, dA$, where D is the region bounded by the parabolas $y = 2x^2$ and $y = 1 + x^2$.

SOLUTION

The parabolas intersect when $2x^2 = 1 + x^2$, that is, $x^2 = 1$, so $x = \pm 1$. The region D is sketched in Figure 8 and we can write

$$D = \left\{ (x, y) \mid -1 \leqslant x \leqslant 1, \ 2x^2 \leqslant y \leqslant 1 + x^2 \right\}$$

Since the lower boundary is $y = 2x^2$ and the upper boundary is $y = 1 + x^2$,

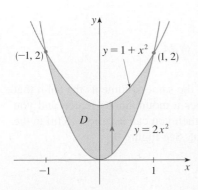

FIGURE 8

Equation 7 gives

$$\iint\limits_{D} (x + 2y)\, dA = \int_{-1}^{1} \int_{2x^2}^{1+x^2} (x + 2y)\, dy\, dx$$

$$= \int_{-1}^{1} \left[xy + y^2 \right]_{y=2x^2}^{y=1+x^2} dx$$

$$= \int_{-1}^{1} [x(1 + x^2) + (1 + x^2)^2 - x(2x^2) - (2x^2)^2]\, dx$$

$$= \int_{-1}^{1} (-3x^4 - x^3 + 2x^2 + x + 1)\, dx$$

$$= -3\,\frac{x^5}{5} - \frac{x^4}{4} + 2\,\frac{x^3}{3} + \frac{x^2}{2} + x \Bigg]_{-1}^{1} = \frac{32}{15}$$ ■

NOTE: When we set up a double integral as in Example 3, it is essential to draw a diagram. Often it is helpful to draw a vertical arrow as in Figure 8. Then the limits of integration for the *inner* integral can be read from the diagram as follows: The arrow starts at the lower boundary $y = g_1(x)$, which gives the lower limit in the integral, and the arrow ends at the upper boundary $y = g_2(x)$, which gives the upper limit of integration.

▪ Average Value

Recall from Section 5.3 that the average value of a function f of one variable defined on an interval $[a, b]$ is

$$f_{\text{ave}} = \frac{1}{b - a} \int_{a}^{b} f(x)\, dx$$

In a similar fashion we define the **average value** of a function f of two variables defined on a rectangle R to be

$$f_{\text{ave}} = \frac{1}{A(R)} \iint\limits_{R} f(x, y)\, dA$$

where $A(R)$ is the area of R.

If $f(x, y) \geq 0$, the equation

$$A(R) \times f_{\text{ave}} = \iint\limits_{R} f(x, y)\, dA$$

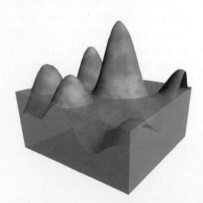

says that the box with base R and height f_{ave} has the same volume as the solid that lies under the graph of f. [If $z = f(x, y)$ describes a mountainous region and you chop off the tops of the mountains at height f_{ave}, then you can use them to fill in the valleys so that the region becomes completely flat. See Figure 9.]

FIGURE 9

▪ **EXAMPLE 4**

A manufacturer has modeled its output by a Cobb-Douglas production function

$$P(L, K) = 70L^{0.6}K^{0.4}$$

where L is the number of monthly labor hours and K is the monthly capital investment (in units of $1000). If L varies roughly evenly from 5000 to 6000 and monthly capital investment varies evenly between \$20,000 and \$30,000, find the average monthly output.

SOLUTION

We compute the average value of the function $P(L, K)$ over the rectangular region R defined by

$$5000 \leqslant L \leqslant 6000 \qquad 20 \leqslant K \leqslant 30$$

The area of R is

$$A(R) = (6000 - 5000)(30 - 20) = 10,000$$

So the average value is

$$P_{ave} = \frac{1}{A(R)} \iint\limits_R P(L, K) \, dA$$

$$= \frac{1}{10,000} \int_{5000}^{6000} \int_{20}^{30} P(L, K) \, dK \, dL$$

$$= \frac{1}{10,000} \int_{5000}^{6000} \int_{20}^{30} 70L^{0.6}K^{0.4} \, dK \, dL$$

$$= \frac{1}{10,000} \int_{5000}^{6000} \left[70L^{0.6} \frac{K^{1.4}}{1.4} \right]_{K=20}^{K=30} dL$$

$$= \frac{1}{10,000} \int_{5000}^{6000} 50L^{0.6}(30^{1.4} - 20^{1.4}) \, dL$$

$$= \frac{(30^{1.4} - 20^{1.4})}{200} \left[\frac{L^{1.6}}{1.6} \right]_{5000}^{6000}$$

$$= \frac{(30^{1.4} - 20^{1.4})(6000^{1.6} - 5000^{1.6})}{320} \approx 44,427.0$$

The average monthly output is about 44,427 units. ▪

▪ Exercises D

1–2 ▪ Find $\int_0^5 f(x, y) \, dx$ and $\int_0^1 f(x, y) \, dy$.

1. $f(x, y) = 12x^2y^3$

2. $f(x, y) = y + xe^y$

3–12 ▪ Calculate the iterated integral.

3. $\int_1^3 \int_0^1 (1 + 4xy) \, dx \, dy$

4. $\int_0^1 \int_1^2 (4x^3 - 9x^2y^2) \, dy \, dx$

5. $\int_0^2 \int_0^1 (2x + y)^8 \, dx \, dy$

6. $\int_0^1 \int_1^2 \frac{xe^x}{y} \, dy \, dx$

7. $\int_1^4 \int_1^2 \left(\frac{x}{y} + \frac{y}{x} \right) dy \, dx$

8. $\int_0^1 \int_0^1 \sqrt{s + t} \, ds \, dt$

9. $\int_0^4 \int_0^{\sqrt{y}} xy^2 \, dx \, dy$

10. $\int_0^1 \int_{2x}^2 (x - y) \, dy \, dx$

11. $\int_0^1 \int_{x^2}^x (1 + 2y) \, dy \, dx$

12. $\int_0^2 \int_y^{2y} xy \, dx \, dy$

13–20 ▪ Calculate the double integral.

13. $\iint\limits_R (6x^2y^3 - 5y^4) \, dA$,

$\quad R = \{(x, y) \mid 0 \leqslant x \leqslant 3, 0 \leqslant y \leqslant 1\}$

14. $\iint\limits_{R} (y + xy^{-2})\, dA,$

$R = \{(x, y) \mid 0 \leqslant x \leqslant 2, 1 \leqslant y \leqslant 2\}$

15. $\iint\limits_{R} xye^{x^2 y}\, dA, \quad R = \{(x, y) \mid 0 \leqslant x \leqslant 1, 0 \leqslant y \leqslant 2\}$

16. $\iint\limits_{R} \dfrac{x}{1 + xy}\, dA, \quad R = \{(x, y) \mid 0 \leqslant x \leqslant 1, 0 \leqslant y \leqslant 1\}$

17. $\iint\limits_{R} xy^2\, dA, \quad D$ is the triangular region with vertices $(0, 0)$, $(1, 0)$, and $(1, 1)$

18. $\iint\limits_{D} \dfrac{y}{x^5 + 1}\, dA, \quad D = \{(x, y) \mid 0 \leqslant x \leqslant 1, 0 \leqslant y \leqslant x^2\}$

19. $\iint\limits_{D} x^3\, dA, \quad D = \{(x, y) \mid 1 \leqslant x \leqslant e, 0 \leqslant y \leqslant \ln x\}$

20. $\iint\limits_{D} (x + y)\, dA, \quad D$ is bounded by $y = \sqrt{x}$ and $y = x^2$

---

21. Find the volume of the solid that lies under the plane $3x + 2y + z = 12$ and above the rectangle $R = \{(x, y) \mid 0 \leqslant x \leqslant 1, -2 \leqslant y \leqslant 3\}$.

22. Find the volume of the solid that lies under the surface $z = 4 + x^2 - y^2$ and above the square $R = \{(x, y) \mid -1 \leqslant x \leqslant 1, 0 \leqslant y \leqslant 2\}$.

23. Find the volume of the solid that lies under the surface $z = xy$ and above the triangle with vertices $(1, 1)$, $(4, 1)$, and $(1, 2)$.

24. Find the volume of the solid that is enclosed by the coordinate planes and the plane $3x + 2y + z = 6$.

25–27 ▪ Find the average value of f over the given region.

25. $f(x, y) = x^2 y, \quad R$ is the rectangle with vertices $(-1, 0)$, $(-1, 5)$, $(1, 5)$, $(1, 0)$

26. $f(x, y) = e^y \sqrt{x + e^y}$,
$R = \{(x, y) \mid 0 \leqslant x \leqslant 4, 0 \leqslant y \leqslant 1\}$

27. $f(x, y) = xy, \quad D$ is the triangle with vertices $(0, 0)$, $(1, 0)$, and $(1, 3)$

---

28. A company models its monthly production by the function

$$P(x, y) = 200x^{3/4} y^{1/4}$$

where x is the number of workers and y is the monthly operating budget in thousand of dollars. The company uses between 50 and 60 workers and its operating budget varies from \$40,000 to \$50,000 per month. Estimate the average monthly output.

29. The state of Colorado is in the shape of a rectangle that measures 388 miles west to east and 276 miles south to north. Suppose the function

$$f(x, y) = 4.6 + 0.02x - 0.01y + 0.0001xy$$

approximates the snowfall, in inches, left during a storm at a location x miles east and y miles north of the southwest corner of the state. According to the model, what was the average snowfall for the entire state during the storm?

30. Researchers assessed the level of airborne pollution, in parts per million (ppm), created by a manufacturing facility throughout a nearby rectangular plot of farmland. The boundaries of the land run 4 miles from east to west and 2 miles from south to north. The level of pollution, measured in parts per million (ppm), at a location x miles west and y miles north of the southeast corner of the plot was modeled by $f(x, y) = 2.7e^{-0.1x - 0.4y}$. Use the model to find the average pollution level over the entire plot of land.

E Answers to Odd-Numbered Exercises

Chapter 1

▪ EXERCISES 1.1 ▪ page 13

1. (a) \$4.00; The price of a 10-lb bag of potting soil is \$4.00.
(b) Domain: $\{4, 10, 50\}$; range: $\{1.60, 4.00, 20.00\}$
3. The population of the city on January 1, 2008, was 64,300;
the population of the city on July 1, 2004.
5. The car averages 24.7 miles per gallon when it is driven
at 65 mi/h.
7. (a) -2 **(b)** 2.8   **(c)** $-3, 1$ **(d)** $-2.5, 0.3$
(e) $[-3, 3], [-2, 3]$
9. $[-85, 115]$ **11.** Diet, exercise, or illness
13.

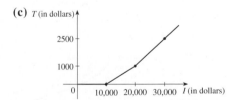

15.

17.

19.

21. (a)

(b) 126 million; 207 million

23. $12, 16, 3a^2 - a + 2, 3a^2 + a + 2, 3a^2 + 5a + 4,$
$6a^2 - 2a + 4, 12a^2 - 2a + 2, 3a^4 - a^2 + 2,$
$9a^4 - 6a^3 + 13a^2 - 4a + 4, 3a^2 + 6ah + 3h^2 - a - h + 2$
25. $8 + h$ **27.** $-3 - h$ **29.** $-1/(ax)$
31. $\left\{ x \mid x \neq \frac{1}{3} \right\} = \left(-\infty, \frac{1}{3} \right) \cup \left(\frac{1}{3}, \infty \right)$
33. $\{ t \mid t \geq -3 \} = [-3, \infty)$
35. No; the scatter plot fails the Vertical Line Test (at $x = 5$
and $x = 12$)
37. No   **39.** Yes, $[-3, 2], [-3, -2] \cup [-1, 3]$
41. $-1, 1, -1$ **43.** $-2, 0, 4$

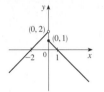

45. $A(L) = 10L - L^2, 0 < L < 10$
47. $S(x) = x^2 + (8/x), x > 0$
49. $V(x) = 4x^3 - 64x^2 + 240x, 0 < x < 6$
51. (a)   **(b)** \$400, \$1900

(c)

53. f is odd, g is even   **55. (a)** $(-5, 3)$ **(b)** $(-5, -3)$
57. Odd   **59.** Neither   **61.** Even
63. Even; odd; neither (unless $f = 0$ or $g = 0$)

▪ EXERCISES 1.2 ▪ page 25

1. The total number of students that attended a math class on
day t of this year
3. The value, in dollars, of gold in the bank's vault at the end
of the nth day of this year
5. The average number of bushels of corn yielded per acre
during year x
7. (a) $f(t) = 58.4 + 2.4t$ **(b)** $f(4) = 68$; the employee
earned a total of \$68,000 in 2004
9. (a) $A(x) = x^2 - 2x + 12$ **(b)** $B(x) = x^2 - 8x - 12$
(c) $C(x) = 3x^3 - 3x^2 - 60x$
(d) $D(x) = (x^2 - 5x)/(3x + 12)$

11. 101, 38   **13.** 20, 25

15. $p(x) = 4x^2 + 4x$, $q(x) = 2x^2 - 1$

17. $p(x) = (1 - \sqrt{x})^3 + 2 - 2\sqrt{x}$, $q(x) = 1 - \sqrt{x^3 + 2x}$

19. $p(x) = x + 2 + \dfrac{1}{x + 2}$, $q(x) = x + \dfrac{1}{x} + 2$

21. The profit, in thousands of dollars, the manufacturer earns during year t

23. $f(g(t))$ gives the average percentage of commuters who carpooled t months after January 1, 2011.

25. (a) $A(m) = 14.7 + 0.2165m + 1.299\sqrt{m}$ is the pressure in PSI that Paul experiences m minutes after he starts his dive.

(b) $A(25) = 26.6075$; Paul experiences a pressure of about 26.6 PSI, 25 minutes after the start of his dive.

27. (a) 4   **(b)** 3   **(c)** 0   **(d)** −2

29. $g(x) = x^2 + 1$, $f(x) = x^{10}$

31. $g(x) = 2x^2 + 5$, $f(x) = \sqrt{x}$

33. (a) $y = f(x) + 4$ **(b)** $y = f(x) - 4$
(c) $y = f(x - 4)$ **(d)** $y = f(x + 4)$ **(e)** $y = -f(x)$
(f) $y = f(-x)$ **(g)** $y = 3f(x)$ **(h)** $y = \frac{1}{3}f(x)$

35. (a) 3   **(b)** 1   **(c)** 4   **(d)** 5   **(e)** 2

37. (a)   **(b)**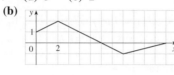

(c) 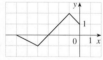 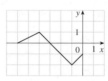   **(d)**

39. 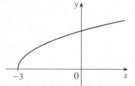   **41.**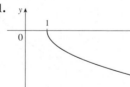

43. 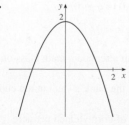   **45.**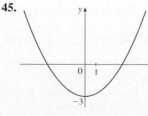

47. (a) $y = -\sqrt{x} + 4$ **(b)** $y = \sqrt{x - 4} - 1$

49. (a) The second reservoir's depth is always 15 feet lower than that of the first reservoir.

(b) The second reservoir's depth always matches the depth of the first reservoir from two months prior.

(c) The first reservoir's depth always matches the depth of the second reservoir from two months prior.

(d) The second reservoir's depth is always 0.8 times, or 80% of, that of the first reservoir.

51. (a) The number of songs sold by the rival service is always 1.3 times the number (30% more) sold by the original service.

(b) The rival service sells 23,000 more songs each month than the original service.

(c) The rival service sells 5000 more songs each month than the original service sold the previous month.

53. (a) $s = \sqrt{d^2 + 36}$ **(b)** $d = 30t$

(c) $f(g(t)) = \sqrt{900t^2 + 36}$; the distance between the lighthouse and the ship as a function of the time elapsed since noon

55. (a) $r(t) = 60t$ **(b)** $A(r(t)) = 3600\pi t^2$; the area of the circle as a function of time

57. $p(x) = \sqrt{x^2 + 6x + 10}$ **59.** $y = -f\left(\frac{1}{2}x\right) + 3$

61. $y = -f(x) - 1$ **63.** $y = f(2x) - 1$

65. $y = -\sqrt{-x^2 - 5x - 4} - 1$ **67.** Yes; $m_1 m_2$

■ **EXERCISES 1.3** ▪ page 37

1. $\frac{3}{2}$ **3.** −20   **5.** $y = 3x - 2$ **7.** $y = 6x - 15$

9. $y = -5x + 11$ **11.** $y = -44x + 260$

13. $y = 3x - 3$

15.

17. $m = -\frac{2}{5}$, $b = 3$ **19.** $m = \frac{5}{6}$, $b = 7$

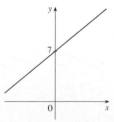 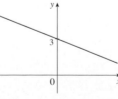

21. $m = -2$, x-intercept 7, y-intercept 14

23. $m = 0.2$, t-intercept 20, y-intercept −4

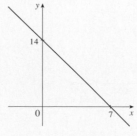 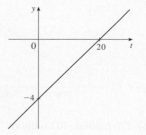

25. $h(x) = 56x - 63$

27. 0.76; The program gains 0.76 million additional viewers each week.

29. -3300; The value of the machine will decrease by \$3300 each year.

31. 0.26; The company pays \$260 per thousand dollars of profit.

33. (a) 8.34, change in mg for every 1 year change
(b) 8.34 mg

35. (a)

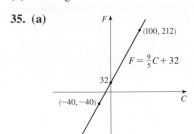

(b) $\frac{9}{5}$, change in °F for every 1°C change; 32, Fahrenheit temperature corresponding to 0°C

37. (a) $T = \frac{1}{6}N + \frac{307}{6}$ (b) $\frac{1}{6}$, change in °F for every chirp per minute change   (c) 76°F

39. (a) $P = 0.434d + 15$ (b) ≈ 196 ft

41. (a)

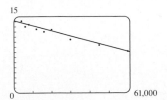

Linear model is appropriate.

(b) $y = -0.0001x + 14.2$

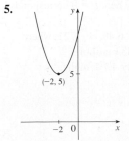

(c) 5.2 per 100 population; extrapolation   (d) No

43. (a)

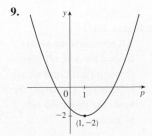

$y = -0.081875x + 15,882.5$ (using first and last data points)
(b) \$14,900   (c) $\approx 193,985$; no

45. (a) $y \approx -0.00009979x + 13.951$
(b) About 11.5 per 100 population   (c) About 6%

47. (a)

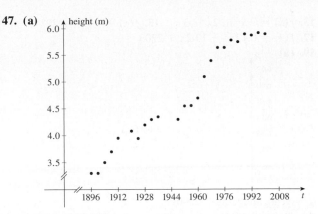

Linear model is appropriate.
(b) $y \approx 0.0265x - 46.8759$ (c) 6.27 m; higher   (d) No

49. (a) A slope of 3   (b) A y-intercept of 3

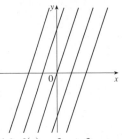
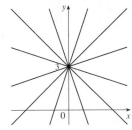

(c) $f(x) = 3x + 3$

51. g

53. $H(t) = \begin{cases} -0.05t + 13.4, & t < 5 \\ 0.667t + 9.97, & t \geqslant 5 \end{cases}$

where $t = 0$ corresponds to 1995

■ **EXERCISES 1.4** ▪ **page 49**

1. (a) Power, polynomial (degree 4)   (b) Power
(c) Polynomial (degree 7)   (d) Rational

3. (a) h (b) f (c) g

5.

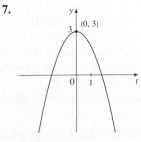

7.

9.

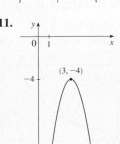

11.

13. $f(x) = -x^2 + 2x - 3$ **15.** $f(x) = -\frac{1}{4}x^2 + 22$
17. $f(x) = 0.12x^2 - 13.2x + 2203$
19. (a)

P: 642, 640, 638, 636, 634, 632; axis 0, 1, 2, 3, 4 t

(t = years after July 1, 2000)
(b) $P(t) = 0.856(t - 3)^2 + 633.4$ **(c)** 688,200
21. (a)

P: 13, 12, 11, 10, 9; axis 0, 2, 4, 6, 8 t

($t = 0$ corresponds to 1995)
(b) $P(t) = 0.2125(t - 4)^2 + 9.1$ **(c)** 11.0%
23. $a(x) = \dfrac{131{,}000}{x} + 0.41$; $2.43 per magazine
25. (a) $R(p) = 142p - 91.4p\sqrt{p}$ thousand dollars
(b) $1.07; 47,455
27. Increasing: $(-5, 8)$; decreasing: $(-\infty, -5)$, $(8, \infty)$
29. Increasing: $(-\infty, -2)$, $(6, 9)$; decreasing: $(-2, 6)$, $(9, \infty)$
31. Increasing: $(-\infty, 0)$, $(1.065, \infty)$; decreasing: $(0, 1.065)$
33. $A = 3.2t$; 115.2 **35.** 225 **37.** $\frac{8}{15}$
39. (a) $D = 0.1875t$ **(b)** 2.625 miles
41. 0.7 dB **43.** The frequency is halved.
45. $H(t) = 15.93t^2 - 5.79t + 1577$, where $t = 0$ corresponds to the year 2000; 2,550,000
47. $P(t) = -0.000285t^3 + 0.522t^2 - 6.40t + 1721$, where $t = 0$ corresponds to the year 1900; 1883 million
49. Production would be approximately 4.7% higher.
51. Four times brighter
53. (a) $N = 3.105x^{0.3080}$ **(b)** 18

■ **EXERCISES 1.5** ▪ page 60

1. (a) $f(x) = a^x, a > 0$ **(b)** $\mathbb{R}$ **(c)** $(0, \infty)$
(d) **(i)** See Figure 3(b). **(ii)** See Figure 3(a).
3.

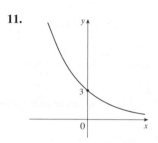

5 $y = 20^x$ $y = 5^x$ $y = e^x$
$y = 2^x$
-1 ⎯ 0 ⎯ 2

All approach the x-axis toward the left, all pass through $(0, 1)$, and all are increasing. The larger the base, the faster the rate of increase.

5.

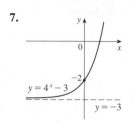

$y = \left(\frac{1}{3}\right)^x$ $y = \left(\frac{1}{10}\right)^x$ 5 $y = 10^x$ $y = 3^x$
-2 0 2

The functions with base greater than 1 are increasing and those with base less than 1 are decreasing. The latter are reflections of the former about the y-axis.

7.

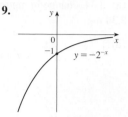

$y = 4^x - 3$
$y = -3$

9.

$y = -2^{-x}$
-1

11.

3

13. (a) $y = e^x - 2$ **(b)** $y = e^{x-2}$ **(c)** $y = -e^x$
(d) $y = e^{-x}$ **(e)** $y = -e^{-x}$

15. x^8 **17.** u^8 **19.** $\dfrac{p^9}{8}$ **21.** $\sqrt[3]{16} = 2\sqrt[3]{2}$

23. $\sqrt[4]{e}$ **31.** Exponential; $f(x) = 5 \cdot 2^x$ **33.** Neither
35. (a) 3200 **(b)** $100 \cdot 2^{t/3}$ **(c)** 10,159
(d) 60,000 $t \approx 26.9$ h

37. $f(x) = 3 \cdot 2^x$ **43.** At $x \approx 35.8$
45. $P = 2614.086(1.01693)^t$, where $t = 0$ corresponds to 1950; 5381 million; 8466 million
47. (a) $f(t) = 0.06698(1.3516)^t$ **(b)** About 2.3 years
(c) The model predicts 92.5 million transistors.
49. (a) 10,853 **(b)** 23,700
(c) 25

7.8 years after January 1, 2000 (October 2007)

40

51. (a) $y = -2^x + 6$ **(b)** $y = 2^{-(x+8)}$

■ EXERCISES 1.6 ▪ page 69

1. (a) It's defined as the inverse of the exponential function with base a, that is, $\log_a x = y \iff a^y = x$.
(b) $(0, \infty)$ **(c)** $\mathbb{R}$
3. (a) 6   **(b)** -2 **5. (a)** 3   **(b)** 7
7. 4.6052   **9.** 2.4037
11. (a) $8^{2/3} = 4$ **(b)** $6^v = u$
13. (a) $\log 1000 = 3$ **(b)** $\log_4 y = x$

15.
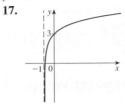
$y = -\ln x$

17.

19. (a) $y = \ln x + 3$ **(b)** $y = \ln(x + 3)$
(c) $y = -\ln x$ **(d)** $y = \ln(-x)$
21. About 68 billion miles   **23.** False   **25.** False
27. $\ln 8$ **29.** $\ln(u^3/25)$
31. (a) By Law 3 of logarithms, $\ln(x^3) = 3 \ln x$ and the domain of both functions is $(0, \infty)$.
(b) The domain of $y = \ln(x^2)$ is $\{x \mid x \neq 0\}$ but the domain of $y = 2 \ln x$ is $\{x \mid x > 0\}$.
33. (a) $\sqrt{e} \approx 1.6487$ **(b)** $-\ln 5 \approx -1.6094$
35. 1.8614   **37.** 6.5850   **39.** 0.4515   **41.** 6.4820
43. May 2019   **45.** ≈ 86.64 ft
47. After about 26.9 hours
49. (a) $Y(v) = -203.94 + 22.14 \ln v$ **(b)** 20.3 years
51. Mid-November 2005

■ CHAPTER 1 REVIEW EXERCISES ▪ page 73

1. (a) 2.7   **(b)** 2.3, 5.6   **(c)** $[-6, 6]$ **(d)** $[-4, 4]$
3. (a)   **(b)** 150 ft

5. $\left(-\infty, \frac{2}{5}\right]$ **7.** $\mathbb{R}$
9. $10, x^2 - 13x + 40, a + 1, 2x + h - 3$
11. (a) Neither   **(b)** Odd   **(c)** Even
13. The amount, in dollars, David contributed to a retirement account during year t
15. (a) $A(x) = 3x^2 - 2^x + 9$ **(b)** $B(x) = 3(2^x - 5)^2 + 4$
(c) $C(x) = 2^{3x^2+4} - 5$
17. (a) Shift the graph 8 units upward.
(b) Shift the graph 8 units to the left.
(c) Stretch the graph vertically by a factor of 2, then shift it 1 unit upward.
(d) Shift the graph 2 units to the right and 2 units downward.
(e) Reflect the graph about the x-axis.

19.

21.

23. (a) -12; 12,000 fewer units will be purchased for each dollar price increase.
(b) $L(x) = -12x + 461$
25. $L(t) = 0.2467t + 50.27$ where $t = 0$ corresponds to 1900 [using the data points for 1920 and 1980]; about 78.6 years
27. $y = \frac{1}{50}(x + 8)^2 + 2$ **29.** $A = 3136/x$
31. Increasing: $(-\infty, -8), (5, \infty)$; decreasing: $(-8, 5)$
33. (a) $9x^2 y^8$ **(b)** 9   **(c)** 2
35. $f(x) \approx 8.3(1.2597)^x$
37. (a) 324,000   **(b)** $P(t) = 4000 \cdot 3^{t/5}$
(c) About 25.13 years
39. About 1.585 hours
41. $y \approx 1.0607t + 21.96$, where $t = 0$ corresponds to 2000; about $34.69

Chapter 2

■ EXERCISES 2.1 ▪ page 82

1. 9   **3.** $\frac{1}{7}$ **5.** 0.115   **7.** 47.432
9. (a) $\frac{143}{63} \approx 2.27$ dollars per day; from March 15 to May 17 the price of gold increased at an average rate of $2.27 per day
(b) $\approx$$0.16/day
11. ≈ 121.4; If spending on advertising is increased from 1.8 to 2.5 million dollars, then the number of vehicles sold increases at an average rate of about 121,400 vehicles per million dollars spent.
13. ≈ 732.72; From 2.5 to 4.5 years after being opened, the balance of the account increased at an average rate of $732.72 per year.
15. $\frac{20}{3} \approx 6.67$; From 3 to 6 hours of charging time, the percentage of full charge increases at an average rate of about 6.67 percentage points per hour.
17. (a) -0.15

(b) Positive   **(c)** $[1, 2]$
19. (a) (i) 20 ft/s   **(ii)** 12 ft/s   **(iii)** 5.6 ft/s
(iv) 4.16 ft/s   **(b)** 4 ft/s
21. (a) (i) 32 ft/s   **(ii)** 25.6 ft/s   **(iii)** 24.8 ft/s
(iv) 24.16 ft/s   **(b)** 24 ft/s

■ **PREPARE YOURSELF 2.2** ▪ **page 92**

1. 0.4091   **2.** 0.4988

3. (a) $(x - 8)(x + 3)$ **(b)** $(a + 5)(a - 5)$

(c) $(2w + 3)(w - 5)$ **(d)** $(b + 1)(b^2 - b + 1)$

4. (a) $\dfrac{x + 1}{x - 4}$ **(b)** $c + 4$ **(c)** $-\dfrac{1}{3q}$

5. $\dfrac{1}{\sqrt{x + 1} + 2}$

6. (a) 8   **(b)** -3 **(c)** 2

■ **EXERCISES 2.2** ▪ **page 92**

1. Yes   **3.** 1.5   **5.** 0.25   **7.** 0.6

9. (a) ≈ 0.41421, ≈ 0.49242, ≈ 0.49980, ≈ 0.49999

(b) 0.5   **(c)** ≈ 0.499999, 0.5, 0, 0; no

(d) The calculator will eventually produce round-off errors; no

(e)

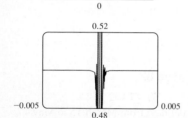

Yes

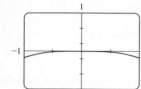

11. 17   **13.** -4 **15.** 4.155   **17.** 0.173

19. (a) $\{x \mid x \neq 2\}$ **(b)** 3   **(c)** 4

21. (a) $\{z \mid z \neq 2, z \neq 3\}$ **(b)** -1 **(c)** 2

23. 5   **25.** $\frac{7}{4}$ **27.** 5   **29.** $\frac{6}{5}$

31. 8   **33.** $\frac{1}{12}$ **35.** $\frac{1}{6}$ **37.** $-\frac{1}{16}$

39. Function values do not approach any particular number as $x \to 0$

41. Function values do not approach any particular number as $t \to 0$

43. (a) 2   **(b)** 3   **(c)** Does not exist   **(d)** 4

(e) Does not exist

45. $f(x)$ approaches 3 as x approaches 1 from the left; $f(x)$ approaches 7 as x approaches 1 from the right; no

47. (a) (i) 0   **(ii)** 0   **(iii)** 1   **(iv)** 1   **(v)** 0

(vi) Does not exist

(b)

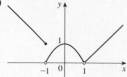

49. (a) 1   **(b)** -1 **(c)** No

51. (a)

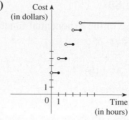

(b) Discontinuities at $t = 1, 2, 3, 4$

53. (a), (b) $\frac{2}{3}$

55. (a) 1   **(b)** 0   **(c)** Does not exist

57. $\frac{2}{3}$

■ **PREPARE YOURSELF 2.3** ▪ **page 108**

1. (a) $a - 3$ **(b)** $v + 1$

2. (a) $-\dfrac{1}{5(5 + t)}$ **(b)** $-\dfrac{1}{4r}$

3. (a) 8   **(b)** 10   **(c)** 12   **(d)** $\frac{1}{6}$

4. $y = \frac{3}{4}x - \frac{13}{2}$ **5.** 6

■ **EXERCISES 2.3** ▪ **page 109**

1. 10   **3.** $-\frac{1}{9}$ **5.** 440 ft/s

7. (a) $\dfrac{f(x) - f(3)}{x - 3}$ **(b)** $\lim\limits_{x \to 3} \dfrac{f(x) - f(3)}{x - 3}$

9. 12   **11.** 13   **13.** 23   **15.** $\frac{1}{4}$

17. (a) 12   **(b)** $4a$

19. (a) $-2 + 8a$ **(b)** 38   **(c)** -10

21.

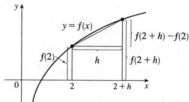

The line from $(2, f(2))$ to $(2 + h, f(2 + h))$

23. $g'(0), 0, g'(4), g'(2), g'(-2)$

25.

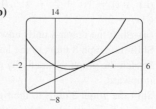

Greater (in magnitude)

27. 7; $y = 7x - 12$

29. (a) 2; $y = 2x - 4$ **(b)**

31. $y = 0.2x + 4$ **33.** $y = -x + 5$

35. -20 m/s; 20 m/s

37. **(a) (i)** \$20.25/unit   **(ii)** \$20.05/unit   **(b)** \$20/unit

39. **(a)** The rate at which the cost is changing per ounce of gold produced; dollars per ounce

(b) When the 800th ounce of gold is produced, the cost of production is \$17/oz.

(c) Decrease in the short term; increase in the long term

41. **(a)** The rate at which the fuel consumption is changing with respect to speed; (gal/h)/(mi/h)

(b) The fuel consumption is decreasing by 0.05 (gal/h)/(mi/h) as the car's speed reaches 20 mi/h.

43. **(a)** The rate at which daily heating costs change with respect to temperature when the temperature is 58°F; dollars/°F

(b) Negative; If the outside temperature increases, the building should require less heating.

45. **(a)** The shop loses \$125 when it sells 80 mugs in a week.

(b) Weekly profit increases at a rate of \$1.50 per mug after 80 mugs are sold.

47. **(a)** 3.296   **(b)** 3.3

49. The rate at which the temperature is changing at 8:00 AM; 3.75°F/h

51. **(a) (i)** 4.65 m/s   **(ii)** 5.6 m/s   **(iii)** 7.55 m/s
(iv) 7 m/s   **(b)** 6.3 m/s

53. **(a) (i)** 9.375 percentage points/year
(ii) 10.35 percentage points/year
(iii) 4.65 percentage points/year
(b) 7.5 percentage points/year
(c) 6.8 percentage points/year

55. **(a)** The rate at which the oxygen solubility changes with respect to the water temperature; (mg/L)/°C

(b) $S'(16) \approx -0.25$; as the temperature increases past 16°C, the oxygen solubility is decreasing at a rate of 0.25 (mg/L)/°C.

57. **(a)** $V(t) = 1.11t^2 - 66.6t + 999$

(b) $V'(15) = -33.3$; the water volume is decreasing at a rate of 33.3 gallons per minute after 15 minutes

59.

61. **(a)** ≈ 296 billion dollars/year   **(c)** -0.84

■ **PREPARE YOURSELF 2.4** ▪ **page 124**

1. $6 - 6t - 3h$ **2.** $-1/[x(x + c)]$ **3.** $2x$

4. $3x^2 + 2$ **5.** $\dfrac{2}{(x + a + 2)(x + 2)}$ **6.** $\frac{1}{4}$

■ **EXERCISES 2.4** ▪ **page 124**

1. **(a)** 1.5   **(b)** 1   **(c)** 0   **(d)** -4
(e) 0   **(f)** 1   **(g)** 1.5

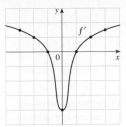

3. **(a)** II   **(b)** IV   **(c)** I   **(d)** III

5.   **7.**

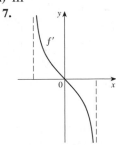

$f'(x) = e^x$

9.   1963 to 1971

11.

13. **(a)** The rate at which the unemployment rate is changing, in percent unemployed per year

(b)

t	$U'(t)$	t	$U'(t)$
1999	-0.2	2004	-0.45
2000	0.25	2005	-0.45
2001	0.9	2006	-0.25
2002	0.65	2007	0.6
2003	-0.15	2008	1.2

15. **(a)** 0, 1, 2, 4   **(b)** $-1, -2, -4$ **(c)** $f'(x) = 2x$

17. $f'(x) = \frac{1}{2}$ **19.** $h'(v) = 8v$ **21.** $f'(x) = 3x^2 - 3$

23. $B(p) = -3/p^2$ **25.** $\dfrac{dy}{dx} = \dfrac{1}{2\sqrt{x}}$

27. $G'(t) = \dfrac{4}{(t + 1)^2}$, $(-\infty, -1) \cup (-1, \infty)$,

$(-\infty, -1) \cup (-1, \infty)$

29. $g'(x) = 1/\sqrt{1 + 2x}, \left[-\frac{1}{2}, \infty\right), \left(-\frac{1}{2}, \infty\right)$

31.

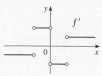

33. (a) The rate at which the percentage of power produced by solar panels is changing, in percentage points per year
(b) On January 1, 2002, the percentage of power produced by solar panels was increasing at a rate of 3.5 percentage points per year.

35. $f'(x) = 4 - 2x,$
$f''(x) = -2$

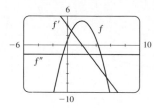

37. (a) (a) Increasing on $(1, 5)$; decreasing on $(0, 1)$ and $(5, 6)$
(b) Local maximum at $x = 5$, local minimum at $x = 1$
(c)

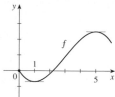

39. Increasing on $(2, 5)$; decreasing on $(-\infty, 2)$ and $(5, \infty)$

41.

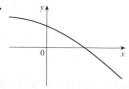

43. If $D(t)$ is the size of the deficit as a function of time, then at the time of the speech $D'(t) > 0$, but $D''(t) < 0$.
45. (a) The rate starts small, grows rapidly, levels off, then decreases and becomes negative.
(b) $(1932, 2.5)$ and $(1937, 4.3)$; the rate of change of population density starts to decrease in 1932 and starts to increase in 1937.
47. $v(t) = 3.4t - 3.1, a(t) = 3.4$
49. $K(3) - K(2)$; concave downward
51. $f'(x) = 4x - 3x^2, f''(x) = 4 - 6x$

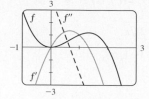

53.

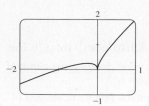

Differentiable at -1; not differentiable at 0

55. (a)

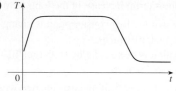

(b) dT/dt is large, positive at first, then $dT/dt = 0$ when the water reaches the temperature of the tank. dT/dt later becomes small, negative as the hot water in the tank is exhausted, eventually reaching zero.
(c)

57. $a = f, b = f', c = f''$

▪ CHAPTER 2 REVIEW EXERCISES ▪ page 129

1. $\dfrac{\sqrt{17} - 2\sqrt{2}}{3} \approx 0.432$

3. 8.5; From the end of the 6th month through the 9th month the revenue increased at an average rate of \$8500 per month.
5. 5.545 **7.** 6 **9.** $\frac{3}{2}$ **11.** 4 **13.** 3
15. (a) (i) 3 **(ii)** 0 **(iii)** Does not exist
(iv) 2 **(v)** Does not exist **(b)** $-3, 0, 2, 4$
17. (a) (i) 3 m/s **(ii)** 2.75 m/s **(iii)** 2.625 m/s
(iv) 2.525 m/s **(b)** 2.5 m/s
19. 3 **21.** $0, f'(5), f'(2), 1, f'(3)$
23. (a) The rate at which the cost changes with respect to the interest rate; dollars/(percent per year)
(b) As the interest rate increases past 10%, the cost is increasing at a rate of \$1200/(percent per year).
(c) Always positive
25. (a) (i) 0.824 **(ii)** 0.8 **(b)** $y = 0.824x + 0.825$
27. $C'(2000) \approx 34.95$; in 2000 the value of US currency in circulation was increasing at a rate of \$34.95 billion/year
29. (a) -0.9 percentage points per year
(b) $P'(2002) \approx 0.85$; in 2002 the percentage of children living below the poverty level was increasing at a rate of approximately 0.85 percentage points per year
31. -4 (discontinuity), -1 (corner), 2 (discontinuity), 5 (vertical tangent)

33.

35. $f'(x) = m$ **37.** $g'(x) = 4x - 3$

39. $A'(w) = -\dfrac{3}{(2w - 1)^2}$ **41.** $g''(x) = 4$

43. (a) About 35 ft/s   **(b)** About $(8, 180)$
(c) The point at which the car's speed is maximized
45. (a) Increasing on $(-2, 0)$ and $(2, \infty)$;
decreasing on $(-\infty, -2)$ and $(0, 2)$
(b) Maximum at 0; minimum at -2 and 2
(c) Concave upward on $(-\infty, -1)$ and $(1, \infty)$; concave downward on $(-1, 1)$

Chapter 3

■ PREPARE YOURSELF 3.1 ▪ page 143

1. (a) $f(x) = x^{1/3}$ **(b)** $g(w) = w^{-1}$ **(c)** $A(t) = 4t^{-1/2}$
(d) $B(v) = 8v^{-3}$ **(e)** $y = x^{3/4}$
2. $y = -2x + 14$

■ EXERCISES 3.1 ▪ page 143

1. $f'(x) = 0$ **3.** $y' = 9x^8$ **5.** $g'(t) = \frac{7}{2}t^{5/2}$
7. $y' = -\frac{2}{5}x^{-7/5}$ **9.** $L'(t) = \frac{1}{4}t^{-3/4}$ **11.** $y' = 48x^5$
13. $f'(x) = -14/x^3$ **15.** $y' = 5$ **17.** $f'(x) = 3x^2 - 4$
19. $y' = 2.8x^3 - 5.4x^2 + 5.1$ **21.** $q' = e^r$
23. $G'(x) = 3x^2 - 4e^x$ **25.** $f'(t) = t^3$
27. $f'(q) = -\dfrac{6}{q^2} + \dfrac{6}{q^3}$ **29.** $y' = \frac{3}{2}\sqrt{x} + \dfrac{1}{2\sqrt{x}}$
31. $F'(x) = \frac{5}{32}x^4$ **33.** $y' = 7 - \dfrac{5}{x^2}$
35. $f'(y) = -10A/y^{11} + Be^y$ **37.** $v' = 2t + \frac{3}{4}t^{-7/4}$
39. $\frac{35}{3}$ **41.** $y = 2x + 2$
43. $y = \frac{3}{2}x + \frac{1}{2}$

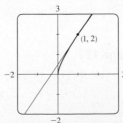

45. 1.76 (dollars per barrel)/week
47. (a) $s(t) = -2.597t^3 + 69.88t^2 + 0.7817t + 1.778$
(b) $s'(4) \approx 435.2$ ft/s
49. 3   **51.** $f'(x) = e^x - 5$ **53.** $f'(x) = 1 - x^{-2/3}$

55. (a)

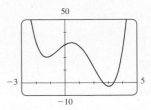

(c) $f'(x) = 4x^3 - 9x^2 - 12x + 7$

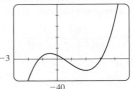

57. $f'(x) = 4x^3 - 9x^2 + 16$, $f''(x) = 12x^2 - 18x$
59. $f'(x) = 0.8x^3 - 3.3x^2 + 1.2x + 2.2$,
$f''(x) = 2.4x^2 - 6.6x + 1.2$
61. (a) 222.8 ft/s   **(b)** 158 ft/s^2
63. (a) $v(t) = 3t^2 - 3$, $a(t) = 6t$ **(b)** 12 m/s^2
(c) $a(1) = 6$ m/s^2
65. $(-2, 21)$, $(1, -6)$ **69.** $a = -\frac{1}{2}$, $b = 2$

■ PREPARE YOURSELF 3.2 ▪ page 155

1. (a) $f'(x) = 2.6 + 0.04x$ **(b)** 4.6
2. 540   **3.** -3.75 **4.** 353.55
5. (a) 1.6   **(b)** 2.0

■ EXERCISES 3.2 ▪ page 156

1. $C(q) = 2000 + 15q$
3. (a) $\dfrac{C(q)}{q} = \dfrac{1800}{q} + 0.12 + 0.003q$
(b) $C'(q) = 0.12 + 0.006q$
(c) \$5.22/bar, \$3.12/bar, \$3.123
5. (a) $C'(x) = 3 + 0.02x + 0.0006x^2$
(b) \$11/pair, the rate at which the cost is changing as the 100th pair of jeans is being produced; the cost of the 101st pair
(c) \$11.07
7. (a) \$146.67   **(b)** $C'(40) \approx 0.085$; cost is increasing at a rate of 0.085 million dollars per 1000 units, or \$85 per unit, when 40,000 units have been produced   **(c)** \$85
9. (a) 390   **(b)** 369   **11.** \$87.50
13. (a) 100 units   **(b)** 7.5   **(c)** 400 units
15. 158   **17.** 351
19. (a) $P(q) = -0.007q^2 + 12q - 2500$
(b) $C'(q) = 4 + 0.01q$, $R'(q) = 16 - 0.004q$ **(c)** 857
21. (a) \$2.15   **(b)** 2418   **23.** 333 units
25. (a) 400 bracelets   **(b)** 1.95 thousand dollars per bracelet, \$1500 per bracelet   **(c)** No
27. (a) \$2800   **(b)** $C'(1400) \approx 0.02$ thousand dollars/unit, $R'(1400) \approx 0.01$ thousand dollars/unit; no   **(c)** 1200 units
29. 175

■ **PREPARE YOURSELF 3.3** ■ **page 164**

1. (a) $f'(x) = 15x^2 + 3$ (b) $g'(x) = -2/x^3$

(c) $r'(x) = \dfrac{1}{2\sqrt{x}}$ (d) $U'(t) = e^t$

2. (a) $\dfrac{5}{t^2}$ (b) $\dfrac{1}{2\sqrt{x}}$ (c) $4\sqrt[3]{x}$

3. (a) 16.4 ft/min   (b) 2.4 ft/min²

■ **EXERCISES 3.3** ■ **page 164**

1. $y' = 5x^4 + 3x^2 + 2x$ **3.** $f'(x) = (x^2 + 2x)e^x$

5. $y' = (x - 2)e^x/x^3$ **7.** $g'(x) = 5/(2x + 1)^2$

9. $F'(y) = 5 + \dfrac{14}{y^2} + \dfrac{9}{y^4}$ **11.** $y' = \dfrac{2t(1 - t)}{(3t^2 - 2t + 1)^2}$

13. $y' = (r^2 - 2)e^r$ **15.** $P' = \dfrac{5(3t^2 - 6t + 2)e^t}{(2 + 3t^2)^2}$

17. $y' = 2v - 1/\sqrt{v}$ **19.** $f'(x) = -ACe^x/(B + Ce^x)^2$

21. -0.556 **23.** $y = 2x$

25. (a) $A'(x) = [xp'(x) - p(x)]/x^2$; The average productivity increases as new workers are added.

27. $-\frac{1}{18} \approx -0.0556$ ft/s²

29. (a) $y = \frac{1}{2}x + 1$ (b)

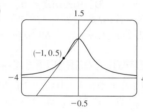

31. (a) $e^x(x - 3)/x^4$ **33.** (a) xe^x, $(x + 1)e^x$

35. $\frac{1}{4}$ **37.** (a) -16 (b) $-\frac{20}{9}$ (c) 20

39. 7   **41.** (a) 0   (b) $-\frac{2}{3}$

43. \$1.627 billion/year

45. (a) $y' = xg'(x) + g(x)$

(b) $y' = \dfrac{g(x) - xg'(x)}{[g(x)]^2}$ (c) $y' = \dfrac{xg'(x) - g(x)}{x^2}$

47. $g^{(n)}(x) = (x + n)e^x$

■ **PREPARE YOURSELF 3.4** ■ **page 175**

1. $f(x) = \sqrt{x}$ **2.** $f(x) = e^x$

3. (a) $2/\sqrt{x}$ (b) $30x^5$ (c) $-\dfrac{2}{3t^{4/3}}$

(d) $-8e^t$ (e) $(x^2 + 2x)e^x$ (f) $\dfrac{1 - w^2}{(w^2 + 1)^2}$

4. $\frac{19}{2}$

■ **EXERCISES 3.4** ■ **page 175**

1. $x/\sqrt{x^2 + 4}$ **3.** $-20x(1 - x^2)^9$ **5.** $e^{\sqrt{x}}/(2\sqrt{x})$

7. $f'(x) = -\dfrac{x}{\sqrt{9 - x^2}}$ **9.** $F'(x) = \dfrac{2 + 3x^2}{4(1 + 2x + x^3)^{3/4}}$

11. $y' = 7(2x^4 - 8x^2)^6(8x^3 - 16x)$

13. $g'(t) = -\dfrac{12t^3}{(t^4 + 1)^4}$ **15.** $A'(x) = 4.24e^{0.8x}$

17. $y' = (1 - 2x^2)e^{-x^2}$ **19.** $P'(t) = 6^t \ln 6$

21. $A' = 4500(\ln 1.124)(1.124^t) \approx 526.02(1.124^t)$

23. $g'(x) = 4(1 + 4x)^4(3 + x - x^2)^7(17 + 9x - 21x^2)$

25. $P'(t) = \frac{1}{3}(\ln 4)(4^{2+t/3})$ or $\frac{16}{3}(\ln 4)(4^{t/3})$

27. $L'(t) = 3(\ln 2)(2^t)e^{3\cdot2^t}$

29. $F'(z) = \dfrac{1}{(z - 1)^{1/2}(z + 1)^{3/2}}$ **31.** $y' = (r^2 + 1)^{-3/2}$

33. $y' = \dfrac{6e^{-0.3t}}{(1 + 2e^{-0.3t})^2}$ **35.** $Q'(x) = \dfrac{3e^{3x} + 1}{2\sqrt{e^{3x} + x}}$

37. $y' = \frac{2}{3}x(x^2 + 2)^{-2/3}e^{\sqrt[3]{x^2+2}}$

39. (a)

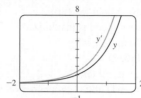

The graph of y' is a vertical stretch of the graph of y.

(b)

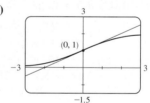

No

41. $y' = -0.5e^{-0.5x}$, $y'' = 0.25e^{-0.5x}$

43. $f'(x) = 30x^4(3x^5 + 1) = 90x^9 + 30x^4$ **45.** -6.888

47. $y = 20x + 1$

49. (a) $y = \frac{1}{2}x + 1$ (b)

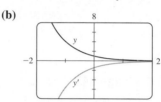

51. (a) $f'(x) = \dfrac{2 - 2x^2}{\sqrt{2 - x^2}}$ **53.** ≈ 0.0757 (μg/mL)/min

55. $A'(3.5) = 1876e^{0.245} \approx 2396.82$; after 3.5 years the value of the account is increasing at a rate of \$2396.82 per year

57. $\approx -24.5°$F/h

59. (a) $f(t) = 100.01(0.00004515^t)$ (b) -670.6 μA

61. (a) $F'(x) = e^x f'(e^x)$ (b) $G'(x) = e^{f(x)}f'(x)$

63. (a) $\frac{3}{4}$ (b) Does not exist   (c) -2

65. $1{,}073{,}741{,}824e^{2x}$

■ **PREPARE YOURSELF 3.5** ■ **page 187**

1. (a) $3 + f'(x)$ (b) $xf'(x) + f(x)$ (c) $3[f(x)]^2 \cdot f'(x)$

(d) $e^{f(x)} \cdot f'(x)$

2. (a) $1 + 4y^3\dfrac{dy}{dx}$ (b) $\dfrac{1}{2\sqrt{x}} + \dfrac{1}{2\sqrt{y}}\dfrac{dy}{dx}$

(c) $2x^2y\dfrac{dy}{dx} + 2xy^2 + \dfrac{dy}{dx}$ (d) $e^x - e^y\dfrac{dy}{dx}$

3. (a) $\dfrac{dN}{da} = 5 + 2a\sqrt{b}$ (b) $\dfrac{dN}{db} = -2 + \dfrac{a^2}{2\sqrt{b}}$

■ **EXERCISES 3.5** ▪ page 187

1. (a) $\dfrac{dy}{dx} = -\dfrac{y + 2 + 6x}{x}$

(b) $y = \dfrac{4}{x} - 2 - 3x, \dfrac{dy}{dx} = -\dfrac{4}{x^2} - 3$

3. $\dfrac{dy}{dx} = -\dfrac{x}{y}$ **5.** $\dfrac{dy}{dx} = -\dfrac{x(3x + 2y)}{x^2 + 8y}$

7. $\dfrac{dy}{dx} = \dfrac{3 - 2xy - y^2}{x^2 + 2xy}$ **9.** $\dfrac{dy}{dx} = \dfrac{1 - 2xye^{x^2y}}{x^2e^{x^2y} - 1}$

11. $-\frac{16}{13}$ **13.** 3   **15.** $y = -x + 2$

17. $y = -\frac{9}{13}x + \frac{40}{13}$

19. ≈ -0.336; When the price increases from \$40, the number of copies that will sell decreases at a rate of 336 copies per \$1 price increase.

21. $f'(x) = 3 - (2/x)$ **23.** $y' = 1.5 + (1/x)$

25. $y' = \dfrac{5(\ln x)^4}{x}$ **27.** $f'(x) = \dfrac{1}{5x(\ln x)^{4/5}}$

29. $\dfrac{dy}{dx} = \dfrac{2(\ln x + 1)}{x} + 2e^x(e^x + 1)$

31. $F'(t) = \dfrac{6}{2t + 1} - \dfrac{12}{3t - 1}$

33. $f'(u) = \dfrac{1 + \ln 2}{u[1 + \ln(2u)]^2}$ **35.** $y' = -\dfrac{x}{1 + x}$

37. $y' = \dfrac{1}{x \ln x}$ **39.** $f'(x) = \dfrac{2x}{x^2 - 5}, f''(x) = \dfrac{-2x^2 - 10}{(x^2 - 5)^2}$

41. $f'(x) = e^x[(1/x) + \ln x], f''(x) = e^x\left(\ln x + \dfrac{2}{x} - \dfrac{1}{x^2}\right)$

43. $y = 4x - 8$ **45.** $f'(x) = (4/x) - 1$

49. 140 dB, $\dfrac{1}{10 \ln 10} \approx 0.0434$ dB/(W/m^2)

51. $\dfrac{dy}{dx} = \dfrac{2x}{x^2 + y^2 - 2y}$

53. (a) $y = \frac{9}{2}x - \frac{5}{2}$ **(b)**

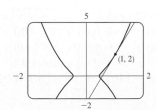

55. (b) $f'(x) = \dfrac{2x - e^{-x}}{x^2 + e^{-x}}$

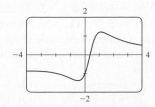

57. $b^2x_0x - a^2y_0y = b^2x_0^2 - a^2y_0^2$ **59.** $\left(\pm\sqrt{3}, 0\right)$

61. $y' = x^x(1 + \ln x)$

63. $\dfrac{dy}{dx} = \dfrac{e^{x^2}\sqrt{x^3 + x}}{(2x + 3)^4}\left(2x + \dfrac{3x^2 + 1}{2(x^3 + x)} - \dfrac{8}{2x + 3}\right)$

■ **PREPARE YOURSELF 3.6** ▪ page 199

1. (a) 5.69   **(b)** $t \approx 5.67$

2. (a) 6.11   **(b)** 2.14   **(c)** 2.12

3. (a) 216.63   **(b)** $-2.5 \ln 0.7 \approx 0.89$

■ **EXERCISES 3.6** ▪ page 199

1. (a) $46{,}500(1.024)^t$ **(b)** $46{,}500(0.976)^t$

3. (a) $A(t) = 5000(1.062)^t$ **(b)** \$7850.62

(c) ≈ 7.81 years

5. (a) $V(t) = 28.6(0.85)^t$ million dollars   **(b)** ≈ 4.27 years

(c) $V'(5) \approx -2.06$; After five years the equipment's value is decreasing at a rate of \$2.06 million/year.

(d) After about 9.45 years

7. (a) (i) \$3828.84   **(ii)** \$3846.11   **(iii)** \$3850.08

(iv) \$3851.61   **(v)** \$3852.01   **(vi)** \$3852.08

(b) ≈ 13.95 years   **(c)** ≈ 13.86 years

9. (a) $V(t) = 16{,}000\left(1 + \dfrac{0.043}{12}\right)^{12t}$

(b) $V'(3.5) \approx 798.10$; After 3.5 years the value of the account is increasing at a rate of \$798.10/year.

11. 5.25% compounded quarterly

13. $1300e^{0.172t}$; ≈ 4723 **15.** About 235

17. (a) $100(4.2)^t$ **(b)** ≈ 7409

(c) $\approx 10{,}632$ bacteria/hour   **(d)** $(\ln 100)/(\ln 4.2) \approx 3.2$ h

19. (a) $P(t) = 260e^{0.6298t}$ **(b)** $\approx 21{,}360$

(c) $\approx 13{,}453$ viruses/day

21. (a) 1508 million, 1871 million   **(b)** 2161 million

(c) 3972 million; wars in the first half of century, increased life expectancy in second half

23. (a) $100e^{-0.02310t}$ or $100\left(\frac{1}{2}\right)^{t/30}$ mg   **(b)** ≈ 9.92 mg

(c) ≈ 199.3 years   **(d)** ≈ -0.229 mg/year

25. ≈ 2500 years

27. (a) ≈ 14.97 hours   **(b)** ≈ 91.2 g   **(c)** ≈ 64.7 hours

29. (a) $\approx 137°$F   **(b)** ≈ 116 min   **(c)** $\approx 0.96°$F/min

31. (a) 1680; 323   **(b)** 930   **(c)** April, 2042

(d) $P'(12) \approx 46.0$; On January 1, 2022, the number of mountain lions is increasing at a rate of about 46 lions/year.

33. $P'(8) \approx 1.18$; On January 1, 2008, the population was increasing at a rate of about 1180 animals/year.

35. (a) 3.23×10^7 kg   **(b)** ≈ 1.55 years

37. (a) ≈ 0.00377 **(b)** $P(t) = \dfrac{100}{1 + 17.87e^{-0.00377t}}$ billion people, where $t = 0$ corresponds to 1990; 5.49 billion

(c) 7.81 billion, 27.68 billion

39. (a)   About 7.4 hours

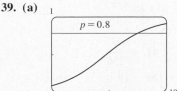

(b) About 11.1 percent of the population per hour

41. Exponential: $P(t) = 1578.3(1.0933)^t + 94{,}000$;

logistic: $P(t) = \dfrac{32{,}658.5}{1 + 12.75e^{-0.1706t}} + 94{,}000$

($t = 0$ corresponds to 1960)

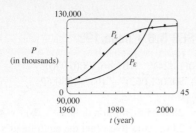

43. $B'(4) < B'(8)$ **45. (b)** 3:36 PM

47. (a)

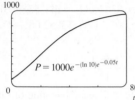

$P = 1000e^{-(\ln 10)e^{-0.05t}}$

(b) ≈ 732, ≈ 11.41 members/year

■ **CHAPTER 3 REVIEW EXERCISES** ▪ **page 206**

1. $f'(x) = 15x^2 - 7$ **3.** $\dfrac{dq}{dr} = \dfrac{1}{3\sqrt[3]{r^2}} - \dfrac{6}{r^2}$

5. $h'(u) = 3e^u - \dfrac{1}{2u^{3/2}}$ **7.** $E'(x) = 2.3(\ln 1.06)(1.06)^x$

9. $B'(t) = 4/t$ **11.** $C'(a) = e^a\sqrt{a} + \dfrac{e^a + 1}{2\sqrt{a}}$

13. $y' = \dfrac{t^2 + 1}{(1 - t^2)^2}$ **15.** $y' = 3(x^4 - 3x^2 + 5)^2(4x^3 - 6x)$

17. $A' = -32e^{-2t}$ **19.** $y' = \dfrac{3x^2}{x^3 + 5}$

21. $y' = \dfrac{2(2x^2 + 1)}{\sqrt{x^2 + 1}}$ **23.** $\dfrac{dz}{dt} = \dfrac{4 - t^2}{2\sqrt{t}\,(t^2 + 4)^{3/2}}$

25. $y' = e^{-1/x}(1/x + 1)$

27. $f'(x) = 10^{x\sqrt{x-1}}(\ln 10)\left(\dfrac{x}{2\sqrt{x - 1}} + \sqrt{x - 1}\right)$

29. $A'(r) = \dfrac{24(\ln r)^3}{r}$ **31.** $y' = 3^{x\ln x}(\ln 3)(1 + \ln x)$

33. $y' = \dfrac{6x[\ln(x^2 + 1)]^2}{x^2 + 1}$

35. $f'(t) = 325e^{0.65t}$, $f''(t) = 211.25e^{0.65t}$

37. $f'(x) = 2 - \frac{15}{4}x^{-1/4}$, $f''(x) = \frac{15}{16}x^{-5/4}$

39. $\dfrac{dy}{dx} = \dfrac{1 - y^4 - 2xy}{4xy^3 + x^2 - 3}$ **41.** $-\dfrac{9}{256}$

43. $y = -x + 2$ **45.** $y = 15x - 14$

47. (a) $\dfrac{10 - 3x}{2\sqrt{5 - x}}$ **(b)** $y = \frac{7}{4}x + \frac{1}{4}$, $y = -x + 8$

(c)

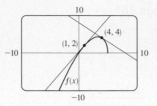

49. (a) $\dfrac{C(q)}{q} = \dfrac{920}{q} + 2 - 0.02q + 0.00007q^2$; 130.11

(b) $C'(q) = 2 - 0.04q + 0.00021q^2$

(c) $C'(100) = 0.1$; Cost is increasing at a rate of $0.10 per unit when 100 items have been produced.

(d) $C(101) - C(100) = 0.10107$

51. (a) $P(q) = -380 + 1.04q - 0.0003q^2$

(b) $\dfrac{R(q)}{q} = 1.36 - 0.0001q$, $R'(q) = 1.36 - 0.0002q$

(c) About 1733

53. (a) $2.6(1.046)^t$ million dollars after t years; ≈ 3.04 million

(b) $2.6(0.954)^t$ million dollars after t years; about 5.57 years

55. (a) $200e^{1.1756t}$ **(b)** $\approx 22{,}040$

(c) $\approx 25{,}910$ bacteria/hour   **(d)** 1.1756 percent/hour

(e) ≈ 3.33 h

57. ≈ 100 h

59. (a) ≈ 4.77 years   **(b)** ≈ 15.84 years

61. (a) 285,000; 59,375   **(b)** $\approx 188{,}000$ **(c)** May 2037

(d) $P'(30) \approx 4.35$; On January 1, 2040, the population will be increasing at a rate of about 4350 animals per year.

63. $(-3, 0)$ **65.** $\left(2/\sqrt{6}, -1\sqrt{6}\right), \left(-2/\sqrt{6}, 1/\sqrt{6}\right)$

67. (a) 2   **(b)** 44

Chapter 4

■ **PREPARE YOURSELF 4.1** ▪ **page 214**

1. ≈ 72.1 miles   **2.** $\sqrt{x^2 + y^2}$ miles

3. (a) $y' = 4[f(x)]^3 f'(x)$ **(b)** $y' = 2x + xf'(x) + f(x)$

4. (a) $y' = A(t)\,B'(t) + A'(t)\,B(t)$

(b) $y' = 2A(t)\,A'(t) + 2B(t)\,B'(t)$

(c) $y' = [A(t)]^2\,B'(t) + 2A(t)\,A'(t)\,B(t)$

■ **EXERCISES 4.1** ▪ **page 215**

1. $dV/dt = 3x^2\,dx/dt$ **3.** 48 cm²/s   **5.** 70   **7.** $\frac{46}{13}$

9. $160/week   **11.** ≈ -118 bottles/year

13. (a) The rate of decrease of the surface area is 1 cm²/min.

(b) The rate of decrease of the diameter when the diameter is 10 cm

(c)

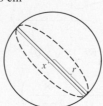

(d) $S = \pi x^2$

(e) $1/(20\pi)$ cm/min

15. (a) The plane's altitude is 1 mi and its speed is 500 mi/h.
(b) The rate at which the distance from the plane to the station is increasing when the plane is 2 mi from the station
(c) 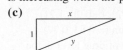   **(d)** $y^2 = x^2 + 1$
(e) $250\sqrt{3}$ mi/h

17. 65 mi/h   **19.** $\frac{720}{13} \approx 55.4$ km/h   **21.** 80 cm³/min
23. 360 ft/s   **25.** $6/(5\pi) \approx 0.38$ ft/min
27. $\approx 1.97°$C/h   **29.** -1.6 cm/min   **31.** $\frac{4}{5}$ ft/min
33. 5 m   **35.** $10/\sqrt{133} \approx 0.87$ ft/s

■ **PREPARE YOURSELF 4.2** ▪ **page 225**

1. (a) $\frac{3}{4}, -1$ **(b)** $0, 2$ **(c)** $1, -1$ **(d)** $-\frac{1}{3}$
2. $\frac{1}{4} \pm \frac{1}{4}\sqrt{41}$ **3.** $(1 + x^2)^{-1/2}(4x^2 + 3)$
4. $-2t^{-1/3}(t + 2)(2t + 5)$
5. (a) $-\frac{2}{3}$ **(b)** e^2 **(c)** $0, \pm 3$

6. (a) $f'(x) = (5x + 1)e^{5x}$ **(b)** $f'(t) = \dfrac{-t^2 - 6t + 4}{(t^2 + 4)^2}$

(c) $\dfrac{dy}{dx} = \dfrac{1}{2x\sqrt{1 + \ln x}}$

(d) $\dfrac{dy}{dx} = \dfrac{1}{2\sqrt{1 + \ln x}} + \sqrt{1 + \ln x}$

■ **EXERCISES 4.2** ▪ **page 225**

Abbreviations: abs, absolute; loc, local; max, maximum; min, minimum

1. Abs min: smallest function value on the entire domain of the function; loc min at c: smallest function value when x is near c
3. Abs max at b, loc max at b and e, abs min at d, loc min at d and s
5. Abs max $f(4) = 4$; abs min $f(7) = 0$; loc max $f(4) = 4$ and $f(6) = 3$; loc min $f(2) = 1$ and $f(5) = 2$
7. 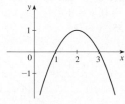   **9.**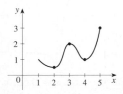
11. (a) 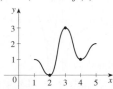   **(b)**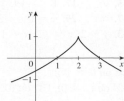

(c)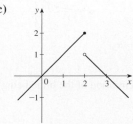

13. (a) 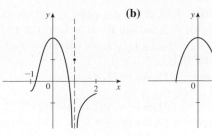   **(b)**

15. Abs max $f(1) = 5$ **17.** None
19. Abs max $f(2) = \ln 2$ **21.** Abs max $f(0) = 1$
23. $-\frac{2}{5}$ **25.** $-4, 2$ **27.** $0, -\frac{1}{2} \pm \frac{1}{2}\sqrt{5}$ **29.** $1/e$
31. $0, 2$ **33.** $0, \frac{8}{7}, 4$ **35.** $f(0) = 5, f(2) = -7$
37. $f(-1) = 8, f(2) = -19$ **39.** $f(3) = 66, f(\pm 1) = 2$
41. $f(\sqrt{2}) = 2, f(-1) = -\sqrt{3}$
43. $f(2) = 2/\sqrt{e}, f(-1) = -1/\sqrt[8]{e}$
45. $f(1) = \ln 3, f(-\frac{1}{2}) = \ln \frac{3}{4}$ **47.** $-2.83, -1.50, 2.83$
49. (a) $2.19, 1.81$ **(b)** $\frac{6}{25}\sqrt{\frac{3}{5}} + 2, -\frac{6}{25}\sqrt{\frac{3}{5}} + 2$
51. (a) $0.37, 0.05$ **(b)** $1/e, 4.5/e^{4.5}$
53. $1.359, 0.837$ **55.** $1.710, -1.077$

57. $\dfrac{6}{13e} \approx 0.170$ mg/mL; after $\dfrac{1}{2.6} \approx 0.385$ h (about 23 min)

59. $\approx 3.9665°$C
61. Cheapest, $t \approx 0.855$ (June 1994); most expensive, $t \approx 4.618$ (March 1998)

63. $f\left(\dfrac{a}{a + b}\right) = \dfrac{a^a b^b}{(a + b)^{a+b}}$

■ **PREPARE YOURSELF 4.3** ▪ **page 235**

1. (a) $(-3, 6)$ **(b)** $(-3, 0) \cup (3, \infty)$
(c) $(-\infty, 0) \cup (4, \infty)$ **(d)** $(-\infty, 4)$ **(e)** (e^6, ∞)
2. (a) $1/e$ **(b)** $-2 \pm \sqrt{2}$

3. (a) $B''(t) = (12t - 12)e^{-2t}$ **(b)** $\dfrac{d^2 y}{dx^2} = -\dfrac{3x^4 + 1}{(x^3 + x)^2}$
4. Negative

■ **EXERCISES 4.3** ▪ **page 236**
Abbreviations: CU, concave upward; CD, concave downward; dec, decreasing; inc, increasing; IP, inflection point; loc, local; max, maximum; min, minimum

1. (a) $(0, 6), (8, 9)$ **(b)** $(6, 8)$ **(c)** $(2, 4), (7, 9)$
(d) $(0, 2), (4, 7)$ **(e)** $(2, 3), (4, 4.5), (7, 4)$
3. (a) I/D Test   **(b)** Concavity Test
(c) Find points at which the concavity changes.
5. Loc min $-\frac{73}{12}$ at $x = \frac{11}{6}$
7. Loc max $M(-\frac{2}{3}) = \frac{385}{27} \approx 14.26$;
loc min $M(\frac{5}{2}) = -\frac{197}{4} = -49.25$
9. Loc max $f(e^2) = 2/e$
11. CU on $(-\infty, \infty)$; no IP
13. CU on $(-\infty, 0.1875)$; CD on $(0.1875, \infty)$;
IP $(0.1875, \approx -5.9664)$
15. (a) Inc on $(-\infty, -2), (2, \infty)$; dec on $(-2, 2)$
(b) Loc max $f(-2) = 17$; loc min $f(2) = -15$
(c) CU on $(0, \infty)$; CD on $(-\infty, 0)$; IP $(0, 1)$

17. (a) Inc on $(-1, 0)$, $(1, \infty)$; dec on $(-\infty, -1)$, $(0, 1)$
(b) Loc max $f(0) = 2$; loc min $f(-1) = 1$, $f(1) = 1$
(c) CU on $\left(-\infty, -1/\sqrt{3}\right)$, $\left(1/\sqrt{3}, \infty\right)$; CD on $\left(-1/\sqrt{3}, 1/\sqrt{3}\right)$;
IP $\left(-1/\sqrt{3}, 13/9\right)$, $\left(1/\sqrt{3}, 13/9\right)$
19. (a) Inc on $(-\infty, 5)$; dec on $(5, \infty)$
(b) Loc max $(5, 25/e)$
(c) CU on $(10, \infty)$; CD on $(-\infty, 10)$; IP $(10, 50/e^2)$
21. (a) Inc on $(-1, \infty)$; dec on $(-\infty, -1)$
(b) Loc min $f(-1) = -1/e$
(c) CU on $(-2, \infty)$; CD on $(-\infty, -2)$; IP $(-2, -2e^{-2})$
23. (a) Inc on $(0, e)$; dec on (e, ∞)
(b) Loc max $f(e) = 1/e$
(c) CU on $(e^{3/2}, \infty)$; CD on $(0, e^{3/2})$; IP $(e^{3/2}, 3/(2e^{3/2}))$

25. (a) Inc on $(-\infty, -1)$, $(2, \infty)$;
dec on $(-1, 2)$
(b) Loc max $f(-1) = 7$;
loc min $f(2) = -20$
(c) CU on $\left(\frac{1}{2}, \infty\right)$; CD on $\left(-\infty, \frac{1}{2}\right)$;
IP $\left(\frac{1}{2}, -\frac{13}{2}\right)$
(d) See graph at right.

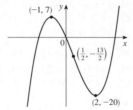

27. (a) Inc on $(-\infty, -1)$, $(1, \infty)$;
dec on $(-1, 1)$
(b) Loc max $h(-1) = 5$;
loc min $h(1) = 1$
(c) CU on $\left(-1/\sqrt{2}, 0\right)$, $\left(1/\sqrt{2}, \infty\right)$;
CD on $\left(-\infty, -1/\sqrt{2}\right)$, $\left(0, 1/\sqrt{2}\right)$;
IP $(0, 3)$, $\left(1/\sqrt{2}, 3 - \frac{7}{8}\sqrt{2}\right)$,
$\left(-1/\sqrt{2}, 3 + \frac{7}{8}\sqrt{2}\right)$
(d) See graph at right.

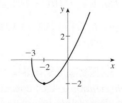

29. (a) Inc on $(-2, \infty)$;
dec on $(-3, -2)$
(b) Loc min $A(-2) = -2$
(c) CU on $(-3, \infty)$
(d) See graph at right.

31. (a) $3, 5$ **(b)** $2, 4, 6$ **(c)** $1, 7$
33. (a) Inc on $(0, 2)$, $(4, 6)$, $(8, \infty)$;
dec on $(2, 4)$, $(6, 8)$
(b) Loc max at $x = 2, 6$;
loc min at $x = 4, 8$
(c) CU on $(3, 6)$, $(6, \infty)$;
CD on $(0, 3)$
(d) 3 **(e)** See graph at right.

35. Loc max $f(-1) = 7$; loc min $f(1) = -1$
37. (a) f has a local maximum at 2.
(b) f has a horizontal tangent at 6.
39. (a) Very unhappy   **(b)** Unhappy   **(c)** Happy
(d) Very happy
41. $P'(t)$ is positive, $P''(t)$ is negative

43. (a) Loc and abs max $f(1) = \sqrt{2}$, no min
(b) $\frac{1}{4}\left(3 - \sqrt{17}\right)$
45. (b) CU on $(-\infty, -2.83)$, $(-0.20, \infty)$;
CD on $(-2.83, -0.20)$; IP $(-2.83, -5.72)$, $(-0.20, 1.21)$
47. CU on $(-\infty, -0.83)$, $(0.77, \infty)$; CD on $(-0.83, 0.77)$
49. 28.57 min, when the rate of increase of drug level in the bloodstream is greatest; 85.71 min, when rate of decrease is greatest
51. 3.06 years after midyear 2000
53. $(3, \infty)$ **57.** $f(x) = \frac{1}{9}(2x^3 + 3x^2 - 12x + 7)$

■ **PREPARE YOURSELF 4.4 ▪ page 247**

1. (a) 1000 **(b)** $-1,000,000$
2. (a) Become larger negative   **(b)** Approach zero
(c) Approach zero   **(d)** Become larger   **(e)** Approach zero
3. They grow larger and larger if $c > 5$, larger and larger negative if $c < 5$

4. They grow larger and larger negative.   **5.** $\dfrac{4 + \dfrac{1}{x} - \dfrac{2}{x^3}}{3 + \dfrac{3}{x}}$

■ **EXERCISES 4.4 ▪ page 247**

1. (a) As x approaches 2, $f(x)$ becomes large.   **(b)** As x approaches 1 from the right, $f(x)$ becomes large negative.
(c) As x becomes large, $f(x)$ approaches 5.
(d) As x becomes large negative, $f(x)$ approaches 3.
3. (a) ∞ **(b)** ∞ **(c)** $-\infty$ **(d)** 1 **(e)** 2
(f) $x = -1$, $x = 2$, $y = 1$, $y = 2$
5.   **7.**

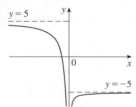

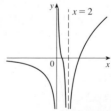

9. $x \approx -1.62$, $x \approx 0.62$, $x = 1$; $y = 1$ **11.** ∞ **13.** $-\infty$
15. ∞ **17.** 0 **19.** $\frac{1}{2}$ **21.** 0 **23.** 2 **25.** $-\infty$
27. ∞ **29.** $y = 0$ **31.** $y = 0$, $y = 1$
33. $y = 2$; $x = -2$, $x = 1$ **35.** $y = 3$

37. $y = 1$, $y = -1$;
inc on $(-\infty, \infty)$;
CU on $(-\infty, 0)$, CD on $(0, \infty)$

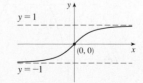

39. (a) $\lim\limits_{x \to \infty} P(x) = \lim\limits_{x \to \infty} Q(x) = \infty$,
$\lim\limits_{x \to -\infty} P(x) = \lim\limits_{x \to -\infty} Q(x) = -\infty$
41. (b) It approaches the concentration of the brine being pumped into the tank.

■ **PREPARE YOURSELF 4.5** ▪ **page 254**

1. $\{x \mid x \neq \pm 2\}$ **2.** $-1, 3$
3. No loc max, loc min $g(0) = 0$; inc on $(0, \infty)$, dec on $(-\infty, 0)$
4. IP $(0, 1)$, $\left(\frac{1}{3}, \frac{26}{27}\right)$; CU on $(-\infty, 0)$, $\left(\frac{1}{3}, \infty\right)$; CD on $\left(0, \frac{1}{3}\right)$
5. $t = 0, t = -7, y = 2$ **6.** Loc max at $x = 2$
7. (a) Loc min at $x = 1$ **(b)** IP at $x = 1$
8. (a) Vertical asymptote $x = 5$
(b) Horizontal asymptote $y = 5$

■ **EXERCISES 4.5** ▪ **page 254**

1. (a) Inc on $(2, \infty)$; dec on $(-\infty, 2)$
(b) Loc min $f(2) = -5$
(c) CU on $(-\infty, \infty)$; no IP
(d) $\lim\limits_{x \to -\infty} f(x) = \infty$, $\lim\limits_{x \to \infty} f(x) = \infty$
(e) See graph at right.

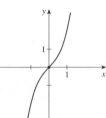

3. (a) Inc on $(-\infty, \infty)$
(b) None
(c) CU on $(0, \infty)$; CD on $(-\infty, 0)$; IP $(0, 0)$
(d) $\lim\limits_{x \to -\infty} f(x) = -\infty$, $\lim\limits_{x \to \infty} f(x) = \infty$
(e) See graph at right.

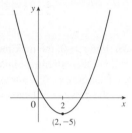

5. (a) Inc on $(1, 5)$; dec on $(-\infty, 1)$, $(5, \infty)$
(b) Loc max $f(5) = 27$; loc min $f(1) = -5$
(c) CU on $(-\infty, 3)$; CD on $(3, \infty)$; IP $(3, 11)$
(d) $\lim\limits_{x \to -\infty} f(x) = \infty$, $\lim\limits_{x \to \infty} f(x) = -\infty$
(e) See graph at right.

7. (a) Inc on $(-3, \infty)$; dec on $(-\infty, -3)$
(b) Loc min $f(-3) = -27$
(c) CU on $(-\infty, -2)$, $(0, \infty)$; CD on $(-2, 0)$; IP $(0, 0)$, $(-2, -16)$
(d) $\lim\limits_{x \to -\infty} f(x) = \infty$, $\lim\limits_{x \to \infty} f(x) = \infty$
(e) See graph at right.

9. Domain $\{x \mid x \neq 1\}$; dec on $(-\infty, 1)$, $(1, \infty)$; no local extreme points; CU on $(1, \infty)$; CD on $(-\infty, 1)$; no IP; asymptotes $x = 1, y = 1$

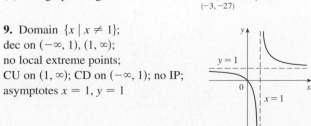

11. Domain $\{x \mid x \neq \pm 3\}$; inc on $(-\infty, -3)$, $(-3, 0)$; dec on $(0, 3)$, $(3, \infty)$; loc max $f(0) = -\frac{1}{9}$; CU on $(-\infty, -3)$, $(3, \infty)$; CD on $(-3, 3)$; no IP; asymptotes $x = -3, x = 3, y = 0$

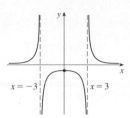

13. Domain $\{x \mid x \neq 0\}$; inc on $(0, 2)$; dec on $(-\infty, 0)$, $(2, \infty)$; loc max $f(2) = \frac{1}{4}$; CU on $(3, \infty)$; CD on $(-\infty, 0)$, $(0, 3)$; IP $\left(3, \frac{2}{9}\right)$; asymptotes $x = 0, y = 0$

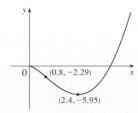

15. Domain $[0, \infty)$; inc on $\left(\frac{12}{5}, \infty\right)$; dec on $\left(0, \frac{12}{5}\right)$;
loc min $\left(\dfrac{12}{5}, -\dfrac{192\sqrt{3}}{25\sqrt{5}}\right) \approx (2.4, -5.95)$;
CU on $\left(\frac{4}{5}, \infty\right)$; CD on $\left(0, \frac{4}{5}\right)$; IP $\left(\dfrac{4}{5}, -\dfrac{128}{25\sqrt{5}}\right) \approx (0.8, -2.29)$;
$\lim\limits_{a \to \infty} h(a) = \infty$

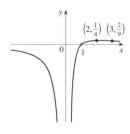

17. Domain $\mathbb{R}$; inc on $\mathbb{R}$; no local extreme points; CU on $(-\infty, 0)$; CD on $(0, \infty)$; IP $\left(0, \frac{1}{2}\right)$, horizontal asymptotes $y = 0$ (at left), $y = 1$ (at right)

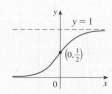

19. $\lim_{x \to \infty} f(x) = 0$, $\lim_{x \to -\infty} f(x) = 0$;
domain $\mathbb{R}$; inc on $\left(-1/\sqrt{2}, 1/\sqrt{2}\right)$; dec on $\left(-\infty, -1/\sqrt{2}\right)$,
$\left(1/\sqrt{2}, \infty\right)$; loc min $f\left(-1/\sqrt{2}\right) = -1/\sqrt{2e}$;
loc max $f\left(1/\sqrt{2}\right) = 1/\sqrt{2e}$; CU on $\left(-\sqrt{3/2}, 0\right)$, $\left(\sqrt{3/2}, \infty\right)$;
CD on $\left(-\infty, -\sqrt{3/2}\right)$, $\left(0, \sqrt{3/2}\right)$;
IP $(0, 0)$, $\left(\sqrt{3/2}, \sqrt{3/2}\,e^{-3/2}\right)$, $\left(-\sqrt{3/2}, -\sqrt{3/2}\,e^{-3/2}\right)$;
horizontal asymptote $y = 0$

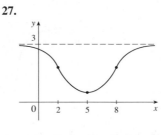

21.

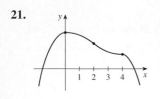

23.

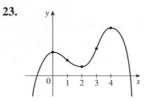

25.

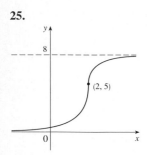

(2, 5)

27.

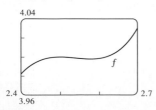

29. Inc on $(0.92, 2.5)$, $(2.58, \infty)$; dec on $(-\infty, 0.92)$, $(2.5, 2.58)$;
loc max $f(2.5) = 4$; loc min $f(0.92) \approx -5.12$,
$f(2.58) \approx 3.998$; CU on $(-\infty, 1.46)$, $(2.54, \infty)$;
CD on $(1.46, 2.54)$; IP $(1.46, -1.40)$, $(2.54, 3.999)$

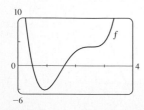

31. Inc on $(-15, 4.40)$, $(18.93, \infty)$;
dec on $(-\infty, -15)$, $(4.40, 18.93)$;
loc max $f(4.40) \approx 53,800$; loc min $f(-15) \approx -9,700,000$,
$f(18.93) \approx -12,700,000$; CU on $(-\infty, -11.34)$, $(0, 2.92)$,
$(15.08, \infty)$; CD on $(-11.34, 0)$, $(2.92, 15.08)$;
IP $(0, 0)$, $\approx(-11.34, -6,250,000)$, $(2.92, 31,800)$,
$(15.08, -8,150,000)$

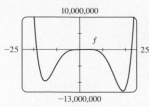

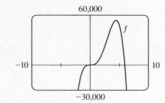

33. For $c \geq 0$, there is no IP and only one extreme point, the origin. For $c < 0$, there is a maximum point at the origin, two minimum points, and two IPs, which move downward and away from the origin as $c \to -\infty$.

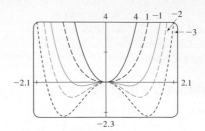

35. For $c > 0$, the maximum and minimum values are always $\pm\frac{1}{2}$, but the extreme points and IPs move closer to the y-axis as c increases. $c = 0$ is a transitional value: when c is replaced by $-c$, the curve is reflected in the x-axis.

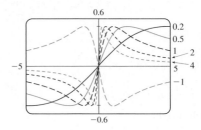

37. For $c > 0$, there is a loc max and IP (to the right of the y-axis) that move closer to the origin as c increases. For $c < 0$, there is a loc min and IP (to the left of the y-axis) that move closer to the origin as c decreases. $c = 0$ is a transitional value: when c is replaced by $-c$, the curve is reflected about the origin.

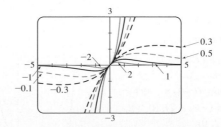

39. All the members have the same basic shape, and are CU on $(0, \infty)$, CD on $(-\infty, 0)$. But as a increases, the four key points shown in the figure move farther away from the origin.

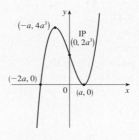

41. (a) When $t = (\ln a)/k$ **(b)** When $t = (\ln a)/k$
(c)

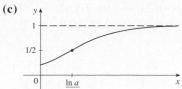

■ PREPARE YOURSELF 4.6 ▪ page 261

1. $\sqrt{x^2 + (20 - x)^2}$ **2.** $\sqrt{(2x + 5)^2 - (x - 2)^2}$
3. $840/[x(x + 4)]$ inches   **4.** b/a **5.** $x = \sqrt[3]{4/3}$
6. $\dfrac{dA}{dm} = \dfrac{-cm^2 - 2pcm + bp}{(cm^2 - bm)^2}$ **7.** $\dfrac{x}{3} + \dfrac{x - 2}{1.5}$ hours

■ EXERCISES 4.6 ▪ page 261

1. (a) 11, 12   **(b)** 11.5, 11.5   **3.** 10, 10
5. 25 m by 25 m
7. (a)

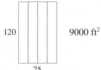

(b)

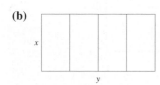

(c) $A = xy$ **(d)** $5x + 2y = 750$ **(e)** $A(x) = 375x - \frac{5}{2}x^2$
(f) 14,062.5 ft^2
9. 4000 cm^3
13. Radius $\sqrt[3]{231/\pi} \approx 4.19$ in, height $\sqrt[3]{231/\pi} \approx 4.19$ in
15. $N = 1$ **17.** $E^2/(4r)$ **19.** (0.703, 1.667)
21. $128\pi/(9\sqrt{3}) \approx 25.80$ in^3
23. (a) Use all of the wire for the square
(b) $40\sqrt{3}/(9 + 4\sqrt{3}) \approx 4.35$ m for the square
25. ≈ 4.85 km east of the refinery
27. Width $60/(4 + \pi) \approx 8.40$ ft; rectangle height
$30/(4 + \pi) \approx 4.20$ ft
29. $y = -\frac{5}{3}x + 10$
31. (a) About 5.1 km from B **(b)** C is close to B;
C is close to D; $W/L = \sqrt{25 + x^2}/x$, where x is the distance
from B to C **(c)** ≈ 1.07; no such value   **(d)** $\sqrt{41}/4 \approx 1.6$
33. $x = 6$ in.

■ PREPARE YOURSELF 4.7 ▪ page 273

1. $f'(x) = 4.2 - 0.6x + 0.006x^2$, $f''(x) = -0.6 + 0.012x$
2. $P'(t) = -4.8e^{-0.4t}$, $P''(t) = 1.92e^{-0.4t}$

3. $g'(x) = 60/x$, $g''(x) = -60/x^2$
4. $h'(a) = \dfrac{3}{2\sqrt{3a}}$ or $\dfrac{\sqrt{3}}{2\sqrt{a}}$, $h''(a) = -\dfrac{9}{4(3a)^{3/2}}$ or $-\dfrac{\sqrt{3}}{4a^{3/2}}$
5. $g(x) = -\frac{1}{300}x + 36$
6. (a) 62.12   **(b)** 4.21

■ EXERCISES 4.7 ▪ page 273

1. (a) $C(0)$ represents fixed costs, which are incurred even
when nothing is produced.
(b) The marginal cost is a minimum there.
(c)

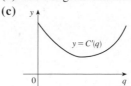

3. $17.40/unit; The cost of producing the 1001st unit is
about $17.40.
5. (a) \$342,491; \$342.49/unit; \$389.74/unit
(b) 400   **(c)** \$320/unit
7. (a) $c(q) = 3700/q + 5 - 0.04q + 0.0003q^2$,
$C'(q) = 5 - 0.08q + 0.0009q^2$
(b) Between 208 and 209 units   **(c)** $c(209) \approx \$27.45$/unit
(d) \$3.22/unit
9. 725 units   **11.** 300 units
13. (a) About 200 yd   **(b)** 192 yd
15. (a) $D(q) = 19 - \frac{1}{3000}q$ **(b)** \$9.50
17. (a) $D(q) = 550 - \frac{1}{10}q$ **(b)** \$175   **(c)** \$100
19. (a) 2571   **(b)** ≈ 0.72; changes in demand are proportionally about 72% of changes in price; inelastic
21. Yes; Lower prices would likely cause a proportionally
greater increase in demand.
23. ≈ 0.49; inelastic   **25.** ≈ 1.23; elastic
27. \$16.55; elastic for $p > \$16.55$, inelastic for $p < \$16.55$
31. (a) ≈ 0.93 **(b)** Raise   **(c)** \$29,000
33. \$50, 625 purses
35. (a) $D(q) = (418.6)(0.9768^q)$ **(b)** About 33,900
(c) ≈ 1.26; lower   **(d)** In millions: \$6.45, \$6.51, \$6.58,
\$6.32, $\approx \$5.92$; about \$175   **(e)** About \$154
37. 200   **39.** 289 boxes; about every 4.7 months

■ CHAPTER 4 REVIEW EXERCISES ▪ page 277

1. Abs max $f(4) = 5$, abs and loc min $f(3) = 1$
3. Abs max $g(-\frac{1}{2}) = g(1) = 3$, loc max $g(-\frac{1}{2}) = 3$,
abs min $g(-1) = g(\frac{1}{2}) = 1$, loc min $g(\frac{1}{2}) = 1$
5. Abs max $f(2) = \frac{2}{5}$, abs and loc min $f(-\frac{1}{3}) = -\frac{9}{2}$
7. Abs max $f(1.73) \approx 0.54$, abs min $f(4) \approx 0.28$
9. (a) Inc on $(-\infty, -2)$, $(4, \infty)$; dec on $(-2, 4)$
(b) Loc max $N(-2) = 33$, loc min $N(4) = -75$
(c) CU on $(1, \infty)$, CD on $(-\infty, 1)$; IP $(1, -21)$
11. (a) Inc on $(-\infty, \frac{3}{4})$, dec on $(\frac{3}{4}, 1)$ **(b)** Loc max $f(\frac{3}{4}) = \frac{5}{4}$
(c) CD on $(-\infty, 1)$; no IP

13. (a) Inc on $\left(-\frac{1}{4}, \infty\right)$, dec on $\left(-\infty, -\frac{1}{4}\right)$
(b) Loc min $f\left(-\frac{1}{4}\right) = -1/(4e)$ **(c)** CU on $\left(-\frac{1}{2}, \infty\right)$,
CD on $\left(-\infty, -\frac{1}{2}\right)$; IP $\left(-\frac{1}{2}, -1/(2e^2)\right)$
15. (a) Inc on $(-\infty, -1)$, $(1, 4)$; dec on $(-1, 1)$, $(4, \infty)$
(b) Loc max at $x = -1$, $x = 4$; loc min at $x = 1$
(c) CU on $(0, 2.5)$; CD on $(-\infty, 0)$, $(2.5, \infty)$
(d) $x = 0$, $x = 2.5$
17. $-\infty$ **19.** 4   **21.** ∞
23. (a) None   **(b)** Dec on $(-\infty, \infty)$ **(c)** None
(d) CU on $(-\infty, 0)$; CD on $(0, \infty)$; IP $(0, 2)$
(e)

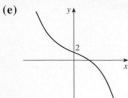

25. (a) Vertical asymptote $x = 2$, horizontal asymptote $y = \frac{3}{2}$
(b) Dec on $(-\infty, 2)$, $(2, \infty)$ **(c)** None
(d) CU on $(2, \infty)$, CD on $(-\infty, 2)$; no IP
(e)

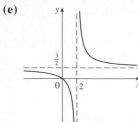

27. (a) None   **(b)** Inc on $(0, \infty)$, dec on $(-\infty, 0)$
(c) Loc min $f(0) = \ln 4$ **(d)** CU on $(-2, 2)$;
CD on $(-\infty, -2)$, $(2, \infty)$; IP $(-2, \ln 8)$, $(2, \ln 8)$
(e)

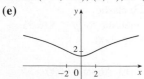

29.

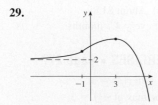

31. Inc on $(-1.73, 0)$, $(0, 1.73)$;
dec on $(-\infty, -1.73)$, $(1.73, \infty)$;
loc max $f(1.73) \approx 0.38$,
loc min $f(-1.73) \approx -0.38$;
CU on $(-2.45, 0)$, $(2.45, \infty)$;
CD on $(-\infty, -2.45)$, $(0, 2.45)$;
IP $(2.45, 0.34)$, $(-2.45, -0.34)$

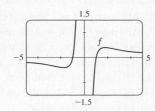

33. Inc on $(-0.23, 0)$, $(1.62, \infty)$; dec on $(-\infty, -0.23)$, $(0, 1.62)$;
loc max $f(0) = 2$; loc min $f(-0.23) \approx 1.96$, $f(1.62) \approx -19.2$;
CU on $(-\infty, -0.12)$, $(1.24, \infty)$; CD on $(-0.12, 1.24)$;
IP $(-0.12, 1.98)$, $(1.24, -12.1)$

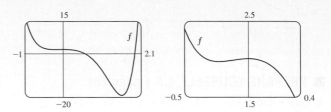

35.

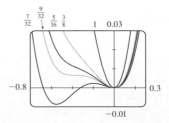

$(\pm 0.82, 0.22)$; $\left(\pm\sqrt{2/3}, e^{-3/2}\right)$

37. If $c = \frac{9}{32}$, f has two critical numbers. It has one for $c > \frac{9}{32}$
and three for $c < \frac{9}{32}$ (except when $c = 0$). If $c \geqslant \frac{3}{8}$, f has no IP;
it has two if $c < \frac{3}{8}$.

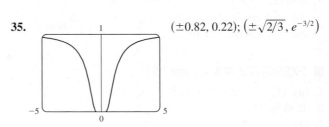

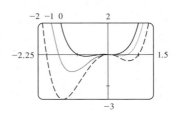

39. About 4.86 years after December 31, 1990 (November 1995)
41. $L = C$ **43.** ≈ -3.19 thousand discs per month
45. 13 ft/s   **47.** 500 and 125
49. 1050 ft (parallel to building) by 350 ft
51. (a) \$28,800, \$15.30/helmet, \$14.40/helmet, \$21,200
(b) 1761   **(c)** 2370
53. \$11.50
55. (a) $E(q) = (7392 - q)/q$
(b) $E(q) < 1$ for $D(q) < \$154$, $E(q) > 1$ for $D(q) > \$154$
(c) \$154   **(d)** $R(q) = 308q - \frac{1}{24}q^2$; \$569,184; yes

■ **EXERCISES 5.1** ▪ **page 293**

1. (a) $30,220 **(b)** $15,600

3. 44.8 ft

5. (a) 12,475 ft ≈ 2.36 mi

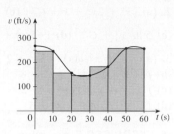

(b) 13,022 ft ≈ 2.47 mi

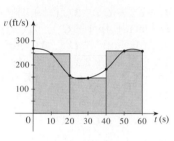

7. 70 L using left endpoints, 63.2 L using right endpoints

9. (a) 41; higher   **(b)** 33; lower

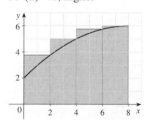

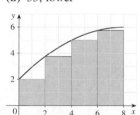

(c) ≈39.1

11. (a) $\frac{77}{60}$, underestimate   **(b)** $\frac{25}{12}$, overestimate

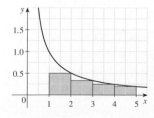

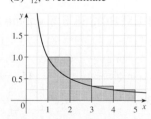

13. (a) 8, 6.875   **(b)** 5, 5.375

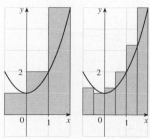

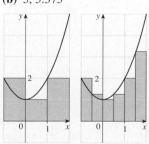

(c) 5.75, 5.9375

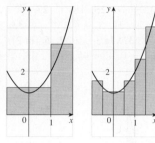

(d) Midpoints with six rectangles

15. (a) Positive   **(b)** Negative

(c) Negative   **(d)** Positive

17. (a) 0   **(b)** −3

19. (a) 4   **(b)** 10   **(c)** −3   **(d)** 2

21. 5   **23.** 2.5   **25.** 124.1644   **27.** 1.6498

29. Lower: −64 (using left endpoints),
upper: 16 (using right endpoints)

31. (a) 4   **(b)** 6   **(c)** 10

33. (a) Yes   **(b)** No, if f is the function graphed in
Exercise 19, then $f(x) < 0$ for $5 \le x \le 6$ but $\int_2^6 f(x) > 0$.

35. B < E < A < D < C   **37. (a)** 8

■ **PREPARE YOURSELF 5.2** ▪ **page 306**

1. $16t^2 + 24t + 9$ **2.** $2x^{5/2} - 3x^{1/2}$

3. (a) $x^{1/2}$ **(b)** $x^{2/3}$ **(c)** x^{-2} **(d)** $x^{-3/2}$

4. $2x^2 + 5x + 2x^{-2}$

5. (a) $dy/dx = x^3$ **(b)** $B'(t) = t^{1/2} = \sqrt{t}$

(c) $L'(u) = 1/u$ **(d)** $dP/dt = 7.3e^t$

(e) $g'(t) = e^{-0.2t}$ **(f)** $f'(v) = 1/\sqrt{v}$

(g) $h'(x) = -3/x^4$ **(h)** $dA/dt = 5^t \ln 5$

■ **EXERCISES 5.2** ▪ **page 306**

1. $F(x) = 2x^3 - 4x^2 + 3x + C$

3. $F(x) = 4x^{5/4} - 4x^{7/4} + C$

5. $F(x) = 2x^{3/2} - x^{-5} + C$ (on any interval that doesn't
include 0)

7. $Q(s) = 5e^s + 3.7s + C$ **9.** $F(q) = q + 2.5e^{0.8q} + C$

11. $G(x) = -\dfrac{1}{x^5} + \dfrac{2}{x^2} + 2x + C$ (on any interval that doesn't
include 0)

13. $f(x) = \frac{8}{3}x^3 - 3\ln|x| + C$

15. $f(x) = 4x^{3/2} + 2x^{5/2} + 4$

17. (a) $f'(x) = 8x^3 + x^2 + 10x - 22$

(b) $f(x) = 2x^4 + \frac{1}{3}x^3 + 5x^2 - 22x + \frac{59}{3}$

19. (a)   **(b), (d)**

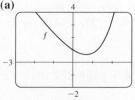

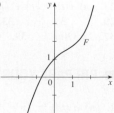

(c) $F(x) = e^x - x^2$

21. 28   **23.** $\frac{364}{3}$ **25.** 18   **27.** $\frac{208}{3}$
29. $-2 + 1/e$ **31.** $\frac{22}{3} - \frac{10}{3}e^{-2.4}$ **33.** 212
35. $\ln 2 + 7$ **37.** $\ln 3$ **39.** $5/(\ln 2)$
41. $30 + 0.3e^4 - 0.3e \approx 45.56$ **43.** $\frac{80}{3} \approx 26.67$
45. 0, 1.32; 0.84
47. 3.75   **49.** 9

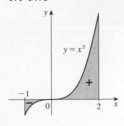

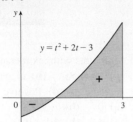

51. $3x^4 + C$ **53.** $\frac{1}{3}t^3 + \frac{3}{2}t^2 + 4t + C$
55. $0.0025q^4 + 0.2q^3 + 1.75q^2 + 14.9q + C$
57. $\frac{5}{2}e^{2t} + C$ **59.** $2u^3 - 2u^{3/2} + C$
61. $2t - t^2 + \frac{1}{3}t^3 - \frac{1}{4}t^4 + C$
63. $\frac{2}{5}x^{5/2} + C$

65. (a) $\frac{1}{8}x^4 + 2x^2 - x + C$ (b) 8
69. 122   **71.** $\int_{-1}^{5} f(x)\,dx$ **73.** $\frac{4}{3}$ **75.** $\frac{413}{6} \approx 68.83$
77. 3

■ EXERCISES 5.3 ▪ page 314

1. The change in cost, in thousands of dollars, when guitar production is increased from 300 to 500
3. The increase in the child's weight (in pounds) between the ages of 5 and 10
5. Number of gallons of oil leaked in the first 2 hours
7. The change in elevation between 3 and 5 miles (measured horizontally) from the start of the trail
9. 18,350 mB (18.35 GB) of data flowed through the router between 6:00 AM and 8:00 AM
11. $58,000   **13.** 142,956   **15.** 16,442
17. ≈ 4512 L   **19.** Thousands of people
21. ≈ 587.6 kilowatt-hours   **23.** 3.54 ft/s
25. (a) $-\frac{3}{2}$ m   (b) $\frac{41}{6}$ m
27. (a) $v(t) = \frac{1}{2}t^2 + 4t + 5$ m/s   (b) $416\frac{2}{3}$ m
29. $\frac{8}{3}$ **31.** $\frac{45}{28}$ **33.** $\approx 57.2°F$ **35.** ≈ 4.06 mg/L
37. 28,320 L   **39.** ≈ 38.3

■ PREPARE YOURSELF 5.4 ▪ page 321

1. (a) $dy/dx = 3x^2 e^{x^3+1}$ (b) $Q'(t) = \dfrac{3 + 2t}{3t + t^2}$

(c) $f'(x) = 16x(2x^2 + 3)^3$ (d) $g'(z) = \dfrac{e^z + 5}{2\sqrt{e^z + 5z}}$
(e) $dr/dt = 3^{2t+2}(2 \ln 3)$
2. (a) $\frac{1}{2}x^6 + 2x^2 - x + C$ (b) $\frac{16}{3}t^{3/2} + C$
(c) $5 \ln|v| + C$ (d) $(-5/v) + C$ (e) $\dfrac{1}{\ln 4}4^x + C$
3. (a) $f(x) = x^4$, $g(x) = 3x^2 + 2$
(b) $f(x) = \sqrt{x}$, $g(x) = x^3 + 8$
(c) $f(x) = 1/x$, $g(x) = x^3 - 2$
(d) $f(x) = e^x$, $g(x) = x^2 + 1$

■ EXERCISES 5.4 ▪ page 321

1. $-e^{-x} + C$ **3.** $\frac{2}{9}(x^3 + 1)^{3/2} + C$
5. $\frac{1}{8}\ln(1 + 4p^2) + C$ **7.** $-\frac{1}{10}(3 - t^2)^5 + C$
9. $\frac{1}{63}(3x - 2)^{21} + C$ **11.** $\frac{1}{3}(q^2 + 3.1)^{3/2} + C$
13. $\frac{5}{3}\sqrt{0.4x^3 + 2.2} + C$ **15.** $-5/[6(z^3 + 2)^2] + C$
17. $\frac{2}{3}(1 + e^x)^{3/2} + C$ **19.** $\frac{1}{3}(\ln x)^3 + C$
21. $\frac{1}{4}e^{2t^2} + C$ **23.** $-\frac{1}{3}\ln|5 - 3x| + C$
25. $2e^{\sqrt{t}+1} + C$ **27.** $-\dfrac{1}{4 \ln 2}2^{3-4t} + C$
29. $\frac{1}{15}(x^3 + 3x)^5 + C$
31. $\frac{1}{8}(x^2 - 1)^4 + C$

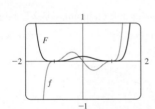

33. $\frac{45}{28}$ **35.** $\frac{182}{9}$ **37.** $\frac{1}{4}$ **39.** $2(e^8 - 1)$
41. $2(e^2 - e)$ **43.** 2   **45.** $\frac{16}{15}$ **47.** 0
49. About 34,200   **51.** About 142.4
53. $\frac{1}{40}(2x + 5)^{10} - \frac{5}{36}(2x + 5)^9 + C$ **55.** 5   **57.** 6π
59. Yes

■ EXERCISES 5.5 ▪ page 327

1. $\frac{1}{2}x^2 \ln x - \frac{1}{4}x^2 + C$ **3.** $2(r - 2)e^{r/2} + C$
5. $\frac{1}{4}x^4 \ln 2x - \frac{1}{16}x^4 + C$ **7.** $(2z + 1)e^{-z} + C$
9. $x \ln \sqrt[3]{x} - \frac{1}{3}x + C$ **11.** $-e^{-3r}\left(\frac{1}{3}r^2 + \frac{2}{9}r + \frac{2}{27}\right) + C$
13. $\frac{3}{16}e^4 + \frac{1}{16}$ **15.** $\dfrac{1}{4} - \dfrac{3}{4e^2}$ **17.** $\frac{1}{2} - \frac{1}{2}\ln 2$
19. $2(\ln 2)^2 - 4 \ln 2 + 2$
21. $-\frac{1}{2}xe^{-2x} - \frac{1}{4}e^{-2x} + C$

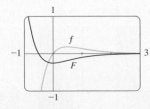

23. 14.133   **25.** 1.936   **27.** 8.592
29. (a) $2 - e^{-t}(t^2 + 2t + 2)$ m   (b) ≈ 1.99 m
31. $\frac{1}{2}(x^2 - 1)\ln(1 + x) - \frac{1}{4}x^2 + \frac{1}{2}x + \frac{3}{4} + C$

35. $x[(\ln x)^3 - 3(\ln x)^2 + 6 \ln x - 6] + C$
37. 2

■ CHAPTER 5 REVIEW EXERCISES ▪ page 329

1. (a) $820,200 **(b)** $242,350
3. (a) 8 **(b)** 5.7

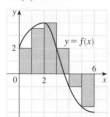

5. $\frac{1}{2} + \pi/4$ **7.** ≈ 56.98
9. $F(x) = \frac{1}{2}x^4 + 3x^2 - 7x + C$
11. $P(r) = 4r + 5 \ln |r| + C$ **13.** 37 **15.** $\frac{9}{10}$
17. $\frac{5}{2}(e^4 - 1)$ **19.** $\frac{45}{2}$ **21.** $2.4t^3 - 2.3t^2 + 18.1t + C$
23. $6 \ln |x| + \frac{1}{2}x^2 + C$
25. Number of barrels of oil consumed from January 1, 2000, through January 1, 2008
27. The change in cost when production is increased from 500 to 1000 laptops
29. $780,000 **31.** $7(1 - e^{-3}) \approx 6.65$ gal
33. (a) $29.1\overline{6}$ m **(b)** 29.5 m
35. $-\frac{37}{3}(e^{-0.6} - 1) \approx 5.56$ oz **37.** $\frac{1}{21}(1 + x^3)^7 + C$
39. $\frac{1}{2} \ln 2$ **41.** $\frac{1}{2}(e^4 - 1)$ **43.** $\frac{2}{3}(e^x + 2)^{3/2} + C$
45. $\frac{1}{2 \ln 5} 5^{x^2} + C$ **47.** $\frac{64}{5} \ln 4 - \frac{124}{25}$
49. $-\frac{1}{e^{4t}}\left(\frac{1}{4}t^2 + \frac{1}{8}t + \frac{1}{32}\right) + C$ **51.** $26/(\ln 3)$
53. $\sqrt{x^2 + 4x} + C$
55. $-4e^{-x^2} + C$

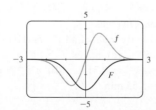

57. 144.26 **59.** 3

■ PREPARE YOURSELF 6.1 ▪ page 337

1. (a) $-\frac{328}{3}$ **(b)** $e^2 - 3$ **(c)** $\ln 4 - \frac{28}{3}$
(d) $\frac{1.5}{\ln 2.5} - \frac{3}{2}$ **(e)** $\ln 2 - 2$
2. ≈ 48.76 **3.** $(-4, 24), (3, 10)$
4. The cost (beyond the fixed costs), in thousands of dollars, of producing the first 1500 units
5. Number of quarts of oil that drain from the end of the 5th minute to the end of the 20th minute

■ EXERCISES 6.1 ▪ page 338

1. $\frac{32}{3}$ **3.** $2e^2 - 8$ **5.** 19.5 **7.** $e - (1/e) + \frac{4}{3}$
9. $\frac{1}{6}$ **11.** $\ln 2 - \frac{1}{2}$ **13.** $\frac{1}{3}$ **15.** 72 **17.** $e - 2$
19. (a) 39 **(b)** 15
21. ≈ 6.32 **23.** 8.38 **25.** 2.80
27. ≈ 393.33; The company's profit decreases by $393.33 after producing and selling the first 200 heaters.
29. ≈ 12.979; In 2011 the factory produced 12,979 more hard drives than in 2010.
31. ≈ 0.206; The value of one of the buildings increased about $206,000 more than the other, 10 years after the purchase date.
33. (a) A **(b)** The distance by which A is ahead of B after 1 minute **(c)** A **(d)** ≈ 2.2 minutes
35. $117\frac{1}{3}$ ft **37.** 4232 cm² **39.** $\ln 2$ **41.** $4^{2/3}$

■ EXERCISES 6.2 ▪ page 348

1. $400,000 **3.** $100,266.67
5. $195,392.69 **7.** $1066.67
9. $346,953.73

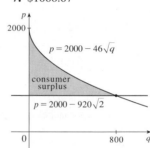

11. $407.25 **13.** $p = -\frac{1}{30}q + 25$; $1500 **15.** $3750
17. $18,304.26 **19.** $12,000 **21.** 3727; $37,753
23. (a) $160 **(b)** $324,900 **(c)** $324,900
25. $112.15 **27.** $233,539.66
29. (a) $433,107.37 **(b)** $267,999.65
31. $498,814.68; no **33.** $346,884.18; $159,013.79
35. $\frac{2}{3}\left(16\sqrt{2} - 8\right) \approx 9.75 million
37. $\dfrac{(1 - k)(b^{2-k} - a^{2-k})}{(2 - k)(b^{1-k} - a^{1-k})}$

■ EXERCISES 6.3 ▪ page 353

1. (a) 1480 **(b)** 9440 **(c)** Not all new members survive
3. 21,046 **5.** 6265 **7.** ≈ 48.3 mg
9. $\approx 12,417$ gallons **11.** 0.00019 cm³/s
13. 6.60 L/min **15.** 5.69 L/min
17. $0.4L$ units per hour, where L is the desired level

■ PREPARE YOURSELF 6.4 ▪ page 362

1. $-3e^{-3x}$; $-2e^{-3x}$ **2.** $(x^3 + 3x^2)e^x$
3. (a) $\ln |x| + C$ **(b)** $-1/x + C$ **(c)** $\ln |x + 4| + C$
(d) $\frac{1}{2} \ln(x^2 + 4) + C$ **(e)** $-\frac{1}{2}e^{-2t} + C$ **(f)** $\frac{2}{3}t^{3/2} + C$

■ **EXERCISES 6.4** ■ page 362

5. $y = \dfrac{2}{K - x^2}, y = 0$ **7.** $y^2 - 2y = x^2 + 2x + K$

9. $y = Kx$ **11.** $y = K\sqrt{x^2 + 1}$ **13.** $u = Ae^{2t+t^2/2} - 1$

15. $y = -\sqrt{x^2 + 9}$ **17.** $y = \sqrt[3]{\frac{3}{2}e^{2x} + \frac{51}{2}}$

19. $A = \sqrt{2 \ln t + 16}$

21. (a) It must be either 0 or decreasing.
(c) $y = 0$ **(d)** $y = 1/(x + 2)$

23. $y = e^{x^2/2}$ **25.** $\dfrac{dB}{dt} = \dfrac{k}{B}$

27. $dP/dt = \frac{1}{100}P$; $P = 20,000e^{t/100}$

29. $dP/dt = \frac{1}{50}P + 500$; 388,156

31. (a) At the beginning; stays positive, but decreases
(c)

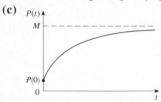

(d) $P(t) = M - Me^{-kt}$; M
33. $y = \pm\sqrt{[3(te^t - e^t + C)]^{2/3} - 1}$
35. $y = Ke^x - x - 1$
37. (a) $C(t) = (C_0 - r/k)e^{-kt} + r/k$
(b) r/k; The concentration approaches r/k regardless of the value of C_0.
39. $dA/dt = k\sqrt{A}\,(M - A)$

■ **PREPARE YOURSELF 6.5** ■ page 368

1. (a) 0   **(b)** ∞ **(c)** ∞ **(d)** ∞ **(e)** 1
(f) 0   **(g)** ∞

2. (a) $-\dfrac{2}{3}\left(\dfrac{1}{6^{3/2}} - \dfrac{1}{3^{3/2}}\right)$ **(b)** $3e^{-1/3} - 3e^{-w/3}$

(c) $\frac{1}{2}e^{t^2} + C$ **(d)** $\ln|\ln x| + C$ **(e)** $-2.5/t^{0.4} + C$

■ **EXERCISES 6.5** ■ page 368

1. $\frac{1}{2} - 1/(2t^2)$; 0.495, 0.49995, 0.4999995; 0.5
3. 1   **5.** 2   **7.** Divergent   **9.** $2e^{-2}$ **11.** 0
13. Divergent   **15.** $\frac{1}{2}$
17. e **19.** Infinite area

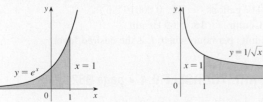

21. $96,000   **23.** $7142.86   **25.** $177,777.78

27. (a)

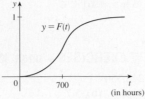

(b) The rate at which the fraction $F(t)$ increases as t increases
(c) 1; All bulbs burn out eventually.

■ **EXERCISES 6.6** ■ page 376

1. (a) The probability that a randomly chosen tire will have a lifetime between 30,000 and 40,000 miles
(b) The probability that a randomly chosen tire will have a lifetime of at least 25,000 miles
3. (b) $1 - \frac{3}{8}\sqrt{3} \approx 0.350$

5. (a) $c = 2$ **(b)** $\dfrac{1}{e} - \dfrac{1}{e^{16}} \approx 0.368$

7. (a) $f(x) \geq 0$ for all x and $\int_{-\infty}^{\infty} f(x)\, dx = 1$ **(b)** 5
11. (a) $e^{-4/2.5} \approx 0.202$ **(b)** $1 - e^{-2/2.5} \approx 0.551$ **(c)** If you aren't served within 10 minutes, you get a free hamburger.
13. $\approx 44\%$
15. (a) 0.0668   **(b)** $\approx 5.21\%$
17. ≈ 0.9545

■ **CHAPTER 6 REVIEW EXERCISES** ■ page 378

1. $\frac{28}{3}$ **3.** $\frac{32}{3}$ **5.** $\frac{8}{3}$ **7.** $\frac{7}{12}$
9. ≈ 3208.33; the manufacturer's profit increases $3208.33 after producing and selling the first 500 backpacks
11. $7166.67   **13.** $19,445.07   **15.** $174,438.57

17. 54,916   **19.** 10,836   **21.** $\dfrac{dA}{dt} = \dfrac{k}{A^2}$

23. $B = -\frac{1}{2}\ln(K - 4\sqrt{t}\,)$ **25.** $y = \sqrt{1/(\frac{1}{8} - x^2)}$
27. $\frac{1}{36}$ **29.** Diverges
31. (a) $f(x) \geq 0$ for all x and $\int_{-\infty}^{\infty} f(x)\, dx = 1$
(b) $\frac{7}{27} \approx 0.259$ **(c)** 6; yes
33. (a) $1 - e^{-3/8} \approx 0.313$ **(b)** $e^{-5/4} \approx 0.287$
(c) $8 \ln 2 \approx 5.55$ min

■ **PREPARE YOURSELF 7.1** ■ page 391

1. (a) $[-5, \infty)$ **(b)** $[-2, 2]$
(c) $\{a \mid a \neq 2\}$ **(d)** $\left(-\infty, \frac{1}{2}\right)$
2. (a)   **(b)**

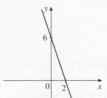
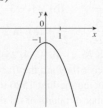

(c)

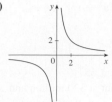

(d)

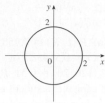

23. $y = 2x - k$

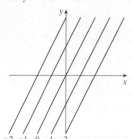

25. $y = k/x$

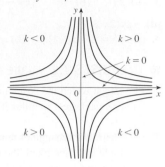

■ EXERCISES 7.1 ▪ page 391

1. **(a)** 37   **(b)** -63 **(c)** -26 **(d)** $8x - 11$
3. **(a)** 9   **(b)** $9e$ **(c)** $1/e^3$ **(d)** $4e^{3y}$
5. **(a)** 3   **(b)** $\frac{3}{2}$ **(c)** 1   **(d)** $-x$
7. **(a)** $e^{(x+h)y} - e^{xy}$ **(b)** $e^{x(y+h)} - e^{xy}$
9. **(a)** 4   **(b)** The entire xy-plane
11. $\{(x, y) \mid y \geq -x\}$

13. $\{(x, y) \mid y \neq x\}$

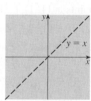

15. $\{(x, y) \mid x^2 + y^2 < 4\}$

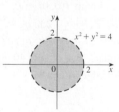

17. $(4, 0, -3)$
19.

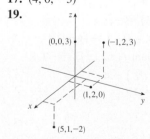

21. $x + y + z = 1$, plane

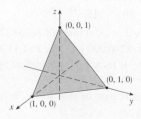

27. $y = ke^{-x}$

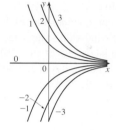

29. **(a)** B   **(b)** III
31. **(a)** A   **(b)** IV
33. ≈ 94.2; When the manufacturer invests $20 million and 120,000 labor hours are spent, its yearly production is about $94.2 million.
35. **(a)** $C = 8000 + 2.5x + 4y + 4.5z$ dollars
(b) 53,500; It costs the company $53,500 to make 3000 small boxes, 5000 medium boxes, and 4000 large boxes.
(c) $\{(x, y, z) \mid x \geq 0, y \geq 0, z \geq 0\}$
37. **(a)** ≈ 20.52; a person who weighs 160 pounds and is 70 inches tall has surface area about 20.52 square feet
39.

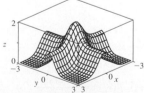

41.

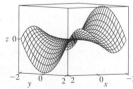

■ PREPARE YOURSELF 7.2 ▪ page 404

1. Rain is falling at a rate of 0.6 inches per hour 1.5 hours after the start of the storm.
2. **(a)** $g'(x) = 15x^2 - 16x + 13$ **(b)** $f'(x) = 8(x + 2)^7$
(c) $K'(v) = 3^v \ln 3$ **(d)** $B'(u) = (u^3 + 3u^2)e^u$
(e) $H'(t) = (t - 3)e^t/t^4$ **(f)** $f'(x) = \dfrac{7 - 7x^2}{(x^2 + 1)^2}$
(g) $g'(y) = \dfrac{1}{2\sqrt{y}} + \dfrac{1}{y}$ **(h)** $dy/dx = 2xe^{x^2+2}$
(i) $\dfrac{dy}{dt} = \dfrac{2t - 5}{t^2 - 5t}$ **(j)** $A'(t) = \dfrac{3t^3}{2\sqrt{t^3 - 1}} + \sqrt{t^3 - 1}$

3. (a) $3x^2$ (b) ae^x (c) $\dfrac{1 + 2ax}{x + ax^2 + b}$

(d) $-c/(2x - c)^2$

4. (a) $y'' = 2/(x - 1)^3$ (b) $y'' = 1/(x^2 + 1)^{3/2}$

▪ EXERCISES 7.2 ▪ page 404

1. The rate at which the cost changes, in dollars per pound, to deliver a package 150 miles if the weight changes from 80 pounds

3. (a) The rate of change of temperature as longitude varies, with latitude and time fixed; the rate of change as only latitude varies; the rate of change as only time varies.

(b) Positive, negative, positive

5. (a) $f_T(-15, 30) \approx 1.3$; For a temperature of $-15°C$ and wind speed of 30 km/h, the wind-chill index rises by 1.3°C for each degree the temperature increases. $f_v(-15, 30) \approx -0.15$; For a temperature of $-15°C$ and wind speed of 30 km/h, the wind-chill index decreases by 0.15°C for each km/h the wind speed increases.

(b) Positive, negative   (c) 0

7. (a) Positive   (b) Negative

(c) Negative   (d) Negative

9. $f_x(1, 2) = -8 =$ slope of C_1, $f_y(1, 2) = -4 =$ slope of C_2

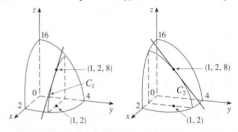

11. (a) $f(x, -2) = -2x^2 - 20x$, $f_x(x, -2) = -4x - 20$

(b) $f(3, y) = 9y - 15y^2$, $f_y(3, y) = 9 - 30y$

13. $f_x(x, y) = -3y$, $f_y(x, y) = 5y^4 - 3x$

15. $f_x(x, y) = 4x^3 + 2xy^2$, $f_y(x, y) = 2x^2y + 4y^3$

17. $\partial z/\partial x = 20(2x + 3y)^9$, $\partial z/\partial y = 30(2x + 3y)^9$

19. $f_x(x, y) = 2y/(x + y)^2$, $f_y(x, y) = -2x/(x + y)^2$

21. $f_r(r, s) = \dfrac{2r^2}{r^2 + s^2} + \ln(r^2 + s^2)$, $f_s(r, s) = \dfrac{2rs}{r^2 + s^2}$

23. $\partial u/\partial t = e^{w/t}(1 - w/t)$, $\partial u/\partial w = e^{w/t}$

25. $f_s(s, t) = -\dfrac{3s}{\sqrt{2 - 3s^2 - 5t^2}}$, $f_t(s, t) = -\dfrac{5t}{\sqrt{2 - 3s^2 - 5t^2}}$

27. $f_x = z - 10xy^3z^4$, $f_y = -15x^2y^2z^4$, $f_z = x - 20x^2y^3z^3$

29. $\partial w/\partial x = 1/(x + 2y + 3z)$, $\partial w/\partial y = 2/(x + 2y + 3z)$, $\partial w/\partial z = 3/(x + 2y + 3z)$

31. $f_x = 1/(y + z)$, $f_y = -x/(y + z)^2$, $f_z = -x/(y + z)^2$

33. $f_x = yzx^{yz-1}$, $f_y = zx^{yz} \ln x$, $f_z = yx^{yz} \ln x$

35. $f_x = y^2z^3t^4$, $f_y = 2xyz^3t^4$, $f_z = 3xy^2z^2t^4$, $f_t = 4xy^2z^3t^3$

37. -27 **39.** $\frac{1}{5}$ **41.** $\frac{1}{4}$

43. $f_{xx} = 6xy^5 + 24x^2y$, $f_{xy} = 15x^2y^4 + 8x^3 = f_{yx}$, $f_{yy} = 20x^3y^3$

45. $w_{uu} = v^2/(u^2 + v^2)^{3/2}$, $w_{uv} = -uv/(u^2 + v^2)^{3/2} = w_{vu}$, $w_{vv} = u^2/(u^2 + v^2)^{3/2}$

51. (a) $P_L(L, K) = 0.9555L^{-0.35}K^{0.35}$, $P_K(L, K) = 0.5145L^{0.65}K^{-0.65}$

(b) $P_L(120, 20) \approx 0.51$, $P_K(120, 20) \approx 1.65$

(c) If capital investment is held constant at $20 million and labor is increased from 120,000 labor hours, then production would increase at a rate of about 0.51 units per thousand labor hours. If labor is held constant at 120,000 labor hours and capital investment is increased from $20 million, then production would increase at a rate of about 1.65 units per million dollars.

53. Complementary   **55.** R^2/R_1^2

▪ EXERCISES 7.3 ▪ page 412

1. (a) f has a local minimum at $(1, 1)$.

(b) f has a saddle point at $(1, 1)$.

3. Maximum $f\left(-1, \frac{1}{2}\right) = 11$ **5.** Minimum $f(-1, 2) = -4$

7. Minimum $f\left(\frac{1}{3}, -\frac{2}{3}\right) = -\frac{1}{3}$

9. Saddle points at $(1, 1)$, $(-1, -1)$

11. Maximum $f\left(\frac{1}{3}, \frac{1}{3}\right) = \frac{1}{27}$, saddle points at $(0, 0)$, $(1, 0)$, $(0, 1)$

13. Maximum $f(0, 0) = 2$, minimum $f(0, 4) = -30$, saddle points at $(2, 2)$, $(-2, 2)$

15. Minimum $f(2, 1) = -8$, saddle point at $(0, 0)$

17. Maximum $f\left(\frac{1}{2}, 0\right) = \frac{1}{2}e^{-1/2}$, minimum $f\left(-\frac{1}{2}, 0\right) = -\frac{1}{2}e^{-1/2}$

19. Minimum $f(0, 0) = 0$, saddle points at $(1, 0)$, $(-1, 0)$

21. $\frac{100}{3}, \frac{100}{3}, \frac{100}{3}$ **23.** $q_A = 2.4$, $q_B = 2.8$

25. $p_1 = 27$, $p_2 = 28$; $q_1 = 6$, $q_2 = 12$

27. Square base of side 40 cm, height 20 cm

29. All edges are 2.5 ft

▪ EXERCISES 7.4 ▪ page 419

1. No maximum, minimum $f\left(\frac{1}{2}, -\frac{1}{4}\right) = -\frac{1}{8}$

3. No maximum, minimum $f(1, 1) = f(-1, -1) = 2$

5. Maximum $f(\pm 2, 1) = 4$, minimum $f(\pm 2, -1) = -4$

7. Maximum $f(\pm\sqrt{2}, 0) = f(0, \pm\sqrt{2}) = 4$, minimum $f(1, \pm 1) = f(-1, \pm 1) = 2$

9. Maximum $f(1, 3, 5) = 70$, minimum $f(-1, -3, -5) = -70$

11. Maximum $f(2, 2, 1) = 9$, minimum $f(-2, -2, -1) = -9$

13. 50, 50   **15.** $\frac{100}{3}, \frac{100}{3}, \frac{100}{3}$

17. $25\sqrt{2} \approx 35.4$ units of A, $20\sqrt{2} \approx 28.3$ units of B

19. (a) $240,000 on labor and $120,000 on capital

(b) $\lambda = 4$; Increasing the budget by $1 will increase the maximum production by 4 units.

21. Square base of side 40 cm, height 20 cm

23. All edges are 2.5 ft

■ CHAPTER 7 REVIEW EXERCISES ■ page 420

1.

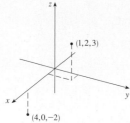

3. (a) e^2 **(b)** $\{(x, y) \mid x \neq 1\}$

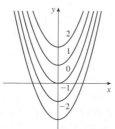

(c) $g_x(2, 1) = 0, g_y(2, 1) = 2e^2$
5. $y = x^2 + k$

7. $f_x(x, y) = 2xy^4 - 2y^5, f_y(x, y) = 4x^2y^3 - 10xy^4$
9. $h_x(x, y) = (2xy + 1)e^{2xy}, h_y(x, y) = 2x^2e^{2xy}$
11. $F_\alpha(\alpha, \beta) = \dfrac{2\alpha^3}{\alpha^2 + \beta^2} + 2\alpha \ln(\alpha^2 + \beta^2),$

$F_\beta(\alpha, \beta) = \dfrac{2\alpha^2\beta}{\alpha^2 + \beta^2}$

13. $G_x = \dfrac{1}{y + 2z}, G_y = -\dfrac{x}{(y + 2z)^2}, G_z = -\dfrac{2x}{(y + 2z)^2}$
15. $f_{xx} = 24x, f_{xy} = -2y = f_{yx}, f_{yy} = -2x$
17. $\approx 3.5°C/m, -3.0°C/m$
19. (a) $P_L(L, K) = 400L^{-1/3}K^{1/3}, P_K(L, K) = 200L^{2/3}K^{-2/3}$
(b) $P_L(100, 80) \approx 371.3, P_K(100, 80) \approx 232.1.$
If capital investment is held constant at 80 units and labor is increased from 100 units, then production would increase at a rate of about 371.3 units per unit labor. If labor is held constant at 100 units and capital investment is increased from 80 units, then production would increase at a rate of about 232.1 units per unit of capital.
(c) Labor (assuming that the units of labor and capital have the same monetary value)
21. Minimum $f(-4, 1) = -11$
23. Maximum $f(1, 1) = 1$; saddle points $(0, 0), (0, 3), (3, 0)$
25. Maximum $f(\pm\sqrt{2/3}, 1/\sqrt{3}) = 2/(3\sqrt{3}),$
minimum $f(\pm\sqrt{2/3}, -1/\sqrt{3}) = -2/(3\sqrt{3})$
27. Maximum 1, minimum -1

■ EXERCISES A ■ page A15

1. $-3a^2bc$ **3.** $2x^2 - 10x$ **5.** $-8 + 6a$
7. $-x^2 + 6x + 3$ **9.** $12x^2 + 25x - 7$
11. $4x^2 - 4x + 1$ **13.** $30y^4 + y^5 - y^6$
15. $2x^3 - 5x^2 - x + 1$ **17.** $1 + 4x$
19. $\dfrac{3x + 7}{x^2 + 2x - 15}$ **21.** $\dfrac{u^2 + 3u + 1}{u + 1}$ **23.** $\dfrac{x}{yz}$
25. $\dfrac{rs}{3t}$ **27.** $\dfrac{c}{c - 2}$ **29.** $2x(1 + 6x^2)$
31. $(x + 6)(x + 1)$ **33.** $(x - 4)(x + 2)$
35. $9(x - 2)(x + 2)$ **37.** $(3x + 2)(2x - 3)$
39. $(t + 1)(t^2 - t + 1)$ **41.** $(2t - 3)^2$ **43.** $x(x + 1)^2$
45. $\dfrac{x + 2}{x - 2}$ **47.** $\dfrac{x + 1}{x - 8}$ **49.** $\dfrac{x - 2}{x^2 - 9}$
51. $(x + 1)^2 + 4$ **53.** $(x - \frac{5}{2})^2 + \frac{15}{4}$ **55.** $(2x + 1)^2 - 3$
57. $1, -10$ **59.** $\dfrac{-9 \pm \sqrt{85}}{2}$ **61.** $\dfrac{-5 \pm \sqrt{13}}{6}$
63. Irreducible   **65.** Not irreducible
67. $a^6 + 6a^5b + 15a^4b^2 + 20a^3b^3 + 15a^2b^4 + 6ab^5 + b^6$
69. 8   **71.** $4a^2b\sqrt{b}$ **73.** $3p^2r^2\sqrt{10p}$ **75.** 3^{26}
77. $16x^{10}$ **79.** $\dfrac{a^2}{b}$ **81.** $\dfrac{1}{\sqrt{3}}$ **83.** 25
85. $2\sqrt{2}\,|x|^3y^6$ **87.** $t^{-5/2}$ **89.** $\dfrac{1}{\sqrt{x} + 3}$
91. $\dfrac{x^2 + 4x + 16}{x\sqrt{x} + 8}$ **93.** $\dfrac{3 + \sqrt{5}}{2}$ **95.** False
97. True   **99.** False   **101.** False
103. $(-2, \infty)$ **105.** $[-1, \infty)$
107. $(0, 1]$ **109.** $(-\infty, 1) \cup (2, \infty)$
111. $(-\sqrt{3}, \sqrt{3})$ **113.** $(-\infty, 1]$
115. $(-1, 0) \cup (1, \infty)$ **117.** $[10, 35]$
119. (a) $T = 20 - 10h, 0 \leq h \leq 12$
(b) $-30°C \leq T \leq 20°C$
121. $1, -7$

■ EXERCISES B ■ page A22

1. 5   **3.** $\sqrt{74}$ **5.** $2\sqrt{37}$ **7.** 2   **9.** $-\frac{9}{2}$
11.

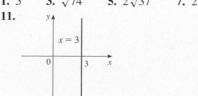

13. $y = 4x - 12$ **15.** $2x - 3y + 19 = 0$

17. $5x + y = 23$ **19.** $y = 6x - 4$ **21.** $y = 2x - 4$
23. $y = 5$ **25.** $x + 2y + 11 = 0$
27. $5x - 2y + 1 = 0$
29. $m = -\frac{1}{3}$,   **31.** $m = 0$,
    $b = 0$ $b = -2$

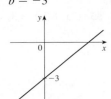   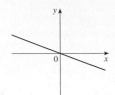

33. $m = \frac{3}{4}$,
    $b = -3$

35.

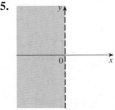

37.

39.

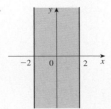

41.

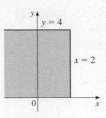

43.

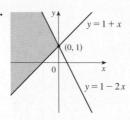

45. $(1, -2)$

■ **EXERCISES C** ■ page A30

1. **(a)** $L_2 = 6, R_2 = 12, M_2 \approx 9.6$
(b) L_2 is an underestimate, R_2 and M_2 are overestimates.
(c) $T_2 = 9 < I$ **(d)** $L_n < T_n < I < M_n < R_n$
3. **(a)** $M_{10} \approx 0.806598, E_M \approx -0.001879$
(b) $S_{10} \approx 0.804779, E_S \approx -0.000060$
5. **(a)** 1.506361   **(b)** 1.518362   **(c)** 1.511519
7. **(a)** 2.660833   **(b)** 2.664377   **(c)** 2.663244
9. **(a)** 2.591334   **(b)** 2.681046   **(c)** 2.631976
11. **(a)** 8.363853   **(b)** 8.163298   **(c)** 8.235114
13. **(a)** **(i)** 3.853518   **(ii)** 3.868367   **(iii)** 3.858416
(b) 3.858471
15. **(a)** 19.8   **(b)** 20.6   **(c)** $20.5\overline{3}$
17. $14.\overline{4}$ **19.** 64.4°F
21. $37.7\overline{3}$ ft/s   **23.** 10,177 megawatt-hours
25.

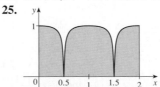

■ **EXERCISES D** ■ page A39

1. $500y^3, 3x^2$ **3.** 10   **5.** 261,632/45   **7.** $\frac{21}{2} \ln 2$
9. 32   **11.** $\frac{3}{10}$ **13.** $\frac{21}{2}$ **15.** $\frac{1}{2}(e^2 - 3)$ **17.** $\frac{1}{15}$
19. $\frac{3}{16}e^4 + \frac{1}{16}$ **21.** 47.5   **23.** $\frac{31}{8}$ **25.** $\frac{5}{6}$ **27.** $\frac{3}{4}$
29. About 9.8 inches

Index

1. (a) What is a function? What are its domain and range?

A function f is a rule that assigns to each input x exactly one output, $f(x)$. The domain of a function is the set of all allowable inputs, and the range is the set of all possible output values as x varies throughout the domain.

(b) What is the graph of a function? What is a scatter plot?

The graph of a function f is the set of input-output pairs $(x, f(x))$ plotted as points for all x in the domain of f. A scatter plot is a collection of individual points and occurs when the domain consists of isolated values.

(c) How can you tell whether a given curve is the graph of a function?

Use the Vertical Line Test: A curve in the xy-plane is the graph of a function of x if and only if no vertical line intersects the graph more than once.

2. Discuss four ways of representing a function. Illustrate your discussion with examples.

The four ways to represent a function are: verbally, numerically, visually, and algebraically. An example of each is given below.

Verbally: An assignment of students to chairs in a classroom (a description in words)

Numerically: A tax table that assigns an amount of tax to an income (a table of values)

Visually: A graphical history of the Dow Jones average (a graph)

Algebraically: A relationship between distance, rate, and time: $d = rt$ (an explicit formula)

3. What is a mathematical model?

A mathematical model is a mathematical description (often by means of a function or an equation) of a real-world phenomenon. (See the discussion on pages 10–11.)

4. What is a piecewise defined function? Give an example.

A piecewise defined function is given by different formulas in different parts of its domain. An example is

$$f(x) = \begin{cases} x^3 - x & \text{if } x \le 2 \\ 2x - 5 & \text{if } x > 2 \end{cases}$$

5. (a) What is an even function? How can you tell if a function is even by looking at its graph?

A function f is even if it satisfies $f(-x) = f(x)$ for every number x in its domain. If the graph of a function is symmetric with respect to the y-axis, then f is even. Examples of an even function: $f(x) = x^2$, $f(x) = x^4 + x^2$, $f(x) = |x|$.

(b) What is an odd function? How can you tell if a function is odd by looking at its graph?

A function f is odd if it satisfies $f(-x) = -f(x)$ for every number x in its domain. If the graph of a function is symmetric with respect to the origin, then f is odd. Examples of an odd function: $f(x) = x^3$, $f(x) = x^3 + x^5$, $f(x) = \sqrt[3]{x}$.

6. Suppose that a function f has domain $(-5, 5)$ and a function g has domain $[0, \infty)$.

(a) What is the domain of $f + g$?

The domain of $f + g$ consists of all the numbers that appear in both the domain of f and the domain of g: $[0, 5)$

(b) What is the domain of fg?

The domain of fg is also $[0, 5)$.

(c) What is the domain of f/g?

The domain of f/g must exclude values of x that make g equal to 0; that is, $\{x \mid 0 \le x < 5, g(x) \ne 0\}$.

7. How is the composition of functions f and g defined? What is its domain?

Given two functions f and g, the composite function h is defined by $h(x) = f(g(x))$. The domain of h is the set of all x in the domain of g such that $g(x)$ is in the domain of f.

8. Suppose the graph of f is given. Write an equation for each of the graphs that are obtained from the graph of f as follows.

(a) Shift 2 units upward: $y = f(x) + 2$

(b) Shift 2 units downward: $y = f(x) - 2$

(c) Shift 2 units to the right: $y = f(x - 2)$

(d) Shift 2 units to the left: $y = f(x + 2)$

(e) Reflect about the x-axis: $y = -f(x)$

(f) Reflect about the y-axis: $y = f(-x)$

(g) Stretch vertically by a factor of 2: $y = 2f(x)$

(h) Shrink vertically by a factor of of 2: $y = \frac{1}{2}f(x)$

(i) Stretch horizontally by a factor of 2: $y = f(\frac{1}{2}x)$

(j) Shrink horizontally by a factor of of 2: $y = f(2x)$

9. Give an example of each type of function.

(a) Linear function: $f(x) = 2x + 1$, $f(x) = ax + b$

(b) Quadratic function: $f(x) = x^2 + x + 1$, $f(x) = ax^2 + bx + c$

(c) Polynomial of degree 5: $f(x) = x^5 + 2x^4 - 3x^2 + 7$

(d) Power function: $f(x) = 2^x$, $f(x) = a^x$

(continued)

(e) Rational function: $f(x) = \dfrac{x}{x^3 + 2}$,

$f(x) = \dfrac{P(x)}{Q(x)}$ where $P(x)$ and $Q(x)$ are polynomials

(f) Exponential function: $f(x) = 2^x$, $f(x) = a^x$

10. What is the slope of a line? How do you compute it? What is the rate of change of a linear function?

The slope of a line is a measure of its steepness. The slope of the line through the points (x_1, y_1) and (x_2, y_2) is

$$m = \frac{\text{rise}}{\text{run}} = \frac{\Delta y}{\Delta x} = \frac{y_2 - y_1}{x_2 - x_1}$$

The rate of change of a linear function is $\Delta y / \Delta x$, which is equal to the slope of its graph.

11. How do you write an equation for a linear function if you know the slope and a point on the line?

If its graph has slope m and passes through the point (x_1, y_1), then its equation is $y - y_1 = m(x - x_1)$.

12. What is a regression line?

A regression line is a line that "best fits" a set of data points and is determined by the method of least squares (see page 35). Normally a graphing calculator or computer software is used to determine the equation of the line.

13. What is the difference between interpolation and extrapolation?

Interpolation means using a model to estimate values *between* given data, whereas extrapolation means estimating values *beyond* the range of the data.

14. What is the shape of the graph of a quadratic function? What is the vertex?

The graph of a quadratic function has the shape of a parabola. It changes direction at its vertex.

15. Sketch by hand, on the same axes, the graphs of the following functions.

(a) $f(x) = x$ (b) $g(x) = x^2$ (c) $h(x) = x^3$

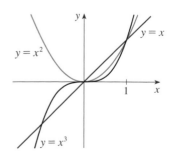

16. Draw, by hand, a rough sketch of the graph of each function.

(a) $y = \sqrt{x}$ (b) $y = 1/x$

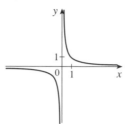

17. (a) Write an equation for a function whose output varies directly with x.

$$f(x) = kx$$

(b) Write an equation for a function whose output varies inversely with x.

$$f(x) = k/x$$

18. Draw, by hand, a rough sketch of the graph of each function.

(a) $y = e^x$ (b) $y = \ln x$

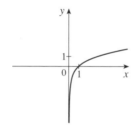

19. (a) What is an inverse function?

An inverse function reverses the inputs and ouputs of the original function.

(b) What is the inverse function of $f(x) = 3^x$?

Its inverse function is $g(x) = \log_3 x$.

2 Summary: Concept Check Answers

1. How is the average rate of change of a function on an interval defined? What are the units of the average rate of change? How can you interpret it geometrically?

   The average rate of change of a function $y = f(x)$ over an interval $[x_1, x_2]$ is the ratio of the change in output to the change in input:

   $$\frac{\Delta y}{\Delta x} = \frac{f(x_2) - f(x_1)}{x_2 - x_1}$$

   The units are output units per input unit. Geometrically, the average rate of change is the slope of the secant line through the points (x_1, y_1) and (x_2, y_2).

2. Explain what $\lim_{x \to a} f(x) = L$ means.

   $\lim_{x \to a} f(x) = L$ means that the values of $f(x)$ approach L as the values of x approach a.

3. State the following Limit Laws.

   (a) Sum Law

   The limit of a sum is the sum of the limits:
   $$\lim_{x \to a} [f(x) + g(x)] = \lim_{x \to a} f(x) + \lim_{x \to a} g(x)$$

   (b) Difference Law

   The limit of a difference is the difference of the limits:
   $$\lim_{x \to a} [f(x) - g(x)] = \lim_{x \to a} f(x) - \lim_{x \to a} g(x)$$

   (c) Constant Multiple Law

   The limit of a constant times a function is the constant times the limit of the function: $\lim_{x \to a} [cf(x)] = c \lim_{x \to a} f(x)$

   (d) Product Law

   The limit of a product is the product of the limits:
   $$\lim_{x \to a} [f(x) g(x)] = \lim_{x \to a} f(x) \cdot \lim_{x \to a} g(x)$$

   (e) Quotient Law

   The limit of a quotient is the quotient of the limits, provided that the limit of the denominator is not 0:
   $$\lim_{x \to a} \frac{f(x)}{g(x)} = \frac{\lim_{x \to a} f(x)}{\lim_{x \to a} g(x)} \quad \text{if } \lim_{x \to a} g(x) \neq 0$$

   (f) Power Law

   The limit of a power is the power of the limit:
   $$\lim_{x \to a} [f(x)]^n = \left[\lim_{x \to a} f(x)\right]^n \quad \text{(for } n \text{ a positive integer)}$$

4. (a) What does it mean for f to be continuous at a?

   A function f is continuous at a number a if the value of the function at $x = a$ is the same as the limit when x approaches a; that is, $\lim_{x \to a} f(x) = f(a)$.

   (b) What does it mean for f to be continuous on the interval $(-\infty, \infty)$? What can you say about the graph of such a function?

   A function f is continuous on the interval $(-\infty, \infty)$ if f is continuous at every real number a. The graph of such a function has no break in it.

5. If we know that a function is continuous at $x = a$, what is the value of $\lim_{x \to a} f(x)$?

   If f is continuous at $x = a$, then $\lim_{x \to a} f(x) = f(a)$.

6. What is the difference between the limit $\lim_{x \to a} f(x)$ and the one-sided limits $\lim_{x \to a^+} f(x)$ and $\lim_{x \to a^-} f(x)$? What is the connection between them?

   When determining $\lim_{x \to a^+} f(x)$ we consider only values of x that are larger than a, whereas for $\lim_{x \to a^-} f(x)$ we consider only values of x smaller than a. For $\lim_{x \to a} f(x)$ we consider values of x both larger and smaller than a; this limit exists only if the two one-sided limits exist and are equal, and then $\lim_{x \to a} f(x) = \lim_{x \to a^+} f(x) = \lim_{x \to a^-} f(x)$.

7. If $y = f(x)$ and x changes from x_1 to x_2, write expressions for the following.

   (a) The average rate of change of y with respect to x over the interval $[x_1, x_2]$:

   $$\frac{f(x_2) - f(x_1)}{x_2 - x_1}$$

   (b) The instantaneous rate of change of y with respect to x at $x = x_1$:

   $$\lim_{x_2 \to x_1} \frac{f(x_2) - f(x_1)}{x_2 - x_1}$$

8. What is a tangent line? Write an expression for the slope of the tangent line to the curve $y = f(x)$ at the point $(a, f(a))$.

   A tangent line is the unique line that touches a curve at a given point and has the same direction as the curve there. The slope of the tangent line to the graph of $y = f(x)$ at $(a, f(a))$ is given by

   $$\lim_{x \to a} \frac{f(x) - f(a)}{x - a} \quad \text{or} \quad \lim_{h \to 0} \frac{f(a + h) - f(a)}{h}$$

9. Give two ways we can estimate an instantaneous rate of change if its value cannot be determined precisely.

   One way is to estimate the value of the limit given in 7(b). Another way is to estimate the slope of a tangent line to the graph of the function.

(continued)

10. Define the derivative $f'(a)$. Discuss two ways of interpreting this number.

$$f'(a) = \lim_{h \to 0} \frac{f(a + h) - f(a)}{h}$$

or, equivalently,

$$f'(a) = \lim_{x \to a} \frac{f(x) - f(a)}{x - a}$$

The derivative $f'(a)$ is the instantaneous rate of change of f (with respect to x) when $x = a$ and also represents the slope of the tangent line to the graph of f at $x = a$.

11. If $f(t)$ is the position function of a moving object, how can you interpret the derivative?

The derivative $f'(t)$ is the velocity of the object.

12. Given a graph of a function $y = f(x)$, how do you graph the derivative f'?

Draw tangent lines to the graph of f at several locations and estimate their slopes. These slope values are graphed as output values (for the same value of x) on the graph of f'. The graph of f' is above the x-axis where f is increasing and below the x-axis where f is decreasing.

13. What does it mean for a function to be differentiable at a?

A function f is differentiable at a if the derivative $f'(a)$ exists.

14. Describe several ways in which a function can fail to be differentiable. Illustrate with sketches.

A function is not differentiable at any value where the graph has a corner, where the graph has a discontinuity, or where it has a vertical tangent line.

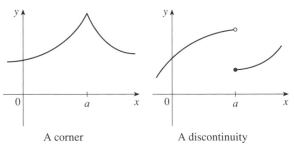

A corner   A discontinuity

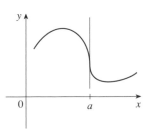

A vertical tangent

15. Define the second derivative of f. If $f(t)$ is the position function of a particle, how can you interpret the second derivative?

The second derivative of f, f'', is the derivative of f'. If $f(t)$ is the position function of a particle, then the first derivative is velocity and the second derivative is the derivative of velocity, namely acceleration.

16. (a) What does the sign of $f'(x)$ tell us about f?

If $f'(x) > 0$ on an interval, then f is increasing on that interval.

If $f'(x) < 0$ on an interval, then f is decreasing on that interval.

(b) What does the sign of $f''(x)$ tell us about f?

If $f''(x) > 0$ on an interval, then f is concave upward on that interval.

If $f''(x) < 0$ on an interval, then f is concave downward on that interval.

17. What does it mean to say that a curve is concave upward? Concave downward?

If a curve is concave upward, then the slopes of tangent lines to the curve increase (from left to right) and the curve bends upward. If the curve is concave downward, then the slopes decrease and the curve bends downward.

3 Summary: Concept Check Answers

1. State each differentiation rule both in symbols and in words.

   (a) The Power Rule

   If n is any real number, then $\frac{d}{dx}(x^n) = nx^{n-1}$.

   The derivative of a variable base raised to a constant power is the power times the base raised to the power minus one.

   (b) The Constant Multiple Rule

   If c is a constant and f is a differentiable function, then

   $$\frac{d}{dx}[cf(x)] = c\frac{d}{dx}f(x)$$

   The derivative of a constant times a function is the constant times the derivative of the function.

   (c) The Sum Rule

   If f and g are both differentiable, then

   $$\frac{d}{dx}[f(x) + g(x)] = \frac{d}{dx}f(x) + \frac{d}{dx}g(x)$$

   The derivative of a sum of functions is the sum of the derivatives.

   (d) The Difference Rule

   If f and g are both differentiable, then

   $$\frac{d}{dx}[f(x) - g(x)] = \frac{d}{dx}f(x) - \frac{d}{dx}g(x)$$

   The derivative of a difference of functions is the difference of the derivatives.

   (e) The Product Rule

   If f and g are both differentiable, then

   $$\frac{d}{dx}[f(x)g(x)] = f(x)\frac{d}{dx}[g(x)] + g(x)\frac{d}{dx}[f(x)]$$

   The derivative of a product of two functions is the first function times the derivative of the second function plus the second function times the derivative of the first function.

   (f) The Quotient Rule

   If f and g are both differentiable, then

   $$\frac{d}{dx}\left[\frac{f(x)}{g(x)}\right] = \frac{g(x)\frac{d}{dx}[f(x)] - f(x)\frac{d}{dx}[g(x)]}{[g(x)]^2}$$

   The derivative of a quotient of functions is the denominator times the derivative of the numerator minus the numerator times the derivative of the denominator, all divided by the square of the denominator.

   (g) The Chain Rule

   If g is differentiable at x and f is differentiable at $g(x)$, then the composite function defined by $F(x) = f(g(x))$ is differentiable at x and F' is given by the product $F'(x) = f'(g(x))\,g'(x)$. The derivative of a composite function is the derivative of the outer function evaluated at the inner function times the derivative of the inner function.

2. State the derivative of each function.

   (a) $y = c$, where c is a constant: $dy/dx = 0$
   (b) $y = x^n$: $dy/dx = nx^{n-1}$
   (c) $y = e^x$: $dy/dx = e^x$
   (d) $y = a^x$: $dy/dx = a^x \ln a$
   (e) $y = \ln x$: $dy/dx = 1/x$

3. (a) How is the number e defined?

   e is the number such that $\lim_{h \to 0} \frac{e^h - 1}{h} = 1$.

   (b) Why is the natural exponential function $y = e^x$ used more often in calculus than the other exponential functions $y = a^x$?

   The differentiation formula for $y = a^x$ $[dy/dx = a^x \ln a]$ is simplest when $a = e$ because $\ln e = 1$.

   (c) Why is the natural logarithmic function $y = \ln x$ used more often in calculus than the other logarithmic functions $y = \log_a x$?

   The differentiation formula for $y = \log_a x$ $[dy/dx = 1/(x \ln a)]$ is simplest when $a = e$ because $\ln e = 1$.

4. Give a formula to compute the balance on an interest-earning investment account after t years if the interest rate is r and the interest is compounded

   (a) annually: $A = P(1 + r)^t$, where P is the principal

   (b) n times per year: $A = P\left(1 + \frac{r}{n}\right)^{nt}$

   (c) continuously: $A = Pe^{rt}$

5. What is the general formula for a quantity that grows at a rate proportional to itself?

   A quantity that grows at a rate proportional to itself exhibits exponential growth and is described by $A(t) = Ce^{kt}$, where C is the initial quantity and k is the relative growth rate.

(continued)

6. **(a)** What does the graph of a logistic function look like?

The graph is "S-shaped." The graph of an increasing logistic function is concave upward at first but changes to concave downward at an inflection point and levels off.

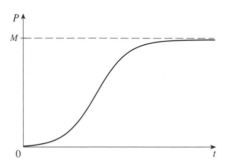

(b) Under what circumstances is this an appropriate model for population growth?

A logistic function is an appropriate model when the population grows at a rate proportional to the size of the population in the beginning, but eventually levels off and approaches its carrying capacity because of limited resources.

7. Explain how implicit differentiation works.

The method of implicit differentiation consists of differentiating both sides of an equation with respect to x, treating y as a function of x. Then we solve the resulting equation for dy/dx.

8. **(a)** What are fixed costs for a business? Variable costs?

Fixed costs are expenses that are incurred whether or not any goods are produced, such as the purchase of machinery or building rental. Variable costs change with the number of goods produced.

(b) How do we compute the average cost of producing q units of a good or service?

Average cost is computed by dividing the total cost by the number of units produced: $C(q)/q$.

(c) What is marginal cost? How do we compute it?

Marginal cost is the instantaneous rate of change of cost with respect to the number of units produced. It is the derivative of the cost function: $C'(q)$. We can think of marginal cost as approximately the cost to produce one additional unit.

(d) How do we determine the minimum average cost?

The minimum average cost occurs when marginal cost and average cost are equal.

9. **(a)** What is marginal revenue? How is it computed?

Marginal revenue is the instantaneous rate of change of revenue with respect to the number of units produced. It is the derivative of the revenue function: $R'(q)$.

(b) How is the profit function defined?

The profit function is defined as the revenue function minus the cost function: $P(q) = R(q) - C(q)$.

(c) How do we determine when profit is maximized?

Profit is maximized when marginal revenue and marginal cost are equal, that is, $R'(q) = C'(q)$ [assuming that $R'(q) > C'(q)$ initially].

10. What is a demand function? Why are demand functions typically decreasing functions?

A demand function $p = D(q)$ gives the price p per unit that a producer can charge in order to sell q units. It is typically decreasing because in order to sell more units, the price normally must be lowered.

11. What is a linear approximation? What is the connection with the tangent line?

A linear approximation is an estimate for a function value found by using the values from a tangent line to the graph of the function as approximations to the function values: $f(a + \Delta x) \approx f(a) + f'(a)\,\Delta x$.

1. Explain the difference between an absolute maximum and a local maximum. Illustrate with a sketch.

The function value $f(c)$ is the absolute maximum value of f if $f(c)$ is the largest function value on the entire domain of f, whereas $f(c)$ is a local maximum value if it is the largest function value when x is near c.

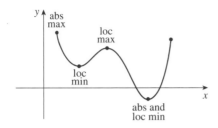

2. (a) What does the Extreme Value Theorem say?

The Extreme Value Theorem states that if f is a continuous function on a closed interval $[a, b]$, then it always attains an absolute maximum and an absolute minimum value on that interval.

(b) Explain how the Closed Interval Method works.

To find the absolute maximum and minimum values of a continuous function f on a closed interval $[a, b]$, we follow these three steps:

- Find the critical numbers of f in the interval (a, b) and compute the values of f at these numbers.
- Find the values of f at the endpoints of the interval.
- The largest of the output values from the previous two steps is the absolute maximum value; the smallest of these values is the absolute minimum value.

3. (a) Define a critical number of f.

A critical number of a function f is a number c in the domain of f such that either $f'(c) = 0$ or $f'(c)$ does not exist.

(b) What is the connection between critical numbers and local extreme values?

If f has a local maximum or minimum at c, then c is a critical number of f. Note that if c is a critical number, it may or may not correspond to a local maximum or minimum.

4. (a) State the Increasing/Decreasing Test.

If $f'(x) > 0$ on an interval, then f is increasing on that interval.
If $f'(x) < 0$ on an interval, then f is decreasing on that interval.

(b) What does it mean to say that f is concave upward on an interval?

If f is concave upward on an interval, then f' is an increasing function on that interval.

(c) State the Concavity Test.

If $f''(x) > 0$ on an interval, then the graph of f is concave upward on that interval.
If $f''(x) < 0$ on an interval, then the graph of f is concave downward on that interval.

(d) What are inflection points? How do you find them?

Inflection points on the graph of a continuous function f are points where the curve changes from concave upward to concave downward or from concave downward to concave upward. They can be found by determining the values at which the second derivative changes sign.

5. (a) State the First Derivative Test.

Suppose that c is a critical number of a continuous function f.

- If f' changes from positive to negative at c, then f has a local maximum at c.
- If f' changes from negative to positive at c, then f has a local mimimum at c.
- If f' does not change sign at c, then f has no local maximum or minimum at c.

(b) State the Second Derivative Test.

Suppose f'' is continuous near c.

- If $f'(c) = 0$ and $f''(c) > 0$, then f has a local minimum at c.
- If $f'(c) = 0$ and $f''(c) < 0$, then f has a local maximum at c.

(c) What are the relative advantages and disadvantages of these tests?

The Second Derivative Test is sometimes easier to use, but it is inconclusive when $f''(c) = 0$ and fails if $f''(c)$ does not exist. In either case the First Derivative Test must be used.

6. Explain what each of the following means and illustrate with a sketch. What does the limit tell us about asymptotes of the graph?

(a) $\lim\limits_{x \to 2^-} f(x) = \infty$

This limit means that the values of $f(x)$ become arbitrarily large as x approaches 2 from the left. The line $x = 2$ is a vertical asymptote of the graph of f.

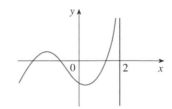

(continued)

(b) $\lim_{x \to \infty} f(x) = 5$

This limit means that the values of $f(x)$ approach 5 as x becomes arbitrarily large. The line $y = 5$ is a horizontal asymptote (at the right) of the graph of f.

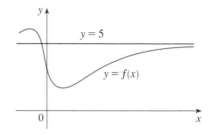

7. (a) Give an example of a familiar function whose graph has a vertical asymptote.

Both the graphs of $y = \ln x$ and $y = 1/x$ have the y-axis as a vertical asymptote.

(b) Give an example of a familiar function whose graph has a horizontal asymptote. Can you think of one that has two different horizontal asymptotes?

The graph of $y = e^x$ has the x-axis as a horizontal asymptote (at the left). The graph of any logistic function has two different horizontal asymptotes, the x-axis on one side and a horizontal line at the carrying capacity on the other.

8. If you have a graphing calculator or computer, why do you need calculus to graph a function?

Calculus reveals all the important aspects of a graph, such as local extreme values and inflection points, that can be missed when relying solely on technology. In many cases we can find exact locations of these key points rather than approximations. Using derivatives to identify the behavior of the graph also helps us choose an appropriate viewing window and alerts us to where we may wish to zoom in on a graph.

9. (a) What are related rates problems? Describe a strategy for solving them.

In a related rates problem, the rate of change of one quantity is calculated from the rate of change of one or more other quantities. To solve them, we first find an equation that relates the different quantities, each of which is a function of time. Then we differentiate both sides of the equation with respect to time. We can determine the desired rate of change by substituting the known values of the quantities and their rates of change. (See page 211.)

(b) What are optimization problems? Describe a strategy for solving them.

Optimization problems are problems in which we are trying to maximize or minimize some quantity, such as cost or profit. To solve them we introduce notation and identify an equation that expresses the quantity we wish maximized or minimized as a function of a single variable. After determining an appropriate domain, we compute the absolute maximum or minimum value using the methods of Sections 4.2 and 4.3. In particular, if the domain is a closed interval we can use the Closed Interval Method. (See page 256.)

10. What is the difference between average cost and marginal cost? How are they related when average cost is minimized?

If the total cost (including fixed costs) is $C(q)$, then the average cost is the cost per unit: $C(q)/q$. Marginal cost $C'(q)$ is the rate of change of the cost function at a particular production level and can be interpreted as approximately the cost of producing one additional unit. When average cost is minimized, the average cost and marginal cost are equal.

11. What is a demand function? How do we use it to compute revenue?

A demand function $p = D(q)$ relates the price per unit of a good to the number of units that can be sold. Revenue is the number of units sold times the price per unit: $R(q) = qp = qD(q)$.

12. (a) What is elasticity of demand? How do we compute it?

Elasticity of demand E is a measure of how much a change in price affects a change in demand for a good. It is computed by the formula

$$E(q) = -\frac{D(q)}{qD'(q)}$$

(b) What does it mean when elasticity of demand is equal to one? Greater than one? Less than one?

When elasticity of demand is equal to one, changes in price and demand are proportional and revenue remains constant. If elasticity of demand is greater than one, then the demand is called elastic and changes in demand are proportionally greater than changes in price. In this case raising prices would decrease revenue. If elasticity of demand is less than one, then the demand is called inelastic and changes in demand are proportionally smaller than changes in price. Raising prices would increase revenue.

(c) How do we use elasticity of demand to determine maximum revenue?

Maximum revenue occurs when $E(q) = 1$, so prices should be adjusted until the elasticity of demand is one.

1. (a) How do we estimate the area under a curve?

One way is to approximate the region by rectangles whose areas are easily computed.

(b) Write an expression for a Riemann sum of a function f.

If we divide a closed interval into n subintervals of equal width Δx and label the endpoints of the subintervals $x_0, x_1, x_2, \ldots, x_n$, then an example of a Riemann sum is $f(x_1)\,\Delta x + f(x_2)\,\Delta x + \cdots + f(x_n)\,\Delta x$.

(c) If $f(x) \geq 0$, what is the geometric interpretation of a Riemann sum? Illustrate with a diagram.

If f is positive, then a Riemann sum can be interpreted as the sum of areas of approximating rectangles, as shown in the figure.

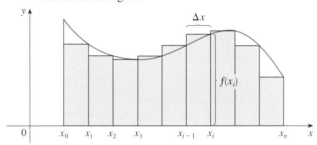

2. (a) Write the definition of the definite integral of a continuous function from a to b.

If f is a continuous function on the interval $[a, b]$, then we divide $[a, b]$ into n subintervals of equal width $\Delta x = (b - a)/n$. We let $x_0\,(= a), x_1, x_2, \ldots, x_n\,(= b)$ be the endpoints of these subintervals. Then

$$\int_a^b f(x)\,dx = \lim_{n \to \infty} \left[f(x_1)\,\Delta x + f(x_2)\,\Delta x + \cdots + f(x_n)\,\Delta x \right]$$

(b) What is the geometric interpretation of $\int_a^b f(x)\,dx$ if $f(x) \geq 0$ on $[a, b]$?

If f is positive, then $\int_a^b f(x)\,dx$ can be interpreted as the area under the graph of $y = f(x)$ and above the x-axis for $a \leq x \leq b$.

(c) What is the geometric interpretation of $\int_a^b f(x)\,dx$ if $f(x)$ takes on both positive and negative values on $[a, b]$? Illustrate with a diagram.

In this case $\int_a^b f(x)\,dx$ can be interpreted as a "net area," that is, the area of the region above the x-axis and below the graph of f (labeled "+" in the figure) minus the area of the region below the x-axis and above the graph of f (labeled "−").

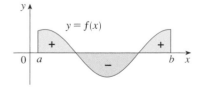

3. What does the Midpoint Rule say?

The Midpoint Rule states that if f is a continuous function on the interval $[a, b]$ and we divide $[a, b]$ into n subintervals of equal width $\Delta x = (b - a)/n$, then

$$\int_a^b f(x)\,dx \approx \Delta x \left[f(\bar{x}_1) + f(\bar{x}_2) + \cdots + f(\bar{x}_n) \right]$$

where $x_i =$ midpoint of $[x_{i-1}, x_i] = \frac{1}{2}(x_{i-1} + x_i)$.

4. (a) What is an antiderivative of a function f?

A function F is an antiderivative of f if $F'(x) = f(x)$.

(b) How do we find the most general antiderivative of f?

If F is an antiderivative of f, then the most general antiderivative is $F(x) + C$, where C is an arbitrary constant.

5. (a) State the Fundamental Theorem of Calculus.

The Fundamental Theorem of Calculus says that if f is continuous on the interval $[a, b]$, then $\int_a^b f(x)\,dx = F(b) - F(a)$, where F is any antiderivative of f.

(b) State the Net Change Theorem.

The Net Change Theorem says that the integral of a rate of change is the net change:
$\int_a^b F'(x)\,dx = F(b) - F(a)$.

6. What does the area from a to b under a marginal cost curve represent?

Marginal cost $C'(q)$ is positive, so the area under its graph from a to b is $\int_a^b C'(q)\,dq$ and represents the increase in cost when production is increased from $q = a$ units to $q = b$ units.

7. If $r(t)$ is the rate at which water flows into a reservoir, what does $\int_{t_1}^{t_2} r(t)\,dt$ represent?

$\int_{t_1}^{t_2} r(t)\,dt$ represents the change in the amount of water in the reservoir between time t_1 and time t_2.

8. Suppose an object moves back and forth along a straight line with velocity $v(t)$, measured in feet per second, and acceleration $a(t)$.

(a) What is the meaning of $\int_{60}^{120} v(t)\,dt$?

$\int_{60}^{120} v(t)\,dt$ represents the change in position of the object from $t = 60$ to $t = 120$ seconds, in other words, in the second minute.

(b) What is the meaning of $\int_{60}^{120} |v(t)|\,dt$?

$\int_{60}^{120} |v(t)|\,dt$ represents the total distance traveled by the object in the second minute.

(continued)

(c) What is the meaning of $\int_{60}^{120} a(t)\, dt$?

$\int_{60}^{120} a(t)\, dt$ represents the change in velocity of the object in the second minute.

9. (a) Explain the meaning of the indefinite integral $\int f(x)\, dx$.

The indefinite integral $\int f(x)\, dx$ is another name for an antiderivative of f, so $\int f(x)\, dx = F(x)$ means that $F'(x) = f(x)$.

(b) What is the connection between the definite integral $\int_a^b f(x)\, dx$ and the indefinite integral $\int f(x)\, dx$?

The connection is given by the Fundamental Theorem: $\int_a^b f(x)\, dx = \left[\int f(x)\, dx\right]_a^b$ if f is continuous.

10. If t is measured in minutes and $g(t)$ is measured in gallons per minute, what are the units for $\int_{10}^{30} g(t)\, dt$?

The units for $\int_{10}^{30} g(t)\, dt$ are the units for $g(t)$ times the units for t: gallons/minute $\times$ minutes = gallons.

11. How do we find the average value of a function f on an interval $[a, b]$?

The average value of f on $[a, b]$ is given by

$$f_{\text{ave}} = \frac{1}{b - a}\int_a^b f(x)\, dx$$

12. (a) State the Substitution Rule. In practice, how do you use it?

The Substitution Rule says that if $u = g(x)$ is a differentiable function and f is continuous on the range of

g, then $\int f(g(x))\, g'(x)\, dx = \int f(u)\, du$. In practice, we make the substitutions $u = g(x)$ and $du = g'(x)\, dx$ in the integrand to make the integral simpler to evaluate.

(b) State the rule for integration by parts. In practice, how do you use it?

The rule for integration by parts states that

$$\int f(x)\, g'(x)\, dx = f(x)\, g(x) - \int g(x)\, f'(x)\, dx$$

In practice, we try to choose $u = f(x)$ to be a function that becomes simpler when differentiated (or at least not more complicated) as long as $dv = g'(x)\, dx$ can be readily integrated to give v. Then the original integral $\int u\, dv$ becomes $uv - \int v\, du$.

13. Explain exactly what is meant by the statement that "differentiation and integration are inverse processes."

The Fundamental Theorem of Calculus (or equivalently the Net Change Theorem) states that

$$\int_a^b F'(x)\, dx = F(b) - F(a)$$

This says that if we take a function F, first differentiate it, and then integrate the result, we arrive back at the original function, but in the form $F(b) - F(a)$. Also, the indefinite integral $\int f(x)\, dx$ represents an antiderivative of f, so

$$\frac{d}{dx}\int f(x)\, dx = f(x)$$

6 Summary: Concept Check Answers

1. How do you find the area between two curves? Does it make a difference if one of the curves lies below the *x*-axis?

If $f(x) \geqslant g(x)$ on the interval $[a, b]$, then the area between the graphs of f and g for $a \leqslant x \leqslant b$ is given by $\int_a^b [f(x) - g(x)] \, dx$. It makes no difference whether the curves are above or below the *x*-axis.

2. What does the area between marginal revenue and marginal cost curves represent?

It represents the change in profit. If $R'(q) > C'(q)$ on a particular interval, then the area between the graphs for that interval represents an increase in profit.

3. Suppose that Sue runs faster than Kathy throughout a 1500-meter race. What is the physical meaning of the area between their velocity curves for the first minute of the race?

It represents the number of meters by which Sue is ahead of Kathy after 1 minute.

4. (a) Given a demand function $p = D(q)$, explain what is meant by the consumer surplus when the amount of a commodity currently available is Q and the current selling price is P. Illustrate with a sketch. How do we compute it?

The consumer surplus is the difference between what a consumer is willing to pay and what the consumer actually pays for a good. The total consumer surplus for the good is the sum of all such "savings" and is given by the integral $\int_0^Q [D(q) - P] \, dq$.

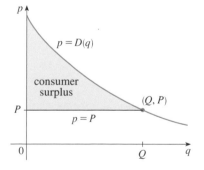

(b) Given a supply function $p = S(q)$, explain what is meant by the producer surplus when the amount of a commodity currently available is Q and the current selling price is P. How is it computed?

The producer surplus is the difference between the price a producer is able to sell a good at and the producer's minimum acceptable price. The total producer surplus for the good is the sum of all such differences and is given by the integral $\int_0^Q [P - S(q)] \, dq$.

(c) What is the total surplus? When is it maximized?

The total surplus is the sum of the consumer surplus and the producer surplus for a good. It is maximized when the market is at equilibrium, that is, when the selling price and quantity produced correspond to the point of intersection of the supply curve and the demand curve.

5. (a) What is the future value of a continuous income stream? How is it computed?

If income is continuously earned at a rate of $f(t)$ dollars per year and then invested at an interest rate r, then the future value of the stream is the total amount of earnings after a specified period of time. It is computed using the formula $\text{FV} = e^{rT} \int_0^T f(t) e^{-rt} \, dt$, where T is the time in years.

(b) How do we find the present value of a continuous income stream? What does this mean?

The present value of an income stream is computed using the formula $\text{PV} = \int_0^T f(t) e^{-rt} \, dt$. It represents the amount of money that would have to be invested now at the interest rate r to match the future value of the income stream.

6. If we have survival and renewal functions for a population, how do we predict the size of the population T years from now?

If a population begins with P_0 members, new members are added at a rate given by the renewal function $R(t)$, where t is measured in years, and the proportion of the population that remains after t years is given by the survival function $S(t)$, then the population T years from now is given by $P(T) = S(T) \cdot P_0 + \int_0^T S(T - t) R(t) \, dt$.

7. (a) What is the cardiac output of the heart?

The cardiac output of the heart is the volume of blood pumped by the heart per unit time.

(b) Explain how the cardiac output can be measured by the dye dilution method.

Dye is injected into part of the heart and a probe measures the concentration of the dye leaving the heart over a time interval $[0, T]$ until the dye has cleared. (See Figure 4 in Section 6.3.) If $c(t)$ is the concentration of the dye at time t, then the cardiac output is given by

$$\frac{A}{\int_0^T c(t) \, dt}$$

where A is the amount of dye used.

(continued)

8. What is a differential equation? What is an initial-value problem?

A differential equation is an equation that contains an unknown function and one or more of its derivatives. An initial-value problem is a differential equation that includes an initial condition such as $y(x_0) = y_0$ that must also be satisfied.

9. What is a separable differential equation? How do you solve it?

It is a differential equation in which the expression for dy/dx can be factored as a function of x times a function of y, that is, $dy/dx = g(x) f(y)$. We can solve the equation by integrating both sides of the equation $[1/f(y)] \, dy = g(x) \, dx$ and solving for y.

10. (a) Write a differential equation for a quantity that grows at a rate proportional to its size.

If $P(t)$ is a quantity that grows at a rate proportional to its size, then $dP/dt = kP$, where k is the constant of proportionality.

(b) Under what circumstances is this an appropriate model for population growth?

It is an appropriate model for population growth under ideal conditions, where there is enough room and nutrition to support unrestricted growth.

(c) What is the solution of this equation?

If P_0 is the initial population, that is, $P_0 = P(0)$, then the solution is $P(t) = P_0 e^{kt}$.

11. (a) Write the logistic differential equation.

The logistic differential equation is $dP/dt = kP(1 - P/M)$, where M is the carrying capacity.

(b) Under what circumstances is this an appropriate model for population growth?

It is an appropriate model for population growth if the population grows at a rate proportional to the size of the population in the beginning, but eventually levels off and approaches its carrying capacity because of limited resources.

12. Define the following improper integrals.

(a) $\int_a^\infty f(x) \, dx = \lim\limits_{t \to \infty} \int_a^t f(x) \, dx$

(b) $\int_{-\infty}^b f(x) \, dx = \lim\limits_{t \to -\infty} \int_t^b f(x) \, dx$

(c) $\int_{-\infty}^\infty f(x) \, dx = \int_{-\infty}^a f(x) \, dx + \int_a^\infty f(x) \, dx$, where a is any real number (assuming that both integrals are convergent)

13. What is a probability density function? What properties does such a function have?

Given a random variable X, its probability density function f is a function such that $\int_a^b f(x) \, dx$ measures the probability that X lies between a and b. The function f has the properties that $f(x) \geq 0$ for all x, and $\int_{-\infty}^\infty f(x) \, dx = 1$.

14. Suppose $f(x)$ is the probability density function for the weight of a female college student, where x is measured in pounds.

(a) What is the meaning of the integral $\int_0^{130} f(x) \, dx$?

It represents the probability that a randomly chosen female college student weighs less than 130 pounds.

(b) Write an expression for the mean of this density function.

The mean is given by $\mu = \int_{-\infty}^\infty x f(x) \, dx = \int_0^\infty x f(x) \, dx$ [since $f(x) = 0$ for $x < 0$].

(c) How can we find the median of this density function?

The median of f is the number m such that $\int_m^\infty f(x) \, dx = \frac{1}{2}$.

15. What is a normal distribution? What is the significance of the standard deviation?

A normal distribution corresponds to a random variable X that has a probability density function with a bell-shaped graph and equation given by

$$f(x) = \frac{1}{\sigma \sqrt{2\pi}} e^{-(x-\mu)^2/(2\sigma^2)}$$

where the mean is μ and σ is the standard deviation. (See Figure 4 in Section 6.6.) The standard deviation measures how spread out the values of X are.

1. (a) What is a function of two variables?

It is a rule that assigns to each input pair (x, y) exactly one output number $f(x, y)$.

(b) What is the graph of a function of two variables?

The graph of a function f of two variables is the surface consisting of all points (x, y, z) in space such that $z = f(x, y)$.

2. What is a linear function of two variables? What type of surface is its graph?

A linear function f of two variables is of the form $f(x, y) = ax + by + c$, where a, b, and c are constants. Its graph is a plane.

3. (a) What is a level curve?

A level curve of a function f of two variables is the curve in the xy-plane given by the equation $f(x, y) = k$, where k is a constant (in the range of f).

(b) What is a contour map?

A contour map is a collection of level curves corresponding to various values of k.

4. What is a function of three variables?

A function f of three variables is a rule that assigns to each input triple (x, y, z) exactly one output number $f(x, y, z)$.

5. (a) Write expressions for the partial derivatives $f_x(a, b)$ and $f_y(a, b)$ as limits.

$$f_x(a, b) = \lim_{h \to 0} \frac{f(a + h, b) - f(a, b)}{h}$$

$$f_y(a, b) = \lim_{h \to 0} \frac{f(a, b + h) - f(a, b)}{h}$$

(b) How do you interpret $f_x(a, b)$ and $f_y(a, b)$ geometrically? How do you interpret them as rates of change?

The partial derivative $f_x(a, b)$ is the slope of the tangent line at the point $(a, b, f(a, b))$ to the vertical trace in the plane $y = b$, and $f_y(a, b)$ is the slope of the tangent line to the trace in the plane $x = a$. Also $f_x(a, b)$ represents the rate of change of f with respect to x (at $x = a$) when y is fixed ($y = b$), and $f_y(a, b)$ is the rate of change of f with respect to y (at $y = b$) when x is fixed ($x = a$).

(c) If $f(x, y)$ is given by a formula, how do you calculate f_x and f_y?

To find f_x, regard y as a constant and differentiate $f(x, y)$ with respect to x. To find f_y, regard x as a constant and differentiate $f(x, y)$ with respect to y.

(d) How do you calculate f_{xy}?

First differentiate with respect to x, to find f_x, and then differentiate the result with respect to y.

6. What do the following statements mean?

(a) f has a local maximum at (a, b).

f has a local maximum at (a, b) if $f(a, b) \geqslant f(x, y)$ when (x, y) is near (a, b).

(b) f has an absolute maximum at (a, b).

f has an absolute maximum at (a, b) if $f(a, b) \geqslant f(x, y)$ for all points (x, y) in the domain of f.

(c) f has a local minimum at (a, b).

f has a local minimum at (a, b) if $f(a, b) \leqslant f(x, y)$ when (x, y) is near (a, b).

(d) f has an absolute minimum at (a, b).

f has an absolute minimum at (a, b) if $f(a, b) \leqslant f(x, y)$ for all points (x, y) in the domain of f.

(e) f has a saddle point at (a, b).

f has a saddle point at (a, b) if $f(a, b)$ is a local maximum in one direction but a local minimum in another.

7. (a) If f has a local maximum at (a, b), what can you say about its partial derivatives at (a, b)?

If f has a local maximum at (a, b) and f_x and f_y exist there, then $f_x(a, b) = 0$ and $f_y(a, b) = 0$.

(b) What is a critical point of f?

A point (a, b) is a critical point of f if $f_x(a, b) = 0$ and $f_y(a, b) = 0$, or if one of these partial derivatives does not exist.

8. State the Second Derivatives Test.

Suppose that $f_x(a, b) = 0$ and $f_y(a, b) = 0$, and let $D = f_{xx}(a, b) f_{yy}(a, b) - [f_{xy}(a, b)]^2$.

- If $D > 0$ and $f_{xx}(a, b) > 0$, then $f(a, b)$ is a local minimum.

- If $D > 0$ and $f_{xx}(a, b) < 0$, then $f(a, b)$ is a local maximum.

- If $D < 0$, then f has a saddle point at (a, b).

- If $D = 0$, then the test gives no information.

9. Explain how the method of Lagrange multipliers works in finding the extreme values of $f(x, y)$ subject to the constraint $g(x, y) = k$.

We first solve the system of equations $f_x(x, y) = \lambda g_x(x, y)$, $f_y(x, y) = \lambda g_y(x, y)$, $g(x, y) = k$ for the unknowns x, y, and λ. Then we evaluate f at all the solution points (x, y); the largest of these values is the maximum value of f, and the smallest is the minimum.